Hot Carriers in Semiconductors

Hot Carriers in Semiconductors

David K Ferry
School of Electrical, Computer, and Energy Engineering, Arizona State University, Tempe, AZ 85287-5706, USA

IOP Publishing, Bristol, UK

ISBN 978-0-7503-3947-6 (ebook)
ISBN 978-0-7503-3945-2 (print)
ISBN 978-0-7503-3948-3 (myPrint)
ISBN 978-0-7503-3946-9 (mobi)

DOI 10.1088/978-0-7503-3947-6

Version: 20211201

IOP ebooks

British Library Cataloguing-in-Publication Data: A catalogue record for this book is available from the British Library.

Published by IOP Publishing, wholly owned by The Institute of Physics, London

IOP Publishing, Temple Circus, Temple Way, Bristol, BS1 6HG, UK

US Office: IOP Publishing, Inc., 190 North Independence Mall West, Suite 601, Philadelphia, PA 19106, USA

Contents

Preface

When I was a young, just-learning scientist, the semiconductor world was populated with new and exciting ideas and discoveries. LEDs were just becoming commercial and the first laser diode was coming from the research lab. The Gunn effect, and a number of unique devices and materials for microwave generation were being discussed. The excitement was multiplied when Lester Eastman (Cornell) convened the first workshop on microwave materials and devices at IEEE headquarters in the Spring of 1964. This brought together a number of outstanding semiconductor scientists (and a few lucky graduate students), to discuss the science of the Gunn diode. Esther Conwell's seminal book *High Field Transport in Semiconductors* had just appeared as well. As I arrived in Vienna for my postdoctoral position, working with Karl-Heinz Seeger, the first papers on the Monte Carlo technique and the first measurements of spin-flip, single-particle Raman scattering of distribution functions were appearing. Working with this group certainly focused on the transport of far-from-equilibrium carriers in semiconductors, and it settled a life-long interest in hot carrier physics.

This may lead one to conclude that this book is a history of the early days, but that is far from the truth. One also could easily point out that the first all-transistor computer was only a few years old as was the real MOSFET and the integrated circuit itself. Thus microelectronics was a growing field already, but was still in its infancy. In fact, Moore's Law had just appeared (1965) and it would be almost another two decades before Christopher Evans' *The Micro Millenium* (1982) would predict the arrival of the information age made possible by microelectronics. So, it would be fair to say that I was a participating witness to the entire formation of the information age, and the technological revolution that has made our present world possible. As such a participant, it is fair to conclude that the field of far-from-equilibrium carrier dynamics has been the driving force for almost all of this technological growth. For example, it was discovered in the late 1980s that highly excited Si FETs could emit light with photon energies down to the order of 0.4 eV, which is less than the band gap. While exotic theories were proposed, it was realized that hot electrons accelerated in the Δ conduction bands could cross the Brillouin zone boundaries and arrive in the second conduction band. The direct transition back to the lowest conduction band would provide a wide spectral range of light emission with the desired range of energies; that is, the light probably was from a hot electron intra-band transition.

Optically-excited hot carriers arrived about this time and have been heavily studied ever since. And, in 2008, the first concept of a hot carrier solar cell was put forward by Robert Ross and Arthur Nozik. Now, the importance of far-from-equilibrium carriers was apparent in an entirely new field. Hence, the so-called 'old' technology was once again 'new' and important. But, it has to be admitted that the new view was far beyond that of the old view. A very wide range of what can be called the science of hot carriers in semiconductors was now involved: hot electrons, hot phonons, phonon lifetimes, proper materials, and so on. While they were interconnected (before as well as now), each had a rich history of its own. This was complicated by a new, nagging problem. Far too many of the scientists working in the solar cell area come from the optical physics realm and aren't acquainted with the rich history of the far-from-equilibrium semiconductor world. In addition, there was the observation that far too many young scientists have the impression that (science) history began the year they entered graduate school. Any science done earlier was apparently irrelevant to their field. While this is my own perception of their view, discussions with other 'senior' scientists suggest that it is a widely-held view. And this brings us to the real reason for this book.

With hot carriers having a new era of interest, it is timely to have a book that summarizes the wide range of topics (including the history) that contribute to hot carriers. And, that is the purpose of this book—to bring together a summary of the science that is important to far-from-equilibrium carriers in semiconductors. Since Conwell's book, written more than half a century ago, there have been some more recent books, but these tend to be edited books, which do not provide a comprehensive and coherent picture of the field. Moreover, even Conwell's book was too early to cover the topics that were new and exciting when I was young. Hopefully, this current effort will rectify (at least some of) that lack, and current scientists working in the area will appreciate this new discourse on the topic.

Over the years, my graduate students and post-docs (and I) have been fortunate to contribute to the growth of the field. The students and post-docs of course did all the heavy lifting, and I have to thank them for their efforts at the time, and their continuing efforts. Some have become colleagues and continue to interact with me, and I owe special thanks to these. In addition, I have been fortunate to have a number of collaborations with other groups, which have provided new opportunities and more insight and understanding, so special thanks also go to them. I want to especially thank Professor Ian Sellers of the University of Oklahoma, who has diligently read each chapter and provided invaluable feedback. This has led to many additions and modifications to the work. However, any mistakes are still my fault, for which I apologize in advance.

David K Ferry

Author biography

David K Ferry

David K Ferry is Regents' Professor Emeritus in the School of Electrical, Computer, and Energy Engineering at Arizona State University. He was also graduate faculty in the Department of Physics and the Materials Science and Engineering program at ASU, as well as Visiting Professor at Chiba University in Japan. He came to ASU in 1983 following shorter stints at Texas Tech University, the Office of Naval Research, and Colorado State University. In the distant past, he received his doctorate from the University of Texas, Austin, and spent a postdoctoral period at the University of Vienna, Austria. He enjoyed teaching (which he refers to as 'warping young minds') and continues active research. The latter is focused on semiconductors, particularly as they apply to nanotechnology and integrated circuits, as well as quantum effects in devices. In 1999, he received the Cledo Brunetti Award from the Institute of Electrical and Electronics Engineers, and is a Fellow of this group as well as the American Physical Society and the Institute of Physics (UK). He has been a Tennessee Squire since 1971 and an Admiral in the Texas Navy since 1973. He is the author, co-author, or editor of some 40 books and about 900 refereed scientific contributions. More about him can be found on his home pages http://ferry.faculty.asu.edu/ and http://dferry.net/.

IOP Publishing

Hot Carriers in Semiconductors

David K Ferry

Chapter 1

Introduction

Many a young experimenter, when first studying electrical things, has seen their subject of investigation go 'poof'. Indeed, some more senior experimenters have been the victim of the same results. But, some have gone on to make the study of how and why the poof happened their lifelong passion. Apparently, the first studies of what is now called breakdown in solids and gasses appears to be that of Eric von Hippel in 1933 [1] and Herbert Fröhlich in 1937 [2]. This was less than a decade from the introduction of quantum mechanics by Heisenberg [3] and Schrödinger [4], and only a decade since electrons had been shown to have wave properties [5, 6]. It would be after the war when the breakdown in silicon was beginning to be studied [7].

The idea of 'hot' carriers may actually have begun with von Hippel [1], but it was surely supported by the later work of Fröhlich [8]. By the term hot carriers, we mean that the electron system has acquired a temperature, from the electric field to which it is subjected, that is larger than the lattice temperature. In this sense, the electron system is no longer in thermal equilibrium, especially with the lattice, and one has already made an assumption that a temperature can describe the electron system. In general, the transport of these hot carriers is quite nonlinear in that the conductance of the material is now a function of the electric field, or voltage, applied to it. Generally, the conductance is related to the field by a conductivity and a mobility, the latter of which depends upon the average energy of the carrier system. The dominant factor in the actual nonlinear conductance does not arise directly from higher-order terms in the field itself, but is usually derived from the fact that the distribution describing the occupancy of momentum states is itself a function of the electric field. In most cases, the determination of this non-equilibrium distribution function is one of the primary problems in describing the behavior of hot carriers.

There are rather standard driving forces which compel the situation described above. As an example, consider two energy states that are described by an energy difference given by the energy of an optical phonon $\hbar\omega_0$. That is, the carrier distribution functions, when a Maxwellian form can be assumed, have the values

doi:10.1088/978-0-7503-3947-6ch1

$f(E)$ and $f(E + \hbar\omega_0)$. When this Maxwellian is a good assumption, then these two values of the distribution function are related by the factor

$$\frac{f(E + \hbar\omega_0)}{f(E)} = e^{-\hbar\omega_0/k_B T_c}, \tag{1.1}$$

where k_B is Boltzmann's constant and T_c is the carrier temperature. There is a dynamic between these two energy levels that is determined by the emission and absorption of phonons [9]. In the steady-state situation, at equilibrium there must be a balance between emission of phonons from the upper level and absorption of phonons in the lower level. But, in non-equilibrium, when the carrier temperature has increased, emission must be larger than absorption in order to transfer energy from the carriers to the phonons (and hence to the lattice). Thus, we require

$$f(E + \hbar\omega_0)(N + 1) > f(E)N, \tag{1.2}$$

where N is the number of phonons at energy $\hbar\omega_0$, and is given by the Bose–Einstein distribution *when the phonons are in equilibrium.* Using the properties of the Bose–Einstein distribution, it is easily shown that

$$\frac{N + 1}{N} = e^{\hbar\omega_0/k_B T_p}, \tag{1.3}$$

where T_p is the temperature of the phonon distribution—the lattice temperature when this distribution is in equilibrium. Using (1.1) and (1.3) in (1.2), we arrive at the condition

$$\exp\left[\frac{\hbar\omega_0}{k_B}\left(\frac{1}{T_p} - \frac{1}{T_c}\right)\right] > 1. \tag{1.4}$$

Hence, if the energy is to flow from the carriers to the optical phonons, then the carrier temperature must be larger than the phonon temperature. One could say that this is just a simple thermodynamic rule. But, it gets more complicated.

It was not long after studies into material breakdown began, that it was observed that the phonons were likely to be out of equilibrium [10]. That is, the emission of a large number of optical phonons by the carriers was quite likely to drive the phonon distribution out of equilibrium and in fact into a form that was not very much like a Bose–Einstein distribution. This non-equilibrium distribution had to have its own decay path, likely through a three-phonon interaction that resulted in the generation of acoustic phonons [11]. By the simple thermodynamic arguments above, the temperature of the optical phonons must be larger than the temperature of the acoustic phonons. But, the latter could also be driven out of equilibrium, so the temperature of the acoustic phonons must be larger than that of the bath—the environment. This all leads to the situation that

$$T_c > T_p > T_a > T_0, \tag{1.5}$$

where T_a is the temperature of the acoustic phonons and T_0 is the bath temperature. So, a very complicated situation has arisen, and it is not a simple task to understand everything that is occurring either experimentally or theoretically.

Much about this complicated situation was known by the first measurements on breakdown in silicon [7]. Indeed, it was about this time that scientists were doing detailed studies of the transport in silicon and germanium, not the least because the first transistor had been created just a couple of years earlier [12]. And, the MOSFET and the integrated circuit were only a few years in the future [13–15]. The period following these developments was a very active time for investigating the properties of various semiconductor materials.

Now, the range of temperatures in (1.5) is all well and good, provided that the carriers and all the various phonons can be described by a temperature. When the system becomes far-from-equilibrium, none of this can be assured. Indeed, when a steady state is achieved, such as in the presence of a very high electric field, any Maxwellian behavior is limited to very low energies. At higher energies, the distribution function takes on much higher values than expected from the thermal distribution [16]. In spite of the obvious, there have been a great many attempts to define temperatures even in these far-from-equilibrium situations. The carriers can be very nonthermal and especially the phonons will behave in this way, as phonon emission by the carriers is within a limited range of phonon wave vectors [9]. Thus, the attempt to describe temperatures can be very misleading in trying to understand the physics of the system. Yet, it occurs.

1.1 Some general observations

When high electric fields are applied to a semiconductor, the resulting current usually becomes sublinear, and is generally accompanied by a reduction in the mobility. The distribution function, characterizing the carriers, is quite far from its low-field Maxwellian form, as suggested above [16]. In this nonlinear transport regime, the ergodic approximation fails [17]. In particular, it fails because the distribution function itself is not a stationary process. It varies in time and its form arises from a delicate balance between the driving forces, the electric field usually, and the dissipative forces, the scattering.

Early studies of high-electric field transport in solids focused mainly on the breakdown studies of dielectrics [7], although the electrical properties of silicon had been studied much earlier [18]. Studies of the variation of the velocity with high electric fields began with Ryder [19], who discovered that the velocity in Si and Ge saturated at high electric fields. The saturated (or nearly saturated) velocity is an important parameter for electron device considerations. For example, with a modern nanoscale MOSFET of say 20 nm gate length, an applied voltage of 1 V produces an average electric field of 0.5 MV cm^{-1} in the channel, which is an incredible electric field, especially when one considers that the saturation velocity occurs at about 20 kV cm^{-1} in Si at room temperature. In many so-called figures of merit for high speed, high frequency, or high power devices, the saturated velocity is an important parameter as it affects the frequency response and power handling properties of the device. It is also important as a mirror into the electron–phonon interactions in the material. But, it needs to be pointed out that breakdown in Si occurs at lower average fields than that mentioned, so such small devices must be operated at lower voltages.

1.1.1 Time scales

In the study of this far-from-equilibrium behavior, one must deal with many characteristic time scales. The importance of these scales is obvious due to the necessity of evaluating the time dependence of the various measurable properties. At the minimum, each of the 'temperatures' in (1.5) is likely to have its own characteristic response time. The non-equilibrium distribution function also evolves with the various time scales mentioned. In a semiconductor, numerous collisions occur and it is these collisions that provide the mechanism of exchange of energy and momentum (to name just a few of the quantities) and relax these toward their equilibrium values. There are collisions between the carriers which randomize the energy and momentum *within* an ensemble but do not relax either of these quantities for the ensemble as a whole. There are also elastic collisions between the carriers and impurities or acoustic phonons which relax the momentum but do little to relax the energy. Finally, there are inelastic collisions between the carriers and lattice vibrations which relax both the energy and the momentum. In general, four generic time scales can be identified [20, 21]:

$$\tau_c < \tau < \tau_R < \tau_H. \tag{1.6}$$

Here the average *duration* of a collision is denoted by τ_c. Generally, this time scale is quite short (perhaps on the fs scale) and not of importance to most considerations. However, on the scale of fast femtosecond laser experiments, this may no longer be true. This collision duration is considered to be the time required to establish the energy-conserving delta function (in the Fermi golden rule for scattering). In distinction, the average time *between* collisions, the mean free time, is denoted by the simple τ. For time scales such that $t \leqslant \tau$, the evolution of the system depends strongly upon the details of the initial state. Generally, $\tau \gg \tau_c$, but this is not always the case at high electric fields, and the breakdown of this inequality can lead to new transport effects, which must be treated in a quantum mechanical manner.

The establishment of equilibrium, or a *non-equilibrium steady state*, can be achieved within a few or a few tens of τ, and the characteristic time associated with this process is the *relaxation time* τ_R. Typical situations described by a relaxation time are the momentum relaxation, and in high fields the energy relaxation. As we will see in later chapters, the momentum relaxation time is often associated with a shift in the distribution function which provides an average velocity or momentum of the carrier system. The corresponding energy relaxation time is associated with the spreading of the distribution which leads to an elevated electron temperature. Now, the phrase '*non-equilibrium steady state*,' may not be familiar to those who have not dealt with far-from-equilibrium systems. But, certainly, one may need only consider Ohm's Law. Current flowing in a semiconductor means that the system is *not* in equilibrium (there is no current flow in equilibrium). Rather, when the current is constant (at a d.c. value), then the system is in a non-equilibrium steady state. Steady-state should not be confused with equilibrium, as they are quite different. In addition, it is also possible for a system to be in a non-steady state while in its non-equilibrium state. One only has to think of an oscillator. It certainly is out of equilibrium, in that it is converting energy (from the bias) from d.c. to some frequency (of the oscillator) while also dissipating energy. The fact that it is oscillating means that the total current is not constant. Another example that will be met further below is that of the GaAs Gunn device, which can also be an oscillator.

Finally, let us consider the final time scale, τ_H. If configuration-space gradients exist, the situation becomes more complex. Relaxation in momentum space proceeds on the scale of τ_R and establishes a 'local' equilibrium over regions smaller than a macroscopic scale, perhaps only a few mean free paths in extent. The achievement of a uniform steady-state of non-equilibrium requires a longer time, the hydrodynamic time $\tau_H > \tau_R$. Only for times large compared to this hydrodynamic time can the ensemble truly be said to be stationary. Examples of the hydrodynamic time scale are diffusion times, arising from recombination processes for excess carriers as well as from local non-homogeneous carrier distributions and sometimes energy relaxation times. Another such hydrodynamic time arises when the actual temperature of the lattice is inhomogeneous (and certainly may be different than that of any heat sink). Examples of the latter are thermoelectric effects, one example of which is the temperature differential that arises from applying an electric field and current to a semiconductor.

1.1.2 Velocity and mobility

One of the most studied observables of hot carriers, at least in homogeneous semiconductors, is the observation of their velocity saturation and the consequent degradation of the mobility. When the carriers are heated by the field, the average energy rises and the distribution spreads with more carriers at higher energy [16]. Since the scattering rate generally increases with energy, there is more carrier scattering (to accommodate the relaxation of the energy input from the field), the mobility is reduced and the velocity does not increase as rapidly. In fact, in Si, the velocity appears to actually saturate at a value near to 10^7 cm s^{-1} at a field of 20 kV cm^{-1} at the lattice temperature of 300 K. We show this in figure 1.1, where the results of an ensemble Monte Carlo simulation are plotted (the calculation procedure is discussed in a later chapter). In this figure, the velocity and *chordal* mobility are plotted for a range of electric field. The non-linearity of the velocity and the mobility are clearly seen. For the range of electric field of about 1–10 kV cm^{-1}, the behavior is usually described as the *warm* electron regime. Above $\sim$ 20 kV cm^{-1}, true saturation of the velocity begins to occur and this is the *hot* electron regime. The chordal mobility is the ratio of the drift velocity to the electric field. One can also compute a differential mobility that is the derivative of the velocity curve, and this is important for microwave applications.

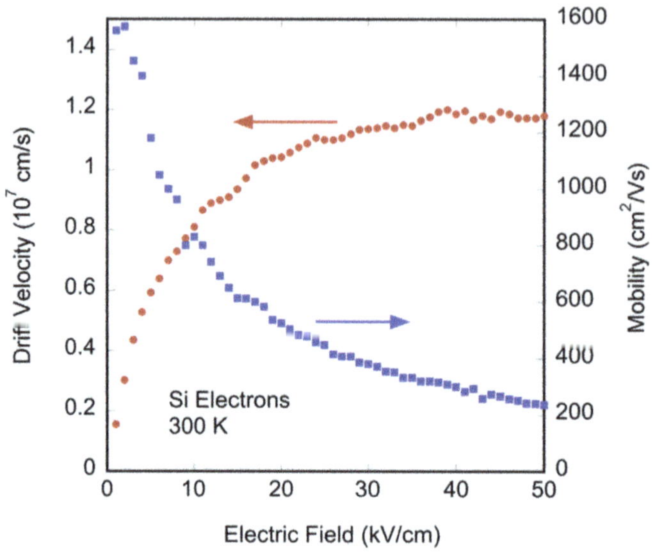

Figure 1.1. The drift velocity and chordal mobility for electrons in Si at room temperature from an ensemble Monte Carlo simulation. The experimental velocity at high fields is usually found to be lower than shown here, $\sim 1 \times 10^7$ cm s^{-1}, but the higher velocity is typical for ensemble Monte Carlo calculation.

Originally, in the early studies, Fröhlich [16] had assumed that breakdown would occur because of a runaway of carriers in a high electric field. He had assumed that the carriers would gain energy from the field, which would increase their velocity and mean free path. It was thought that the latter would increase because there would not be an increase in the scattering rate (shorter relaxation time). But, curves like those of figure 1.1 are clearly stable, so that runaway does not occur for Si. It is thought to occur in some other, wide band gap materials, and we will return to it in chapter 2. It is the 'warm' electron regime that is of interest at this point. In this regime, Ryder found that the velocity (actually the current density in Si and Ge with a fixed carrier density) increased as the square root of the electric field [19, 22]. In view of these results, Shockley had earlier worked out a theory, using a Maxwellian distribution, which fit this behavior [23]. In this approach, he had shown that the mobility would decrease with the carrier temperature, as it was dependent through the mobility on the scattering time, which decreased with the thermal velocity. The latter was dependent upon the square root of the temperature, which was thought to increase linearly in the electric field in this regime. This would give the observed behavior. This observation set in motion the study of the field dependence of the mobility that continued for the next few decades. It was later that attention turned to further studies of the velocity itself, rather than the mobility—the velocity is the critical response to the electric field, and the mobility is derived from this velocity response. So, it is more important to focus upon the velocity rather than just its derivative.

While Shockley's model is over-simplified, it can actually give some information on what is happening in the velocity saturation regime. If the carrier temperature begins to increase as the square of the electric field, then the above argument would lead to the mobility decreasing linearly in the field, and this would give a saturated velocity. In reality, however, the processes are much more complicated and the temperature assumption doesn't work any longer, but the results are the same—the velocity is saturated in most material dominated by phonons other than the polar LO modes. The latter phonons are driven by the Fröhlich interaction which depends upon a lattice polarization and is Coulombic in nature [24]. Other phonons interact with the carriers via a form of deformation potential, and this can lead to a saturated velocity, as will be seen in chapter 2.

1.1.3 Inter-valley scattering

Transfer between e.g. different valleys of the conduction band is one of the older and most interesting aspects of high-electric-field transport in semiconductors. There are two possible scenarios by which transfer between sets of valleys can occur. In one, the symmetry-breaking properties of the high electric field are used to break the symmetry between valleys that are normally equivalent. This occurs, for example, in Si at low temperatures when the electric field is not oriented at the same angle with all six minima of the conduction band. This leads to breaking the symmetry of the equivalent valleys. In equilibrium, each of the six valleys of the conduction band has an equal number of carriers in it, and the cubic symmetry of the crystal is maintained. However, if the electric field is oriented so that it makes a different angle with the principal axes of the six ellipsoids, then each valley can be heated differently by the field. This is because each ellipsoid is characterized by two different effective masses: a longitudinal mass along the major axis of the ellipsoid and a transverse in the directions normal to this axis. Hence, carriers being accelerated by the field will see a different average effective mass depending upon the angle the field makes with the principal axis. We will delve into this in more detail in chapter 2.

The second method by which inter-valley transfer can play a major role is when only a single conduction band minimum, with a small effective mass, is normally occupied, such as in the case of GaN or GaAs, or any of the direct gap compound semiconductors (the principal minimum is at the Γ point). With the application of a high electric field, the carriers may be heated to relatively high energies. Then, a fraction of the carriers will actually be at energies above the minima of a secondary set of valleys (which lie along the (111) directions at the set of L points of the Brillouin zone in GaAs, and are some 0.3 eV above the Γ minimum) of the conduction band. They may even be higher than the third set of minima at the X point in the zone. Since, the set of equivalent L or X minima have relatively high values of the effective mass, their density of states is much larger than that of the central minimum, and inter-valley scattering will cause electrons to move to the satellite valleys of the conduction band. Consider, for example, the conduction band of GaAs shown in figure 1.2. There is a central valley characterized by a small effective mass of $0.067m_0$. In addition, there are subsidiary minima located at both the L point, lying some 0.29 eV higher, and near the X point, lying some 0.5 eV

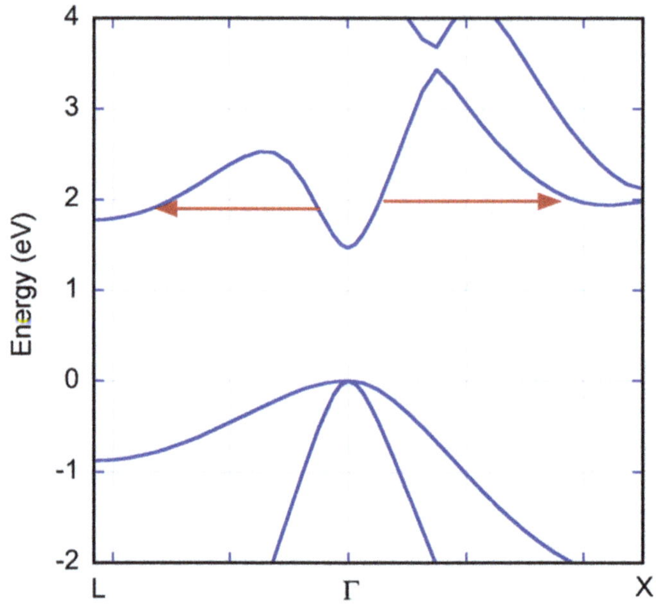

Figure 1.2. The band structure around the principal energy gap, calculated by a local empirical pseudo-potential method. Inter-valley transfers from the Γ valley to the satellite L and X valleys are shown by the red arrows.

higher. These valleys have a considerably greater effective mass and density of states, as the former are Ge-like while the latter are similar to Si. Under normal circumstances, the central valley is the only one occupied. However, for an applied field of some 2.5–3.0 kV cm^{-1} at 300 K, electrons begin to transfer to the L valleys, a process first observed by Gunn in GaAs [25]. It was identified clearly as inter-valley transfer by Ridley and Watkins [26, 27]. In transport, this is observed as a drop in the velocity, as the carriers in the Γ have a low mass and hence a high velocity, while those carriers in the L and X valleys have a high mass and hence a low velocity. This drop in velocity gives rise to what is called negative differential conductance (NDC), as the velocity decreases as the field increases. If the material is biased in this region of NDC, it is unstable against the electric field breaking up into inhomogeneous regions. As the differential conductivity is negative, any fluctuation in the electric field will grow, rather than be damped, and this leads to an instability characterized by different values of the electric field at different positions, usually a very high electric field region that can move through the semiconductor, and a low field outside this region. This leads to current oscillations (and will be dealt with in much more depth in chapter 2).

Inter-valley transfer is seen in a great many semiconductors, particularly those with a direct bandgap, so that the lowest lying conduction band is at the center of the Brillouin zone. While it hasn't been mentioned, creation of high-energy electrons in photo-excitation can also produce inter-valley transfer, and is discussed next. Recent examples include observations of this effect in atomically thin transition-metal dichalcogenides [28, 29], and in the absorber material on AlInAs hot-carrier solar cells [30].

1.2 Optically excited hot carriers

The properties of photo-excited carriers in semiconductors are some of the more interesting aspects, and indeed are important in nearly all optically motivated devices today, from the vast and growing field of solar cells to the advanced lasers used in most scanning devices. The role of these photo-generated carriers is crucial to the operation of such devices. Here, we will introduce just the optically generated carriers that are produced by direct transitions in materials such as GaAs, with a more extensive treatment in chapter 4.

At sufficiently high frequencies, such as in the optical region, we achieve the situation where $\hbar\omega > E_G$, and the electromagnetic field can induce electrons from the valence band into the conduction band. Each photon can create a new electron–hole pair that can contribute to the conduction process. This is the usual optical absorption that provides a method to study the various transitions possible from valence to conduction band, or even between different valence or different conduction bands. The requirement is that the lower energy state must be occupied and the upper energy state empty, to satisfy the Fermi statistics. Of interest here are the direct transitions, such as between the top of the valence band and the bottom of the conduction band in GaAs at Γ, which is known as the E_0 transition (the direct band gap, which is about 1.4 eV in GaAs). Transitions in GaAs for a 2.1 eV photon are shown in figure 1.3, in which the photon energy is considerably larger than E_0. There are three transitions shown as there are three valence bands—the heavy- and light-hole bands and the split off spin–orbit band. This latter band is not shown in figure 1.2, but the maximum lies 340 meV below the valence band maximum in GaAs, and a transition from this band to the conduction band is also shown in figure 1.3.

It is clear that the electrons and holes created by the absorption of these photons are hot carriers; in at least one case, the electrons actually lie above the L valleys, and can rapidly scatter to these latter valleys. In general, these electrons (and holes) decay to the band extrema by the emission of optical phonons. Through the phonon emission process and through interactions between and among the electrons (or holes), the carriers will relax into a non-equilibrium distribution near the band extrema. In general, and more so for these optically created carriers, the distribution will be characterized by a temperature well above the lattice temperature, but this is not always true. As discussed above, this is a steady-state distribution if the illumination is constant, but is *not* an equilibrium distribution. If the optical excitation is pulsed, then the distribution is not even steady-state.

As an example, the excitation from the heavy-hole band puts electrons into the conduction band at about 500 meV above the conduction band minimum. If these decay within this band by the emission of the LO phonons, then about 13 phonons will be emitted by each of these carriers. Still, enough energy is brought into the distribution of the electrons that they will still have a temperature above the lattice

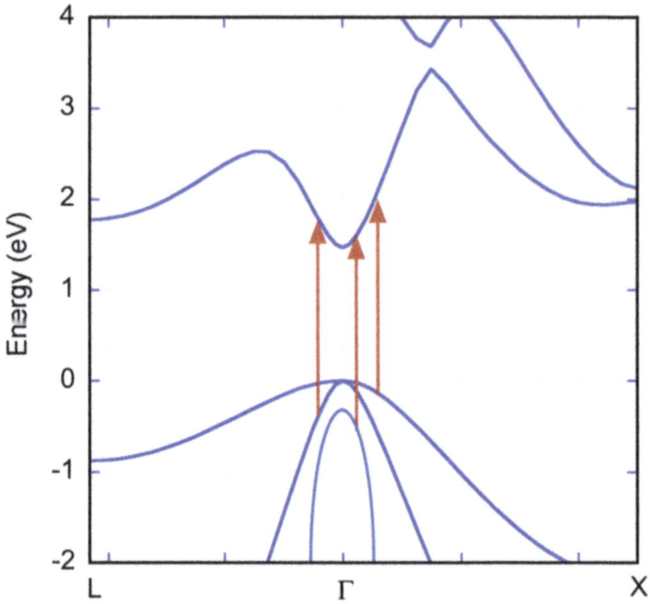

Figure 1.3. The band structure of GaAs and the allowed transitions for a photon of 2.1 eV. This differs from figure 1.2 in that the split spin–orbit band is shown, and a transiton from it to the conduction band (the center arrow) is also allowed.

temperature. Moreover, there is a likelihood that the LO phonon distribution will also be driven out of equilibrium, and will have to decay through three-phonon processes involving other phonon modes. This creates a complicated system of coupled photon–phonon–electron (or/and hole) distributions that remains quite difficult to solve for the dynamical response.

There are many complicated time scales for this coupled system. While the average scattering time for the polar LO mode phonon in GaAs is about 200 fs, it is possible for the electrons to emit these phonons within the first ~ 25 fs after photo excitation [31]. On the other hand, the carrier-carrier interaction builds up on the same time scale. For an intense photon beam in which ~ 10^{18} cm^{-3} electron–hole pairs are generated, the screening plasma is built-up on the 10 fs scale, and the coupled plasmon mode will hybridize with the polar LO phonons, shifting the apparent phonon energy to higher values. In this regime, the phonon replicas of the narrow absorption energy in the conduction band can no longer be distinguished and only a broad distribution is seen already at 50 fs after the pulsed laser source [31]. In analyzing this coupled system completely, it is actually necessary to treat the inter-carrier interactions in real space to fully account for the various parts of the interaction [32, 33].

As remarked above, the photo-excited carriers will relax toward the band extrema by the emission of phonons. Once this occurs, then the electrons and holes can begin to recombine by a radiative recombination process (which is characterized by a hydrodynamic time, in the sense of section 1.1.1). Measurements of the

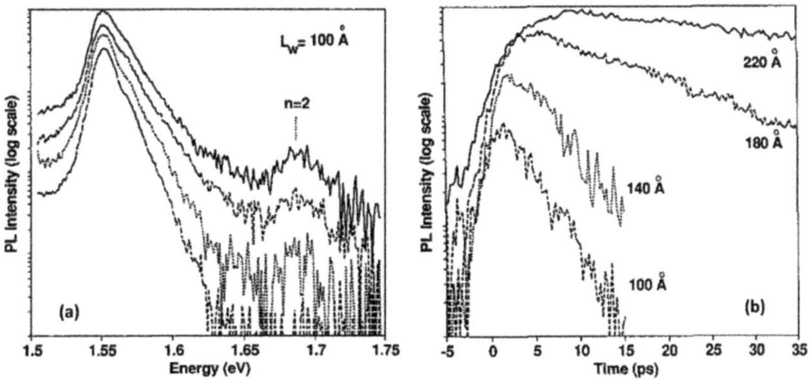

Figure 1.4. (a) Luminescence spectra measured at 100 K, for different decay times after the pump pulse for a multi-quantum well system in GaAs. The quantum wells are 10 nm thickness. The time delays from top to bottom are 1, 2, 5, and 10 ps after the pump pulse. (b) Time-delay behavior of the luminescence for the $n = 2$ peak for samples with different quantum well thicknesses. Reprinted from Morris *et al* [34], copyright (1994) with permission from Elsevier.

recombination radiation is termed photoluminescence, and such emission from GaAs quantum wells is shown in figure 1.4 [34]. The luminescence is collected from a 20 period superlattice of GaAs quantum wells and $Ga_{0.7}Al_{0.3}As$ barriers. Illumination is provided by a dye laser at 2.06 eV, with 0.4 ps pulses. In panel (a) of the figure, the energy dependence of the lowest energy transitions from electron level to hole level are shown for several different times after the pump pulse. Notice that it is the logarithm of the radiation that is plotted, but it may be noticed that each curve nearly has a characteristic slope in panel (a), which have some relation to a 'temperature' of the nonequilibrium electrons and/or holes. The increasing slope with time is an indication of the cooling of the electron–hole system with time. In panel (b), the time dependence of the $n = 2$ transition peak luminescence is shown for different quantum well thicknesses. The faster decay times are likely due to the increased number of phonon decay paths due to the zone folding that occurs in quantum wells and superlattices.

In some cases, the slopes of the luminescence are nearly the same for quite different laser intensities. This is a result of Einstein's explanation of the photoelectric effect [35]. While the electrons and holes generated do not have sufficient energy to overcome the work function and be emitted from the semiconductor, it is still true that the excess energy passed to the photo-generated electrons and holes is a function only of the energy of the photon and not of the intensity of the radiation. This will still be the case after the particles emit their phonons and relax to states near the band edges. Hence, their equivalent temperatures, if such exist, should be nearly the same whatever the intensity of the radiation, to first order. Secondary effects can, however, occur. One can immediately think of electron–electron (or hole–hole and/or electron–hole) scattering as redistributing the energy in the distribution [36]. Then there is the situation in which the phonon decay encounters a bottleneck, that slows the decay.

Pushing particles to higher energy will lead to additional phonon emission and a lowering of the effective temperature. At high densities, however, the self-energy arising from the real part of the carrier–carrier interactions can be large enough to create a band-gap narrowing effect, through the role of the self-energy in lowering the kinetic energy of the carriers.

1.3 Hot carriers in devices

It is hard to underestimate the importance of hot carriers, especially when we consider semiconductor devices. As we remarked above, a modern nanoscale MOSFET of say 20 nm gate length, an applied voltage of 1 V produces an average electric field of 0.5 MV cm^{-1} in the channel. And this is the *average* field. This is 25× the field at which velocity saturation begins in Si. Modern integrated circuits are composed of billions of these transistors, so one must carefully control the way the fields are present in these devices. Impact ionization begins in Si at fields of only 0.2 MV cm^{-1}, so control of the field is essential. Nevertheless, the size of transistors continues to grow smaller.

The growth in the number of transistors on a leading technology integrated circuit has been driven, and in turn is calibrated, by Moore's Law. Moore's Law is not a physical law, but an empirical law coming from economics rather than physical science. Transistors and computer components tend to be laid out on the microchip in a planar, rectilinear fashion much like the houses in any block of a modern city. This allows their length scale to be reduced by a manner such that the area is linearly reduced (approximately by a factor of 2) in each new generation. The main technology utilizes silicon as the semiconductor of choice. This is important, as one finds that it is the cost of silicon area on the chip that drives Moore's Law. Originally, the rule was composed of three parts: (i) reducing the transistor size, (ii) increasing the size of the microchip itself (no longer followed due to heat dissipation problems), and (iii) circuit cleverness, by which the number of transistors needed to perform a function is reduced, thus saving area [37]. Originally, the microcomputers contained a great many peripheral chips and these were laid out on a large circuit. As the transistors were shrunk, more of these peripheral chips could be included on the main processor chip. Since the basic cost of the main chip has not changed much over the intervening decades, this means that the cost per functional block has been greatly reduced. Thus, the cost of computing power has exponentially gone down. It is this economic argument that drives Moore's Law.

1.3.1 FET structures

The field effect transistor (FET) is the oldest known form of transistor, as it was originally patented by Lilienfeld in 1926 [38]. But, a working device was not demonstrated until 1959 [39], as control of the surface was a serious problem which held up development until after the bipolar transistor. Within a few years, the metal-oxide-semiconductor FET (MOSFET) was the preferred device for the integrated circuit due both to its planar technology and its generally lower power dissipation. A schematic drawing of a simple MOSFET is shown in figure 1.5. In this case, it is an

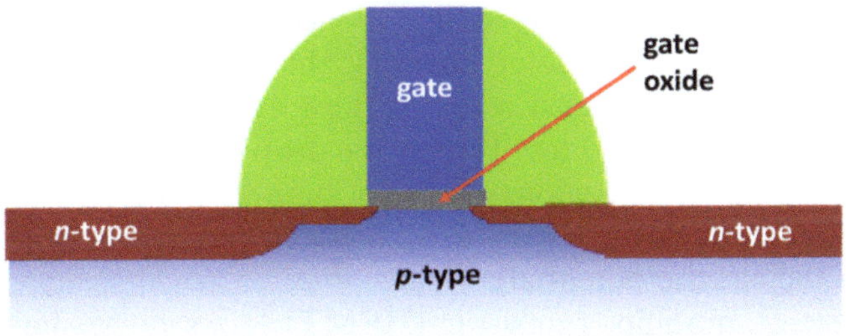

Figure 1.5. Schematic diagram of a planar MOSFET.

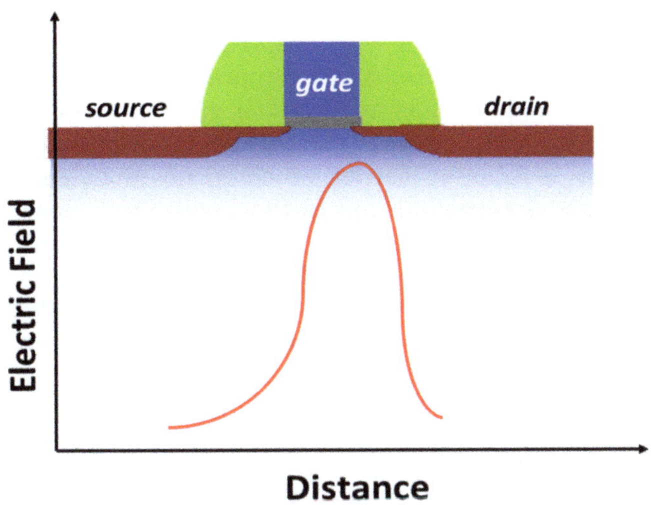

Figure 1.6. Variation of the electric field in the channel region of the MOSFET.

n-channel device, which begins with a *p*-type substrate. Typically, the gate is formed first by the growth of the gate oxide, over which a gate conductor is grown or deposited. Then, the *n*-type source and drain are formed, the (green) sidewall insulators participate in this (the detailed manner in which the transistor is typically fabricated is discussed elsewhere [40]). The device operates on the premise that a positive voltage applied to the gate will lead to an inversion at the surface such that a narrow *n*-type channel will connect the source and drain regions. The source region is usually grounded and positive voltage is applied to the drain. In the sense of this operation, the device is often termed an *enhancement* mode device, because the gate potential opens and increases the size of the *n*-channel, which enhances the conductance between source and drain.

An important aspect of the transistor is the internal electric field. This field is very small at the source end of the channel (at the source contact) and rises throughout the gate region (see figure 1.6). It reaches its peak value at the drain end of the gate, and then drops abruptly in the drain. This is important, as the density in the channel at the source end is set by the inversion density given by the gate voltage, $\sim C_G V_G/e$

(as the source is grounded). This is mainly a sheet density (quasi-two dimensional density). As the carriers move down channel, they are accelerated by the electric field. But, the carrier density must drop, as the total current through the device is constant by Kirchoff's current Law. Hence, the drop in density arises as the carriers spread out into the interior regions (broader than just the channel). Yet, all the carriers are falling through the energy drop between the source and drain (energy is the negative of the electric potential).

The high-electron-mobility transistor (HEMT) looks quite similar in its basic structure, but is crucially different in physics. The *p*-type region of figure 1.5 is replaced, for example, with an undoped GaAs layer (see figure 1.7), and the gate oxide is usually AlGaAs for the typical heterojunction HEMT. The channel will form in the GaAs, or in a narrower band gap material such as InGaAs or GaN, while the undoped 'insulator' is a wider band gap material such as AlInAs or AlGaN, respectively. A band offset barrier appears at the interface between the narrower gap material and the wider gap material, and this barrier keeps the carriers in the channel formed on the narrower gap side of this hetero-interface.

The gate metal is a Schottky barrier which creates the channel in the GaAs between the source and drain without any further applied bias. Then, a negative gate voltage closes this channel. Another difference is that the source and drain are moved some distance from the gate region to avoid breakdown due to the opposite voltages on gate and drain. This is referred to as a *depletion* mode device. But, the operating equations and principles are mostly the same as for a MOSFET.

The electric field in the HEMT closely parallels that of the MOSFET, with most of the field under the gate and peak of the field at the drain end of the gate. However, because of the gaps around the gate, the field can penetrate into the regions at either end of the gate. The gate is spaced further from the drain to reduce any field in this region. In many cases, such as the very high fields in AlGaN/GaN HEMTs, the gap regions are covered with a protective dielectric layer such as Si_3N_4. It is quite natural in either device for the electric field easily to be sufficiently high to cause inter-valley transfer of the electrons. This is not such a problem in Si MOSFETs, as there are not

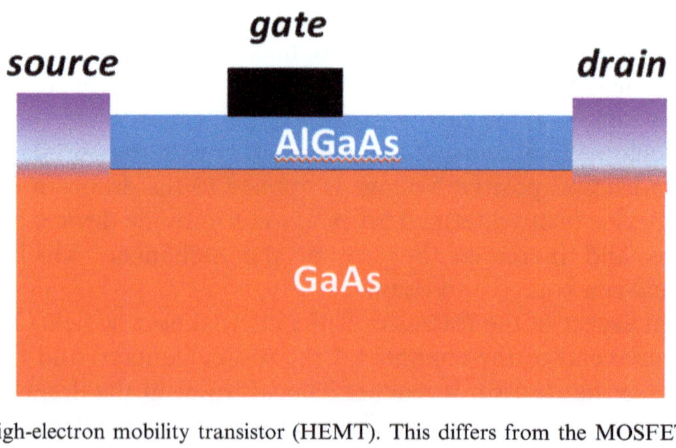

Figure 1.7. The high-electron mobility transistor (HEMT). This differs from the MOSFET in the choice of materials and in the open regions between the source and gate and between the drain and gate.

any satellite valleys close enough to be a problem. In the III–Vs, however, intervalley scattering is a way of life and can offer some advantages for the devices, as will be discussed in chapter 2.

1.3.1.1 Nonequilibrium phonons

Anyone who has felt the case of a modern microchip understands that these things get hot! Although Apple's new M1 chip is designed around a 10 W power envelope, some tests have shown it can dissipate upwards of 20 W or more for a 1–2 cm^2 chip. The details are not as important as the recognition that this is the power being dumped into the actual transistors, and this must eventually be removed by the heat sink. The process of transferring that power from the electrons in the device to the lattice involves the emission of a variety of phonons and these phonons transfer the power to other phonons, and eventually fall to the thermal conductivity to get rid of it. Both the transfer of power from a high energy phonon to low energy phonons and the thermal conductivity depend upon three-phonon interactions. It is a big deal to get this heat out of the devices [41]. It was recognized quite some time ago that the phonons had to be driven out of equilibrium to achieve this task [42]. It was found that the transfer of energy from the electrons was dominated by the phonons connecting valleys along the same axis, and the resulting heating was mainly near the drain end of the channel [43]. Quite similar behavior is observed in HEMTs, and the heating can be quite large in these devices. Consequently, the existence of nonequilibrium phonons in semiconductors can be a major field of study [44].

When carriers are highly excited into states well beyond their normal equilibrium states, as in the devices of the last paragraph, whether through the application of an electric field or by photo excitation, they relax their excess energy by phonon emission. When they are in the Γ valley, such as for direct gap III–Vs, this usually involves the polar LO modes. In figure 1.8, we show the situation for optical excitation. Here, we show the central part of the band-structure for $In_{0.52}Al_{0.48}As$ [45], although this is generic for the direct gap materials of interest, such as GaAs and GaN. The direct gap is 1.5 eV and photons of 2.5 eV are assumed. Most of the excess energy goes into the electrons exciting them into higher energy states in the conduction band. These carriers relax by the emission of the LO modes as shown by the blue cascade in the figure. Only excitation from the heavy hole band is shown, and roughly 90% of the excess energy goes into the photo-electrons. Hence, the average photo-electron will emit something like 20 polar LO phonons in this cascade. It is quite likely that the phonon distribution, normally the Bose–Einstein distribution, will be driven out of equilibrium for those wave numbers that can be excited by the electrons. This will be discussed further in chapter 4.

1.3.1.2 Light emission

Light emission from III–V devices would not surprise anyone. Most of these devices use materials that have a direct band-gap and are naturally used in LEDs and diode lasers. But, light coming from Si devices is a surprise, because Si has an indirect band-gap and is not expected to be an emitter in this manner. To be sure, light emission has been seen in Si *p-n* junctions under conditions of diode breakdown [46, 47].

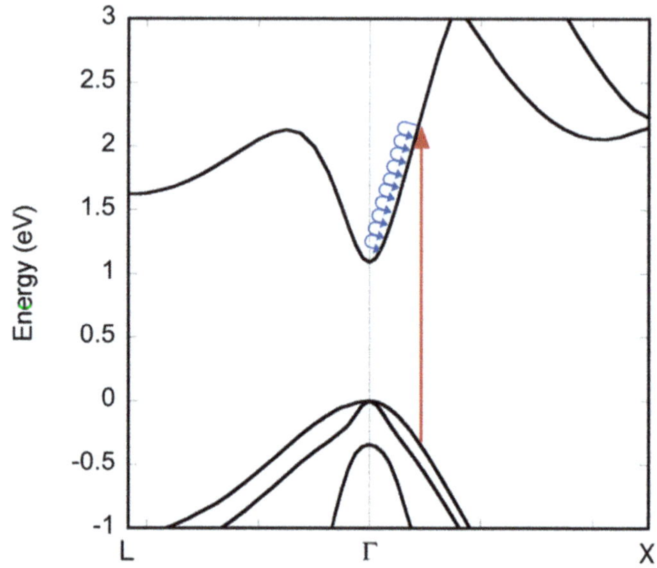

Figure 1.8. Highly excited electrons, such as those shown here being photo-excited from the valence band to the conduction band (red), can relax by emitting a cascade of optical phonons (blue). The large number of emitting phonons can drive the phonon distribution out of equilibrium.

However, light has been observed from MOSFETs operating under hot electron conditions [48]. It was suggested that this radiation was the result of a bremsstrahlung effect of the hot carriers being deflected by impurity atoms [49]. However, the more energy a carrier has, the less it is deflected by the Coulomb field of an impurity, which is contrary to such an explanation. The spectrum of the light was shown to be Maxwellian with a temperature in the range of 1000–1500 K for a 1.5 μm channel length device, with the lower temperatures at higher values of the gate voltage [50]. Spatial studies showed that this light emission was coming from the drain end of the gate. Surprisingly, this emission was observed over the range 0.75–1.55 eV, and that meant some light was sub-band-gap in origin [51]. The latter authors also showed that electron–hole recombination radiation led to additional spectrum different from the back-ground light emission.

Eventually, it was determined that the likely source of the light emission was not the bremsstrahlung effect, but was transitions within the conduction band itself [52–54]. That is, these light-emitting transitions involved only a single type of carrier—the electrons in the *n*-channel devices. Simulations using the full band structure for silicon identified the emission as a direct transition between different bands of the conduction band itself [52]. Then, the observation of peaks near 1.8–2.0 eV were due to reflections in the oxide that signaled the importance of the inter-conduction band process [53]. Further studies indicated that the impurity densities were far too low for effective bremsstrahlung emission and a shoulder at the lower energy range, below the band-gap, was not expected for this type of emission, thus supporting direct transitions within the conduction band [54].

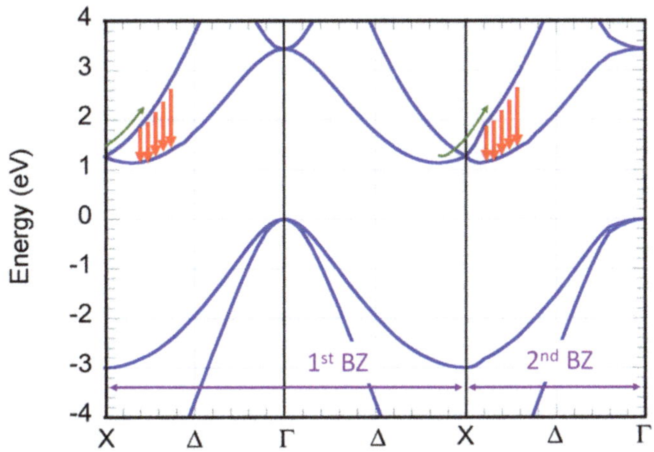

Figure 1.9. The band structure of Si along the (100) axis into the second Brillouin zone. Electrons in the conduction band minimum along Δ can be excited into an upper conduction band as they cross the zone boundary, and then can undergo direct optical transitions.

To understand how this can be the case, reference to the band structure must be made. In figure 1.9, the band structure for Si is shown along the X-Γ-X-Γ (100) axis, passing through the first Brillouin zone into the second zone. Since the momentum is only defined within a reciprocal lattice vector, electrons reaching X can be transported to negative value of X, or appear to progress across the zone boundary into the second zone. As carriers are heated in an electric field, they will move up the band to the zone boundary and then continue into a higher lying band. Here, they can begin to undergo direct optical transitions back to the lower conduction band, and give rise to the observed spectrum. And, there is no reason for this emission to be cutoff at the band gap. The observation of below band-gap radiation is a supporting sign of this inter-conduction band optical transition.

1.3.2 The hot carrier solar cell

Solar cells are a growing renewable energy source. As a result of this, there continues to be interest in improving the technology, especially if one can improve the efficiency attained in these cells. As in many fields, the science that forms the basis for these cells is old. Bequerel discovered the photovoltaic process in 1839 [55], using platinum electrodes on a silver halide. Fritts developed the first solar cell apparently in 1883 [56], using gold electrodes on selenium. The first *p-n* junction solar cell was patented in 1946 [57]. Changing to Si as the material choice provided relatively efficient cells in 1954 [58]. The search for even greater efficiency has brought into play a great many materials today.

It is important to realize that the normal *p-n* junction diode quite differently than a *p-n* junction solar cell. A normal *p-n* junction diode is a *minority* carrier device. When the diode is forward biased, carriers move from high density to low density regions. Electrons are injected into the *p*-region by traversing the *p-n* junction and then recombine with holes. Holes are injected into the *n*-region by traversing the *p-n* junction and then recombine with electrons. Both processes provide a current. and eventually combine with the electrons to contribute to the current. The directions in which these minority carriers move are *opposite* to their movement in response to the built-in electric field, as indicated in figure 1.10(a). This electric field is generally never mentioned in discussions of the diode current.

Modern heterostructure solar cells should be considered a *majority* carrier device. Electron–hole pairs occur wherever photons are absorbed throughout the device. This usually occurs in the depletion region. The much higher absorption coefficients in the compound semiconductors means that this depletion region can occupy most of the cell itself. Usable photogenerated current occurs at voltages less than V_{to}, or more specifically, less than the open circuit voltage V_{OC}. At this voltage, the current is zero, even under illumination. In any solar cell, the carriers need to reach the junction depletion region, simply because that is where the electric field separates the electrons and holes and thus provides a photo-voltage. For this to be efficient, the depletion region should occupy as much of the cell volume as possible, and, in any case, be as close to the illuminated surface as possible. This is, of course, easier to do in a III–V cell than in e.g. silicon. This is depicted in figure 1.10(b).

Within the depletion region itself, the carriers respond to the built-in electric field of the junction. The photo-electrons move toward the *n*-region, acting as excess majority carriers. The holes move toward the *p*-region, where they also will be excess majority carriers. In both cases, the carriers move to their respective contacts entirely as drifting majority carriers. The resulting photocurrent depends entirely upon their movement in response to the electric field.

Just as in the case of optically created hot carriers in section 1.2 above, each particle of the electron–hole pair will usually have a kinetic energy that is away from the band extrema. For example, the solar spectrum peaks at around 2.0–2.5 eV, while the band gap of a typical semiconductor in the solar cell is around 1.0 eV. Thus, the electron and hole have a combined kinetic energy of some 1.0–1.5 eV, which is a considerable excess energy. This has to be relaxed by the emission of phonons. As we will see in chapters 4 and 5, this is likely the optical phonon, which has to then decay into acoustic phonons for the energy to finally go into heating of the lattice.

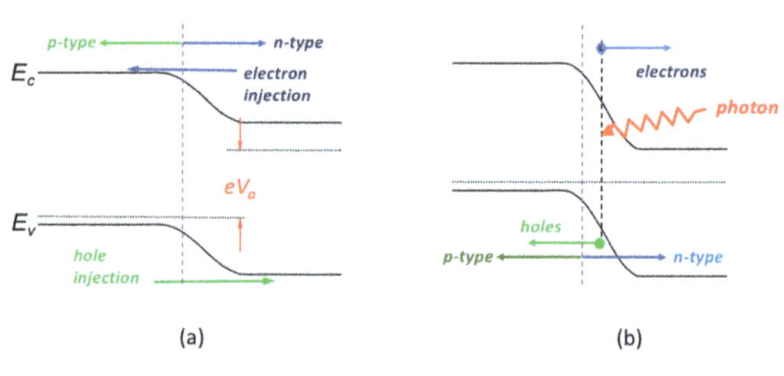

Figure 1.10. (a) Carrier injection in a normal, forward-biased *p-n* junction. The injected carriers become minority carriers in the respective region. (b) The electron and hole pair created by a photon in the depletion region, are separated by the built-in electric field and become majority carriers in their respective regions.

There are several sources of loss in modern solar cells, that still exist in spite of the years of research and development. These are displayed in figure 1.11. The important point is the energy gap of the semiconductor that provides absorber layer(s). Photons whose energy lies below the band gap will not be absorbed in the material, although there has been work on using mid gap levels, and multi-photon absorption to overcome this limit, but the fundamental limit is still with us. Hence, those photons that have energy below the band gap still represent a loss from the solar spectrum. The efficiency of the solar cell, under usual conditions, is thought to be limited by the Shockley–Queisser theory, which is dependent upon some thermodynamic limits evaluated under detailed balance approximations [59], which are probably questionable in their own right. From the figure, it is evident that, for sufficiently low energy photons, the largest single loss mechanism is the cascade of phonons emitted as the hot carriers cool to the band edges (the green section of the figure). For example, with an absorber band gap of 0.75 eV, some 45% of the solar energy is lost to the phonons. If one could significantly reduce this effect, the cells could be far more efficient.

In 1982, Ross and Nozik proposed that a hot carrier solar cell (HCSC) could be made that would reduce the thermal losses and provide a much higher efficiency [60]. To achieve the predicted efficiencies, they pointed out that one has to achieve two objectives: (1) prevent the photo-carriers from thermalizing to the band edges by the emission of optical phonons, and (2) extract only the hot carriers into the contacts through an energy-selective contact. Since then, an extensive effort has been made to achieve these goals with limited success. These two requirements point out the role that non-equilibrium phonons play in the solar cell and point to the use of heterostructures to create the necessary barriers such that only the hot carriers are

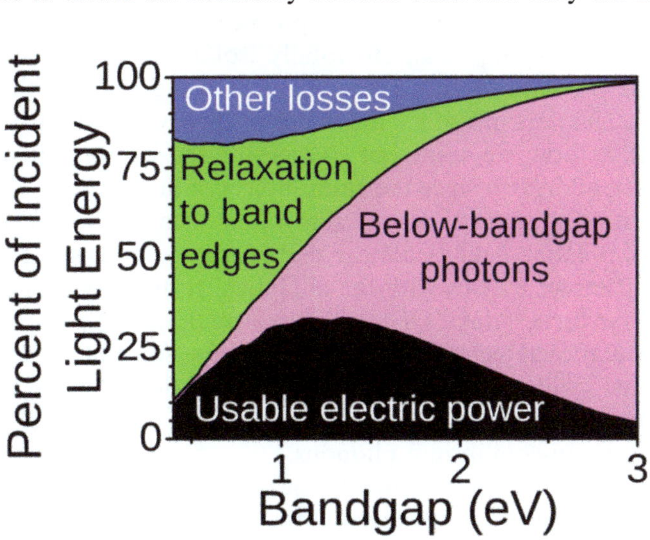

Figure 1.11. The principal losses in solar cells. The black region is the Shockley–Queisser limit which has limited solar cell progress. The green section is loss to the phonon cascade by the absorbed hot carriers. A modified solar spectrum as it appears at the earth's surface has been assumed. (From Wikipedia Commons, attributed to Sbyrnes321, used under the Creative Commons license.)

collected. One approach that has been pursued heavily is to seek a blockade of the phonon emission process, and this is often created by the phonons themselves when they are far out of equilibrium. The blockade can also appear when the decay pathway of the optical phonons is also out of equilibrium, thus slowing the decay itself. More recently, it has been suggested to park the electrons in the upper valleys of the conduction in order to reduce the losses to the phonons [34, 61]. These topics will discussed further in chapter 4.

1.4 What is in this book

This introduction has primarily served to lay out the wide scope of fields in which hot carriers are important. Nevertheless, it cannot have described the full range of these fields, and the detailed study of such fields is to be carried out in the subsequent chapters. The reader has probably already noted that few experimental details have appeared in this chapter; this will hopefully be rectified in the following text. In chapter 2, the nature of hot carriers that arise in high electric fields will be treated beginning with the simplest measurements of velocity and mobility. The wide range of inter-valley scattering and measurements of these events also will be discussed, both for d.c. fields and for microwave fields. In chapter 3, this discussion is repeated for mesoscopic devices and materials at low temperature.

In chapter 4, we turn to optically induced hot carriers and their measurements. Although this chapter focuses upon the optical generation of the carriers, mention of optical measurements of high electric field effects, by nature, will overflow the previous chapter as well as this one. The modifications that appear in quantum wells and heterostructures will also be discussed. In chapter 5, the discussion turns to the issue of non-equilibrium phonons and the problems, and perhaps advantages, that they create. Again, it is impossible to totally isolate this issue from the previous chapters, but the focus here will be the dynamics of the phonons themselves.

The last three chapters introduce the realm of theory and simulation, addressing the question as to how we treat hot carriers in the computational world. This question has been addressed since the earliest days of hot carriers. These treatments will connect with the general knowledge of condensed matter physics, but a knowledge of this latter field is almost certainly a prerequisite for dealing with these chapters. Chapter 6 deals with the analytical forms of simulation and theory that arose from the earliest days. Of course, these analytical approaches can easily be treated by computational techniques, as the latter ease some of the difficulty of some integrals. Chapter 7 will entail the ensemble Monte Carlo technique, and the manner in which it is applied to the wide range of hot carrier physics, including statistical approaches to the non-equilibrium phonons.

Finally, in chapter 8, an introduction to the quantum approaches that extend, and parallel, the semi-classical approaches will be given. By necessity, this cannot be a full and complete treatment. A more extensive treatment may be found in Ferry [62]. While many would want to approach the full quantum transport treatment of hot carriers, it is worth pointing out that such an approach usually brings little new information except on the very short time scale (fs), while at the same time considerable new difficulty. A rationale for this statement will be provided in the chapter, but it is in no way intended to dissuade those who wish to undertake the more difficult path. Nevertheless, those pursuing this path should be encouraged to approach it properly and not stop at relatively easy solutions which avoid the entire rationale of the full quantum approach (especially when using the non-equilibrium Green's functions) [63].

References

[1] von Hippel E 1933 *Zeitschr. Elekrochem.* **39** 506
[2] Fröhlich H 1937 *Proc. R. Soc.* A **160** 230
[3] Heisenberg W 1925 *Z. Phys.* **26** 879
[4] Schrödinger E 1926 *Ann. Phys.* **79** 361
[5] Davisson C and Kunsman C H 1921 *Science* **54** 522
[6] Thompson G P 1927 *Nature* **119** 890
[7] Fröhlich H and Seitz F 1950 *Phys. Rev.* **79** 526
[8] Fröhlich H and O'Dwyer J 1950 *Proc. Phys. Soc.* A **63** 81
[9] Ferry D K 2020 *Semiconductor Bonds and Bands* 2nd edn (Bristol: IOP Publishing)
[10] Van Vleck J H 1941 *Phys. Rev.* **59** 724
[11] Weinrich N 1965 *Solids: Elementary Theory for Advanced Students* (New York: Wiley)
[12] Bardeen J and Brattain W 1948 *Phys. Rev.* **74** 230
[13] Khang D 1960 *U.S. Patent* 3,102,230
[14] Kilby J S 1959 *U.S. Patent* 3,138,743
[15] Noyce R N 1959 *U.S. Patent* 2,981,877
[16] Fröhlich H 1947 *Proc. R. Soc.* A **188** 532
[17] Price P J 1979 *Semiconductors and Semimetals* ed R K Willardson and A C Beer vol 14 (New York: Academic) pp 249–308
[18] Wick F G 2008 *Phys. Rev. (Ser. I)* **27** 11
[19] Ryder E J 1953 *Phys. Rev.* **90** 766
[20] Mori H, Oppenheim I and Ross J 1962 *Studies in Statistical Mechanics* ed J de Boer and G E Uhlenbeck (Amsterdam: North Holland)
[21] Chester G V 1963 *Rep. Prog. Phys.* **26** 411
[22] Ryder E J and Shockley W 1951 *Phys. Rev.* **81** 139
[23] Shockley W 1951 *Bell Sys. Tech. J.* **30** 990
[24] Fröhlich H 1954 *Ann. Phys.* **3** 325
[25] Gunn J B 1963 *Sol. State Commun.* **1** 88
[26] Ridley B K and Watkins T B 1961 *Proc. Phys. Soc. London* **78** 293
[27] Ridley B K and Watkins T B 1961 *J. Phys. Chem. Sol.* **22** 155
[28] Ferry D K 2017 *Semicond. Sci. Technol.* **32** 085003
[29] He G, Nathawat J and Kwan C-P *et al* 2017 *Sci. Rep.* **7** 11256
[30] Ferry D K 2019 *Semicond. Sci. Technol.* **34** 044001
[31] Osman M A and Ferry D K 1987 *Phys. Rev.* B **36** 6018
[32] Ferry D K, Kriman A M, Kann M J and Joshi R P 1991 *Comp. Phys. Commun.* **67** 119
[33] Kriman A M, Kann M J, Ferry D K and Joshi R P 1990 *Phys. Rev. Lett.* **65** 1619
[34] Morris D, Houde D, Deveaud B and Regreny A 1994 *Superlatt. Microstruc.* **15** 309
[35] Einstein A 1905 *Ann. Phys.* **19** 82
[36] Joshi R P, Grondin R O and Ferry D K 1990 *Phys. Rev.* B **42** 5685

[37] Moore G 1965 *Electronics* **5** 114
[38] Lillienfeld J E 1926 Method and apparatus for controlling currents *US Patent* 1,475,175
[39] Atalla M M, Tannenbaum E and Scheibner E J 1959 *Bell Syst. Tech. J.* **38** 749
[40] Ferry D K 2020 *Transport in Semiconductor Mesoscopic Devices* 2nd edn (Bristol: IOP Publishing)
[41] Vasileska D, Raleva K and Goodnick S M 2008 *J. Comp. Electron.* **7** 66
[42] Van Vleck J H 1941 *Phys. Rev.* **59** 724
[43] Aksamija Z and Ravaioli U 2006 *J. Comp. Electron.* **5** 431
[44] Ferry D K 2021 *Appl. Phys. Rev.* **08** 021324
[45] Welland I and Ferry D K 2019 *Semicond. Sci. Technol.* **34** 064003
[46] Chynoweth A G and McKay K G 1956 *Phys. Rev.* **102** 369
[47] Goetzberger A and Stephens C 1961 *J. Appl. Phys.* **32** 2646
[48] Tam S, Hsu F C and Ko P K *et al* 1983 *Electron Dev. Lett.* **4** 386
[49] Tam S and Hu C 1984 *IEEE Trans. Electron Dev.* **31** 1264
[50] Toriumi A, Yoshimi M and Iwase M *et al* 1987 *IEEE Trans. Electron Dev.* **34** 1501
[51] Herzog M and Koch F 1985 *Appl. Phys. Lett.* **53** 2620
[52] Bude J and Yoshii A 1992 *Phys. Rev.* B **45** 5848
[53] Hublitz K and Lyon S A 1992 *Semicond. Sci. Technol.* **7** B567
[54] Lacaita A L, Zappa F, Bigliardi S and Manfredi M 1993 *IEEE Trans. Electron Dev.* **40** 577
[55] Bequerel E 1839 *Comp. Rend.* **9** 561
[56] Fritts C E 1883 *Am. J. Sci.* **26** 465
[57] Ohl R S 1946 *U.S. Patent* 2402662
[58] Chapin D M, Fuller C S and Pearson G L 1954 *J. Appl. Phys.* **25** 676
[59] Shockley W and Queisser H J 1961 *J. Appl. Phys.* **32** 510
[60] Ross R T and Nozik A J 1982 *J. Appl. Phys.* **53** 3813
[61] Ferry D K 2019 *Semicond. Sci. Technol* **34** 044001
[62] Ferry D K 2017 *Quantum Transport in Semiconductors* (Singapore: Pan Stanford Publishing)
[63] Ferry D K, Nedjalkov M and Weinbub J *et al* 2020 *Entropy* **22** 1103

IOP Publishing

David K Ferry

Chapter 2

High electric field transport

In theoretical studies of hot carrier transport, the over-riding problem is the need to obtain the relevant distribution function in the presence of both the high electric field and the varying properties of the scattering processes as the average energy of the carriers increases. For transport properties, the distribution function is not an end in itself, since integrals over this function must be performed in order to evaluate the transport coefficients, the velocity and the mobility, etc. To the experimentalist, however, interest lies in the observable quantities that describe measurable transport. An important point is that meaningful theoretical advances are not usually achieved without a strong understanding of the measurable properties of the system of interest. So, before addressing how the theory can be approached, the description of the experimental observables will begin with this chapter on electric field effects.

As pointed out in chapter 1, the idea of 'hot' carriers may actually have begun with von Hippel [1], but it was surely supported by the later work of Fröhlich [2]. By the term hot carriers, we mean that the electron system has acquired a temperature, from the electric field to which it is subjected, that is larger than the lattice temperature. In this sense, the electron system is no longer in thermal equilibrium, especially with the lattice, and one has already made an assumption that a temperature can describe the electron system. In general, the transport of these hot carriers is quite nonlinear in that the conductance of the material is now a function of the electric field, or voltage, applied to it. But such temperature arguments are good only for the Maxwellian or Fermi–Dirac distribution, something that is often not possible. The statement itself is nothing more than a conservation of energy in the system, although this itself is not guaranteed. This conservation of energy should be maintained in order to establish a nonequilibrium steady-state in the system.

The reason for this non-specificity in the distribution function can be seen from figure 2.1. In the absence of an electric field, the distribution is symmetry in momentum space, due to the symmetry of the cubic lattice. However, when an

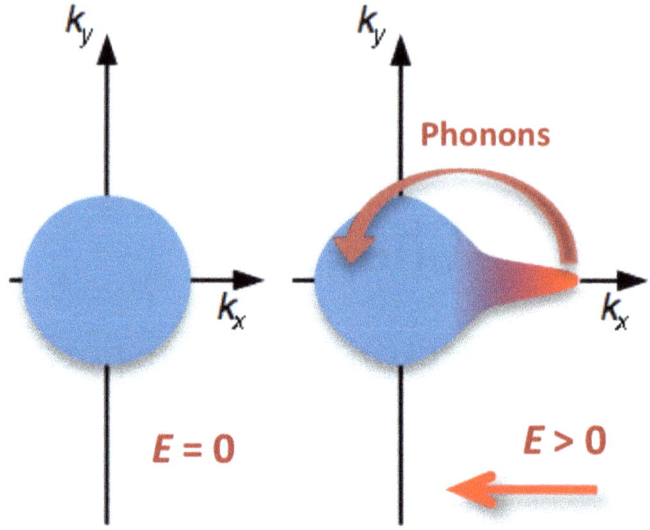

Figure 2.1. Role of the electric field in inducing asymmetry in the distribution function. Then, relaxation by the phonons usually balances the driving force of the electric field.

electric field is applied, the carriers move in response to this field. The direction indicated is for electrons, which move toward the higher voltage. This leads to a streaming of the distribution in the field direction. This streaming is counter-balanced by the emission of phonons, which may be a single one or an entire cascade. In general, a steady state is achieved when the excitation by the field is balanced by the relaxation of the phonons. Quite often, as will be seen in the remaining chapters, the distribution can not be described by a temperature. Then, it may be termed to be athermal.

Sometimes, steady state is not the stable state. The simplest case in which steady-state is *not* obtained comes from electron transfer between non-equivalent valleys, such as transfer from the Γ valley to upper L and X valleys in the Gunn effect [3, 4]. In this case, the electrons transfer from a high mobility central valley to the low mobility satellite valleys, and this causes a negative differential mobility (conductance) to appear in the current-electric field relation. This negative differential mobility is a region of instability and leads to traveling high electric field domains [3, 5]. Hence, no steady-state exists in this situation. These coherent structures (the domains) can be maintained only through a sufficient flow of energy into the system, and have been termed dissipative structures [6]. Even so, the distribution function must be carefully determined if these effects are to be properly treated and understood. The motion of these domains can actually be observed, a point that will be considered further in this chapter.

2.1 Velocity and mobility

Early studies of high electric field transport in solids focused mainly on the breakdown studies of dielectrics [7], although the electrical properties of silicon

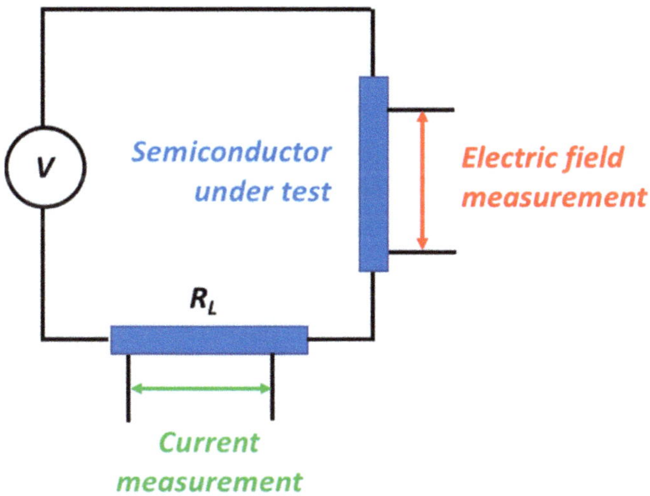

Figure 2.2. A test circuit, first described by Wick [8] allows measurements of the electric field and current through a semiconductor sample while protecting against contact effects.

had been studied much earlier [8]. Studies of the variation of the velocity with high electric fields began with Ryder [9], who discovered that the velocity in Si and Ge saturated at high electric fields. The saturated (or nearly saturated) velocity is an important parameter for electron device considerations. The saturated velocity is also important as a mirror into the electron–phonon interactions in the material. But, this is only one of the many observables that are encountered in the study of semiconductors.

Already, in the early years, Wick [8] had realized that measurement would have many problems. The circuit he described for these measurements has many modern attributes, such as the one shown in figure 2.2. First, of course, is the semiconductor structure that is being measured, which may be taken to have a resistance R_S. Current is introduced to the sample by contacts at the end as shown. But, these current-carrying contacts have so-called 'contact resistance' that disturbs the homogeneity of the electric field within the structure. So, voltage measurements are made with a pair of non-current-carrying contacts placed upon the side of the sample. Knowing the distance between these two contacts allows one to determine the electric field from the voltage measured across them. Then, a small load resistance R_L is placed in series with the semiconductor under test, with the requirement that $R_L \ll R_S$. The voltage across this load resistor allows one to determine the current through the semiconductor sample. Then, from the properties of the sample, or from measurements such as the Hall effect, one knows the carrier density and its dimensions, so that the velocity can be determined from the current measurement. The only modern modification of this approach arises from the realization that the sample is quite likely to heat up under the applied power. To avoid this sample heating, one uses pulsed voltage sources.

One of the first actual measurements of velocity were those of Ryder [9]. Using a pulsed voltage source and a variant of figure 2.2, he measured the velocity of Si and

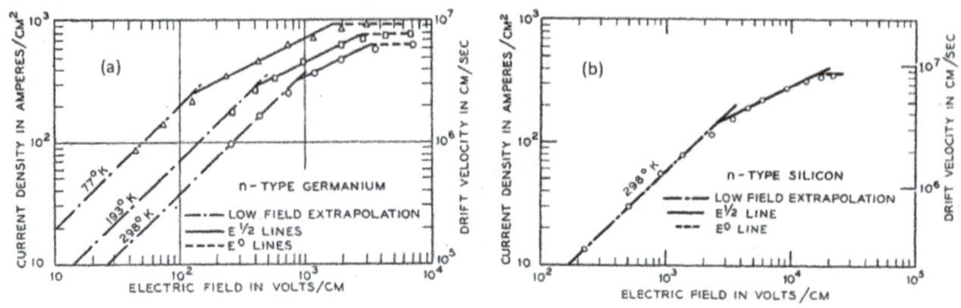

Figure 2.3. (a) Current density, and velocity, in *n*-type Ge as a function of the electric field. (b) Current density, and velocity, in *n*-type Si as a function of electric field. Reprinted with permission from E J Ryder [9], copyright (1953) by the American Physical Society.

Ge at a variety of temperatures. This work identified both the warm electron and the hot electron regime, as is illustrated in figure 2.3. He clearly observed velocity saturation in both materials, and concluded that this saturation was a result of the electrons transferring their energy to the optical phonon modes, and that this process would support the concept of a saturated velocity (we will return to this point below). From the observed velocity as a function of the electric field, one can easily obtain the chordal mobility for these same fields. The author also took care with the proper evaluation of the contacts to avoid the injection of minority electrons.

2.1.1 Saturated velocity

Another method of measuring the velocity is a variation on figure 2.2, and derives from the Haynes–Shockley experiment [10]. In this experiment, a very high resistivity sample is used, and electrons (or holes) are injected at one end of the sample, and the temporal length of the current pulse, due to these injected electrons, is measured. Then, from the physical length of the sample and the transit time given by the temporal length, a velocity can be inferred. Such a measurement was carried out by Canali and Ottaviani on Si to measure the temperature dependence of the saturated velocity [11]. The electrons were created by a short (0.15 ns) pulse of either 40 keV electrons or 5.47 eV natural α particles from ^{241}Am. The electric fields were oriented along either the (111) direction or the (100) direction. The former direction makes an equal angle with all six ellipsoids of the Si conduction band so these ellipsoids should all heat equally. In the case of the (100), the field does not make an angle with all six ellipsoids, but separates the group into two sets: one pair whose major axis is parallel to the field, and a set of four which do make an equal angle with the field. This can lead to some anisotropy in the velocity, especially at lower temperatures, but the effect is still seen at room temperature. We will return to this later. Their results are shown in figure 2.4, and compared with results from Duh and Moll [12] and from Rodriguez and Nicolet [13]. A theoretical comparison was provided by Costato and Reggiani [14], but gave values higher than any of the experiments. The values of the saturated velocity at room temperature were just below 10^7 cm s^{-1}, which agree fairly well with the value obtained by Ryder. This continues to be the accepted value even today, although smaller values are sometimes found in device studies.

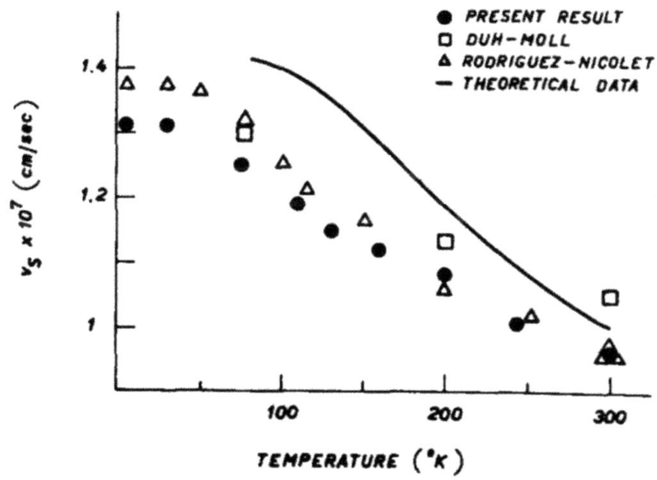

Figure 2.4. Saturation values of the drift velocity in silicon for the (111) direction as a function of the temperature of the lattice. The experimental errors are all here of the order of ±5%. The results are compared with those of Duh and Moll [12] and Rodriguez and Nicolet [13]. The theory curve of Costato and Reggiani was eventually published [14]. Reprinted from Canali and Ottaviani [11], copyright (1970) with permission from Elsevier.

Another result of the time-of-flight studies is the ability to extract a field dependent diffusivity for the hot carriers that result from using a high electric field. A reason for studying hot carrier diffusivity is that the Einstein relation is no longer valid in a nonequilibrium situation. Thus, any relationship between the mobility and the diffusion constant must be obtained experimentally. Usually, the excitation pulse will have a Gaussian shape, and this will spread during the transit of the carriers. As the Gaussian exits the sample, it will induce a fall time of the current pulse which is related to the Gaussian spread. From this, one can often extract a diffusivity for the carriers.

It was remarked above that one expected to achieve a saturated velocity when the energy was effectively transferred to the optical modes of the phonons. This is true when these optical modes are the nonpolar modes, such as those that dominate the energy relaxation in Si. It turns out in the case of nonpolar phonons that the saturation velocity is thought to be a scalable parameter from material to material, precisely because it is a mirror into the carrier-lattice interactions. When we have electrons dominated by relative high energy (nonpolar) phonons, then we can estimate the order of magnitude of the energy loss from equating the energy input to the loss as

$$ev_sE \sim \frac{3}{2}\hbar\omega_0\left[\left(N_q + 1\right) - N_q\right] = \frac{3}{2}\hbar\omega_0(e^x - 1)N_q, \tag{2.1}$$

where the factor in square brackets in the first form arises from the difference between emission and absorption of a phonon and v_s is the saturated velocity. (The equivalent relation for polar modes involves the carrier energy, so that scaling does not hold.) Here, (1.3) has been used to relate the two Bose–Einstein factors and

$x = \hbar\omega_0/k_BT$, and T is the lattice temperature for the phonons in equilibrium. The simple form of this equation (and the next one) arises from the fact that the matrix element for the scattering interaction is basically the same in this equation and the next, so will cancel. On the other hand, the balance of momentum which leads to the saturated velocity may be estimated by

$$eE \sim 2m_c v_s(e^x + 1)N_q, \tag{2.2}$$

where m_c is the effective mass, since both emission and absorption of a phonon cause a loss of momentum. The numerical factors in these two equations are a property of nonpolar optical phonon scattering [15]. These two equations may be solved together to yield an estimate of the saturated velocity as

$$v_s \sim \sqrt{\frac{3\hbar\omega_0}{4m_c}} \tanh^{1/2}\left(\frac{\hbar\omega_0}{2k_BT}\right). \tag{2.3}$$

As remarked (2.3) is strictly correct only for nonpolar optical phonons in their role of dominating the scattering, and even in this case (2.3) has appeared in slightly different forms in different derivations that appear in the literature. Nevertheless, the result suggests that the velocity can be scaled among various semiconductors as it depends only upon material parameters and the lattice temperature. In figure 2.5, we show this scaling by plotting the observed saturated velocity (either experiment or calculated) versus the prediction of (2.3). The well-documented experimental values are shown in red, while the calculated expectations are in blue. The green symbols are the peak velocities seen in material which shows negative differential conductance. These latter materials show electron transfer from the Γ to the L or X minima, and the transport in the Γ valley is typically dominated by the polar optical phonon. Nevertheless, the peak velocity seems to agree in some cases with the expected result (2.3). Many materials lie off the line by a significant distance. A simple argument lies in the fact that C, GaP, Si, SiC and Ge are all indirect gap materials with multiple minima of the conduction band. But, so is AlSb. Perhaps the best argument is that one should not put too much faith into such scaling relationships.

2.1.2 Role of the phonons

The question as to why polar optical phonons differ from their nonpolar comrades lies in the nature of the matrix elements. It was pointed out some time ago that the microscopic theory of the electron–phonon interaction will involve the matrix element as a function of the scattering wave number q [16]. Generally, in dealing with scattering through e.g. the Fermi golden rule, scattering is a long-range interaction, so that q is relatively small. If one expands the generic matrix element in a power series in q, the result varies as

$$M \sim \frac{A}{q^2} + B + Cq + \cdots. \tag{2.4}$$

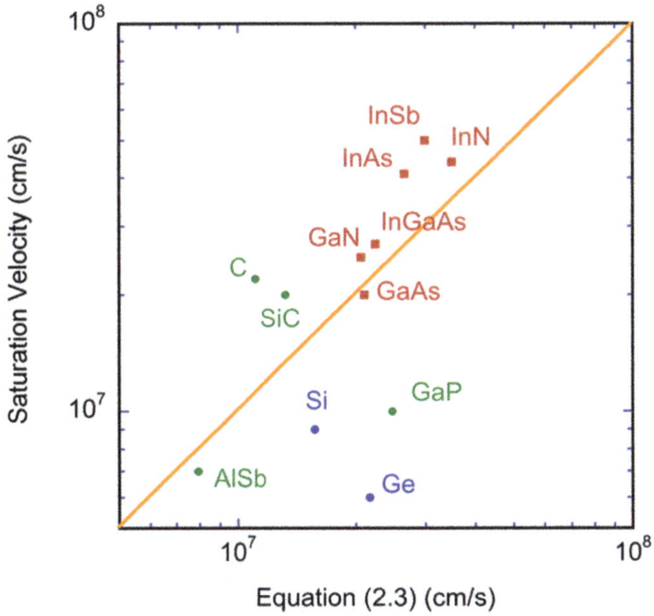

Figure 2.5. Some observed values of the saturated, or peak, velocities compared to the results expected from (2.3). Blue are experimental results for the saturated velocities and red are for the peak velocity. Results expected from simulations are in green.

Generally, the optical phonon scattering is considered in the rigid-ion approximation and this corresponds to the *B* term; that is, the interaction matrix element is independent of the wave number. Acoustic phonon scattering, and first-order optical phonon scattering contribute to the *C* term. The *A* term comprises Coulombic scattering processes. In this *A* term, the elastic contributions arise from impurity scattering and piezoelectric scattering, while the optical contribution is the polar optical mode. These Coulombic processes tend to be anisotropic in the scattering angle and do not reduce the momentum as effectively as the isotropic scattering processes.

The polar optical mode arises from an electromagnetic polarization of the lattice itself, and the strength of the matrix element depends upon the difference between the optical and static permittivities. This interaction is clearly apparent by a splitting of the LO and TO modes at the Γ point in the phonon dispersion in the Brillouin zone. This splitting is coupled to the two dielectric permittivities through the Lydane–Sachs–Teller relationship [17]. Because of the lattice polarization, the LO mode is commonly referred to as the polar LO mode. (There is of course an LO mode in materials like Si, but it has no connected polarization.) This leads to an additional energy dependence in the scattering rate and an energy dependence in (2.3) that is not present in the nonpolar materials. As a result, the velocity does not saturate. This is illustrated in figure 2.6 for three strong polar semiconductors. The data is taken from Glicksman and Hicinbothem [18] and Neukermans and Kino [19] for InSb at 77 K, from Bauer and Kuchar [20] for InAs at 77 K, and from Ruch and

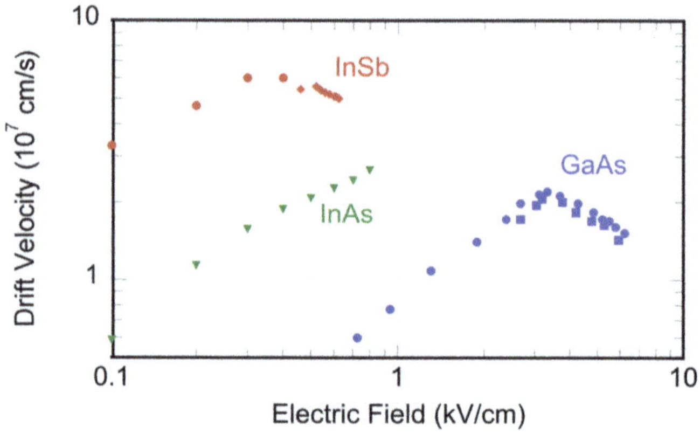

Figure 2.6. Drift velocities for several III–V compounds as a function of the applied electric field. These are all polar materials and do not exhibit any strong saturation of the current. The negative differential conductance in InSb and GaAs are due to inter-valley scattering and the onset of negative differential conductance. (It is known to also occur in InAs, but no measurements have been made to show it.) InSb and InAs are measured at 77 K, while the GaAs data is at room temperature.

Kino [21] for GaAs at 300 K. These materials exhibit quite general behavior for a polar material. While there may be significant reduction in the chordal mobility, saturation is not normally seen while the carriers remain in the Γ valley. The trends in figure 2.6 are that the effective mass gets smaller as the curves lie higher and the mobilities get lower as one moves to the right. As we will see below, the breakdown fields for these three materials are 200 V cm^{-1} for InSb at 77 K, 1 kV cm^{-1} for InAs at 77 K, and about 4×10^5 V cm^{-1} for GaAs at 300 K. The onset of the negative differential conductance in InSb and GaAs is not limited to polar materials as will be shown later in this chapter.

Importantly, the polar LO mode can scatter electrons and holes through the resulting Fröhlich interaction [22, 23],with an effective interaction parameter [24]

$$\frac{1}{\gamma} = \omega_{\mathrm{LO}}^2\left(\frac{1}{\epsilon_\infty} - \frac{1}{\epsilon(0)}\right), \tag{2.5}$$

where ϵ_∞ and $\epsilon(0)$ are the optical and static permittivity. This interaction parameter replaces the normal deformation field or potential in the Fermi golden rule. The field and potential that arise from the charge dipole of the two atomic charges lead to a long-range scattering potential of Coulomb nature. It is important to note that this scattering process falls off very rapidly with wave number q. While the conduction band (or valence band) may have multiple valleys, this scattering process is important really only for intra-valley scattering due to this fact. The inter-valley scattering can appear to be much larger due to the density of states available in the satellite valleys.

2.1.3 Distribution functions

It was remarked at the start of this chapter that determining the distribution function was one of the first problems in studying the hot carriers theoretically. In fact, it is now possible to actually measure some aspects of the distribution function experimentally. This has been done mostly with Raman scattering. Normally, one uses Raman scattering to measure the optical phonons, as will be discussed in chapter 5. This phonon interaction gives rise to the Stokes and anti-Stokes lines arising from emission and absorption of phonons by the photo-excited carriers. But, Raman scattering can also measure parts of the hot carrier distribution [25–27], through single-particle scattering. Generally, one uses a *p-i-n* structure to create a sizable electric field, which is set by an applied bias to the device. Light is sent through the device, parallel to the applied field. The Raman configuration is usually prepared to sense only the single spin scattering [27]. Normally, the sample is illuminated with a picosecond or sub-second pulse from a relatively high energy laser ($\hbar\omega > E_G$) in the back-scattering configuration. The Raman scattering is conducted in an optical configuration such that only the single-particle scattering associated with spin-density fluctuations is detected. This is measured with a standard double spectrometer and photon counting electronics. One important aspect of this is that the Raman scattering cross-section is inversely proportional to the square of the effective mass of the electrons, so that primarily only electrons in the very light mass Γ valleys are selectively probed [28]. The scattering rides on a background signal of luminescence that extends from the Stokes or the anti-Stokes sides of the back-scattering signal. Care has to be taken, as there is often an additional luminescence peak from the spin–orbit split-off valence band that has to be removed from the data.

In figure 2.7, data from a GaAs *p-i-n* structure is shown [29]. Here, a 1.91 eV photon source was used, as this frequency is nearly that of the $E_0 + \Delta$ scattering (from the split-off band) and the data will not be contaminated by this latter luminescence. An electron–hole density of $\sim 10^{17}$ cm^{-3} was generated. If the semiconductor is non-degenerate, as is the case here, the scattered light is approximately given by [30]

$$\hbar\omega = \hbar\mathbf{q} \cdot \mathbf{v} + \frac{\hbar^2 q^2}{2m^*}, \tag{2.6}$$

where **v** is the electron velocity, m^* is the effective mass of the electron *at that velocity*, and **q** is the wave number of the photon. The right-hand side of (2.6) gives

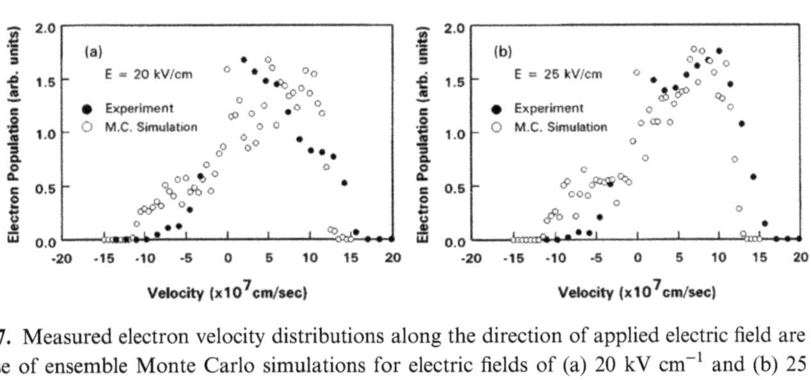

Figure 2.7. Measured electron velocity distributions along the direction of applied electric field are compared with those of ensemble Monte Carlo simulations for electric fields of (a) 20 kV cm^{-1} and (b) 25 kV cm^{-1}, respectively. Qualitatively, the fit is good except for the very high velocity regions. Reprinted from E D Grann *et al* [29], copyright (1995) with the permission of AIP Publishing.

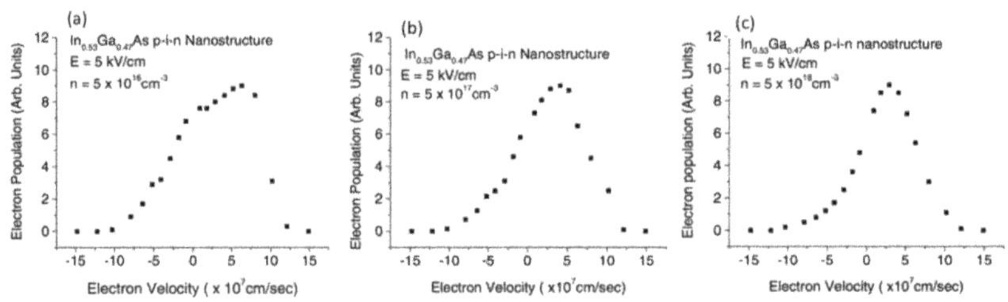

Figure 2.10. Typical nonequilibrium electron distributions for an $In_{0.53}Ga_{0.47}As$ based *p-i-n* nanostructure, taken at an electric field of 5 kV cm^{-1}, with densities of (a) 5×10^{16} cm^{-3}, (b) 5×10^{17} cm^{-3}, and (c) 5×10^{18} cm^{-3}. The electron distribution evolves from a quite non-Maxwellian shape toward a more *shifted* Maxwellian as the density increases. Reprinted from K T Tsen *et al* [33], copyright IOP Publishing. All rights reserved.

One of the points of great interest is the role played by carrier–carrier scattering. In particular, it is known that this scattering is thought to drive the distribution toward a Maxwellian shape, which would make it symmetrical about zero velocity. Experimentally, this is not often the real case. In figure 2.10, this idea is addressed through determining the distribution function at various carrier densities, and hence at varying degrees of carrier–carrier scattering [33]. The sample used is a *p-i-n* structure with InP *n* and *p* regions and an unintentionally doped $In_{0.53}Ga_{0.47}As$ 'intrinsic' region. A single pulse of 0.8 ps width from a mode-locked YAlG laser at 1.92 eV is used (the mode-locking produces these pulses at 76 MHz). All data were taken at room temperature and an electric field of 5 kV cm^{-1} was used at all densities. Three respective densities are shown in the panels of this figure. In panel (a), the density is 5×10^{16} cm^{-3} and it is clear that the distribution cannot be compared in any way with a Maxwellian. In panel (b), the density is 5×10^{17} cm^{-3} and this is likely to be weakly degenerate. Even so, the distribution is not that of a Fermi–Dirac, although it seems to be tended toward a drifted Maxwellian—a Maxwellian centered upon a positive non-zero value of velocity. This becomes even clearer in panel (c), which is for a density of 5×10^{18} cm^{-3}. In this latter panel, it seems to be quite evident that there a carrier 'temperature' that makes the distribution non-degenerate and fairly well described by the drifted Maxwellian format. Another interesting point is that the velocity is only weakly affected by the photo-excited density for this long pulse width, and stays in the range 3–3.4 × 10^7 cm s^{-1} at all densities, with the lower values at the higher densities [33]. While it appears that the changes apparent in the electron distributions are arising from the carrier–carrier scattering, it needs to be pointed out that quite similar effects on the distribution arise from taking into account state filling of the final states of electron–phonon scattering [34]. These two effects seem to have complementary roles in forming the shape of the distribution function. We will return to this discussion in chapter 4.

2.2 Transient transport

As we have previously discussed, there are at least two different relaxation times corresponding to the relaxation of both momentum and energy. In the case of hot carriers, these become important in deciding how the transient behavior occurs when one turns on an electric field. That is, the velocity response to the onset of an electric field is far more complicated than the case at low fields. When we study the transient transport, we have to consider both of these times, as well as the presence of any other relaxation processes. But, at least these two processes are required to see a transient velocity that is greater than the steady-state velocity, which is known as an *overshoot* velocity. Sometimes, this is referred to as ballistic behavior, which was mentioned in the previous section.

We may think about these two times as corresponding to the shift in the carrier distribution function that occurs upon applying the electric field (the momentum relaxation process) and the spread of the distribution to higher energies as the electrons gain energy (the energy relaxation process). The shift in momentum space is often described, at high carrier densities, as leading to a drifted Maxwellian (see figure 2.10 above). The energy relaxation spreads the distribution and leads to the elevated electron temperature in this latter case. If the energy relaxation process is slower than the momentum relaxation process, as is often the case, the velocity can overshoot its steady-state value (the saturation velocity) in high fields. This is because the electrons respond with a momentum relaxation time corresponding to the lattice temperature. As the average energy increases, this leads to a reduction in the momentum relaxation time, which lowers the average velocity.

We can illustrate this overall process with the Langevin equation approach. Normally, it is known that the representative velocity for each particle of an ensemble of electrons may be written as [35]

$$\frac{dv}{dt} = \frac{eE}{m_c} - \frac{v}{\tau_m} + \frac{1}{m_c}R(t), \tag{2.7}$$

where the last term on the right-hand side is the *random force*, and E is the electric field. Scattering causes the relaxation time τ_m, which arises from the *average effect* (over time) of the scattering, while the random force arises from the variations in scattering from its average effect; the time average of $R(t)$ is zero, by definition, so it represents the fluctuations in the scattering time. The random force thus leads to noise from the scattering and this can be inter-valley noise as well as intra-valley noise [36]. When we now average (2.7) over the ensemble of electrons, the resulting drift velocity is given by

$$v_d = \langle v \rangle = \frac{e\tau_m}{m_c}E(1 - e^{-t/\tau_m}). \tag{2.8}$$

The end result of this equation is just the low-field mobility that is common in most transport descriptions. Since the only place where the derivative of the velocity vanishes is the long time limit, this cannot lead to any overshoot. That is, the equation has only two stable points (where $dv/dt = 0$): $t = 0$ and the steady-state drift velocity, $t \to \infty$.

In general, the scattering rate increases as the average energy of the carriers increases. Hence, the momentum relaxation time must get shorter, and a description of this has to be added to the Langevin equation of (2.7). This can be handled in a variety of ways, one of which is to use a retarded Langevin equation [24]. Or, we can recognize that τ_m should really be $\tau_m(\varepsilon)$, where ε here denotes the energy to avoid confusion with the field E. Then we write an equation for the energy as

$$\frac{d\varepsilon}{dt} = evE - \frac{\varepsilon - \varepsilon_0}{\tau_\varepsilon} + R_\varepsilon(t), \tag{2.9}$$

where τ_ε represents the energy relaxation time and R_ε represents the random fluctuations in the energy. In either approach—the retarded Langevin equation or the addition of the energy relaxation equation—the result will be the same. We will have a retarded response of the velocity, and there will be the possibility that the velocity can have an overshoot behavior. We will examine such equations as these in chapter 6 in greater detail.

In figure 2.11, we plot the transient velocities for the carriers in $In_{0.53}Ga_{0.47}As$ at room temperature for several values of the electric field. The curves for the two higher values of field clearly exhibit the property of overshoot, with some rather high values for this quantity. From curves such as these, one could fit the above equations and try to obtain values for the two relaxation times. But, this is not such an important exercise. Now, what one can infer from figure 2.11 is interesting. For example, at 5 kV cm^{-1}, one can estimate that the average carrier will have traveled ~ 1 μm in the first picosecond. If this were a 100 nm gate length transistor, the carrier would have exited the device well before it reached this velocity. In a high-electron mobility transistor, there is usually a space between the actual source and the gate,

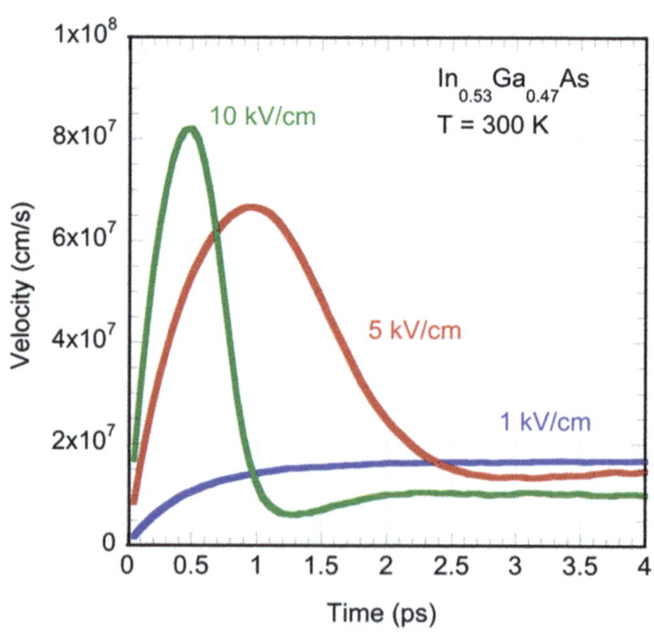

Figure 2.11. The velocity as a function of time for electrons in $In_{0.53}Ga_{0.47}As$ at 300 K, and for three different values of the electric field. The two higher fields are for values of the field above the peak velocity, and exhibit the phenomenon of overshoot, or ballistic, behavior.

and similarly between the gate and the drain, so that the source–drain distance is larger than the gate length. As a result, the carriers are accelerating prior to entering the gate region, and hence the higher overshoot velocities can be used to effect a higher cutoff frequency for the device, as f_T usually depends on the transit time through the gate region [37]. Typically, the field is highest at the drain end of the gate, and carriers will transfer to the satellite valleys already under the gate. In very high frequency devices, the carriers will usually remain in the satellite valleys until they enter the drain. What this all means is that proper engineering of the transistor design means that the overshoot velocities can be used to impact the design and improve the performance of very high frequency transistors. We will return to this inter-valley transfer below.

An important aspect of this discussion is that there are techniques by which the higher transient velocities and the distribution function can be measured experimentally. One such approach is based upon single-particle Raman scattering [25], discussed in the previous section. In the first studies, a sample of GaAs was cooled to 1.6 K, although the pulsed laser heated the sample to about 25 K, as determined by the anti-Stokes TO phonon scattering from a pulsed laser without any electric field applied. The material was illuminated by this pulsed laser, with the scattered light collected at a right angle to the incident pulse, and the probe pulse was timed with the application of the electric field. Significant changes in the free carrier scattering appeared for electric fields above 2 kV cm^{-1} at room temperature, so that most studies were done at the low temperature mentioned. The scattering from the free particles results in a Stokes shift from the distribution when the particle velocity is positive and an anti-Stokes shift when the velocity is negative. Hence, by measuring the frequency shift of these two scattering processes, the population at each velocity can be measured, and the distribution function determined. These early experiments clearly exhibited the electron heating from the applied electric field.

Studies of AlGaAs *p-i-n* structures at 80 K, with a 3 ps laser pulse from a dye laser at 2.17 eV, and photo-generated density of 2×10^{18} cm^{-3} density, showed that the average velocity achieved was about 2×10^7 cm s^{-1} at 18–28 kV cm^{-1} [38]. With a pulse this long, it is unclear whether overshoot is observed, but this velocity is above that expected for the valley velocity. However, studies of a GaAs *p-i-n* structure at this temperature with 1.92 eV photons gave velocities of $\sim 5 \times 10^7$ cm s^{-1} for densities up to 10^{18} cm^{-3} and fields in the 15–35 kV cm^{-1} range [27]. These are clearly above the steady-state values, so it is likely that one is observing strong overshoot behavior. Similarly, velocities as high as 4.7×10^7 cm s^{-1} were observed in InP at room temperature using 0.6 ps pulses from a Nd:YAlG laser at 1.95 eV [39]. When these studies are done with shorter excitation pulses and monitored by a probe pulse, then the role of the delay and time variation can be explored, as indicated above in figure 2.9(b). Similarly, a velocity of 5.4×10^7 cm s^{-1} was measured at 15 kV cm^{-1} for electrons in $In_{0.53}Ga_{0.47}As$ at 300 K at a density of 5×10^{17} cm^{-3} and photons at 1.43 eV [40]. In each case, the observed velocity is an average over the pulse width of the actual transient velocity, and the velocity is deduced from integrating over the electron distribution obtained as shown in the previous section.

In general, overshoot can occur whenever the energy relaxes on a slower time scale than the momentum, as discussed above. But, most of the experimental studies discussed are for the case of inter-valley scattering. Hence, the electrons accelerate to velocities corresponding to the Γ valleys and then transfer to the satellite valleys, which have a much lower velocity not least because of their higher values of the effective mass. The long-time steady-state velocities are dominated by the satellite valley behavior for fields above the peak in the velocity–field curves. So, curves such as those in figure 2.11 exhibit both behaviors. First, there is overshoot to very high velocities, which occurs on the sub-picosecond scale for the electric fields used in the experiments. This first rapid rise occurs as the energy relaxation due to the polar LO modes is quite slower than the momentum relaxation. However, on this energy relaxation scale, inter-valley transfer begins to occur as well, so the intermediate range of the quasi-steady-state velocity for electrons in the Γ valley is not seen in figure 2.11. However, the optical experiment is sensitive primarily to the Γ valley electrons due to the Raman scattering cross-section being sensitive to the inverse square of the effective mass. Hence, the experiments using single-particle back-scattering pick out the quasi-steady-state velocity for electrons in the Γ valley.

2.3 Inter-valley scattering

Inter-valley transfer between different valleys in the full energy band is one of the oldest and most interesting aspects of high field transport in semiconductors. It is already apparent in figure 2.2. that the velocity-field characteristic is dramatically modified by transfer of carriers from the Γ valley to satellite valleys which lie at higher energies. It was also pointed out above that even materials like Si depend upon the direction of the electric field, and valley transfer can occur when the field does not make the same angles with all six equivalent ellipsoids of the conduction band. So, there are at least two types of inter-valley transfer that are important, one between the valleys in a set that are nominally equivalent in equilibrium and a second between sets of non-equivalent valleys within the overall conduction band.

2.3.1 Symmetry breaking in equivalent valleys

The minima of the conduction band in Si lies approximately 85% of the way to X along the (100) direction. As there are six such directions in the cubic unit cell, there are going to be six equivalent minima, and a constant energy surface is an ellipsoid of revolution, as shown in figure 2.12. These ellipsoids are characterized by a heavier longitudinal mass, which is the mass in the direction of the principal axis of the ellipsoid, and a lighter transverse mass, in the two directions normal to the principal axis. The presence of these six ellipsoids of revolution pose an important point when we talk about transport away from thermal equilibrium. For example, if we orient the electric field in the (100) direction, the four ellipsoids lying in a plane that is normal to the electric field exhibit their transverse mass, while the two ellipsoids with their major axis parallel to the electric field exhibit their longitudinal mass. As a result, carriers in the four ellipsoids will be heated to a higher electron temperature than the other two as they exhibit the smaller mass. As a result, there will be a

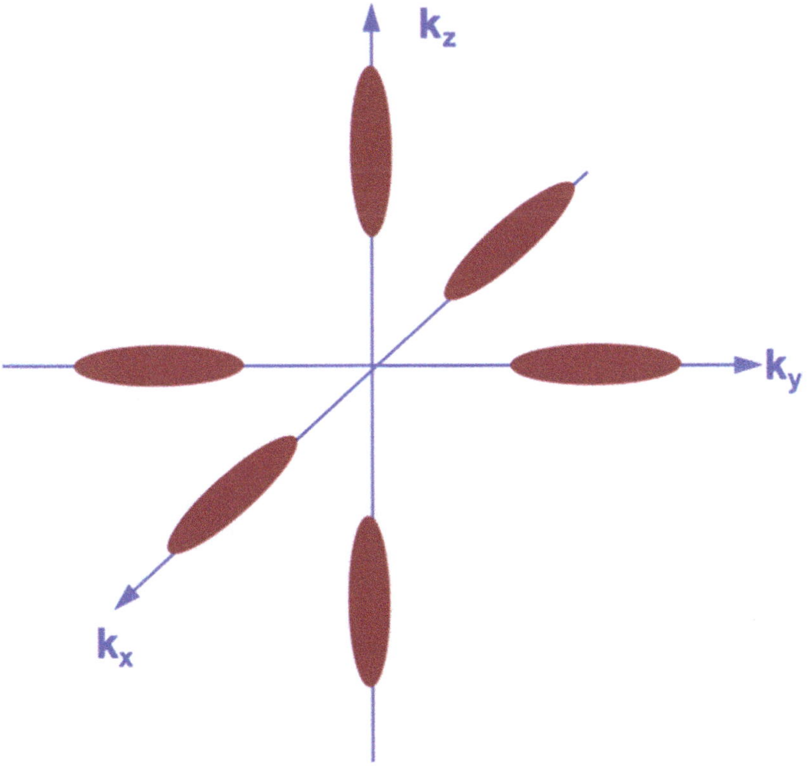

Figure 2.12. The six equivalent ellipsoids of Si, shown as constant energy surfaces, align along the set of [100] directions, which align with the three momentum axes as shown.

repopulation among the 'hot' and 'cold' valleys [15]. If we orient the electric field in a (110) direction, then there will be four valleys in the plane of the electric field that all make the same angle with the field. The two valleys normal to the electric field show the transverse mass, while the other four show a mass arising from the angle between the field and the major axes of the ellipsoids, thus a mixture of the longitudinal and transverse masses. Once again, it is possible for a repopulation between the 'hot' and 'cold' valleys to occur. Finally, if we orient the field in the (111) direction, it will make the same angle with all six of the ellipsoids and no repopulation will occur.

The repopulation was discovered quite early [41–43]. Before discussing this, we want to explore later experimental results using time of flight measurements [44], that confirm these expectations, and are well observable at lattice temperatures below room temperature [45]. In figure 2.13, we show experimental and theoretical results for Si over a range of electric fields and temperatures. It is clear from the experimental data that different curves are obtained for the (111) and (100) directions, even at room temperature. At low temperature, at and below 20 K, the (100) curve even shows negative differential conductance. The authors used Monte Carlo simulations to study the repopulation and drift velocity to illuminate the physics involved in these measurements. The combination of experiment and

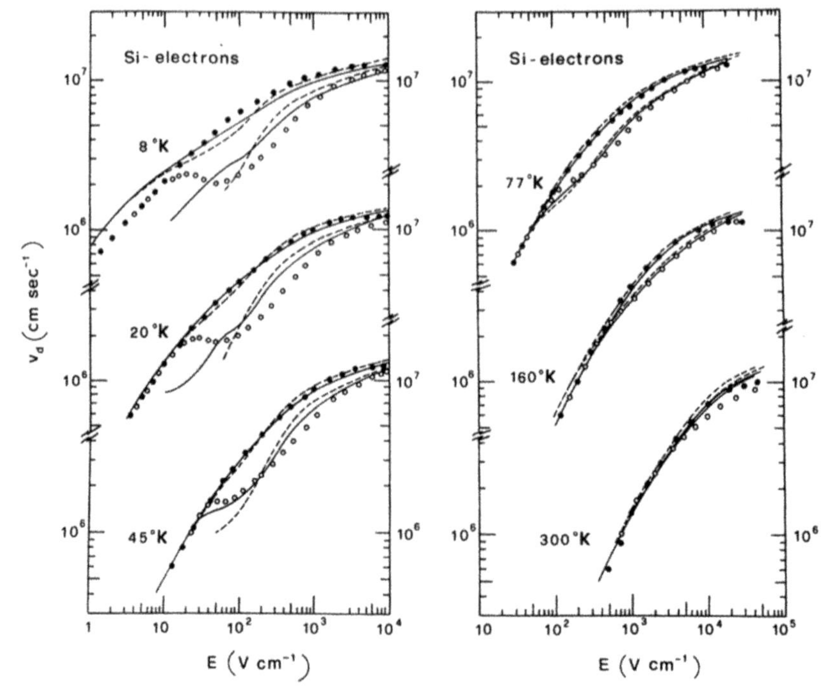

Figure 2.13. Electron drift velocity as a function of the electric field at various temperatures. Closed and open circles refer to experimental data with the field parallel to [111] and [100] directions, respectively. For the same directions, continuous and broken lines indicate the theoretical results from Monte Carlo simulations. The solid line includes the low energy phonons, while the broken lines omit these phonons. Reprinted with permission from Canali *et al* [45]. Copyright (1975) by the American Physical Society.

theory and the resulting outstanding fit between the two is remarkable given that it occurred in what could be called the 'early days' of such studies. When the electric field gets very large and the current saturates, the difference in field direction tends to be washed out by the very high electron temperatures.

The phonons involved for inter-valley scattering in Si are of two types, referred to as the *f* and *g* scattering between the six ellipsoids. The *f*-phonons connect ellipsoids on different axes, while the *g*-phonons connect the two valleys along the same axis. The *f*-phonons thus have a length of 2 × 0.85X along the (110) direction, which reaches into the square face of the second zone near an X point. The *g*-phonons lie at about 0.3X along (100). In each case, there are high energy optical modes and low energy acoustic modes which have these momentum values. For each of these two processes, it was found that a high energy optical mode and a low energy acoustical mode would satisfy the scattering process. Normally, the acoustic modes are considered to be forbidden for these scattering events. Yet, the results of Canali *et al* [45] found that the inclusion of these inter-valley acoustic modes was necessary to fit the drift-velocity data as a function of temperature, field strength, and field orientation at all temperatures. It was shown slightly later that, while forbidden, these inter-valley acoustic modes could be treated by a first-order interaction

[46, 47], and this was shown to yield a much better agreement with the temperature dependence of the conductivity in Si. The important point is that the energy dependence of the first-order interaction is stronger than the normal zero-order interaction. As a result, the first-order interaction, while normally weaker, becomes dominant at energies above ~ 0.5 eV. More recent studies, using a full empirical pseudopotential band structure, coupled to the ensemble Monte Carlo technique, has confirmed that these inter-valley acoustic modes are important, and may even be dominant, in transport in Si [48].

There is another interesting effect that arises from the repopulation among the six equivalent valleys. To illustrate this, consider the [110] plane shown in figure 2.14. This plane contains all three principal axes of the Brillouin zone. If one moves along (110) from Γ, one reaches the second zone X point as shown. Moving along (111) leads through the L point to the Γ point in the second zone. Along (100), there is one of the six ellipsoids. If the electric field is applied in an arbitrary direction (not one of the principle axes), the resulting current is rotated toward the light mass directions. The longitudinal mass is ~ $0.91m_0$, while the two transverse masses are ~ $0.19m_0$. So, this rotation can be significant given the five-fold difference in the masses. At near equilibrium, in low electric fields, the six current vectors sum to a net current that is parallel to the electric field, thus supporting the cubic symmetry of the crystal [49]. However, once the electric field is sufficiently high to begin repopulating the valleys, the cubic symmetry is upset, and the net current is no longer parallel to the electric field. In Ge, the ellipsoids are along the (111) directions and centered on the L points at the edge of the zone, but the same principals apply as it is also cubic.

In an actual long semiconductor sample that is being measured, the current is, of course, along the axis of this sample. It is the electric field that is observed to be rotated relative to the current. This provides a transverse, Hall-like, electric field and

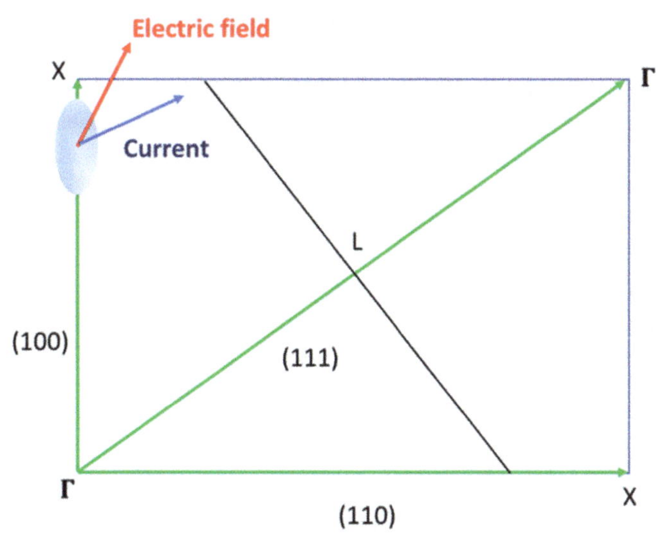

Figure 2.14. Rotation of the current away from the electric field for an ellipsoidal constant energy surface, here in Si. This is shown in a typical [110] plane with various positions in the first and second Brillouin zones.

a resulting transverse voltage, an effect known as the Sasaki–Shibuya effect [41–43]. The transverse voltage is measurable, but vanishes by symmetry for any of the three principal directions in the crystal. In figure 2.15, the angle between the current and the electric field for germanium is shown as a function of the angle between the electric field and the (100) axis, in the [110] plane. It is easy to see that the angle between current and voltage goes to zero at the (111) axis and reverses sign for continuing increases of this device angle toward (110). The dashed line is a calculation from Shibuya [41], which underestimates the effect. The angle between the field and the current increases with electric field until a maximum effect and then decreases at very high fields as the valleys begin to again have equal populations (as their temperatures become the same, in the saturated current regime). This is shown in figure 2.16 for a variety of sample temperatures, for an angle of 30° between the sample current and the (100) axis.

2.3.2 Transfer between non-equivalent valleys

The second major inter-valley transfer that occurs is when only a single conduction band minimum, with a small effective mass, is normally occupied, such as in the case of GaAs. With the application of a high electric field, carriers in this central Γ valley are heated to relatively high energies. Those carriers in the high energy tail of the distribution can even reach energies where they can scatter via inter-valley phonons to the satellite L and X valleys of the conduction band. In GaAs, we know that the L valleys are about 0.29 eV above the conduction band minimum, and the X valleys are some 0.2 eV higher still. Since these satellite valleys tend to have relatively high effective masses (similar to Si for X valleys and Ge for L valleys), the energetic carriers see a higher density of states in these satellite valleys and tend to prefer to scatter to them. Scattering between non-equivalent valleys of the conduction band in a material such as GaAs became of interest with the discovery of microwave emission from GaAs and InP when a high electric field was applied [3]. While the current–voltage relation was approximately linear in these devices when a 10 ns pulse was applied, the resistance was observed to rise as the voltage increased, indicating a sub-linear behavior. Once a critical electric field was surpassed, in the range of 2–4 kV cm^{-1}, the current began to fluctuate in a nearly sinusoidal fashion. This is shown in figure 2.17 [3].

Gunn originally discounted inter-valley transfer as a possible cause in early calculations, and even suggested that it could be a plasma-like interaction between the drifting electrons and the polar optical phonons [50]. However, the work of Ridley and Watkins earlier predicted that inter-valley transfer was likely to occur in this manner [4], and the former subsequently suggested that indeed this process was the cause of the oscillations [51]. Gunn, in fact, made measurements of the velocity-field characteristic for GaAs, and found the presence of negative differential mobility (or conductance) above 3.8 kV cm^{-1}, but the negative differential mobility was only about 300 cm^2 Vs^{-1}, much lower than expected [52]. Subsequent measurements by Ruch and Kino [53] used a time-of-flight technique to observe a better velocity–field relation. These measurements agreed well with Monte Carlo simulations by Butcher and Fawcell [54], and are shown in figure 2.18.

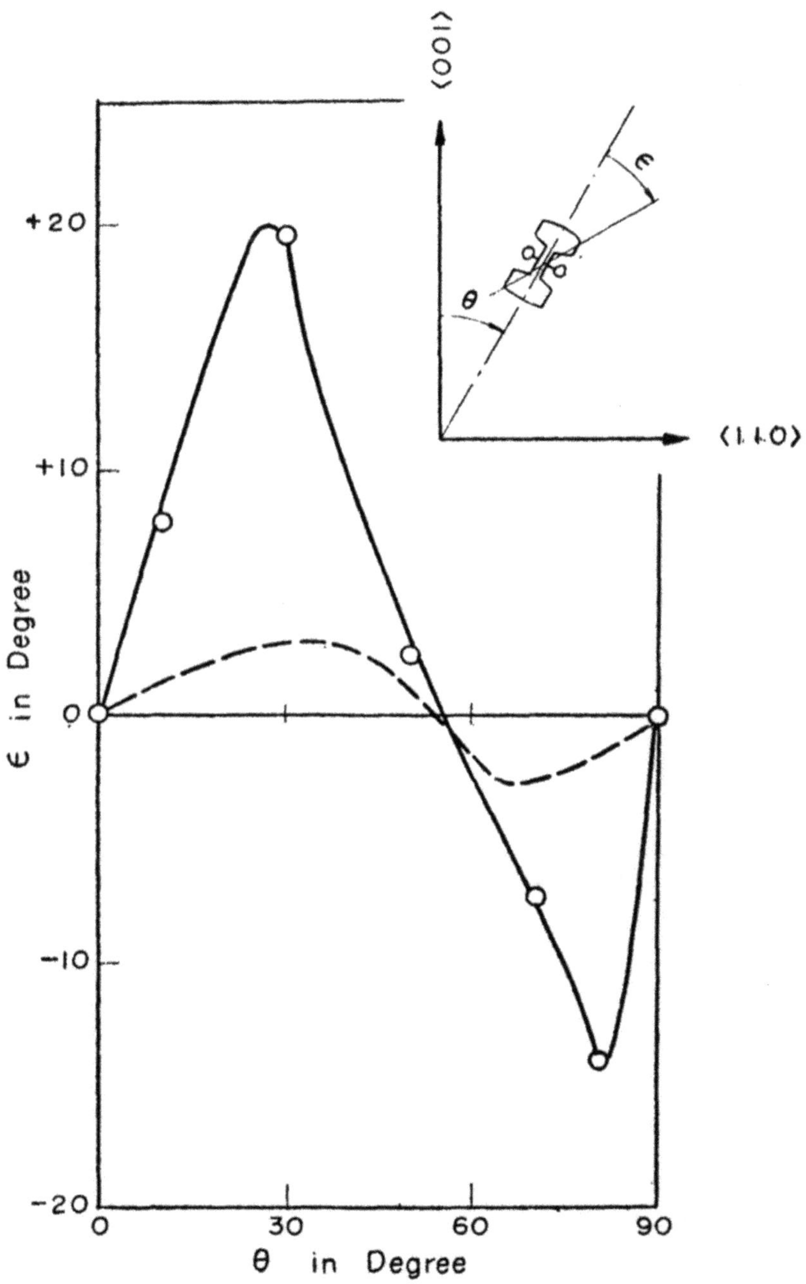

Figure 2.15. The angle between the field and the current (defined in the inset above) as a function of the angle between the sample axis and the (100) direction in the [110] plane. Reprinted from Sasaki *et al* [42], copyright (1958) the Physical Society of Japan.

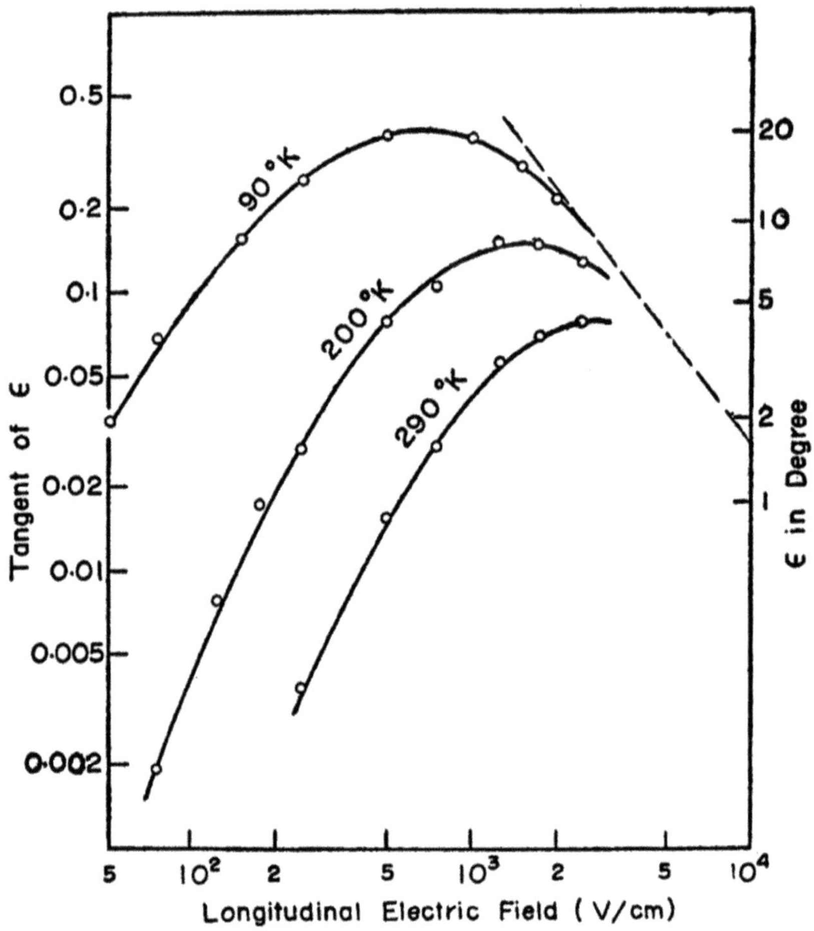

Figure 2.16. The angle between the electric field and the current is plotted as a function of the electric field for various temperatures. Here, the current axis is at an angle of 30° from the (100) axis and the material is *n*-type Ge. Reprinted from Sasaki *et al* [42], copyright (1958) by the Physical Society of Japan.

The current oscillations are a natural by-product of the presence of the negative differential conductance (NDC). For example, if one biases the device into a region where the average electric field is 6–8 kV cm^{-1}, it can be seen from figure 2.18 the device has NDC. Normally, fluctuations die out with the dielectric relaxation frequency $\omega_D = \sigma/\epsilon$. But, if the differential conductance is negative, the fluctuations do not die out, but grow! Biasing the device into the NDC region gives an unstable system, which is quite likely not to have a steady-state. Rather, the fluctuation in electric field will grow, so that there is a high field domain embedded in a very low field background. Since the domain requires charge accumulation (or depletion), it is quite likely to move with the charge itself. A natural consequence is that the oscillation frequency will be correlated with the length of the device, and given by

$$f = v_d/L, \tag{2.10}$$

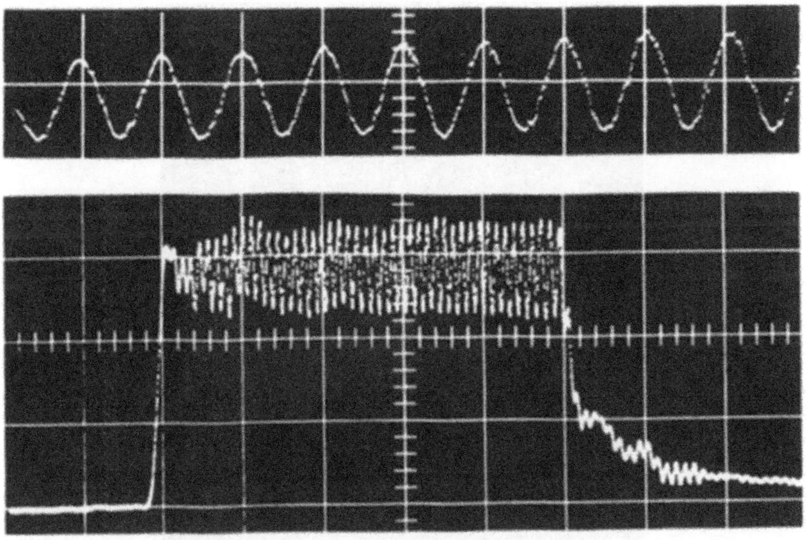

Figure 2.17. Current waveform produced by the application of a voltage pulse of 16 V amplitude and 10 ns duration to a specimen of *n*-type GaAs of 25 μm (average electric field of 6.4 kV cm^{-1}). The frequency of the oscillating component (top trace) is 4.5 GHz. Time scale of the lower trace is 2 ns cm^{-1} (grid lines at 1 cm spacing). Reprinted from J B Gunn [3], copyright (1963) with permission from Elsevier.

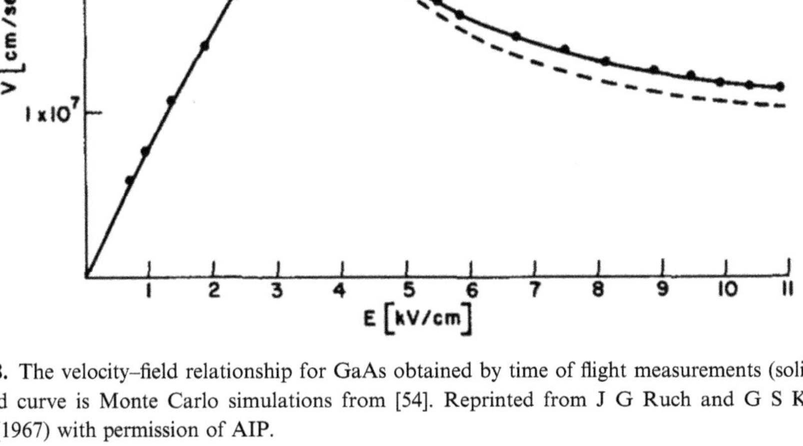

Figure 2.18. The velocity–field relationship for GaAs obtained by time of flight measurements (solid curve). The dashed curve is Monte Carlo simulations from [54]. Reprinted from J G Ruch and G S Kino [53], copyright (1967) with permission of AIP.

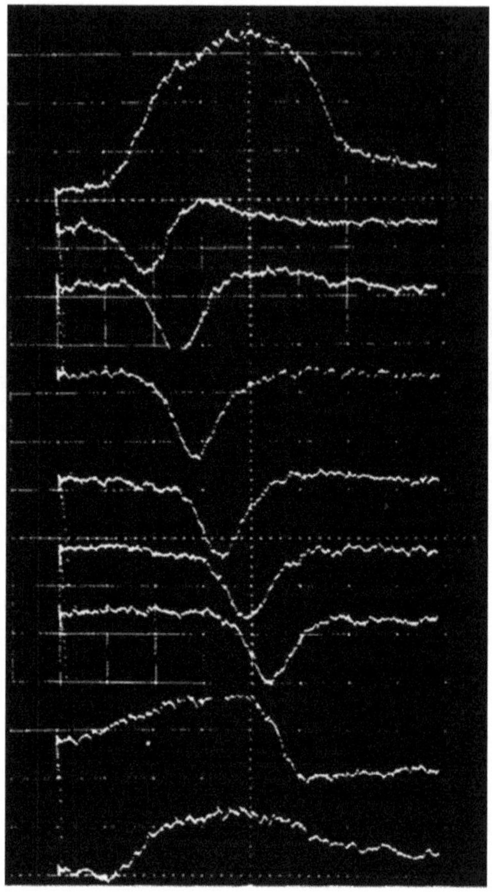

Figure 2.19. Temporal derivative of the voltage as a function of position along the length of the Ge device. The horizontal scale is approximately 50 μm per grid cell. The movement of the high electric field domain can be clearly recognized. Reprinted from B J Elliott *et al* [57], copyright (1967) with permission of the AIP.

where v_d is the drift velocity and L is the length of the device. In Gunn's original devices, the frequency and length suggested a drift velocity of $\sim 1.1 \times 10^7$ cm s^{-1}, which is below that expected from figure 2.18 for the average field of the measurement.

Similar oscillations have been observed in Ge [55, 56]. The electrons in Ge reside in the L valleys and not a Γ valley (which lies some 180 meV higher in energy). Yet, NDC and domain propagation are seen in this material. There are additional valleys along Δ near X, which are about 170 meV higher in energy, so that it might be expected that inter-valley transfer could lead to NDC at low temperature (77 K). The existence of the NDC and the domain propagation was confirmed by Gunn using a capacitive probe which could be moved along the edge of the device's length [57]. Measurements of these domains are shown in figure 2.19. As it is a capacitive probe, the signal is the time derivative of the voltage at the position of the probe along the device length. What is seen in each trace of the figure is the time variation of the local voltage as the domain passes that position. From several traces such as these, on can infer the drift velocity of the domain itself, in this case about 1.3×10^7 cm s^{-1}.

2.3.2.1 An analytical treatment

Understanding why electrons transfer requires knowledge of several different ideas. First, of course, is the scattering rates within the Γ valley and from this valley to the satellite valleys, because the transfer depends upon the relative size of these quantities. In figure 2.20, these scattering rates are plotted for $In_{0.53}Ga_{0.47}As$, a semiconductor of considerable interest as its band gap aligns well with the minimum dispersion energy of quartz fibers, is useful for hot carrier solar cells [58], and is used in high speed HEMTs [59]. There are two curves in each case, and the upper curve is for emission of the appropriate phonon while the lower curve is for the absorption of the phonon during the scattering process. Normally, the inter-valley process is a weaker process than the polar interaction. However, the multiplicity of the satellite valleys, and the much larger mass in the satellite valleys, combine to give a much larger density of final states for the inter-valley scattering process. That leads it to be much more likely than the intra-valley polar LO mode scattering.

The strength of the inter-valley process takes density from the Γ valley and puts it into the L valleys. To understand how this leads to a negative differential

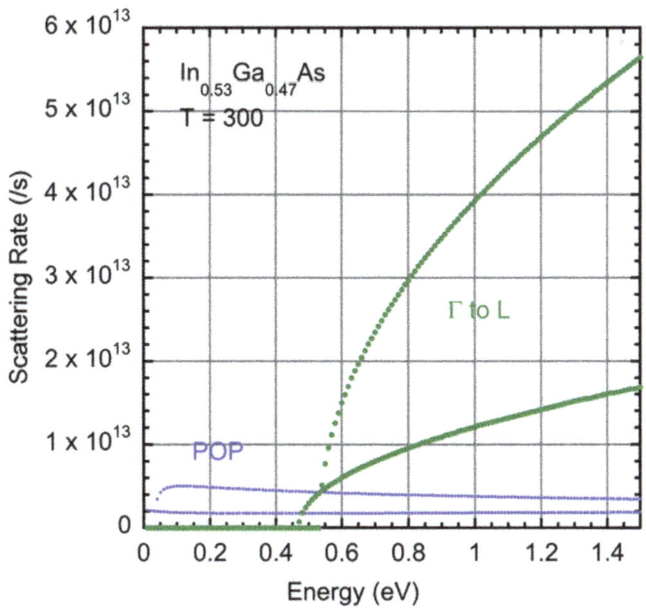

Figure 2.20. The scattering rates for polar optical phonons in the Γ (blue) and for inter-valley scattering to the L valleys from the Γ valley (green). The upper curve, in each case, is for emission of a phonon while the lower curve is for the absorption of a phonon.

conductance, we need only some simple parametric equations, beginning with the basic density of states (parabolic bands are assumed here; the extension to full hyperbolic bands that arise with the spin–orbit interaction does not alter the argument to any great extent)

$$n_{3D}(E) = \frac{1}{2\pi^2}\left(\frac{2m_c}{\hbar^2}\right)^{3/2} E^{1/2}, \tag{2.11}$$

where E is the energy within the particular band and m_c is the effective mass for that band. It is important to note that the latter quantity, the effective mass for a band, is the density of states effective mass which involves the multiplicity of the particular set of valleys [24]. For a non-degenerate Γ valley at an electron temperature $T_{e\Gamma}$, the electron density is given as

$$n_\Gamma = \frac{1}{4}\left(\frac{2m_\Gamma k_B T_{e\Gamma}}{\pi\hbar^2}\right)^{3/2} \exp\left(\frac{E_F}{k_B T_{e\Gamma}}\right). \tag{2.12}$$

Similarly, the electron population in the L valleys may be written as

$$n_L = \frac{1}{4}\left(\frac{2m_L k_B T_{eL}}{\pi\hbar^2}\right)^{3/2} \exp\left(\frac{E_F - \Delta_{\Gamma L}}{k_B T_{eL}}\right), \tag{2.13}$$

where $\Delta_{\Gamma L}$ is the energy separation between the conduction band edge and the bottom of the L valleys. Hence, the ratio of the two densities can be written as

$$\frac{n_L}{n_\Gamma} = \left(\frac{m_L T_{eL}}{m_\Gamma T_{e\Gamma}}\right)^{3/2} \exp\left[\frac{E_F}{k_B}\left(\frac{1}{T_{eL}} - \frac{1}{T_{e\Gamma}}\right)\right] \exp\left(-\frac{\Delta_{\Gamma L}}{k_B T_{eL}}\right). \tag{2.14}$$

For GaAs, the appropriate masses are $m_\Gamma = 0.067m_0$ and $m_L = 1.2m_0$, in which the latter includes a factor of $4^{2/3}$ to account for valley multiplicity. $\Delta_{\Gamma L}$ is approximately 0.29 eV. If the two temperatures are the same, then the ratio of populations is about 70, and the two populations will become comparable at an electron temperature of the order of 950 K. This is not unreasonable for high electric fields.

The condition above is actually more stringent than necessary and should be modified for the respective mobilities when the conductance is being considered. Hence, a simple argument on the conductivity gives a good result. The contribution of the two densities to the conductivity is given as

$$\sigma = e(n_\Gamma \mu_\Gamma + n_L \mu_L). \tag{2.15}$$

The interest is in the NDC regime, which may be addressed with the field derivative of the current density, as

$$\frac{dJ}{dE} = \sigma + E\frac{d\sigma}{dE} = \sigma + eE(\mu_\Gamma - \mu_L)\frac{dn_\Gamma}{dE}, \tag{2.16}$$

where here E is the electric field, and the negative sign arises from the fact that the L valley density is increasing when the Γ valley density is decreasing. For NDC, (2.16) leads to the condition

$$-\frac{\mu_\Gamma - \mu_L}{\mu_\Gamma + (n_L/n_\Gamma)\mu_L}\frac{E}{n_\Gamma}\frac{dn_\Gamma}{dE} > 1. \tag{2.17}$$

For the case of interest, the density in the central valley is a decreasing function of the electric field, so that $\mu_\Gamma > \mu_L$ is required, which is almost always satisfied in the case of a small mass central valley. To simplify further, it will be assumed that the two temperatures are the same, which is likely to be the case near the peak of the velocity–field curve, when few carriers have actually transferred. With the use of (2.12)–(2.14), (2.17) can be rewritten as

$$\frac{n_L}{n_L + n_\Gamma}\left(\frac{\Delta_{\Gamma L}}{k_B T_e}\right)\frac{E}{T_e}\frac{dT_e}{dE} > 1. \tag{2.18}$$

It is not a bad approximation to assume at this point that electron temperature increases linearly with the electric field, which then leads to the result that only an electron temperature of ~ 600 K is needed and a transfer of only 10%–15% of the electrons will give rise to NDC. Thus, the onset of NDC does not require a large fraction of the carriers to be transferred. This is a result of the large differences in mobility in the different valleys.

2.3.2.2 Stabilizing the device

The experimental results discussed earlier in this section could lead one to believe that the non-existence of a steady-state current is the basic mode of the NDC device. But, this is not necessarily the case. If a domain is propagating across the device, as shown in figure 2.19 above, the voltage distribution across the device basically has a step jump across the domain, much like the curve second from bottom in the figure. In order for this step to occur, and for the domain to form, there must be a dipole of charge in the device. Generally, one may think of this in the following way: the velocity found from the domain above was less than the actual carrier velocity found from the velocity–field curve. Hence, carriers ahead of the domain leave the region faster than the movement of the domain. On the other hand, carriers behind the domain also move faster and catch the domain, thus causing an accumulation on the back-side and depletion on the front-side. This helps form the domain, and strengthens it as it moves. Nevertheless, there is a thickness associated with the domain, which is the positional size of the jump in the voltage. If the density of the device is lowered and/or the length is made smaller, the necessary charge for the dipole may not exist in the device. When this is the case, the domain may either fill the entire device, or be pinned near one of the contacts. In either case, there will not be a travelling domain and a steady-state can be obtained. This is useful in making a reflection amplifier that uses the NDC to advantage in a microwave device [60].

In principle, the above argument boils down to requiring the dielectric relaxation time (inverse of the dielectric relaxation frequency) must be longer than the transit time of carriers through the device. This leads to the result that

$$nL < \frac{\epsilon E}{e} \sim 7 \times 10^{10}\,\mathrm{cm}^{-2}, \tag{2.19}$$

where the permittivity of GaAs and a field of 10 kV cm^{-1} have been used. So, for a lightly doped material of approximately 10^{15} cm^{-3}, a sample length of less than 0.7 μm can be tolerated for a steady-state solution.

2.3.2.3 The strange case of InSb

Broad-band microwave emission was observed in InSb quite early [61]. It was subsequently also seen in InAs [62]. Radiation was observed when the electric field reached a threshold value that was dependent upon a magnetic field. In these materials, with their very small band gap, impact ionization is observed at relatively low electric fields, ~ 200 V cm^{-1} in InSb at 77 K and ~1 kV cm^{-1} in InAs at 77 K [63]. The microwave emission set in about 140 V cm^{-1} in both materials. While the magnetic field was relatively high, and would have qualified for the formation of Landau levels at lower temperatures, at 77 K, the thermal broadening of these levels, especially at the relatively high electric field for these materials, was sufficient to wash out the quantization. In any case, as will be seen below, the need for the magnetic field in these early studies was actually misleading. Nevertheless, the orientation of the electric and magnetic fields with respect to crystal orientation was anisotropic, as was the threshold for emission [64].

Due to the close resemblance of InSb and InAs both in band structure and in scattering mechanisms, it was felt that the underlying mechanism for the broad-band microwave emission was the same in both materials. While the source of the emission was obscure at the time, it was apparent that only the electrons, as contrasted to the holes, was significant in producing the emission. Nevertheless, the emission spectrum is quite broad, ranging from a few MHz to more than 10 GHz, as can be seen in figure 2.21 [65], and its general decay is as $1/f$ (common for some noise sources). The source of the oscillations will be discussed in the following.

As commented above, the role of the magnetic field is found only by consideration of the symmetry of the device. For this, consider a sample that is relatively thin and has a width W and a length L, with $L \gg W$. Nearly metallic contacts are applied at the end of the sample across the entire W, and the magnetic field is applied normal to the sample plane. What is not also appreciated is that the metallic contacts short out the Hall field in the sample. Since the current must be rotated with respect to the electric field, this means that the current enters (and leaves) the sample at an angle to the electric field. Since $\mu B > 1$ in these samples, this angle is not small. At some distant from the contacts, the normal Hall field has built up and the current is now directed along the length of the sample. The current at the contacts is thus concentrated into a corner of the sample, and corresponding higher electric fields appear in these corners than in the main body of the sample. Thus, the real threshold

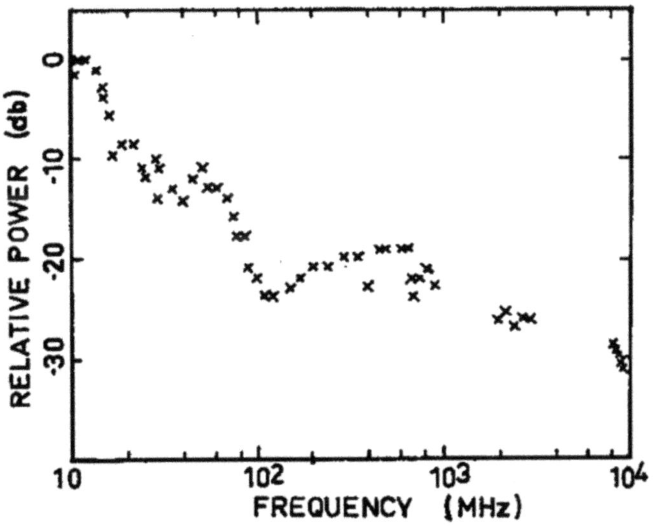

Figure 2.21. Frequency spectrum of the microwave emission from InSb, with an applied field of 5 kG, and an electric field of 100 V cm^{-1}. Reprinted from D K Ferry *et al* [65], under the Creative Commons 3.0 open access.

for the emission appears for fields in the sample corners, and this is different than the average field measured in the sample. This field intensification effect plays an important role in the understanding of the emission.

It was remarked above that impact ionization occurs in InSb for fields above ~200 V cm^{-1} at 77 K. Studies of this show additional interesting behavior, even for pulsed measurements. If a quasi-constant voltage is applied, the current merely rises once the threshold for impact ionization is surpassed. On the other hand, if a quasi-constant current pulse is used the current above threshold exhibits negative differential resistance (NDR), with a set-back in voltage and an S-type instability [66, 67], as shown in figure 2.22. The current level at which the set-back occurs depends upon the magnetic field that is supplied [68]. Such a voltage set-back was previously observed during avalanche breakdown in Si bipolar transistors, and was termed *second* breakdown [69]. It is known that the S-shaped current–voltage curves would lead to current filaments [51], and the heating in these filaments could lead to device failure [70]. Indeed, such filament formation and device destruction has been observed in InSb samples as well [71].

To understand what was happening, one has to understand that different events happen on different time scales. Impact ionization in these materials is a nanosecond event, whereas inter-valley transfer is very much a sub-nanosecond event. It was generally assumed that inter-valley transfer in these materials would not be important as the band gap (0.17 eV) was much smaller than the gap between the satellite valleys and the conduction band minimum (0.5 eV). Nevertheless, it was observed that inter-valley transfer could occur on the nanosecond time scale [72, 73], as shown in figure 2.23. The threshold for inter-valley transfer was observed to be

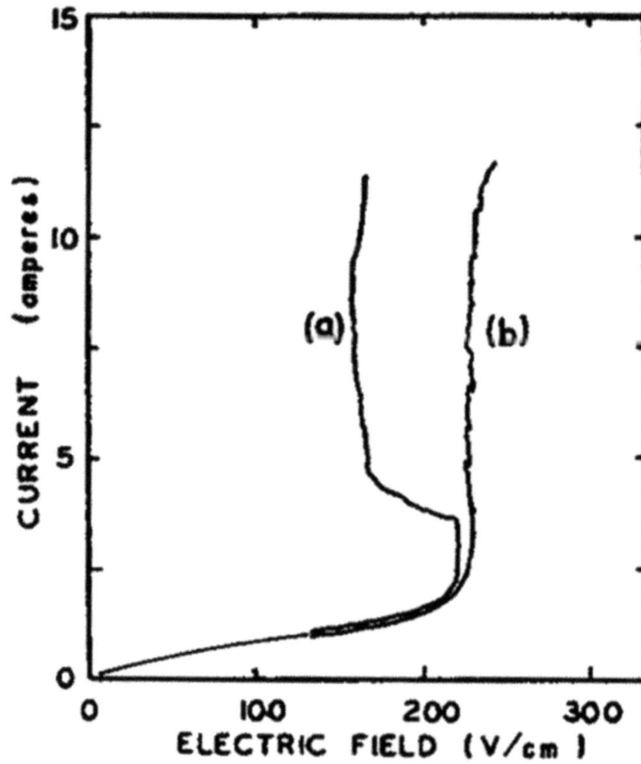

Figure 2.22. A comparison of the current–voltage curves showing instability after the onset of avalanche breakdown. (a) Curves for a quasi-constant current bias, and (b) curves for a quasi-constant voltage bias. The ohmic portion of the curves has been displaced slightly for clarity. The curves are taken 10 ns after the start of the pulses. Reprinted with permission from D K Ferry and H Heinrich [67], copyright (1968) by the American Physical Society.

about 600 V cm^{-1}, and the enhanced electric field in the first domain through the sample would generate the onset of impact ionization. So, on different time scales, both the Gunn effect and impact ionization could be observed.

What finally clarified the broad-band emission from InSb really was observations on GaAs devices. It is generally known that if an oscillator is over-damped, that is, when the load resistance is larger than the magnitude of the negative resistance, the oscillation will break into noise. It was observed that similar behavior occurs in GaAs when the device sees too large a load resistance. This is critical. In the case of InSb, it was observed that even for as small an average electric field of 12.4 V cm^{-1} at 77 K, the field at the cathode contact was sufficiently large to launch high electric field domains [74]. However, these domains could not propagate throughout the device, but died a random distance from the cathode. This meant there was no coherence from one domain to the other and the device emitted broad-band radiation, even while showing domains, as appears in figure 2.24. Here, the sample was doped to 3×10^{15} cm^{-3}, with a mobility of 2.2×10^{5} cm^{2} Vs^{-1} at 77 K, and had a length of approximately 3 mm. The domain propagation indicated a velocity in the

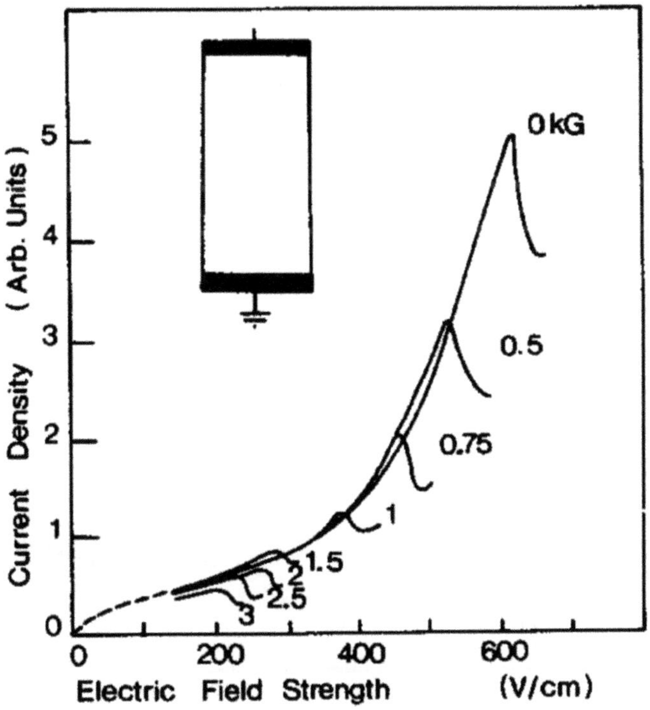

Figure 2.23. In long samples with a length-to-width ratio of 4.3, the onset of the inter-valley transfer is greatly reduced in InSb (at 77 K) by the application of a magnetic field (parameter is the magnetic field in kG). Data is taken 2 ns after onset of the voltage pulse. Reprinted from D K Ferry *et al* [73], copyright (1973) by the American Physical Society.

range of 5–7 × 10^7 cm s^{-1}. With the doping in this sample, the threshold for broad-band microwave emission is relatively independent of any magnetic field, so it is the high field at the cathode that initiates the domains. The randomness of these domains is thought to be the source of this emission, as mentioned previously. The random nucleation of these domains results in a current modulation that has no specific periodicity, and thus has charactistics similar to $1/f$ noise, as seen in figure 2.21.

2.3.2.4 Monolayers

In recent years, the consideration of new materials for electronic devices beyond Si, and the III–Vs, has led to monolayer materials, with graphene being perhaps the most familiar. However, with the absence of a band gap, other materials have been considered, such as the transition metal di-chalcogenides (TMDCs). The most common of these is MoS_2. Several studies of these for electronics applications have appeared. These include mobility [75] and field effect transistors [76]. Of importance here is, while the bulk material has an indirect band gap, the monolayer version is direct. The direct band gap makes the material interesting for optical as well as electronic applications. Here, however, our attention is focused on the inter-valley transfer that affects electronic properties desired for future transistors.

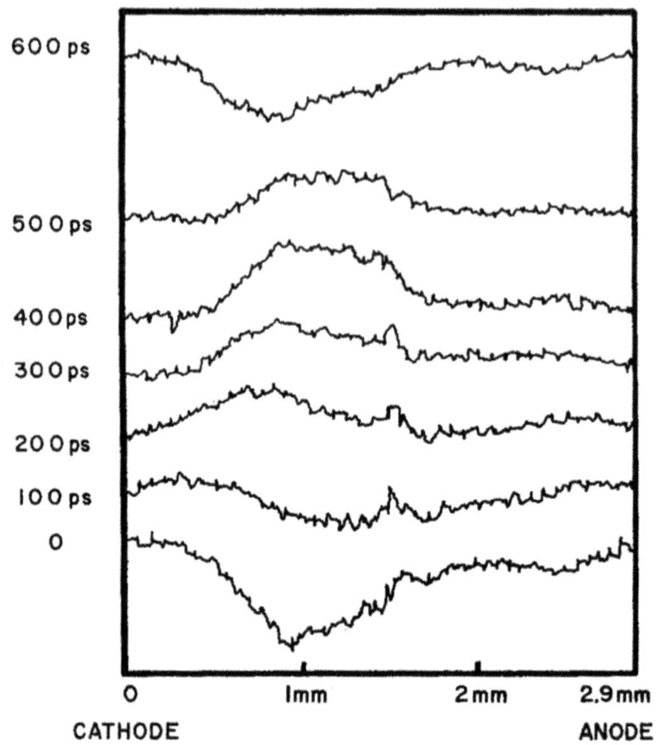

Figure 2.24. Formation and movement of a domain in InSb at 77 K with an average electric field of 12.4 V cm^{-1} across the device. This is above the threshold for the broad-band microwave emission seen from the device. Note that the domain decays rapidly after leaving the region near the cathode contact. Reprinted from W A Porter and D K Ferry [74], copyright (1972) with permission from Elsevier.

MoS_2 is a layered compound in which each monolayer is actually a tri-layer composed of Mo atoms in the center layer and S atom layers displaced above and below this center layer. As mentioned, it has an indirect band gap in bulk form, located at the T points in the Brillouin zone, which are midway along the Γ to K line, where the K points are at the six corners of the hexagonal cell [24]. The gap becomes direct only in the monolayer limit [73], for which the band extrema are located at the K points of the Brillouin zone, so that there are two minima for the conduction band. Subsidiary valleys of the conduction arise from the residual T valleys of what was the indirect gap. These properties are quite general for most of the TMDCs, and inter-valley transfer and NDC have been predicted for both MoS_2 and WS_2 [77]. Here, we will discuss results for the latter material, for which the effective masses are approximately $0.32m_0$ for the K point primary minima of the conduction band and $0.75m_0$ for the subsidiary T minima. The latter lies approximately 80 meV above the K point minima.

For the experiments discussed here, WS_2 was grown by chemical vapor deposition on an oxidized Si wafer. The conducting Si can be used as a gate to change the conductivity of the TMDC layer via capacitive coupling through the oxide.

The conclusion of the growth of a single monolayer is tested by Raman scattering as the relative phonon peaks will shift with the number of layers [78]. The transport through these device layers is usually enhanced by thermal annealing which relaxes strain and improves the contacts to the device [79]. In this case, a modest annealing of about an hour produced an increase in the current through the layer. The defects responsible for this may be traps and random potentials, as it is believed that S vacancies lead to deep traps from the vacancy itself and a random potential from the change in crystal potential the vacancy leads to [80]. By using the Si back gate as part of an FET, an on/off current ratio of better than 10^5 was achieved.

In figure 2.25, results of the TMDC behavior clearly indicate the presence of NDC, which can be inferred to arise from inter-valley transfer [77]. Similar behavior was observed in all devices studied. Notice in panel (a) that the NDC is observed

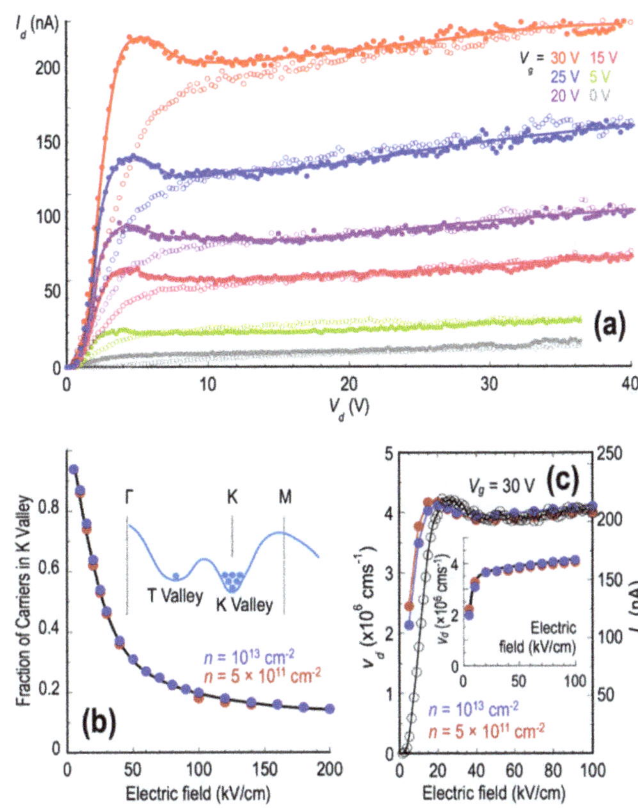

Figure 2.25. (a) Negative differential conductance in the current–voltage curves of one of the devices studied. The filled and open circles refer to up-sweeps and down-sweeps of the applied voltage. The solid line is a guide to the eye. (b) The fraction of electrons that remain in the K valleys, for two densities, as a function of the electric field as determined by a Monte Carlo simulation [77]. The inset shows the positions of the T and K valleys along the line from Γ to K and then K to M (between two adjacent K points on the edge of the hexagonal unit cell). (c) Match between simulated and measured current–voltage curves. The inset is a sample that is annealed for hours and discussed in the text. Reprinted from G He *et al* [78], under the Creative Commons CC BY licensed open access.

when increasing the drain voltage, and not when the voltage is being reduced. This is a natural hysteresis observed in the Gunn effect—when the voltage is being increased the electric field is nearly homogenous, and domains form only after the threshold is increased. In a decreasing voltage, the domain sustains itself until the voltage is reduced well below the threshold, which leads to a clockwise hysteresis in the current–voltage curves. Once NDC is observed, there is a significant increase in current noise in the channel, which is likely due to the domain travel as discussed above. One aspect of the short anneal in these devices is that there will still be strain in the layer which likely raises the T valleys relative to the K valleys. If the T valleys sit too low, there is too small a barrier to the transfer from K to T, and this will occur at relatively small voltages. By raising the T valleys, one gets a more abrupt onset of transfer and the appearance of NDC. In the inset to panel (c), an *I*–*V* curve is shown for a device that is annealed for several hours, so that no obvious NDC is observed. From theoretical studies, it is estimated that the strain in the shorter anneal results in an increase of the valley separation to about 110 meV, which is sufficient to make NDC observable [75].

2.3.3 Inter-valley transfer in devices

It is already clear from the previous paragraph that inter-valley transfer is an important consideration for electronic devices. This effect can be either good or bad in the manner in which it appears in normal device operation. This will be explored here with two examples in which the transfer has been identified and shown to have a significant effect on real devices. The bad, which will be treated first, may be an extreme case, but it has to be understood if one is to avoid such a pitfall in creating useful devices structures for real world applications.

2.3.3.1 The bad—domains in characteristics

The GaAs field effect transistor has been around for quite some time [81]. The structure is very similar to that of figure 1.7, except that the substrate GaAs is usually semi-insulating material with a resistivity above 10^7 ohm-cm, and the AlGaAs layer is replaced with a doped GaAs layer and serves as the active layer in the device. In the devices studied, the epitaxial GaAs active layer was about 0.3 μm thick and doped to 10^{17} cm^{-3}. The source-to-drain distance was 9 μm and the gate length was 3 μm [82]. While this is a massive device by today's standards, it was typical of the time (late 70s). There were a great many studies by this time of the Gunn effect itself, but the dynamic properties of any domains and the conditions for their existence in actual transistors was not well studied. The problem was to determine how the NDC affected the performance of real devices. This device sustained current oscillations as indicated by using a spectrum analyzer in conjunction with the normal electronic curve tracer. The problem is illustrated in figure 2.26. When essentially dc voltage was applied, as would be typical for a curve tracer, the characteristics shown in panel (a) of the figure were obtained. However, when short pulses were used and the *I*–*V* curves extracted with a sampling oscilloscope, the characteristics of panel (b) were obtained.

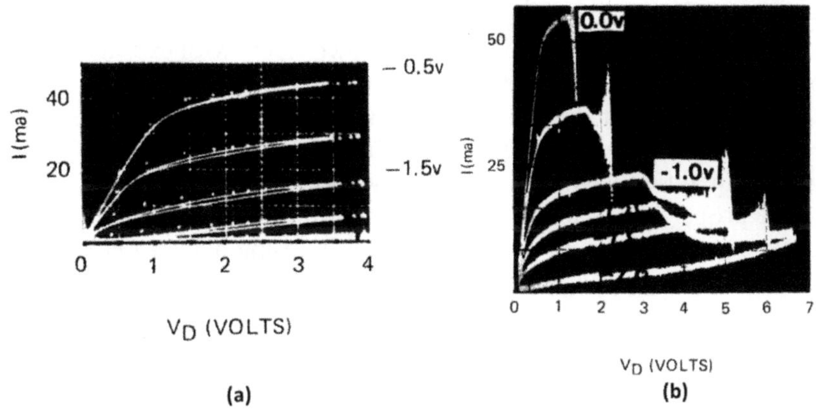

Figure 2.26. (a) Typical characteristics for an FET obtained from a curve tracer. The hysteresis indicates something complicated is occurring. (b) *I–V* curves using short pulses and a sampling oscilloscope clearly indicated the presence of NDC and current oscillations. Reprinted from H L Grubin, D K Ferry, and K R Gleason [82], copyright (1980) with permission from Elsevier.

Understanding of the problem requires extensive simulation of the behavior of such FETs, and this was done as well. It turns out that there is a characteristic number, the ratio of the negative gate voltage for turn-off of the device and the voltage at which the current curves tend to saturate (pinch-off at the drain end of the gate)

$$K = \frac{V_{TO}}{V_{sat}}. \tag{2.20}$$

In these devices, $K \sim 3.9$. In other devices, in which $K \sim 1$, no oscillations were observed, so it was assumed that such devices would be much more stable. The simulations showed that, when K was much larger than unity, domains would form at the drain end of the gate and these domains would propagate mostly *across* the doped layer. On the other hand, for low K devices, it appeared that a single domain would appear at the drain contact and was stable in this position. In order to reduce K, the most likely approach is to thin the active, doped layer of GaAs in order to reduce the gate turn-off voltage V_{TO}. In essence, this is using the thickness of the active layer as the domain 'length' in (2.19) to stabilize the device [82]. The drive to smaller devices to reach the THz domain has thus obviated much of this problem.

2.3.3.2 The good—InP HEMTs

The high-electron mobility transistor (HEMT) has become a favorite device for high frequency operation and low noise applications. Versions of it based upon GaN are of interest for high power applications as well. In modern versions, the active channel is a thin quantum well, so that the problems of the previous device are not encountered here. In fact, we will see that inter-valley transfer is a quite useful part of the operation of such devices. Here, the discussion will center on a particular implementation of the HEMT of figure 1.7. The structure is based upon an InP substrate, particularly as ternary alloys such as $In_{0.53}Ga_{0.47}As$ and $In_{0.52}Al_{0.48}As$ are lattice matched to this substrate and can be grown easily with excellent control. The device of interest is grown with the layer structure (beginning with the substrate) InP, $In_{0.52}Al_{0.48}As$ buffer layer, $In_{0.75}Ga_{0.25}As$ quantum well channel, and $In_{0.52}Al_{0.48}As$ top layer, a structure that has produced very high frequency devices [83]. It should be noted that the quantum well channel is strained in this structure, and that will affect its transport properties. Channel lengths from 70 to 20 nm have been explored in such a structure. This type of device, modified by using a $In_{0.53}Ga_{0.47}As$ quantum well with a narrower InAs layer embedded within it, has produced devices of 25 nm gate length that yielded 9 db gain at 1 THz [59]. These are among the highest operating frequency devices in the world.

Our interest here is how this behavior depends upon inter-valley transfer, and the indicator is that the band gap of the strained $In_{0.75}Ga_{0.25}As$ is 0.58 eV, while that of $In_{0.53}Ga_{0.47}As$ is 0.75 eV. Normally, one would expect these devices to break down, due to avalanche breakdown at voltages less than 1.0 V. However, these devices are stable amplifiers for drain bias up to 2.5 V [84, 85]. To investigate the unusual stability of these devices and their THz response, simulations were undertaken with full band Monte Carlo techniques [86]. This involved determining the full phonon spectrum in the strained quantum well, as well as the full band structure of the material. The full band simulation was developed with what is known as a cellular method in momentum space [87]. As is usual in these devices, the peak value of the electric field occurred at the drain end of the gate. What was unusual is that a vast majority of the carriers had transferred to the satellite valleys by this point in the device, and *remained there until reaching the drain itself*. This condition is reflected in figure 2.27 [88]. In panel (a), the population of electrons located between the source and gate is shown in the Brillouin zone. A few carriers have been excited to higher energies due to the extension of the high electric field into the region before the gate. Panel (b) shows the population of electrons in the Brillouin zone in the region between the gate and the drain. Clearly, most of the carriers are occupying the full set of satellite valleys, both X and L. In the figure, the energy shading of the particles is the total energy measured from the valence band maximum. From panel (b), it is

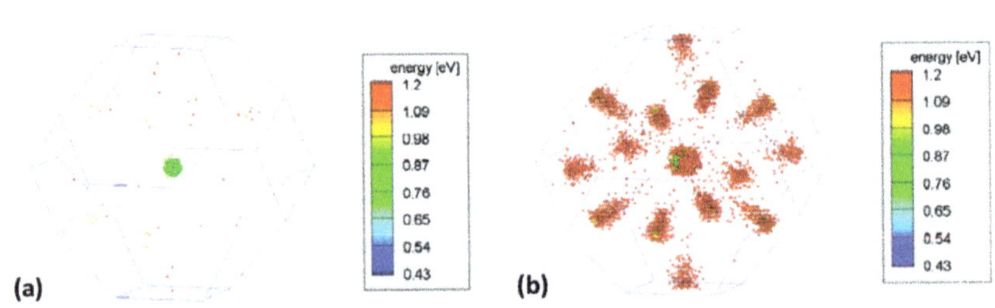

Figure 2.27. The projection of the simulated particles in the HEMT onto the Brillouin zone. (a) The projection for particles in the region between the source and the gate. As the electric field actually reaches into this region, some particles have already been transferred to satellite valleys. (b) The projection for particles in the region between the gate and the drain. In both cases, the shading represents the particle energy, relative to the valence band maximum. Reprinted from R Akis *et al* [88], copyright by IOPP. All rights reserved.

clear that the particles reach not only the L valleys, but also the X valleys of the conduction band. This, in fact, explains the high breakdown voltage observed in these devices. When the carriers are in the satellite valley, the onset of impact ionization is shifted to much higher electric fields, and energies, than expected by the normal band gap. First, the carriers in the satellite valleys have a much higher gap to the valence band, and, secondly, the effective mass is higher in the satellite valleys, and this also raises the threshold for ionization, as will be seen in the next sections. Thus, inter-valley transfer can be quite useful for improving device response. Here, it raises the breakdown voltage and it enhances the speed at which the carriers cross the gate-to-drain spacing in the HEMT, thus producing less series resistance in the devices. The essence of modern devices follow this trend as well, so that such inter-valley transfer lives well in them.

2.4 Impact ionization and breakdown

As was pointed out at the beginning of this chapter, the early studies of high field effects in semiconductors were aimed at understanding the breakdown of the material [1, 2]. Over the years, there have been two basic approaches to characterizing the current rise due to *impact ionization*, or as discussed in devices, the *avalanche breakdown*. There is a connection between these two terms. For example, when an electron in the conduction band has enough energy, it can interact with a bound electron in the valence band, and excite it across the energy gap. This creates an additional electron in the conduction band and a hole in the valence band, and is the impact ionization process. Of course, it can be initiated by a hole as well, producing an additional hole in the valence band and an electron in the conduction band. But, in a reverse biased *p-n* junction, the three particles existing after this process can each be accelerated and produce two more particles each. This exponential growth in electrons and holes is the avalanche breakdown process. What is measured depends upon the difficulty in accessing the critical physical processes occurring in the depletion region.

One of the earliest, but still quite tractable, approaches to measuring the ionization coefficient α comes from McKay [89]. Here, α is the number of electron–hole pairs produced per electron per centimeter of travel in an electric field E (measured in V cm^{-1}). It is first being assumed that electrons and holes produce the same value of α. In the junction, it further is assumed that the electric field is a function only of the position coordinate. Then, one can connect the ionization coefficient to the multiplication factor $M = n/n_0 \sim I/I_0$, where I is the current and I_0 is the reverse saturation current, by

$$1 - \frac{1}{M} = \int_0^W \alpha dx. \tag{2.21}$$

Here, W is the thickness of the depletion region in the junction. This equation illustrates the experimental problem. The ionization coefficient is not directly measurable, but must be extracted from measurements of the multiplication factor. For an abrupt *p-n* junction, the electric field profile is usually triangular, and with some manipulation, one can arise at a formula for the extracted ionization rate as [89]

$$\alpha(E_{\max}) = \frac{2}{W^2}\frac{d(1-\frac{1}{M})}{dE_{\max}} - \frac{4}{W^3}\left(1-\frac{1}{M}\right)\frac{dW}{dE_{\max}}, \tag{2.22}$$

where $E_{\max}$ is the peak electric field in the junction. This approach can of course be extended to cases where the electron ionization coefficient α differs from a hole ionization coefficient β, but this just complicates the equations and the ability to extract these parameters [90].

In figure 2.28, several sets of data [91–93] for electrons in Si are plotted as a function of the inverse electric field. The variation within the experimental data is significant, which is likely due to the difficulty of measuring the ionization coefficient effectively. The data of Lee *et al* [91] appears to give $\alpha \sim E^{-2}$, while the other data has a slightly higher coefficient. This is not unusual, as it will become apparent that there are various complications in the ideas of breakdown.

The problem that causes the determination of the ionization coefficient to be difficult is the very high electric fields that are necessary for impact ionization in wide

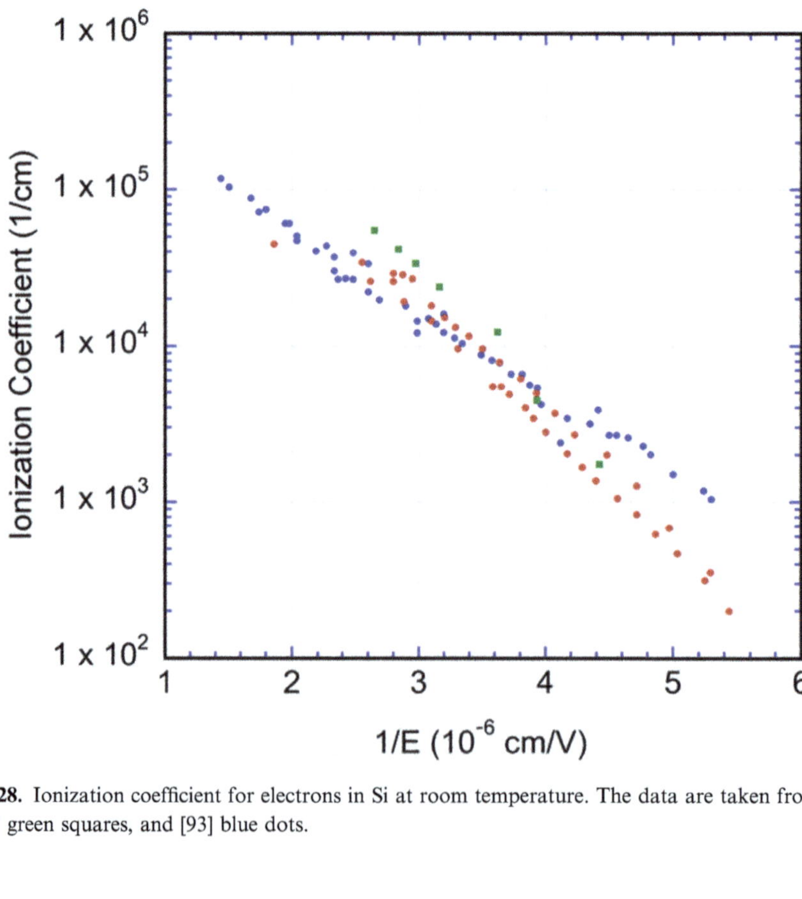

Figure 2.28. Ionization coefficient for electrons in Si at room temperature. The data are taken from: [91] red dots, [92] green squares, and [93] blue dots.

band gap materials, even in Si. The necessary fields are of the order of hundreds of kilovolts per centimeter. In narrow gap materials such as InSb, a different approach is possible. As we noted in the preceding sections, impact ionization in InSb at 77 K sets in around 200 V cm^{-1}, which is considerably smaller than the values found above. This means that we go directly to the determination of the process from the time dependence of the current using pulsed electric fields on the nanosecond scale (the pulse must be sufficiently short to avoid heating the lattice and changing the mobility of the material). Then, in the situation where it is clear that the increasing current is due to carrier multiplication, one can get the generation rate (number of particles per second generated by an incident electron) as [94]

$$g(E) = \frac{1}{I_0}\frac{dI}{dt}\bigg|_{t=0} = \frac{1}{n_0}\frac{dn}{dt}\bigg|_{t=0}. \tag{2.23}$$

The generation rate for InSb has been studied by several groups [94, 95, 96], and the results are summarized in figure 2.29. The data is taken from McGroddy and Nation [94] for an electron density of 7.4×10^{13} cm^{-3} (green squares in the figure), Ferry and Heinrich [95] for a carrier density of 6×10^{13} cm^{-3} (red circles), and

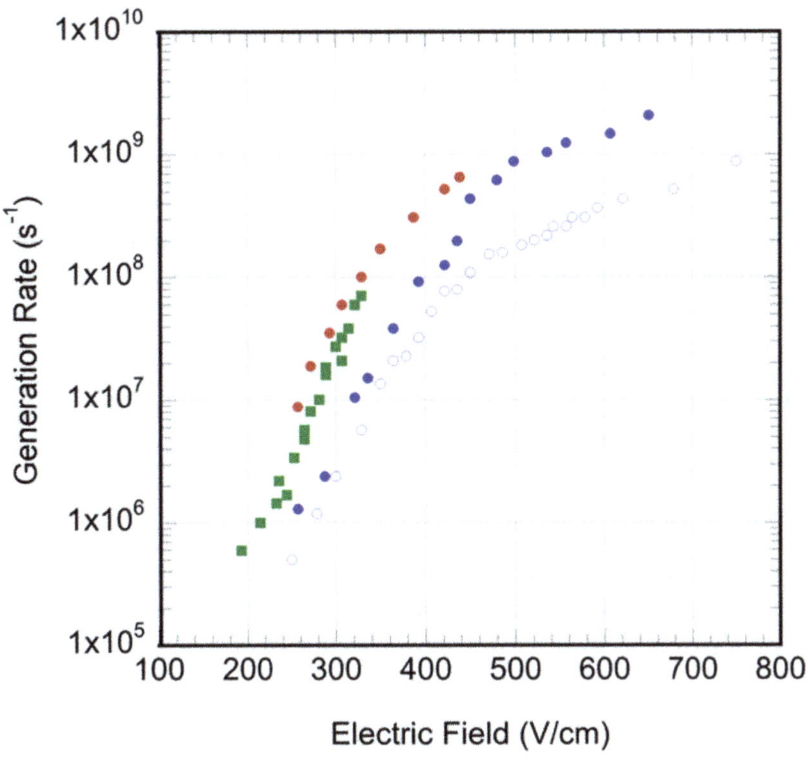

Figure 2.29. Impact ionization generation rate in InSb at 77 K. Data are taken from McGroddy and Nation (green) [94], Ferry and Heinrich [95] (red), and Dick and Ancker-Johnson (blue) [96]. The different curves are discussed in the text.

Dick and Ancker-Johnson [96] for carrier densities of 2.7×10^{15} cm^{-3} (solid blue circles) and 5×10^{14} cm^{-3} (open blue circles), all at 77 K. Simulations with Monte Carlo techniques suggest that the lower generation rates at high fields are expected with increased electron density [97].

Of course the generation rate g is related to the ionization coefficient α, as both are average quantities connected to the distribution function at the high electric field. The first of these is expressed as

$$\alpha = \frac{1}{n_0 v_d} \sum_{E > E_{\mathrm{T}}} \rho(E) \Gamma_{\mathrm{ion}} f(E), \tag{2.24}$$

where Γ_{ion} is the impact ionization scattering rate (discussed below), n_0 is the low field carrier density, and the other parameters have their normal meaning. The generation rate is also described by this form with an added factor, the velocity, as

$$g = \frac{1}{n_0 v_d} \sum_{E > E_{\mathrm{T}}} \rho(E) v(E) \Gamma_{\mathrm{ion}} f(E), \tag{2.25}$$

where here E is the energy. In both of these equations, E_{T} is the threshold energy for the ionizing collision, $\rho(E)$ is the density of states, $f(E)$ is the distribution function, and Γ_{ion} is the probability of an ionizing collision.

2.4.1 The energy threshold

There are two major factors that will affect the shape exhibited by either the generation rate (2.25) or the ionization coefficient (2.24) and their functional dependence upon the electric field. These are the threshold energy for ionization E_{T} and the energy dependence of the ionizing collision. The first of these will be dealt in this section and the second in the next section.

Consider a typical ionizing collision in which an electron in the conduction band (a similar process can occur from a hole in the valence band for p-type material) interacts with an electron in the valence band (e.g., an electron which is not free to travel). Prior to the collision, the incident electron lies at a relatively high energy $E_c(\mathbf{p}_1)$ and the bound electron lies in the valence band at $E_v(\mathbf{p}_2)$. This electron is in the heavy-hole band, as it offers the lowest energy threshold. After the collision, the incident electron has created an electron–hole pair, giving us two electrons in the conduction band and a hole in the valence band. The two electrons in the conduction band now lie at $E_c(\mathbf{p}_3)$ and $E_c(\mathbf{p}_4)$. The hole lies at $E_v(\mathbf{p}_2)$, where the original electron was situated. There are several complications to consider. First, the initial electron and the two final electrons do not have to be in the same conduction band valley, and certainly do not satisfy this for Si, where a phonon has to be involved for the indirect process. Hence, phonons may be involved in the various transitions.

During the ionizing collision, both the total energy and total momentum have to conserved [98]. In general, one writes equations for the total energy and momentum and then minimizes these to find the threshold energy. It appears that this leads to the condition that the velocities at each of the momenta values, including that of the phonon, must be the same at the energy threshold [98]. If phonons are involved, this is a very restrictive condition as it requires the charge particles to have the same velocity as the phonons. Since phonon velocities are usually quite small when compared to electron velocities, this restricts the charge carriers to lie very near the band extrema for phonon involved processes. This also restricts the phonon processes to be for cases in which the incident electron makes a band transition.

In a direct-gap semiconductor with spherical, parabolic bands, the effective masses are taken to be m_e and m_h. The threshold criterion requires that the electron and hole velocities v_e and v_h are equal in the absence of the phonon processes. This leads to

$$k_v = \gamma k_c, \quad \gamma = m_h/m_e, \tag{2.26}$$

where k_v and k_c are valence and conduction band wave numbers, respectively. Then, the conservation of energy can be written as

$$\begin{aligned} E_T = E_c(p_1) &= 2\left(\frac{\hbar^2 k_c^2}{2m_e}\right) + \frac{\hbar^2 k_v^2}{2m_h} + E_G \\ &= (2+\gamma)\frac{\hbar^2 k_c^2}{2m_e} + E_G, \end{aligned} \tag{2.27}$$

the factor of two is for the two electrons, and the momentum becomes

$$p_1 = 2p_c - p_v = \hbar k_c(2+\gamma). \tag{2.28}$$

Here, it is assumed that the two final electron states are in the same band and have the same momentum and velocity. From the last equation, the threshold energy may be determined to be

$$E_T = E_c(p_1) = (2+\gamma)^2 \frac{\hbar^2 k_c^2}{2m_e}. \tag{2.29}$$

This can now be used to eliminate the momentum from (2.27), and finally find the threshold energy to be

$$E_T = E_G\left(\frac{2+\gamma}{1+\gamma}\right). \tag{2.30}$$

This is for the electron initiated process. When $\gamma = 1$, both threshold energies are given by $1.5E_G$. On the other hand, if γ is very large, the ionization threshold for electrons is just the gap itself, while if this parameter is very small, it becomes twice the gap. For the case of hole initiated ionization, γ is replaced by $1/\gamma$, and the two situations are reversed. With the foregoing discussion, it is clear that one can determine the threshold energy quite easily with the equal velocity criterion *if sufficiently accurate band structures are available.* The lowest threshold energies in the various directions for some common semiconductors have been determined [98, 99], and are summarized in [24].

2.4.2 The ionizing collision

The collision rate for the ionizing collision can be calculated in a similar manner to that used for normal momentum scattering and relaxation times. The interaction potential is the screened Coulomb interaction between the incident electron, which we take to be in the conduction band, and the bound electron in the valence band. Further, we will assume that the bands are spherical and parabolic, although the approach is easily extended to a non-parabolic band [97]. It will also be assumed that the dielectric function is just a simple value and the frequency and momentum dependence of this quantity will be ignored. We may then write the probability of scattering as

$$\Gamma_{\text{ion}} = \frac{1}{\tau_{\text{ion}}} = \frac{2\pi}{\hbar}\sum_{k_2,k_4} |M(k_1, k_2, k_3, k_4)|^2\, \delta(E_1 - E_2 - E_3 - E_4). \tag{2.31}$$

Using plane wave assumptions for the wave functions in the matrix elements, one finds that the scattering rate is given as

$$\Gamma_{\text{ion}} = \frac{m_e^3 e^4}{\hbar^3\pi^3\varepsilon^2 m_h^2}\left(1 + \frac{m_h}{m_e}\right)\int_0^{E-E_G} \frac{\sqrt{E'(E - E' - E_G)}}{(E_G + E_D)^2}dE', \tag{2.32}$$

where $E_D = \hbar^2 q_D^2/2m_h$ is an energy corresponding to a hole with the Debye screening wave vector. Equation (2.32) can now be readily integrated to give the resulting total scattering rate for the ionizing collision as

$$\Gamma_{\text{ion}}(E) = \frac{m_e^3 e^4}{8\hbar^3\pi^2\varepsilon^2 m_h^2}\left(1 + \frac{m_h}{m_e}\right)\left(\frac{E - E_T}{E_T + E_D}\right)^2, \tag{2.33}$$

and, of course, the energy must be larger than the energy gap, which here is the threshold energy (2.30).

In low-dimensional systems, the problem can become more complicated. In graphene, for example, there is no real band gap and the nature of the ionizing collision depends more on the effective gap between possible electron and hole states near the Fermi energy. Nevertheless, the impact ionizing collision has been calculated for graphene [100–102], with good results. It is somewhat more problematic in the transition metal di-chalcogenides, where the conduction and valence bands are primarily derived from the metal *d* states. Here, it is found that these principle bands are rather narrow in energy and certainty not as wide as the band gap. As a result, impact ionization is not really expected, yet it occurs [103]. On the other hand, these materials suffer from significant numbers of chalcogenide vacancies [104], that introduce defect states near mid-gap. Thus, impact ionization could excite electrons trapped in these states to the conduction band. Also, there could arise a multi-step process where the valence band electron is excited to the defect level by one hot electron and then is ionized to the conduction band by a second hot electron. This latter two-step process would be the inverse of Shockley–Read–Hall recombination.

2.4.3 Ballistic versus diffusive ionization

Exact calculations for the ionization coefficient (2.24) and the generation rate (2.25) must involve knowledge of the distribution function of the carriers, as is clear from these formulas. As discussed above, this is in itself a difficult task to achieve. Most of the so-called exact theories really boil down to finding the distribution function first. The desire here is really to describe the impact ionization process itself and to understand a reasonable view of how it depends upon energy and/or the electric field. In the early days, there were two competing ideas on how impact ionization occurred. In one of these, it was assumed that 'lucky' electrons occasionally escaped some scattering events and ballistically traveled to high energies where they underwent impact ionization [105]. The second approach assumed that the distribution function was mainly spherically symmetric, with a characteristic temperature, and the diffusion of particles to higher energy provided the source for impact ionization.

Wolff [106], in one of the earliest theories applicable to semiconductors, assumed that a strong scattering interaction was always present in the semiconductor, as did Fröhlich earlier [22, 107], and that this led to a quasi-Maxwellian distribution function, dominated by the spherically symmetric part. Nearly all of the discussion in this chapter contradicts this approach, so it is fair to ask whether Wolff's assumptions could be relevant. Yet, at very high electric fields, and carrier energies, the diffusive effect (in energy space) dominates any anisotropy arising from the scattering processes. So, the symmetric part can become a dominant term and lead to the majority of impact ionization. As we will see, Wolff's approximations are an important part of the result. This diffusion approximation leads to the the ionization coefficient as

$$\alpha(E) \cong A \exp\left(-\frac{C}{E^2}\right), \tag{2.34}$$

where here, E is the electric field and C is an arbitrary constant.

Shockley [105], on the other hand, assumed a very weak interaction between the electron and the lattice, so that the distribution function retained a strongly peaked form arising from the electric field streaming term in the Boltzmann equation. He considered that impact ionization arose from a few 'lucky' electrons (traveling ballistically) in the tail of the asymmetrical part of the distribution that streamed to high enough energy to initialize the impact ionization. So, in his case, the ionization coefficient would vary as

$$\alpha(E) \cong B \exp\left(-\frac{C'}{eEL}\right), \tag{2.35}$$

where L is the scattering mean free path and C' is an arbitrary constant.

It is perhaps more rational to inquire as to whether the two approaches above can be combined in some way. This task was pursued by Baraff [108]. He assumed that the distribution function could be expanded into two terms, one of which is spherically symmetric and the second of which is almost a delta function in space oriented along the electric field (an extreme version of figure 2.1). Hence, the distribution function was assumed to be

$$f(p, \cos\vartheta) = A(p) + B(p)\delta(1 - \cos\vartheta), \tag{2.36}$$

and then the two functions A and B, the distribution function can be expanded in a Legendre polynomial series, as will be discussed in chapter 6. With this approach, Baraff found that the spherically symmetric part is very similar to a well-known distribution, the Druyvesteyn distribution [109], and may be written as

$$f_0(p) = \frac{1}{(p^2)^a}\exp\left(-\frac{p^2/2m_c}{k_B T_e}\right), \tag{2.37}$$

with

$$\begin{aligned} \frac{1}{a} &= \frac{2}{3} + \frac{eEL}{3\hbar\omega_0} \\ T_e &= \frac{(eEL)^2}{3\hbar\omega_0} + \frac{2}{3}eEL \end{aligned}. \tag{2.38}$$

Thus, both forms appear in the solution, and that moving from one to the other is merely due to the size of the electric field. If $eEL = \hbar\omega_0$, a solution quite similar to the Shockley form [105] is found. On the other hand, if $eEL \ll \hbar\omega_0$, a form which is basically the Wolff form [106] is found. So, the initial onset of impact ionization seems to arise from the quasi-ballistic carriers streaming to high energy, while for higher fields the diffusion from the main body of the distribution takes over. It is worth remarking that, especially in Shockley's regime, the band structure details are expected to be exceedingly important. This is especially true in Si, due to the six ellipsoids in a constant energy surface with the anisotropic masses. Hence, not only L, but also the threshold energy E_T, is orientation dependent. In fact, this direction dependence has been observed in GaAs [110] and Si [111]. In these studies, it was found that the highest impact ionization occurred in the (110) direction in GaAs and in the (100) direction in Si. In both cases, it was observed that the directional dependence was greatly reduced as the electric field was increased, in keeping with the above expectations.

2.5 Microwave studies

The idea of studying carriers at microwave frequencies is quite old, and there are two approaches that provide interesting results. First, there is using a small microwave signal to monitor the conductance as a function of the high electric field. For example, replacing the microwave frequency with an optical frequency leads to the curve in figure 2.9(b), where the transient velocity in InP appears to peak at a delay of some 300 fs. This would correspond to a peak in the conductance of the material at a frequency of some 6.7 THz. This peak will also transfer into a peak in the frequency spectrum of the conductance of an InP sample measured, for example, with Fourier Transform Infrared Spectroscopy, in order to reach into the THz regime, which is usually hard to measure. Second, one can use very high power microwave, or THz, signals to excite hot carriers in semiconductors and then measuring the properties with small dc fields or with another optical signal. One approach for this latter case is, of course, free carrier absorption in high microwave fields.

2.5.1 Probing with small microwave fields

Perhaps the first use of microwaves to probe the warm carrier mobility was by Morgan in *n*-type Ge [112], but it was used more effectively by Seeger [113] to study warm and hot carriers in Ge. In general, the idea is to use a pulsed high electric field applied to the sample while monitoring the conductance of the sample with the microwave field, or, as in later experiments, an infrared laser [114], in which the free carrier absorption is monitored. But, it can be modified to heat the carriers with a strong microwave field and probe the results with a weak microwave field [115].

In modern times, however, this methodology is used to study real semiconductor devices under normal operating conditions. The measurement may use an external microwave source or internally generated conditions of the device itself, such as the hot carrier noise generated during operation of the device. Such noise signals have been used for decades as an attempt to characterize normal hot carrier behavior in semiconductors, but has naturally become of interest in devices. For example, it has been shown that noise measurements can give an indication of the effective carrier temperature in MOSFETs, but also demonstrate that the noise temperature is affected by the gate voltage [116]. More recently, device noise has been used to characterize the behavior of GaN HEMTs and to study the role of hot phonons in these devices [117]. These latter authors find that the energy relaxation time deduced from the noise measurements can give information on the LO phonon nonequilibrium state. Beyond noise, the electroluminescence from GaN HEMTs operating in the microwave range has been used to characterize different modes of operation of the device [118].

An $Al_{0.26}Ga_{0.74}N$/GaN HEMT was studied using a vector network analyzer to give microwave frequencies up to 60 GHz [119]. This allows one to study not only the S parameters of the device, but also evaluate the small signal current gain, the unilateral gain, and the maximum available gain of the device under normal operating conditions. This was used to study the role of different passivation techniques on the device (the surface passivation in GaN devices is usually needed to avoid surface breakdown in the high fields of these high devices).

Planar Schottky diodes are used to detect very high frequency signals. Recent studies have used microwave reflectometry to study the THz response of such devices [120]. The results were sufficiently promising, that the authors suggested that an imaging camera at 150 GHz was feasible using arrays of these detectors.

Today, it seems that we no longer describe such traditional microwave measuring systems as being used to study hot electrons, but rather it is now the study of device performance. The presence of the hot electrons is accepted as commonplace to the operation of these high frequency and high power devices.

2.5.2 High microwave fields

Early measurements of microwave heating of carriers in solids used pulsed magnetrons for the exciting source and perhaps a small electric field to monitor the conductance. This was done for direct carrier heating [121] and to study the Hall fields in a magnetic field [122], both in GaAs as a method to gain more information about the Gunn effect. Later, Hess *et al* [123] used strictly high power microwaves to study hot carriers in Ge, directly measuring the incident, reflected and transmitted power levels, deducing the sample impedance from the known electric field strength and the absorbed power. Surprisingly, it was found in all these cases that the heating and mobility were determined to be approximately the same as from dc electric fields. Subsequently, very high frequency (890 GHz) from a submillimeter laser was used to probe the conductivity in GaAs and InP in order to determine directly the energy and momentum relaxation times by probing closer to the frequencies corresponding to these relaxation times [124].

Kawashima and Kataoka used large microwave fields to study the velocity-field characteristics in GaInSb, across the compositional range [125], as a method to avoid contact effects in determining the hot carrier behavior. They were able to observe NDC for compositions above 50% GaSb. Surprisingly, the 35 GHz microwaves they were using did not indicate NDC at lower compositions, which would have been expected, given the discussion in earlier sections, but the 200 ns pulses may well have been sufficiently long to allow impact ionization to dominate the measurements.

Once it was confirmed that hot carriers led to non-ohmic conduction, then it became feasible to begin to study frequency multiplication using such devices. Seeger noted that it was easy to generate third harmonics of a 9.4 GHz signal already with warm electrons [126]. It was an easy extension to use the hot carrier regime for frequency mixing as well [127]. Machi then established that using hot electron bulk materials made far more efficient mixers than e.g. point contact diodes [128]. The process has been extended in recent years to replacing the semiconductor with a superconducting hot electron bolometer for the detection of THz signals in the mixing process [129, 130].

2.5.3 THz transistors

The InP-based HEMT that has exhibited gain at 1 THz has already been mentioned in section 2.3.3. In order to reach such high frequencies, it is necessary to utilize the very high overshoot, or ballistic, velocities that can be reached on a transient basis. The key is the transit time of the electrons (or holes in *p*-channel devices) under the gate itself. But, it is important to note that it is not the *physical* dimension of the gate, but the *effective* gate length. The physical dimension is the drawn length of the actual metal gate, while the effective length is the acceleration length in the region of the channel mainly under the gate. Usually, this has to be established by proper simulations of the entire device to determine the actual velocity as a function of position within the device. An example is shown in figure 2.30 for several gate lengths [88]. At the top of the figure, beginning at about 100 nm, are a series of colored horizontal lines of different length. These are the physical gate dimensions. The main curves are the velocity as function of distance along the channel. At each position, the velocity is integrated across the thickness of the electron density of the

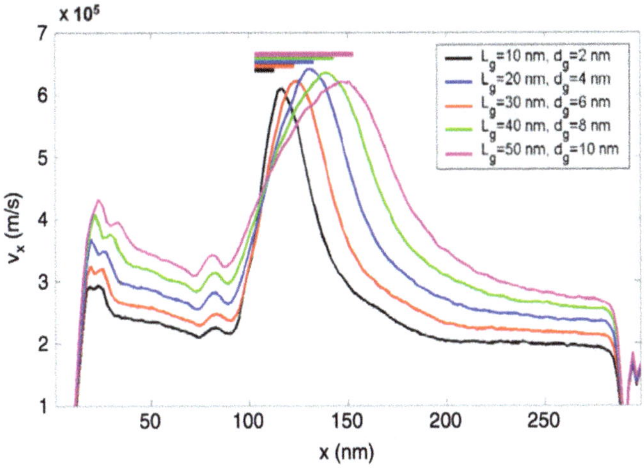

Figure 2.30. The velocity along the channel for a set of scaled HEMT devices. The details are discussed in the text. Reprinted from Akis *et al* [88], copyright IOP Publishing. All rights reserved.

channel, so that this velocity is an average over the cross-section of the channel at each position along the channel. It may be noticed that the velocity begins to increase in the electric field before the drain edge of the gate is reached. This occurs at about the 90 nm point. This is defined as the left edge of the effective gate length [37]. The right edge of the effective gate is at the peak of the velocity, which is also the peak of the electric field in the channel. This peak generally occurs at the drain end of the gate, or ever so slightly beyond in very short gate lengths. Thus, this effective gate length is larger than the physical gate length. There are other definitions for the effective gate length, but this one appears to work better in determining the frequency response of the transistor.

The important factor to determine is the transit time under the effective gate acceleration. First, the average velocity under the gate is determined by integrating the curves in figure 2.30 as

$$v_{\text{av}} = \frac{\int_{y_1}^{y_2} n(y)v(y)dy}{\int_{y_1}^{y_2} n(y)dy}, \tag{2.39}$$

where y is the *transverse* dimension across the channel. The parameters y_1 and y_2 define the top and bottom of the channel. This average is then used to compute the transit time as

$$\tau_T = \frac{1}{2\pi f_T} = \int_0^{L_G^{\text{eff}}} \frac{dx'}{v_{\text{av}}(x')}, \tag{2.40}$$

where f_T is the unity gain frequency, where $h_{21} = 1$. In this last equation, the position is relative to the position where acceleration begins. The advantage of this definition of the effective gate length is found when a plot of the expected f_T as a function of the inverse effective gate length is made, where it is found that a linear behavior is observed.

In the actual device [84], it was found that the observed f_T was only about 610 GHz for a physical gate length of 35 nm. For this physical gate length, the effective gate length in the scaling arguments is about 44 nm, which would lead to a unity gain frequency of 1.5 THz, if there were no mediating effects. Such effects can arise from excessive contact resistance, heating of the device, or parasitic effects from the circuit itself. Nevertheless, it was found the the frequency at which the maximum available gain, $f_{\max}$, occurred was 1.5 THz.

It is possible to create gain above the unity gain frequency by using resonances within the device circuit [131]. This has been demonstrated with both measurements and simulations of the device and circuit. The authors have shown that two peaks in gain can be achieved beyond the normal cutoff. The first of these is due to a resonance involving the intrinsic device capacitance and the circuit inductance. The second is a resonance involving primarily the circuit capacitance and inductance. Before proceeding, however, it is useful to point out that microwave power HEMTs often use a number of gate 'fingers.' In order to increase the power capability, the current through the device must be increased. The normal way to do this is to use a greater gate width (the transverse dimension from the gate length). However, in microwaves the gate is a transmission line, and if the gate width begins to be a significant fraction of the wavelength λ, then parts of the device are out of phase with other parts. To avoid this, the gate width is broken into a number of parallel devices. In figure 2.31, this is illustrated with two such GaN HEMTs. The left panel shows a device with two fingers [132], while the right panel shows a device with eight fingers [133]. Both devices have a similar layout, which is governed by the fact that these devices are connected to coplanar wave guide or strip lines. Hence, the source provides a ground plane and fills the outer edges of the device. Then, the gate comes in from one side and the drain goes out the other side. The study of resonances was

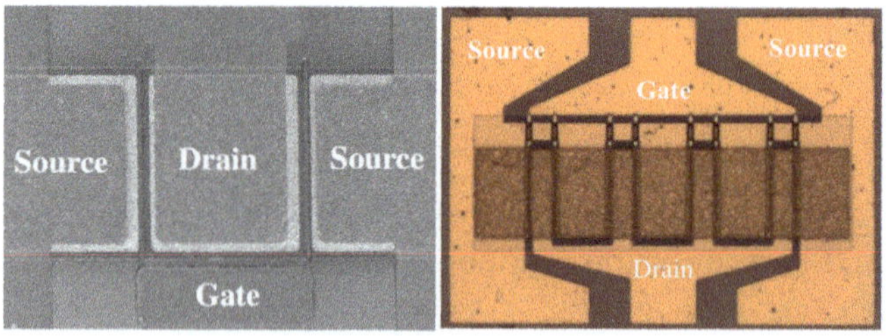

Figure 2.31. Illustration of the use of multiple gate fingers for pseudo-parallel devices so that the width of each individual gate remains much smaller than the microwave wavelength. The left panel has two such fingers, each of which is 50 μm wide. This is reprinted from Lachab *et al* [132]. The right panel has eight fingers, each of which is 125 μm wide, reprinted from Toprak *et al* [133]. Both are reprinted with permission of IOP Publishing.

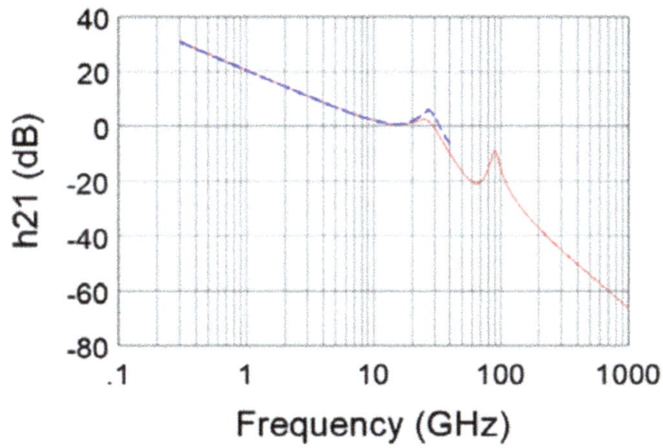

Figure 2.32. Spectral response of a GaN HEMT, showing the resonance in the gain at high frequency. Reprinted from Crupi *et al* [131], copyright (2018) with permission of Wiley.

carried out for a variety of devices [131]. For example, a 0.25 μm GaN HEMT in which the gate is composed of eight fingers of width 50 μm each. This device was operated at $V_{GS} = -2.72$ V and $V_{DS} = 10$ V, and showed the resonances at approximately 30 and 80 GHz, whereas the unity gain frequency was nearer 20 GHz. Similarly, in a 0.15 μm GaAs HEMT with the same gate finger dimensions, and $V_{GS} = -0.6$ V and $V_{DS} = 6$ V, the resonances appeared at about 90 and 400 GHz whereas the unity gain frequency was about 60 GHz. The experimental and theoretical results are shown in figure 2.32 for a 0.7 μm gate length GaN HEMT with a gate composed of two fingers of 400 μm width each, with biases of $V_{GS} = -2.0$ V and $V_{DS} = 19.5$ V. The first resonance was observed at 18 GHz and is clearly seen in the experiment, while a second is expected at about 90 GHz.

2.6 Ballistic devices

The earliest attempt to utilize ballistic, or hot carrier overshoot, velocity for devices may well have been the hot electron transistor, in which two semiconductor electrodes are separated by a thin metal layer [134]. The idea was that this would produce a smaller, but still efficient, bipolar transistor, in which the electrons would traverse the thin metal base quickly enough to not lose any energy. The idea has been pursued for some time, but has not prospered over the years. A variant on this idea was subsequently proposed, in which a metal-oxide-metal structure would be used as an electron injection structure with bias between the two metals in order to extract the electrons from the source metal [135]. Then, a second barrier whose height could be varied would allow spectroscopy on the electrons injected from the primary part of the device.

Somewhat later, studies of tunneling through single heterostructure barriers in a GaAs/AlGaAs/GaAs system, with a heavily doped source region and a lightly doped drain region, showed that electrons which tunneled would emit phonons with precise energy losses that could be measured quite easily at low temperature [136]. This led to an oscillatory current, in which the period corresponded to the LO optical phonon energy, and the oscillations could be significantly enhanced by using a magnetic field to freeze the carriers in the lightly doped drain region. There was much discussion over the exact cause, as self-consistent fields of the electrons in the depleted drain will modulate the potential distribution and affect the current.

The combinations of these above devices appeared shortly afterward. In one version, the metal-oxide-metal structure of Heiblum was replaced by a GaAs/AlGaAs/GaAs structure in which the two GaAs layers were heavily doped [137]. Then, the second barrier was an AlGaAs layer with graded composition, so that the barrier would be raised or lowered depending upon the bias applied across it. In essence, the second barrier acted as a spectrometer to measure the energy distribution of the electrons emitted from the first part of the structure. The second approach created all layers of a similar structure by variation of the doping across the system [138]. The two barriers themselves were very lightly doped regions of GaAs, while the other parts were more heavily doped regions of GaAs. Again, bias across the second barrier allowed its use as an analyzer for the energy spectrum of the electrons emitted over the first barrier. The analyzer allowed the measurement of the scattering rates for these injected electrons [139].

Whether the above work played a role or not, the structure appears in another form in the concept of ballistic electron emission microscopy (BEEM) [140], where the potential landscape looks a lot like that of Heiblum [137]. But, it reaches its most useful form when combined with the scanning tunneling microscope (STM) [141, 142]. The metal tip of the STM and vacuum between the tip and the substrate provides the first barrer, which is the tunneling barrier. This then allows the study of potential steps in the substrate, such as the Schottky barrier in a semiconductor [143]. BEEM has been used to study hot carrier transport in semiconductor Schottky barrier made of Au on Si, and the characteristics of the hot carriers coming over the barrier have been characterized [144]. The potential surface for the BEEM device will look something like that of figure 2.33, taken from Bell *et al* [145]. Here, the tip corresponds to the STM tip and the area between the tip and the base is the vacuum gap of the tip from the surface. The base is the barrier metal on the semiconductor,

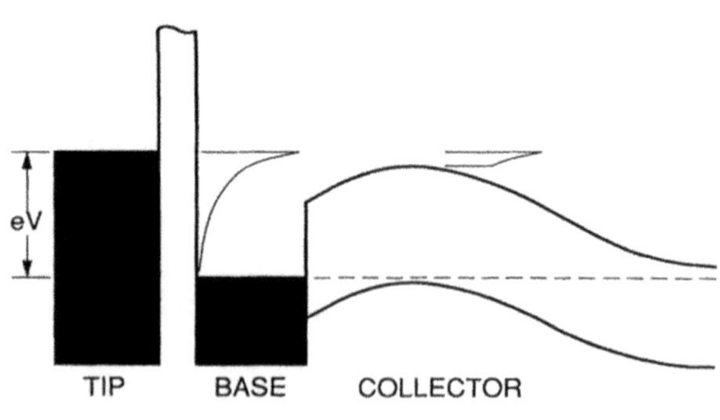

Figure 2.33. Potential profile of a BEEM experiment on a metallized Si *p-n* junction. The tip is that of an STM used to inject the ballistic electrons. Reprinted from Bell *et al* [145], copyright (1993) by the American Physical Society.

in this case Au on *p*-type Si. The STM tip is also Au. In this work, focus was upon metallized Si *p-n* junctions. The *p*-layers varied from 20 to 100 nm, so that the scattering properties of the injected electrons could be studied. The early work on BEEM has been reviewed by Bell *et al* [146].

2.7 Real-space transfer

One usually considers that carriers move through real space in any transport problem, and that is true. What is different about real-space transfer is the modification of this idea with hot carriers. The original concept considers a superlattice of GaAs and GaAlAs [147]. In this case, the GaAlAs was considered to be doped and the GaAs undoped. Thus, carriers would transfer into the GaAs, but be situated quite near each side of the quantum well, next to the hetero-interfaces. This was, in fact, an early application of modulation doping [148]. When a high electric field is applied in the plane of the quantum wells, the carriers there are heated to a high electron temperature. Then, they can diffuse across the interface into the GaAlAs layers where the mobility is much lower. The current across the interface can be estimated as if it were thermionic emission current, but there is an added force from the donors that attracts the carriers to the wide band gap material. Once in the GaAlAs, the transferred carriers would no longer have either a high mobility or a high velocity, and the idea was this would lead to negative differential conductance. Shortly after the idea was conceived, it was confirmed with ensemble Monte Carlo simulations that such negative differential conductivity would occur [149]. While this idea was formulated for the heterostructures, subsequent calculations showed that it could occur in Si/SiO_2 such as charged-coupled devices [150].

Measurements in structures with $Al_{0.17}Ga_{0.83}As$ barriers of 100 nm thickness and GaAs wells that varied from 40 nm to 1.0 μm, at 77 K, clearly established that real-space transfer was occurring, and negative differential conductance was observed [151]. Here, it was found that the strength of the negative region would dramatically increase with the doping level in the AlGaAs. Thinning the barrier thickness to 50 nm also increased the negative values, and the effect could also be seen at room temperature. Further theoretical studies showed that there should be an increase in the noise signal when real-space transfer began to occur [152]. In a similar study of a GaAs/AlGaAs HEMT, it was shown that velocity modulation would occur with varying gate voltages [153], and it was firmly established that real-space transfer could occur from these devices [154]. In the HEMTs, real-space transfer is aided by inter-valley scattering, since transfer from the Γ valleys in GaAs to the L valleys in AlGaAs is quite easy. In these HEMTs, this would be detrimental to their operation, since charge accumulation in the AlGaAs would lead to a loss of gate control on the device, and general detrimental behavior [155]. On the other hand, this transfer could lead to an interesting class of fast charge storage devices [156].

In wide band gap materials, real-space transfer was observed in SiO_2/AlGaN/GaN HEMTs, where the transfer would be from the normal channel at the heteroface of AlGaN on GaN, to the top interface next to the oxide [157]. Such transfer can lead to very sharp drops in the current and hence strong negative differential conductance [158].

When the field is turned normal to the surface, and transport is through the superlattice or heterostructure new behavior is observed. Early on, it was discovered that transport normal to a superlattice enhanced the impact ionization observed in either of the component materials [159, 160]. It was soon confirmed that the carriers responded to both the applied electric field plus any superlattice induced fields [161, 162]. It was later shown that tunneling could also enhance the ionization [163]. This led to the realization that a stronger negative differential conductance could be found with transport normal to the heterostructure, thus creating a new 'hot electron' diode [164, 165].

So far, transfer into a lower mobility region has been discussed. But, it is also possible to transfer into a higher mobility region.

References

[1] von Hippel E 1933 *Zeitschr. Elekrochem.* **39** 506
[2] Fröhlich H and O'Dwyer J 1950 *Proc. Phys. Soc.* A **63** 81
[3] Gunn J B 1963 *Sol.-State Commun.* **1** 88
[4] Ridley B K and Watkins T B 1961 *Proc. Phys. Soc.* **78** 293
[5] Shaw M P, Grubin H L and Solomon P R 1979 *The Gunn-Hilsum Effect* (New York: Academic)
[6] Nicolis G and Prigogine I 1977 *Self-Organization in Nonequilibrium Systems* (New York: Wiley-Interscience)
[7] Fröhlich H and Seitz F 1950 *Phys. Rev.* **79** 526
[8] Wick F G 2008 *Phys. Rev. (Ser. I)* **27** 11
[9] Ryder E J 1953 *Phys. Rev.* **90** 766
[10] Haynes J and Shockley W 1949 *Phys. Rev.* **75** 691
[11] Canali C and Ottaviani G 1970 *Phys. Lett.* **32A** 147
[12] Duh C Y and Moll J L 1968 *Sol. State Electron.* **11** 917
[13] Rodriguez V and Nicolet M A 1969 *J. Appl. Phys.* **40** 496
[14] Costato M and Reggiani L 1970 *Lett. Nuovo Cim.* **3** 728
[15] Conwell E M 1967 *High Field Transport in Semiconductors* (New York: Academic)
[16] Vogl P 1976 *Phys. Rev.* B **13** 694
[17] Lyddane R H, Sachs R G and Teller E 1941 *Phys. Rev.* **59** 673
[18] Glicksman M and Hicinbothem W A Jr 1963 *Phys. Rev.* **129** 1572
[19] Neukermans A and Kino G S 1970 *Appl. Phys. Lett.* **17** 102
[20] Bauer G and Kuchar F 1969 *Phys. Lett.* **30A** 399
[21] Ruch J G and Kino G S 1968 *Phys. Rev.* **174** 921
[22] Fröhlich H 1937 *Proc. R. Soc.* A **160** 280
[23] Fröhlich H 1954 *Adv. Phys.* **3** 325
[24] Ferry D K 2020 *Semiconductors Bonds and Bands* (Bristol: IOP Publishing)
[25] Mooradian A and McWhorter A L 1970 Proc. of the 10th Int. Conf. on Physics of Semiconductors S P Keller, J C Hansel and F Stern (Oak Ridge, TN: US Atomic Energy Commission) p 380
[26] Ralph S E and Wolga G J 1990 *Phys. Rev.* B **42** 11353
[27] Grann E D, Sheih S J and Tsen K T *et al* 1995 *Phys. Rev.* B **51** 1631
[28] Grann E D, Sheih S J and Chia C *et al* 1994 *Appl. Phys. Lett.* **64** 1230
[29] Grann E D, Tsen K T and Sankey O F *et al* 1995 *Appl. Phys. Lett.* **67** 1760

[30] Chia C, Sankey O F and Tsen K T 1993 *Mod. Phys. Lett.* B **7** 331
[31] Tsen K T, Poweleit C and Ferry D K *et al* 2005 *Appl. Phys. Lett.* **86** 222103
[32] Tsen K T, Ferry D K and Wang J-S *et al* 1999 *Physica* B **272** 416
[33] Tsen K T, Kiang J G and Ferry D K 2006 *J. Phys. Cond. Matter* **18** L585
[34] Lugli P and Ferry D K 1985 *IEEE Trans. Electron Dev.* **32** 2431
[35] Kubo R 1974 *Lecture Notes in Physics: Transport Phenomena* vol 31 (Berlin: Springer) p 75
[36] Price P 1960 *J. Appl. Phys.* **31** 949
[37] Akis R, Ayubi-Moak J and Farelli N *et al* 2008 *IEEE Electron Dev. Lett.* **29** 306
[38] Grann E D, Sheih S J and Tsen K T *et al* 1995 *IEEE J. Sel. Topics Quantum Electron.* **1** 1093
[39] Tsen K T, Ferry D K and Wang J-S *et al* 1996 *Appl. Phys. Lett.* **69** 3575
[40] Liang W, Tsen K T and Ferry D K *et al* 2004 *Semicond. Sci. Technol.* **19** S23
[41] Shibuya M 1955 *Phys. Rev.* **99** 1189
[42] Sasaki W, Shibura M and Mizuguchi K 1958 *J. Phys. Soc. Jpn.* **13** 456
[43] Sasaki W, Shibura M, Mizuguchi K and Hatoyama G M 1959 *J. Phys. Chem. Sol.* **8** 250
[44] Norris C B and Gibbons J F 1967 *IEEE Trans. Electron Dev.* **14** 38
[45] Canali C, Jacoboni C, Nava F, Ottaviani G and Alberigi-Quaranta A 1975 *Phys. Rev.* B **12** 2265
[46] Ferry D K 1976 *Phys. Rev.* **B14** 1605
[47] Siegel W, Heinrich A and Ziegler E 1976 *Phys. Stat. Sol. (a)* **35** 269
[48] Fischetti M V, Yoder P D and Khatami M M *et al* 2019 *Appl. Phys. Lett.* **114** 222104
[49] Smith R A 1959 *Semiconductors* (Cambridge: Cambridge University Press)
[50] Gunn J B 1963 *Phys. Lett.* **4** 194
[51] Ridley B K 1963 *Proc. Phys. Soc. London* **82** 954
[52] Gunn J B 1966 *Phys. Lett.* **22** 369
[53] Ruch J G and Kino G S 1967 *Appl. Phys. Lett.* **10** 41
[54] Butcher P and Fawcett W 1966 *Phys. Lett.* **21** 489
[55] McGroddy J C and Nathan M I 1967 *IBM J. Res. Develop.* **11** 337
[56] Guion W G and Ferry D K 1970 *Appl. Phys. Lett.* **16** 269
[57] Elliott B J, Gunn J B and McGroddy J C 1967 *Appl. Phys. Lett.* **11** 253
[58] Hirst L C, Walters R J, Führer M F and Ekins-Daukes 2014 *Appl. Phys. Lett.* **104** 231115
[59] Mei X, Yoshida W and Lange M *et al* 2015 *IEEE Electron Dev. Lett.* **36** 327
[60] Foyt A G 1966 *Int. Sol. State Circuits Conf.* (Piscataway, NJ: IEEE) p 84
[61] Buchsbaum S J, Cynoweth A G and Feldman W L 1965 *Appl. Phys. Lett.* **6** 67
[62] Ferry D K and Dougal A A 1965 *Appl. Phys. Lett.* **7** 318
[63] Bauer G and Kuchar F 1972 *Phys. Stat. Sol. (b)* **13** 169
[64] Ferry D K and Porter W A 1969 *IBM J. Res. Develop.* **13** 621
[65] Ferry D K, Young R W and Dougal A A 1967 *Zeit. Naturforsch.* **22A** 576
[66] Ancker-Johnson B 1967 *J. Phys. Soc. Jpn.* **22** 1156
[67] Ferry D K and Heinrich H 1968 *Phys. Rev.* **169** 670
[68] Müller E and Ferry D K 1970 *J. Phys. Chem. Sol.* **31** 2401
[69] Schafft H A and French J C 1962 *IEEE Trans. Electron Dev.* **9** 129
[70] Ferry D K and Dougal A A 1966 *IEEE Trans. Electron Dev.* **13** 627
[71] Ancker-Johnson B and Drummund J E 1963 *Phys. Rev.* **131** 1961
[72] Müller E and Ferry D K 1970 *Sol. State Commun.* **8** 855
[73] Ferry D K, Heinrich H, Keeler W and Müller E 1973 *Phys. Rev.* B **8** 1538

[74] Porter W A and Ferry D K 1972 *Physica* **60** 155
[75] Kaasberg K, Thygeson K S and Jacobsen K W 2012 *Phys. Rev.* B **85** 115317
[76] Radisavljevic B, Whitwick M B and Kis A 2011 *Nat. Nano.* **6** 147
[77] Ferry D K 2017 *Semicond. Sci. Technol.* **32** 085003
[78] He G, Nathawat J and Kwan C-P *et al* 2017 *Sci. Rep.* **7** 11256
[79] Ovchinnikov D, Allain A and Huang Y S *et al* 2014 *ACS Nano* **8** 8174
[80] Zhu W J, Low T and Lee Y H *et al* 2014 *Nat. Commun.* **5** 3087
[81] Mead C A 1966 *Proc. IEEE* **54** 307
[82] Grubin H L, Ferry D K and Gleason K R 1980 *Sol.-State Electron.* **23** 157
[83] Dawson D, Samoska L and Fung A K *et al* 2005 *IEEE Micro. Wireless Compon. Lett.* **15** 874
[84] Mei X, Yoshida W and Lange M *et al* 2015 *IEEE Electron Dev. Lett.* **36** 327
[85] Radisic V, Mei X B and Deal W R *et al* 2007 *IEEE Micro. Wireless Compon. Lett.* **17** 223
[86] Ayubi-Moak J S, Ferry D K and Goodnick S M *et al* 2007 *IEEE Trans. Electron Dev.* **54** 2327
[87] Saraniti M and Goodnick S M 2000 *IEEE Trans. Electron Dev.* **47** 1909
[88] Akis A, Ayubi-Moak J S and Ferry D K *et al* 2008 *J. Phys. Condens. Matter* **20** 384201
[89] McKay K G 1954 *Phys. Rev.* **94** 877
[90] Sze S M and Gibbons G 1966 *Sol. State Electron.* **9** 831
[91] Lee C A, Logan R A and Latdorf R L *et al* 1964 *Phys. Rev.* **134** A761
[92] Crowell C R and Sze S M 1966 *Appl. Phys. Lett.* **9** 242
[93] van Overstraeten R and DeMan H 1970 *Sol.-State Electron.* **13** 583
[94] McGroddy J C and Nathan M I 1967 *J. Phys. Soc. Jpn., Suppl.* **22** 437
[95] Ferry D K and Heinrich H 1968 *Phys. Rev.* **169** 670
[96] Dick C L and Ancker-Johnson B 1972 *Phys. Rev.* B **5** 526
[97] Curby R C and Ferry D K 1973 *Phys. Stat. Sol. (a)* **15** 319
[98] Anderson C L and Crowell C R 1972 *Phys. Rev.* B **5** 2267
[99] Czajkowski I K, Allam J and Silver M *et al* 1990 *Proc. IEE: J. Optoelectron.* **137** 79
[100] Rana F 2007 *Phys. Rev.* B **76** 155431
[101] Winzer T, Knorr A and Malic E 2010 *Nano Lett.* **10** 4839
[102] Girdhar A and Leburton J P 2011 *Appl. Phys. Lett.* **99** 043107
[103] He G, Nathawat J and Kwan C-P *et al* 2017 *Scientific Rep.* **7** 11256
[104] Zhu W, Low T and Lee Y-H *et al* 2013 *Nat. Commun.* **10** 1038
[105] Shockley W 1961 *Sol.-State Electron.* **2** 36
[106] Wolff P A 1954 *Phys. Rev.* **95** 1415
[107] Fröhlich H 1939 *Rep. Prog. Phys.* **6** 411
[108] Baraff G A 1962 *Phys. Rev.* **128** 2507
[109] Druyvesteyn M J 1930 *Z. Phys.* **64** 790
[110] Pearsall T P *et al* 1978 *Sol.-State Electron.* **21** 297
[111] Tang J Y, Shichijo H, Hess K and Iafrate G J 1981 *J. Phys. Colloques* **42** C7–63
[112] Morgan T N 1959 *J. Phys. Chem. Sol.* **8** 245
[113] Seeger K 1963 *Z. Phys.* **172** 68
[114] Seeger K and Vana H 1979 *Phys. Stat. Sol. (b)* **96** 605
[115] Abe M and Kaneda S 1972 *Jpn. J. Appl. Phys.* **11** 1675
[116] Tanimoto M and Ferry D K 1983 *IEEE Electron Dev. Lett.* **4** 246
[117] Šermukšnis E, Liberis J and Šimukovic *et al* 2017 Int. Conf. Noise Fluc. (Piscataway, NJ: IEEE)

[118] Brazzini T, Casbon M A and Uren M J *et al* 2017 *IEEE Trans. Electron Dev.* **64** 2155
[119] Gassoumi M, Helali A and Gassoumi M *et al* 2019 *Silicon* **11** 557
[120] Shurakov A, Prikhodko A and Mikhailov D *et al* 2020 *J. Phys. Conf. Ser.* **1695** 012156
[121] Hamaguchi C, Kono T and Inuishi Y 1967 *Phys. Lett.* **24A** 500
[122] Acket G A 1967 *Phys. Lett.* **25A** 374
[123] Hess K, Nimtz G and Seeger K 1969 *Sol. State Electron.* **12** 79
[124] Philipp A, Kuchar F and Seeger K 1977 *Phys. Stat. Sol. (b)* **79** 115
[125] Kawashima M and Kataoka S 1979 *Jpn. J. Appl. Phys.* **18** 1311
[126] Seeger K 1963 *J. Appl. Phys.* **34** 1608
[127] Nimtz G and Seeger K 1968 *J. Appl. Phys.* **39** 2263
[128] Machi Y 1969 *IEEE Trans. Microwave Theory Techniq.* **17** 333
[129] Khosropanah P, Zhang W and Hovenier J N *et al* 2008 *J. Appl. Phys.* **104** 113106
[130] Richter H, Semenov A D and Pavlov S G *et al* 2008 *Appl. Phys. Lett.* **93** 141108
[131] Crupi G, Raffo A and Vadala V *et al* 2018 *Microw. Opt. Technol. Lett.* **60** 3023
[132] Lachab M, Sultana M and Fatima H *et al* 2012 *Semicond. Sci. Technol.* **27** 125001
[133] Toprak A, Osmanoglu S and Östürk M *et al* 2018 *Semicond. Sci. Technol.* **33** 125017
[134] Sze S M, Crowell C R, Carey G P and LaBate E E 1966 *J. Appl. Phys.* **37** 2690
[135] Heiblum M 1981 *Sol.-State Electron.* **24** 343
[136] Hickmott T W, Solomon P M and Fang F F *et al* 1984 *Phys. Rev. Lett.* **52** 2053
[137] Heiblum M, Nathan M I, Thomas D C and Knoedler C M 1985 *Phys. Rev. Lett.* **55** 2200
[138] Levi A F J, Hayes J R, Platzman P M and Wiegmann W 1985 *Phys. Rev. Lett.* **55** 2071
[139] Levi A F J, Hayes J R and Bhat R 1986 *Appl. Phys. Lett.* **48** 1609
[140] Kaiser W J and Bell L D 1988 *Phys. Rev. Lett.* **60** 1406
[141] Binnig G, Rohrer H, Gerber C and Weibel E 1982 *Appl. Phys. Lett.* **40** 178
[142] Binnig G, Rohrer H, Gerber C and Weibel E 1982 *Phys. Rev. Lett.* **49** 57
[143] Ludeke R, Prietsch M and Samsavar A 1991 *J. Vac. Sci. Technol.* B **9** 2342
[144] Milliken A M, Manion S J and Kaiser W J *et al* 1992 *Phys. Rev.* B **46** 12846
[145] Bell L D, Manion S J and Hecht M H *et al* 1993 *Phys. Rev.* B **48** 5712
[146] Bell L D, Kaiser W J and Hecht M H *et al* 1993 *Scanning Tunneling Microscopy* ed J A Stroscio and W J Kaiser (Boston, MA: Academic) pp 307–48
[147] Hess K, Morkoç H, Shichijo H and Streetman B G 1979 *Appl. Phys. Lett.* **35** 469
[148] Dingle R, Störmer H L, Gossard A C and Wiegmann W 1978 *App. Phys. Lett.* **33** 665
[149] Glisson T H, Hauser J R and Littlejohn M A *et al* 1980 *J. Appl. Phys.* **51** 5445
[150] Hess K and Shichijo H 1980 *IEEE Trans. Electron Dev.* **27** 503
[151] Keever M, Shichijo H and Hess K *et al* 1981 *Appl. Phys. Lett.* **38** 36
[152] Tang J Y and Hess K 1981 *IEEE Trans. Electron Dev.* **28** 285
[153] Hirakawa K, Sakaki H and Yoshino J 1985 *Phys. Rev. Lett.* **54** 1279
[154] Widiger D J, Kizilyalli I C, Hess K and Coleman J J 1985 *IEEE Trans. Electron Dev.* **32** 1092
[155] Quay R, Hess K and Reuter R *et al* 2001 *IEEE Trans. Electron Dev.* **48** 210
[156] Keever M, Hess K and Ludowise M 1982 *IEEE Electron Dev. Lett.* **3** 297
[157] Sagyi S, Kodymov A and Adivarahan V *et al* 2005 *App. Phys. Lett.* **87** 043505
[158] Shur M 2013 *ECS Trans.* **50** 129
[159] Chin R, Holonyak N and Stillman G E *et al* 1980 *Electron. Lett.* **16** 467
[160] Capasso F, Tsang W J, Hutchinson A L and Williams G F 1982 *Appl. Phys. Lett.* **40** 38
[161] Brennan K, Wang T and Hess K 1985 *IEEE Electron Dev. Lett.* **6** 199

[162] Brennan K, Hess K and Capasso F 1987 *Appl. Phys. Lett.* **50** 1987
[163] Chuang S L and Hess K 1987 *J. Appl. Phys.* **61** 1510
[164] Hess K, Higman T K, Emanuel M A and Coleman J J 1986 *J. Appl. Phys.* **60** 3775
[165] Arnold D, Hess K and Iafrate G J 1988 *Appl. Phys. Lett.* **53** 373

IOP Publishing

Hot Carriers in Semiconductors

David K Ferry

Chapter 3

Carrier heating at low temperature

The study of carrier heating at low temperatures is relatively old, but not quite as old as the study of carrier heating in bulk materials. The latter were described in some detail in chapter 2. The study of carrier heating at low temperatures is often described as a study of mesoscopic devices. This generally began with work on the detailed properties of Si MOSFETs at low temperatures, where electron heating [1] and velocity saturation [2] were observed, just as in bulk materials, but with different characteristics. In many cases, carrier heating was most of interest in the decoherence of the carriers in studies of the quantum behavior, which is more readily addressed at low, and very low, temperatures.

Decoherence is a process where the quantum mechanical interference properties of a mesoscopic system decay, and the system begins to behave classically. The characterization of this phenomenon is based upon measurement of the phase-breaking time τ_φ of the system, which determines the temperature (and length) scales over which quantum interference is observed. Generally, the phase-breaking time arises from the *inelastic* scattering processes in the device. But, these inelastic processes are just those in which the carrier can lose or gain energy from e.g. the phonons. This gives rise to another characteristic time when the system is driven out of equilibrium, for example by passing a large current through it. In this latter case, the system returns to equilibrium by a process called energy relaxation, and this energy relaxation process is exactly that discussed in chapter 2. The energy-relaxation time τ_E is another parameter of importance as it describes the characteristic time for the system to return to equilibrium. There have been many experimental and theoretical studies directed toward understanding the process of dephasing and its variations with lattice temperature. There are few studies, however, in which both the phase-breaking and energy relaxation have been studied in the same system [3–5]. Generally, such measurements provide us with the experimental means to determine the equivalent carrier temperature that arises from the heating by the current or voltage, and to determine the effective relaxation times, as well as studying how these two times differ.

Studies of phase-breaking or energy relaxation involve heating experiments where Joule heating of, for example, a 2DEG is used to extract energy-loss rates and the energy-relaxation time. In order to analyze the heating, and to determine things like the average power input per electron and the energy-relaxation time, one needs to have a 'thermometer', a mechanism by which the actual *electron* temperature can be determined from the measurements. One of the earliest methods was to measure the dependence of the mobility on both the electric field and the lattice temperature (at low electric field) [6]. At low temperatures, particularly in bulk material, the mobility is dominated by scattering from the ionized impurities. In this interaction, only a single temperature—that of the carriers—is involved, so that variation of this temperature by either of the two excitations leads to a scale that can be used to convert the mobility at a given electric field to that at a given lattice temperature. Hence, the carrier temperature is determined. As high mobility material became available, the effect of the impurities was dramatically reduced by the use of modulation doping and the resulting increase in effective screening. Thus, the mobility ceased to be a useful method in mesoscopic studies. Alternatively, it was shown that using the Shubnikov–de Haas oscillations provided an effective thermometer. Here, varying the temperature and applied fields could provide a good measure of obtaining the carrier temperature from the amplitude of the oscillations [6]. This has remained a good method for estimating the carrier temperature. If these are studied at various temperatures and currents, these two sets of data may be used to extract the temperature dependence of the energy-loss rate $P(T)$, or equivalently the power being dissipated within the mesoscopic system. This appears in the literature in two forms, the first being the loss per electron, and the second is just the total loss in the device. Obviously, these differ by the area or volume of the device and the carrier density at which the measurements are made. This energy-loss rate is normally a function of a power law of the hot carrier temperature, and the exponent is often considered to be an indicator of the type of electron–phonon scattering that is dominating the inelastic process in the device. In spite of years of work in both normal semiconductor systems, as well as mesoscopic devices, there is not yet a clear understanding of the entire process.

3.1 Early work

As with the early study of hot carriers in semiconductors, Ryder studied the velocity of electrons in Ge at 20 K, and noticed an increase of the current as the field was raised [7]. This is easily explained by scattering by the ionized impurities. This scattering is Coulombic in nature, so the faster the electrons go, the less they are affected by the impurities. Thus, as one raises the field (or current), and therefore the average energy of the carriers, the mobility will increase, as the scattering is a function only of the carrier temperature. As noted above, this provides a good thermometer for further studies of such behavior at low temperature.

Some of the earliest work was done by Morgan with heating of electrons in Ge [8], and by Seeger with microwave heating in Ge and Si [9]. Hess followed this up with the use of microwave mixing to determine the energy relaxation times for carriers in Ge and Si [10]. However, the work that will be used to illustrate the effect is due to Fang and Fowler [11], who studied the transport properties of Si MOSFETs at low temperature. They measured the conductance in the inversion channel at 4.2 K as a function of the electric field, and then measured it as a function of the base temperature. As, impurity scattering is the only effective scattering at these low temperatures, they could then determine the electron temperature as a function of the electric field, as shown in figure 3.1. The temperature dependence on the field is shown in this figure for several values of the inversion density, and it appears that, from the data, the change in temperature is proportional to the field to the 3/2 power. At very high electric fields, they found that the mobility decreased with field

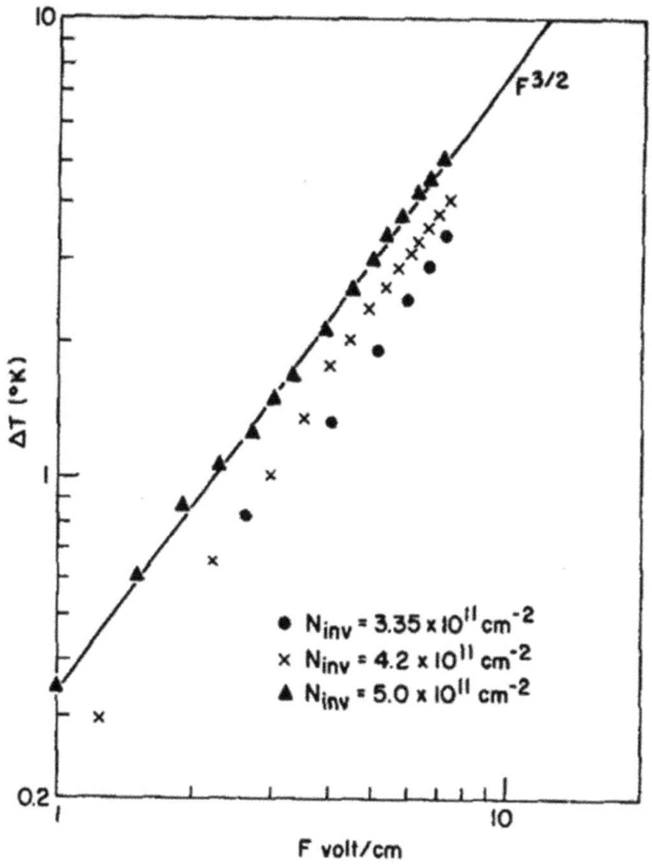

Figure 3.1. The change in electron temperature as a function of the electric field in a Si MOSFET at 4.2 K. Data for three different inversion densities are shown. Reprinted from Fang and Fowler [11], copyright (1970) with permission from AIP Publishing.

at all temperatures and electron densities. They measured the electron drift velocity for different surface orientations, electron densities and temperatures up to 300 K. The device used was a MOSFET with a 10 μm channel length and thick gate oxides, in order to reduce the depletion that occurs at the drain end of the channel. With the thick oxide, the gate voltage could be made larger than the drain voltage and so reduce the depletion. Measurements were made with pulsed drain voltage on the device, using 300 ns pulses. In figure 3.2, the electron velocity at 300 K is shown for devices on the (100) surface of Si, for an inversion density of 6.6×10^{12} cm^{-2}, which is fairly large for a Si device. Nevertheless, a saturation of the velocity was observed in the devices. The observed saturated velocity of ~6.5×10^{6} cm s^{-1} should be compared with the observations of figure 2.3 and the time-of-flight measurements in figure 2.13, both for bulk material. The device values are close to those of figure 2.3 and only slightly smaller than the time-of-flight measurements. Similar measurements on (111) and (110) surfaces yielded saturated velocities of 5.5×10^{6} cm s^{-1} and 4×10^{6} cm s^{-1}, respectively. The values obtained for the saturated velocities seemed to be independent of the substrate doping or the inversion layer density.

Krowne and Holm-Kennedy [12] pursued studies in *n*-type Si at a lattice temperature of 4.2 K, and again where the impurity scattering was used as the thermometer. This allowed the determination of the electron temperature directly from the change in mobility, and this is shown in figure 3.3 for a range of gate

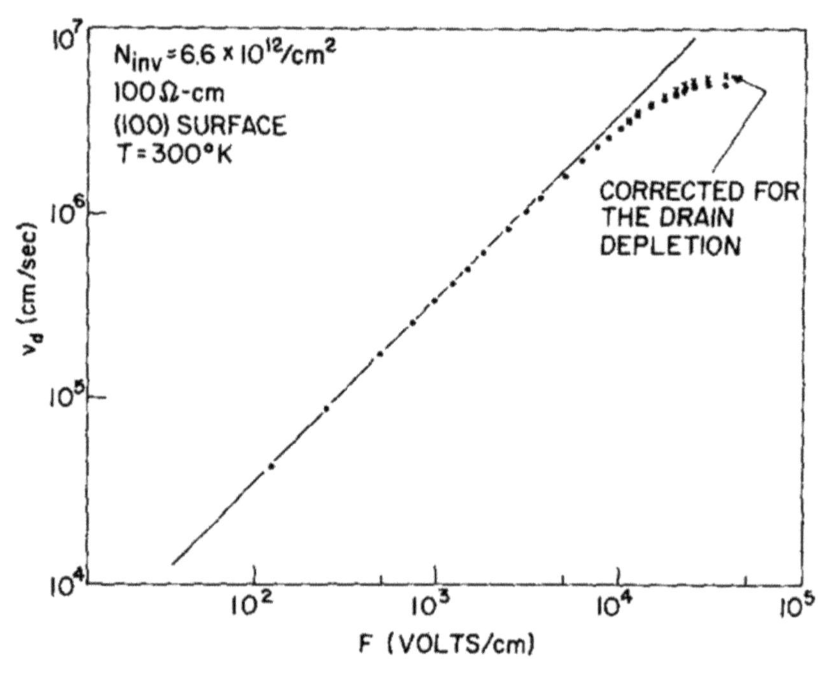

Figure 3.2. Typical drift velocity variation as a function of the electric field for a MOSFET on the (100) surface of Si. The device characteristics are discussed in the text. Reprinted from Fang and Fowler [11], copyright (1970) with permission from AIP Publishing.

induced carrier densities. As discussed above, one compares the mobility with actual lattice temperature over a range of the latter, and then uses the electric field heating to infer the temperature at a mobilty in the field corresponding to the lattice temperature. If we know the electron temperature from experimental measurements, then the energy relaxation time can be found from the relationship

$$P(E) = \frac{k_B(T_e - T)}{\tau_E}, \tag{3.1}$$

with the power input being given by $e\mu E^2$, and E is the electric field. In figure 3.4, the energy relaxation time determined by this method and the measured electron temperature is plotted for the same experiments from these authors. These are the calculated values of the energy relaxation time as a function of the measured electron temperatures. The connection of this temperature to the electric field is given by figure 3.3. The dependence upon the actual inversion density arises from the

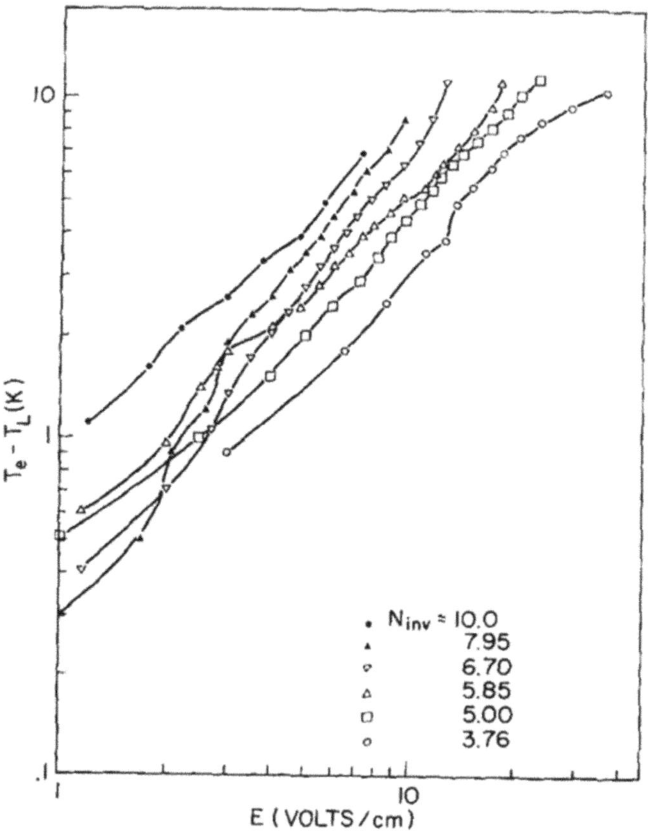

Figure 3.3. The difference in electron temperature and lattice temperature for electrons in n-type Si at a base temperature of 4.2 K. The inversion density in the MOSFET is given in units of 10^{11} cm^{-2}. Reprinted from Krowne and Holm-Kennedy [12], copyright (1974) with permission of Elsevier.

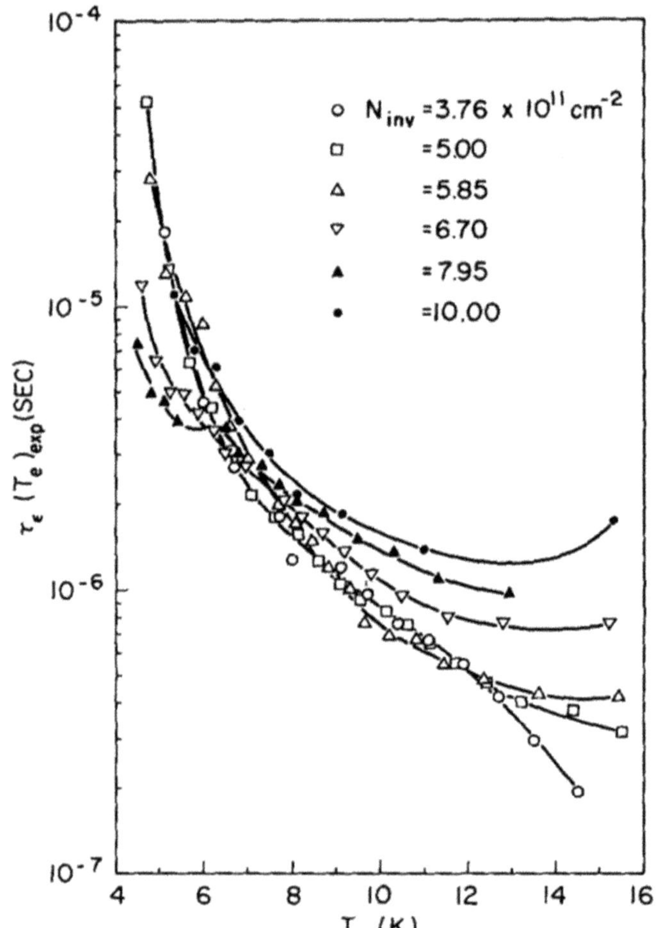

Figure 3.4. The energy relaxation time determined from experiments is plotted versus the electron temperature similarly determined from figure 3.3. The various curves are for different inversion densities as indicated by the labels. Reprinted from Krowne and Holm-Kennedy [12], copyright (1974) with permission of Elsevier.

fact that the mobility itself depends upon the inversion density, but it may be seen that the various curves seem to coalesce for densities below ~6×10^{11} cm^{-2}. It should be remarked that Fermi–Dirac statistics were used in all evaluations on the role played by the density. The variation at higher densities is felt to be a result of degeneracy in the electron gas. In many cases, one can use the variation of the relaxation time with temperature to infer the dominant scattering process, as will be illustrated in later sections. Here, however, the scatter among the curves for the various carrier densities is too large to infer an exponent for the temperature in the curves. The point is that, as stated above, different scattering processes will lead to different dependences on the lattice and electron temperatures.

3.2 Phase-breaking

Transport through small structures at low temperatures is often referred to as mesoscopic behavior [13]. Scattering of carriers in these devices is normal scattering behavior in a near equilibrium situation. As a result, the mobility and the diffusion constant are related. This transport is different from the ballistic or quasi-ballistic behavior discussed previously only in that many scattering events occur. While the actual transport may well be quasi-ballistic in nature, the phase-breaking events depend upon the scattering that does occur, particularly when energy is gained or lost in the process. Hence, the mobility, or the diffusion constant, can be used to define a characteristic phase-breaking length from the inelastic phase-breaking time, via

$$l_\varphi = \sqrt{D\tau_\varphi}, \tag{3.2}$$

where D is the diffusivity of the semiconductor material. One often thinks about this type of transport as diffusive transport, although the majority carriers are the important quantity and their motion is mainly determined by the mobility. But, in equilibrium, the mobility and diffusivity are connected by the Einstein relation, although one needs to use a slightly modified form for the degenerate materials. The modification is a factor given by the ratio of two Fermi integrals, but the exact integrals are determined by the dimensionality of the sample.

In many mesoscopic devices, transport studies are focused upon interference effects and determination of this phase-breaking time. In order to see interference effects, it is necessary that the phase coherence length be comparable to, or larger than, the size of the device under study. It is for this reason that nearly all mesoscopic experiments are carried out at low temperature, where the scattering is much weaker than at room temperature. At non-zero temperatures, an additional source of dephasing arises from the fact that a range of energies, near the Fermi energy, are involved in the transport. The width of energies involved is of order $3.5k_BT$, where T is often used as the lattice temperature but should be the electron temperature. We can therefore define a *thermal broadening time* by using the broadening that scattering induced and relating it to the spread in energy as

$$\frac{k_BT\tau_T}{\hbar} \sim 1 \;\rightarrow\; \tau_T \approx \frac{\hbar}{k_BT}. \tag{3.3}$$

From this, a thermal diffusion length, in analogy to the phase coherence length (3.2), can be defined as

$$L_T = \sqrt{D\tau_T} = \sqrt{\frac{\hbar D}{k_BT}}. \tag{3.4}$$

This last contribution to decoherence is an unavoidable result of operating at a non-zero temperature, as this decoherence effect would exist even if it were possible to eliminate all the inelastic scattering processes. In this section, we want to examine a number of mesoscopic studies which focus upon determining experimentally the phase-breaking time.

3.2.1 Weak localization

There are many situations in which disorder in the crystalline potential, due to defects or to impurities, can lead to a back-scattering effect on the carriers. Disorder can lead to a set of localized scatterers which produce exactly this back-scattering, and provide an addition to the resistance in the sample at low temperatures [14]. In the bulk, this process is known as weak localization. The idea is that there will be a set of scatterers that cause an electron to deflect off them and eventually find its path reversed back to the source of the carriers. Hence, they scatter around a ring in a sense, but a somewhat disordered ring. But, the carrier can go around the ring in either of two directions, so that there are two time-reversed trajectories in which the particles scatter from each site and eventually return to their original direction as a back-scattering. Since we cannot know which direction the particle scatters around the 'ring', both directions are possible and represent the time-reversed paths that can interfere with each other. The onset of weak localization causes a reduction in the conductivity of the sample. That is, weak localization gives a negative contribution to the conductivity that arises from the back-scattering of the carriers. This is usually a small but significant effect if there is sufficient disorder in the system.

In figure 3.5, the classic signature of weak localization is shown for an epilayer of Ge grown on Si [15]. As the temperature is lowered, the resistance decreases due to the reduction of scattering in the two-dimensional layer. However, at the lowest

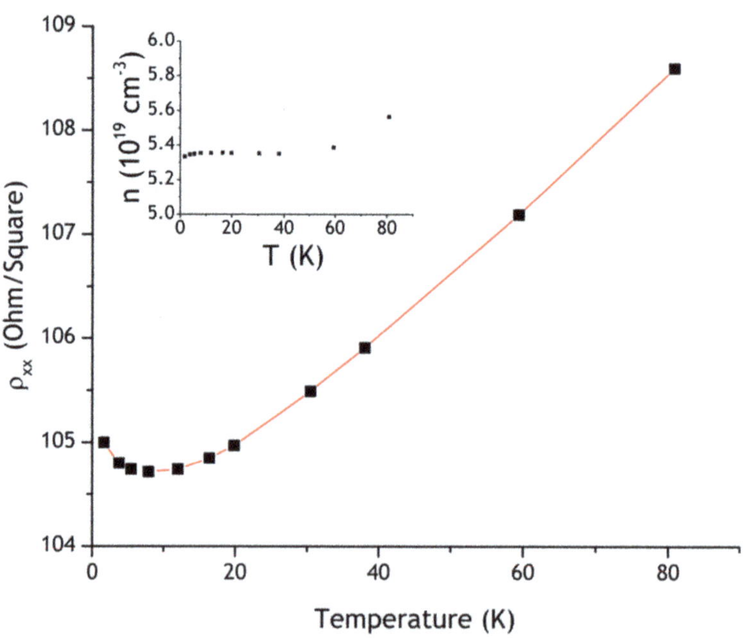

Figure 3.5. The temperature dependence of the zero magnetic field resistance of the Ge epitaxial layer. The inset shows the variation of the carrier density over this same temperature range. The increase in resistance at the lowest temperatures is the usual signal for weak localization. Reprinted from Newton *et al* [15], copyright (2017) with permission of AIP Publishing.

temperatures, the resistance deviates and increases again; this is the weak localization signal. The sample was grown by reduced pressure vapor phase epitaxy. First, a 1.2 μm undoped buffer layer is grown on the Si substrate, and then a 50 nm n-type Ge layer is grown. This layer is heavily doped to 5.3×10^{19} cm^{-3} and has a mobility of 220 cm^2 Vs^{-1}. It may be seen from the inset to the figure that the carrier density remains nearly at the impurity density over the temperature range, increasing only at the highest temperature studied.

The conductivity is determined by the current–current correlation function in the Kubo formula. It can also be found from an approach using the retarded Langevin equation [16], which itself generates this correlation function. Using the correlation function, the conductivity can be expressed as

$$\sigma = \frac{ne^2}{m^*\langle v^2(0)\rangle}\int_0^\infty \langle v(t)v(0)\rangle dt \sim \frac{ne^2}{m^*}\int_0^\infty e^{-t/\tau}dt = \frac{ne^2\tau}{m^*}. \tag{3.5}$$

Here, it has been assumed that the velocity decays with a simple exponential of the scattering time, which is the case in equilibrium. The conductivity, in these conditions, is measured at the Fermi surface in degenerate material. (This is not strictly required for the definitions that are used this equation and throughout the rest of this chapter.) The term in the angle brackets, along with the prefactor squared velocity average compose the velocity autocorrelation function. The exponential relationship in equation (3.5) can be thought of as the probability that a particle moves, without scattering, for a time τ. This time is taken as the mean time between collisions, and will appear again in chapter 7 in discussions of the Monte Carlo technique. This exponential arises from the fact that the autocorrelation function decays in time with this simple exponential when the system is in equilibrium [17]. In one dimension, $D = v_F^2\tau$ (or more specifically, $D = v_F^2\tau/d$, where d is the number of spatial dimensions in the system), $l_e = v_F\tau$, and the one-dimensional density is $n_1 = k_F/\pi$, so equation (3.5) becomes

$$\sigma_1 = \frac{e^2}{\pi\hbar}2D\int_0^\infty \frac{1}{2l_e}e^{-t/\tau}dt. \tag{3.6}$$

In two dimensions, $D = v_F^2\tau/2$ and $n_2 = k_F^2/2\pi$, to give

$$\sigma_2 = \frac{e^2}{\pi\hbar}2D\int_0^\infty \frac{k_F}{2l_e}e^{-t/\tau}dt. \tag{3.7}$$

The integrand has the units L^{-d} in both cases. It is naturally expected that this scaling will continue to higher dimensions. However, the quantity inside the integral is no longer just the simple probability that a particle has escaped scattering. Instead, it now has a prefactor that arises from critical lengths in the problem. To continue, one could convert each of these to a conductance by multiplying by L^{d-2}. This dimensionality couples with the diffusion constant and time integration to produce the proper units of conductance. To be consistent with the remaining discussion, however, this will not be done.

In weak localization, the correlation function is related to the probability of return to the initial position. The integral is analogously related to the above and termed a correlator, specifically known as the particle–particle correlator $W(t)$ (this is just a variance of the correlation function in name, and not in detail). Hence [18],

$$C(\tau) = \int_0^\infty W(t)dt \sim \int_0^\infty \frac{1}{2L^d} e^{-t/\tau} dt, \tag{3.8}$$

The overall expression above has an exact analogy to the previous equations (in fact, the correlation function in the first two equations is just this integral). In the case of weak localization, it will be seen that the drift time of the free carriers is not the important parameter. Rather, since the diffusive transport of the carriers involves considerable scattering, the probability that this scattering returns the particle to its original position is the critical parameter. Thus, weak localization is evaluated by replacing C or W determined from a proper diffusion equation in this strongly scattering regime. Hence, the correlation function of interest will be found to be more complicated than the simple exponential decay.

As discussed above, the weak localization correction factor is defined by the probability that a particle diffuses some distance and then returns to the original position. This latter probability is the quantity $W(t)$. The time t of interest is that required to diffuse around the scattering loop (in returning to the original position). The conductivity correction is then determined in the manner of equation (3.7) as

$$\Delta\sigma = -\frac{2e^2}{h} 2D \int_0^\infty W(t)dt, \; x(t) \to x(0). \tag{3.9}$$

The prefactor is just that occurring in equations (3.6) and (3.7) above. (The negative sign arises because weak localization *reduces* the conductance.) As a rule, the majority of particles will not return to the original position. Only a small fraction will do so. As a result, the conductance correction is small. Although this correction arises from a small fraction of particles, it is this small fraction that is of interest. In the above equations, the resulting correlation function philosophically is a decay of 'knowledge' of the initial state. For the puposes here, however, the 'probability' that particles can diffuse around a closed loop over a time t and return to their initial position and still have enough memory of the wave function to cause interference is the important quantity. Only when there is interference between the initial wave and the returning wave, does weak localization arise. While $W(t)$ has been treated as a probability function, it has the units L^{-d}, characteristic of conductivity, and thus is not a true probability function. In calculating the conductance correction, this dimensionality is treated correctly, and no problems will arise.

It has been assumed up to this point that the transport of the carriers is diffusive, and the motion involves a great many scattering centers, which results in the net motion being characterized as Brownian motion. It is the quantum effects that cause the interference leading to equation (3.9). But the general motion may be described as classical. What is usually meant by this is that $kl_e < 1$, where k is the carrier's wave vector (usually the Fermi wave vector) and l_e is the mean free path between collisions. Normally, this is the elastic mean free path, which is usually shorter than the inelastic mean free path. Near equilibrium, the probability function will be Gaussian (characteristic of Boltzmann transport). In this situation, it may be assumed that $W(t)$ will satisfy the diffusion equation for motion away from a point (at time $t = 0$). The usual form of such a diffusion equation is simply [18]

$$\left(\frac{\partial}{\partial t} - D\nabla^2\right)W(t) = \delta(t)\delta(\mathbf{r}), \tag{3.10}$$

which has the general solution

$$W(\mathbf{r}, t) = \frac{1}{(4\pi Dt)^{d/2}} \exp\left(-\frac{r^2}{4Dt}\right). \tag{3.11}$$

Now, this solution is for motion in an open system without boundaries and is characteristic for a d-dimensional system. If the system has boundaries, the result is more complicated, and the modal solutions determined by these boundaries must be found. Only the simple, open solution will be considered here. As discussed above, it is the probability of return, $r = 0$, that is of interest. Obviously, in this limit, the exponential in (3.11) disappears, but there is another factor to be considered, and that is the probability that the particle retains some phase memory while diffusing through these multiple collisions. This simple probability, which is also an exponential, is added empirically. Finally, the probability of return to $r = 0$ after a time t, without loss of phase, is given by

$$W(\mathbf{r}, t) = \frac{1}{(4\pi Dt)^{d/2}} \exp\left(-\frac{t}{\tau_\varphi}\right). \tag{3.12}$$

It is the phase-breaking time τ_φ that has been introduced to characterize the phase-breaking during the diffusive process. It is seen that the dimensionality of $W(t)$ is L^{-d}, which is the dimensionality of the integrand in (3.6) and (3.7). Thus, this agrees with the discussion above.

A further modification has been suggested by Beenakker and van Houten [19]. They point out that we do not expect diffusive effects in the ballistic transport regimes. On the short-time basis, these effects go away. Here, 'short time' is used in the sense that no collisions are likely to occur. Collisions must occur before diffusive transport arises. When there are no collisions, the particle has no chance to be back-scattered to the original position. These authors suggest changing equation (3.12) to account for this process. They suggest the new form for the probability should be

$$W(\mathbf{r}, t) = \frac{1}{(4\pi Dt)^{d/2}} \exp\left(-\frac{t}{\tau_\varphi}\right)(1 - e^{-t/\tau}). \tag{3.13}$$

The only quantum mechanics in the this approach has been the phase memory that must survive the trip around the scattering loop. This is described by the phenomenological phase-breaking time τ_φ. While no quantum mechanical calculation has been done, we have introduced the necessary phase interference through this phenomenological term. The important point is to recognize where these quantum effects should have entered the discussion.

The dimensionality that appears in $W(t)$ will be corrected when the total conductance, rather than the conductivity, is determined. However, this change will not be made. Instead, equation (3.13) in used in equation (3.9). The conductivity corrections for weak localization in the important dimensionalities, become

$$\Delta\sigma = -\frac{e^2}{\pi\hbar}\begin{cases} \dfrac{1}{2\pi l_\varphi}\left(\sqrt{1+\dfrac{\tau_\varphi}{\tau}}-1\right), & d=3 \\ \dfrac{1}{2\pi}\ln\left(1+\dfrac{\tau_\varphi}{\tau}\right), & d=2 \\ l_\varphi\left(1-\sqrt{\dfrac{\tau}{\tau_\varphi+\tau}}\right), & d=1 \end{cases}. \tag{3.14}$$

The important length in this diffusive description is the phase coherence length $l_\varphi = \sqrt{D\tau_\varphi}$, determined from the phenomenological phase-breaking time. To be sure, this result is the easiest one that is found under reasonable constraints. Nevertheless, the results are quite useful. The weak localization correction is relatively universal, and primarily depends upon the ratio of the two important time scales that arise in this type of transport. The quantum mechanics lies buried in the ad hoc assumption of a phase coherence time τ_φ. Without this coherence time, none of the above formulas is meaningful. The reduction in conductivity given by (3.14) is thought to cause the increase in resistance that was apparent in figure 3.5. From data such as that of figure 3.5, a value for the phase-breaking time can be inferred once the scattering time τ is determined from the measured mobility.

The weak localization effect can be destroyed by the application of a magnetic field. The magnetic field breaks time-reversal symmetry. An alternative version of the effect of weak localization is that the particle can move around the scattering loop in either direction, and the two reverse paths lead to interference continuously around the loop. The magnetic field leads to a Lorentz force, which produces a force normal to both the magnetic field and the local velocity of the particle. Since the two paths have opposite velocities, the Lorentz force on the two will have opposite directions, hence causing the paths to diverge and breaking up the interference effect. It is the measurement of this destruction of the interference by a magnetic field which is normally used to determine the phase-breaking time. In the theory, it is assumed that a critical magnetic field may be defined, as that for which the weak localization correction has been reduced to one-half of its peak value. This field is called the *correlation* magnetic field. It is argued that this value of critical magnetic field is the field value for which exactly one flux quanta (h/e) is enclosed within the area of the phase-coherent loops that lead to the back-scattering.

In the presence of a magnetic field, the diffusive paths that return to their initial coordinates will enclose magnetic flux. To account for this behavior the diffusion equation (3.10) must be modified to include the vector potential that gives rise to the magnetic field. The magnetic field is oriented in the z-direction and, in the Landau gauge, that leads to $\mathbf{A} = (0, Bx, 0)$. Only a thin two-dimensional slab with no z variation is assumed, so that equation (3.10) becomes

$$\left[\frac{\partial}{\partial t} - D\left(\nabla - i\frac{2e\mathbf{A}}{\hbar}\right)^2 + \frac{1}{\tau_\varphi}\right]W(t) = \delta(t)\delta(\mathbf{r}), \tag{3.15}$$

where the vector potential is incorporated via the Peierls' substitution. In addition, the phase-breaking time has been introduced. In the two-dimensional system of the normal heterostructure interface, there will be no z variation. The magnetic field and phase-breaking terms can be combined into the relationship

$$\frac{1}{\tau_\varphi} \to \frac{1}{\tau_\varphi} + Dk^2 = \frac{1}{\tau_\varphi} + (2n + 1)\frac{eBD}{\hbar}, \tag{3.16}$$

where the relevant momentum has been assumed to be the Fermi momentum which is defined by the cyclotron orbits that may result from the magnetic field. This treatment is simplified, and a more extensive exact treatment exists. Such a treatment is beyond the level that we are discussing here, but it has been shown that [20]

$$\Delta\rho \sim \frac{e^2}{\pi\hbar}\left[-F\left(\frac{\tau_\varphi}{\tau_B}\right) + 3F\left(\frac{\tau_\varphi\tau}{\tau_B(\tau_\varphi + \tau)}\right)\right], \tag{3.17}$$

where

$$F(x) = \ln(x) + \Psi(0.5 + 1/x), \tag{3.18}$$

$\tau_B^{-1} = 4eDB/\hbar$ and $\Psi(\cdot)$ is the digamma function.

In figure 3.6, the conductance and resistance for the sample of figure 3.5 are plotted as a function of magnetic field at different temperatures [15]. These data are fit to (3.17) in panel (b), although this equation is modified in order to account for spin–orbit interaction which is strong in Ge [21].

By estimating the magnetic field at which the weak localization correction is reduced by a factor of two, one has an estimate of the value of the phase-breaking time, since it will be equal to the magnetic time defined in the second term of (3.16)

$$B_c \sim \frac{h}{el_\varphi^2}, \tag{3.19}$$

where it is assumed that D is known. This leads one to conclude that the critical magnetic field is that necessary to couple one flux quantum (h/e) through a phase-coherent area defined by the diffusive coherence length.

The weak localization seen in monolayer graphene grown on a Ge (111) surface is shown in figure 3.7 [22]. The growth on this Ge surface leads to poly-crystalline

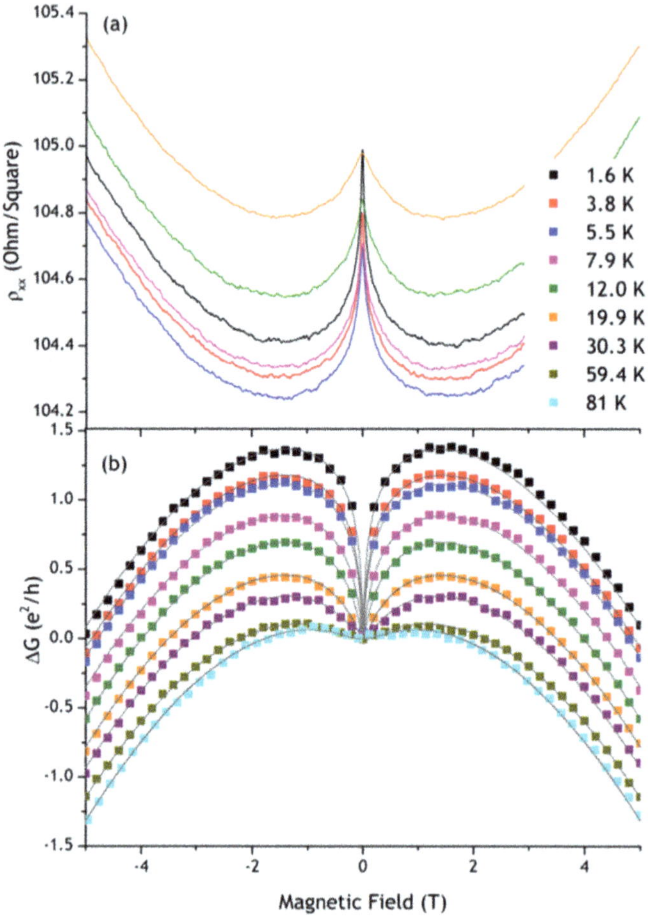

Figure 3.6. The (a) magnetoresistance and (b) differential conductance of the same sample used for figure 3.5. The magnetoresistance includes an arbitrary offset for clarity. The colored lines and data are for the temperatures indicated in panel (a). The lines are fits to (3.17), modified for the spin–orbit interaction in Ge. Reprinted from Newton *et al* [15], copyright (2017) with permission from AIP Publishing.

graphene, which responds to the disorder this causes. The weak localization is not seen for graphene on the (110) surface of Ge. In the poly-crystalline case, it is thought that the grain boundaries induce strong elastic intervalley scattering which contributes to the disorder needed for weak localization. Here, the formula (3.17) is modified for intervalley scattering and for additional effective scattering due to the trigonal warping in the graphene conduction band at higher energies. The various lengths in figure 3.7(c) are the phase-breaking length (3.2), the scattering length $L_i = \sqrt{D\tau}$, and the effective intervalley length $L_* = \sqrt{D\tau_*}$. Here, the scattering time τ_* is defined by $\tau_*^{-1} = \tau_s^{-1} + \tau_W^{-1}$, where τ_s accounts for the intervalley scattering time and τ_W accounts for extra scattering due to the trigonal warping.

If there is a strong many-body effect and significant spin–flip scattering, such as can occur if there is a strong spin–orbit interaction, then weak anti-localization can

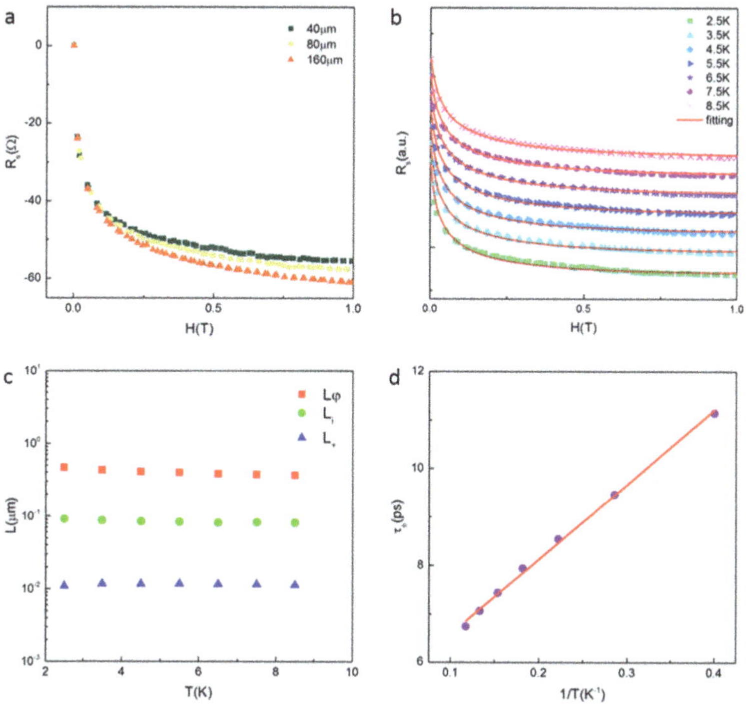

Figure 3.7. The magnetoresisitivity properties for graphene grown on Ge (111). (a) Resistivity for different lengths of the sample as a function of magnetic field. (b) Resistivity at different temperatures for the sample as a function of magnetic field. (c) Various scattering lengths as a function of temperature (see text). (d) The deduced phase-breaking time as a function of the inverse temperature. Reprinted from Sun *et al* [22], via the Creative Commons 4.0 license.

occur, for which there is a reduction in the resistance and an increase in the conductance at low temperature. This weak anti-localization is seen in *p*-type Ge epilayers doped to 8.5×10^{18} cm^{-3} [15].

3.2.2 Shubnikov–de Haas experiments

Studies of the phase-breaking time generally involve varying the lattice temperature T and studying how the phase-breaking process depends upon this temperature. The phase-breaking process is then obtained from various effects, such as weak localization, as was discussed in the previous section. Or it could be deduced from the conductivity as a function of temperature, as in section 3.1. As high mobility material became available, the effect of the impurities was dramatically reduced by the use of modulation doping and the resulting increase in effective screening. Thus, the mobility ceased to be as useful a method in mesoscopic studies. Alternatively, it was shown that using the Shubnikov–de Haas oscillations provided an effective thermometer. Here, varying the temperature and applied fields could provide a good measure of obtaining the carrier temperature from the amplitude of the oscillations [6, 23]. This has remained a good method for estimating the carrier temperature.

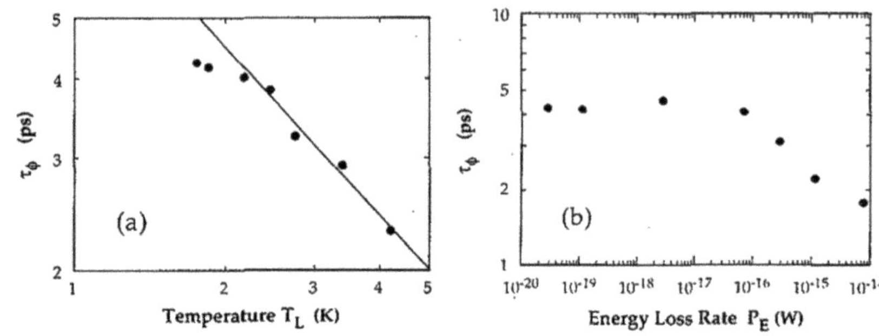

Figure 3.8. Phase-breaking time for one of the Si/SiGe samples (a) as a function of lattice temperature and (b) as a function of the input power to the electrons in the quantum well. The line in panel (a) has a temperature variation as $T^{-0.85}$, although a slightly larger value is expected from theory. Reprinted from Stöger *et al* [23], with permission from IOP Publishing. All rights reserved.

In figure 3.8 are shown results from Stöger *et al* [23] for a Si/GeSi heterostructure grown by vapor phase epitaxy. Here, a 500 nm $Si_{0.7}Ge_{0.3}$ layer was first grown on a Si substrate. Then a 10 nm Si strained layer was grown to provide the quantum well of interest. This was followed by an undoped 4 nm $Si_{0.7}Ge_{0.3}$ layer, followed by doped and top layers of $Si_{0.7}Ge_{0.3}$. This provides remote impurities separated from the channel to reduce the scattering from these impurities, while the electrons move to the channel itself. The doped layer involved a P concentration of 4×10^{18} cm^{-3}, which led to a channel mobility of 7000 to 12 500 cm^2 Vs^{-1} at 4.2 K for various samples, after illumination with a GaAs LED. The measurements discussed here, and in the next section, result from a sample with channel concentration of 0.88×10^{12} cm^{-2} and channel mobility of 13 100 cm^2 Vs^{-1} at 1.8 K. The electron temperature as a function of the current was found from the Shubnikov–de Haas amplitude as mentioned, and further described later in this section. From measurements of weak localization, the phase-breaking time was determined. This is plotted in figure 3.8(a) as a function of the lattice temperature, and presumably the electron temperature. It is also shown as a function of the electron loss rate, discussed in the next section in figure 3.8(b). It is seen that there is a transition to a temperature independent phase-breaking time at lower power inputs and lower sample temperatures, which is quite usual in such studies. We return to this below.

To further illustrate this, and particularly the manner of determining the electron temperature, a study of quantum wells in a heterostructure grown on a lattice matched InP substrate is used. This consists of a 25 nm wide InGaAs quantum well sandwiched between two InAlAs barrier layers [5]. The top barrier layer consists of a 30 nm *n*-doped layer and a 10 nm undoped spacer layer (closest to the InGaAs quantum well). The bottom barrier layer is a 250 nm undoped layer. The doping in the top barrier provides sufficient carriers in the quantum well to populate two subbands, but this does not affect the experiment. The total electron concentration was 1.3×10^{12} cm^{-2} with a mobility of about 140 000 cm^2 Vs^{-1}. Measurements were made at a base temperature of 4.2 K. A micrograph of the sample is shown in the inset to figure 3.9. In the main panel of figure 3.9, the performance of the sample is

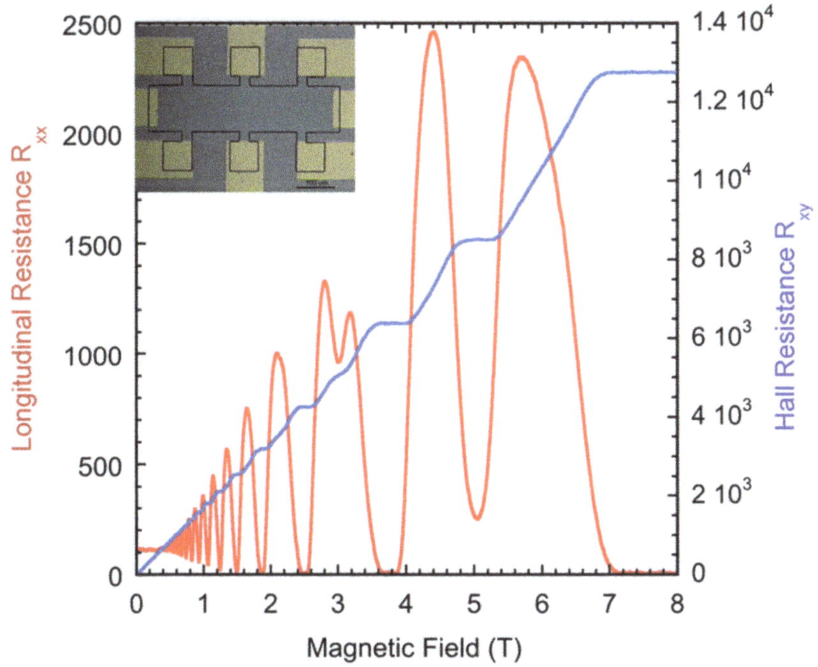

Figure 3.9. Quantum Hall measurements for an InGaAs quantum well with InAlAs barriers. The inset is a optical micrograph of the sample and its contacts.

shown by plotting the quantum Hall resistance and longitudinal resistance. It can be seen that plateaus are readily seen, even for the lower mobility of this sample. As one can see, the sample is configured as a Hall bar, with current passed through the end leads and the voltages measured on the side arms. Data are recorded for various lattice temperatures in the range 4.2–30 K with an excitation current of 500 nA. The measurements for varying temperature are shown in figure 3.10. The curves, of decreasing amplitude as indicated by the black arrow, are for temperatures of 4.2, 6, 8, 10, 15.5, 20, and 30 K. Then sample heating experiments are conducted with currents ranging from 0.5 to 250 μA at the base temperature of 4.2 K. These measurements clearly show a dual frequency (in $1/B$) behavior, by which the density in each of the two sub-bands may be determined. These two frequencies can be either attributed to the presence of two populated sub-bands in the system or the presence of one sub-band that is spin split, wherein this spin splitting extends to zero magnetic field. Discerning which one is the true case can be done by comparing the carrier density found from the Fourier spectra to that found from the Hall measurements. The values for the carrier density of the sub-bands (computed including spin degeneracy) are $7.9 \pm 0.1 \times 10^{11}$ cm^{-2} and $5.0 \pm 0.1 \times 10^{11}$ cm^{-2} for the lowest and second sub-band, respectively, and they add up to a total of $1.29 \pm 0.2 \times 10^{12}$ cm^{-2}. Since the Hall value of the total carrier density is 1.28 ± 10^{12} cm^{-2}, two sub-bands are populated in the quantum well.

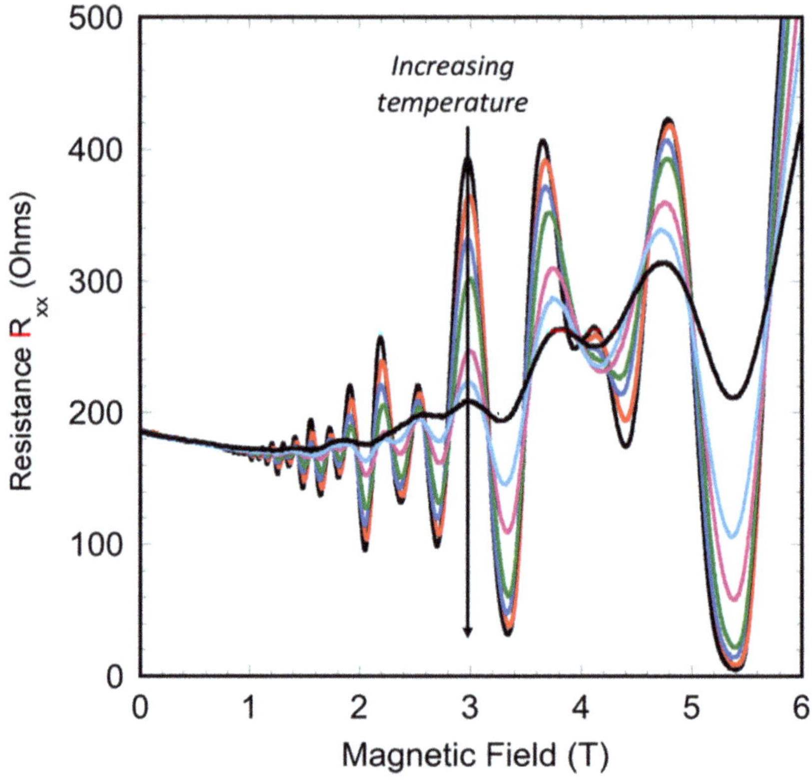

Figure 3.10. Variation of the Shubnikov–de Haas amplitudes as the lattice temperature is raised. Similar variation with current through the device is used to find the electron temperature at a given current. The temperatures for the curves, from top to bottom (according to the arrow are 4.2, 6, 8, 10, 15.5, 20, and 30 K).

The interest is in the major peaks that appear in the plot and how these change with temperature, as indicated by the arrow in the figure. Three of these peaks, for 1.6, 2.0, and 3.2 T, were carefully measured for their temperature dependence. At each magnetic field, the neighboring peak and trough are both measured, and this difference is used as the appropriate measure. Then, the measurements were made at base temperature with varying current. There is a small magnetic field dependence on the carrier heating, so using these three values of magnetic field also in the current-dependent measurements allows one to both gain more data points and to compensate for the small field dependence. The resulting change in the amplitude of the oscillations is used to correlate a particular temperature with a particular current. This comparison is shown in figure 3.11. Here, one can see that at the lowest current level, there is no apparent heating of the sample. The straight line drawn in this figure suggests that the temperature varies approximately as the square root of the current, but should be taken simply as a guide to the eye.

For studies of the phase-breaking time, temperatures in the mK range were employed. The same analysis described above is carried out over a range of temperatures and currents, the latter of which were much lower than those of

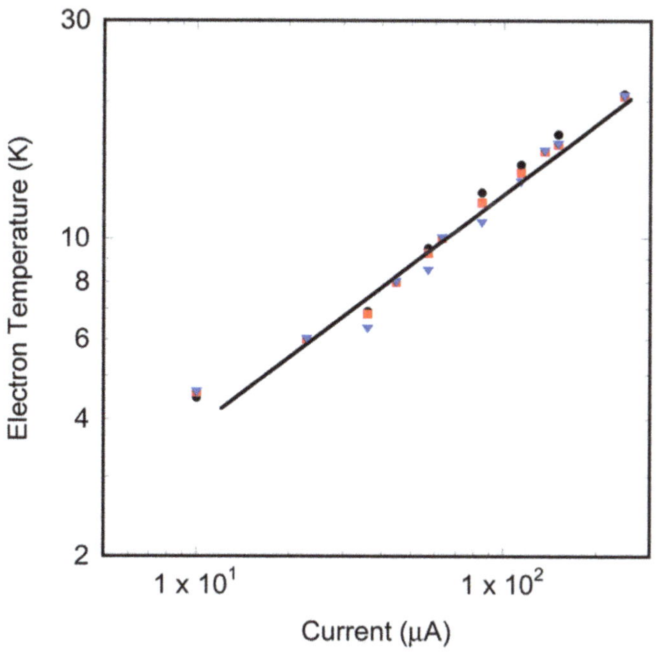

Figure 3.11. The electron temperature as a function of the drive current for three different magnetic fields: black circles (1.6 T), red squares (2.0 T), and blue triangles (3.2 T).

figure 3.11, being in the nA range because of the lower base temperature of 40 mK. The phase-breaking time is determined from looking at the current fluctuations that occur as the magnetic field is swept and determined from the magnetic correlation function for these fluctuations. This is done in the same manner as those for weak localization discussed in the previous section. We can utilize the magnetic correlation function given by

$$C(\Delta B) \sim \sum_i G(B_i + \Delta B)G(B_i), \tag{3.20}$$

where the averaging is done over the initial magnetic field, and this function is normalized to unity at $\Delta B = 0$. At relatively low magnetic fields, this correlation function does not depend upon this initial magnetic field. The critical field is taken to be the value of ΔB at which the correlation function has fallen to 0.5, and this is called the *correlation magnetic field* ΔB_c. But, the correlation magnetic field ΔB_c can become a function of the magnetic field itself when edge states start to form [24, 25]. Then, the slope of this change can be used to experimentally determine the phase-breaking time from the correlation magnetic field as [13]

$$\tau_\varphi = \frac{hB}{E_F \Delta B_c}. \tag{3.21}$$

This phase-breaking time is plotted in figure 3.12 as a function of both the lattice temperature and the current through the sample.

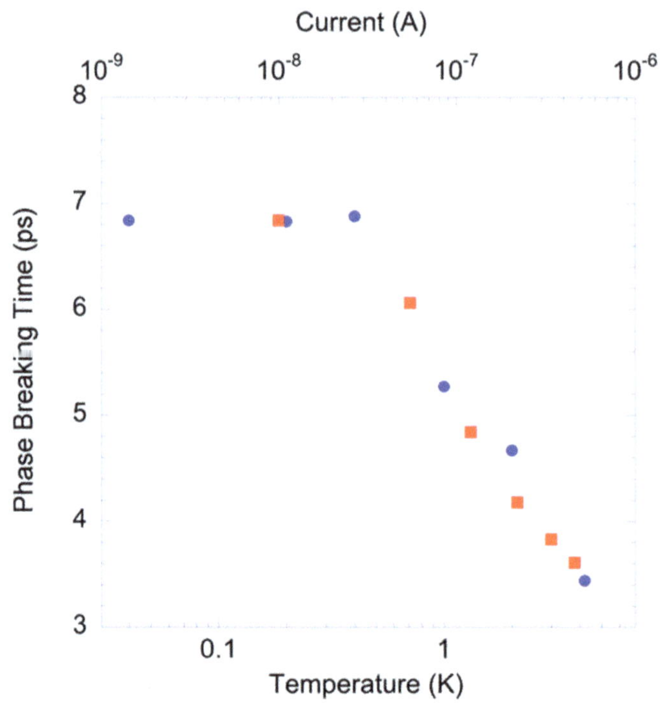

Figure 3.12. Phase-breaking time for a two-dimensional electron gas as a function of the lattice temperature (blue dots) and the current through the device (red squares).

It should be pointed out that there is a theoretical approach based upon the electron–electron interaction that suggests that the phase-breaking time is given in two dimensions by [26, 27]

$$\frac{1}{\tau_\varphi} = \frac{k_B T}{2\pi\rho(E)D\hbar^2}\ln[\pi\rho(E)D\hbar]. \tag{3.22}$$

Here, $\rho(E)$ is the density of states at the Fermi energy and D is the diffusion constant at this energy. This equation certainly predicts the $1/T$ behavior seen in figure 3.12, but does not explain the temperature independent portion of the data. This temperature independence at low temperatures has always been seen in studies of the phase-breaking time, and so can be considered to be intrinsic to the process. And, indeed, other theoretical approaches which also depend upon the electron–electron scattering, but principally deal with the electron–plasmon interaction, to produce the dephasing show both temperature ranges, despite being somewhat controversial. This latter form is given as [28, 29]

$$\frac{1}{\tau_\varphi} = \frac{e^2}{\pi\sigma}\sqrt{\frac{2D}{\tau}}\left[2T\sqrt{\tau\tau_\varphi} + 1\right], \tag{3.23}$$

where σ is the conductivity. The first term arises from the high temperature region and leads to a $1/T^{3/2}$ behavior, while the second term arises from the low

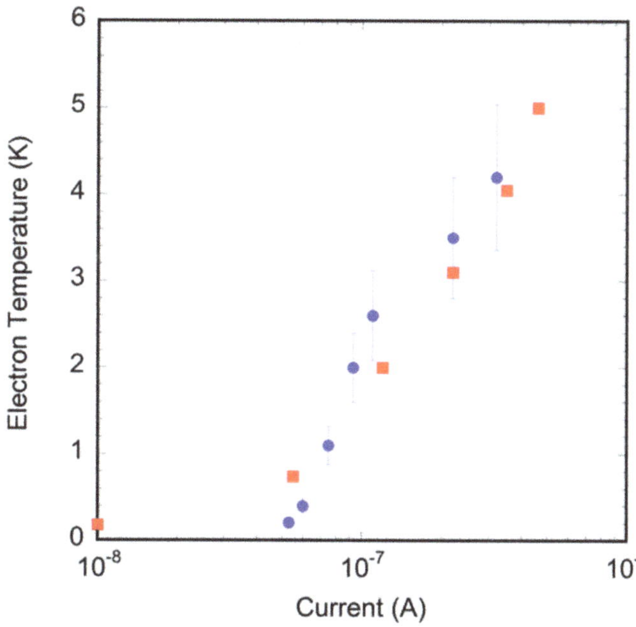

Figure 3.13. A comparison of the apparent electron temperature found from using the amplitude of the Shubnikov–de Haas oscillations (blue circles) and the phase-breaking time (red squares) obtained from the correlation function of the current fluctuations.

temperature region where $k_B T < \hbar\omega$, and ω is the two-dimensional plasmon frequency. The plasmon interaction is the interaction of an electron with the collective plasmon mode. When one introduces a screening effect, this effectively limits the single-particle interaction to long range (small k). To completely include the full electron–electron interaction, one now has to include the scattering of an electron off the collective (short-range interaction) motion [29]. This will reappear below in the treatment of energy relaxation in graphene.

What also appears in figure 3.12 is that the phase-breaking time itself is a valuable thermometer for determining the electron temperature for warm and hot electrons in these mesoscopic systems. In figure 3.13, a comparison of the electron temperature determined from the Shubnikov–de Haas amplitudes and the phase-breaking time is shown as a function of the current passed through the device. The lowest current, 10 nA, apparently causes no observable heating and lies off the general curve. Other than that, there seems to be a very good agreement between the two approaches to determining the electron temperature.

3.3 Energy relaxation time

Generally, the phase-breaking time discussed above arises from the inelastic scattering processes in the device. But let us repeat an earlier comment: these inelastic processes give rise to another characteristic time when the system is driven out of equilibrium, for example by passing a large current through it. In this latter case, the system returns to equilibrium by a process called energy relaxation. The energy-relaxation time τ_E is another parameter of importance as it describes the characteristic time for the system to return to equilibrium. There have been many experimental and theoretical studies directed toward understanding the process of dephasing and its variations with lattice temperature, and these have been discussed in the preceding chapters. There are few studies, however, in which both the phase-breaking and energy relaxation have been studied in the same system [5, 30, 31]. Generally, such measurements provide us with the experimental means to determine the equivalent carrier temperature that arises from the heating by the current or voltage, and to determine the effective relaxation times, as well as studying how these two times differ.

As above, using the Shubnikov–de Haas oscillations provides an effective thermometer. Here, varying the temperature and applied fields can provide a good method of obtaining the carrier temperature from the amplitude of the oscillations [6]. This has remained the method of choice for estimating the carrier temperature. If these are studied at various temperatures and currents, these two sets of data may be used to extract the temperature dependence of the energy-loss rate $P(T)$, or equivalently the power being dissipated within the mesoscopic system. This appears in the literature in two forms, the first being the loss per electron, and the second is just the total loss in the device. Obviously, these differ by the area or volume of the device and the carrier density at which the measurements are made. This energy-loss rate is normally a function of a power law of the hot carrier temperature, and the exponent often considered to be an indicator of the type of scattering that is dominating the inelastic process in the device. In spite of years of work in both normal semiconductor systems, as well as mesoscopic devices, there is not yet a clear understanding of the entire process.

It has been shown that the general formula for the energy-loss rate often can be written as [32]

$$P(T_e) = F(T_e) - F(T) = A(T_e^p - T^p), \tag{3.24}$$

where F is a general function, and A is a combination of material parameters and fundamental constants. The exponent p (and the prefactor A) depend upon the specific scattering mechanism that governs energy relaxation. For example in heterostructures for the III–V compound semiconductors, it has been asserted that one expects $p = 7$ for acoustic phonon interactions, and $p = 5$ for the screened piezoelectric interaction, and $p = 3$ for the unscreened piezoelectric interaction [33] (we will address these parameters beginning in chapter 6, where the theory is discussed). Various values of the exponent depend upon the energy dependence of the scattering rate, as well as the dimensionality of the system. For example, the function F for acoustic mode scattering in a quasi-two-dimensional system is given as [32]

$$F(T_i) = \frac{1}{\hbar E_F}\int dq \mid \psi(q) \mid^2 \left\langle \frac{m^* V}{\pi\hbar^3} S^2(Q) | M_3 |^2 (\hbar\omega)^3 N_\pm \right\rangle, \tag{3.25}$$

where q and Q are the phonon momentum vector parallel to, and normal to, the two-dimensional layer, respectively, ψ is the electron wave function normal to the layer, S is the screening function, M_3 is the three-dimensional scattering matrix element, ω is the phonon radian frequency and N is the Bose–Einstein distribution at T_i. The $\pm$ signs tell us that we use $N + 1$ for emission and N for absorption of the phonon. The triangular brackets indicate an average over the scattering angle in the two-dimensional plane. It is clear that different scattering processes will have different matrix elements and thus different power laws and the integrals are carried out.

When measurements such as these are made at low temperatures, one cannot always assert that the phonons are in the equipartition limit, where the full details of the Bose–Einstein distribution can be ignored and a leading term used. A decision about this involves the Bloch–Gruneisen temperature. In general, the largest q vector for a phonon that can be generated in an electron–phonon scattering event at low temperature is the one that spans the Fermi surface from one side to the other, basically a value of $2k_F$. With this wavelength, the maximum acoustic phonon energy is then $2\hbar k_F s$ where s is the sound velocity. The issue is whether or not the exponential in the denominator of the Bose–Einstein statistic can be expanded to simplify the expression. The argument of the exponential is the ratio of the phonon energy in the collision to the thermal energy. Thus, the critical temperature is said to be the Bloch–Gruneisen temperature

$$T_{BG} = \frac{2\hbar k_F s}{k_B}. \tag{3.26}$$

For temperatures below this, the full Bose–Einstein distribution has to be kept, and this complicates the determination of the loss to phonons. For example, for a two-dimensional density of 10^{12} cm^{-2}, one finds that in a 2DEG for a typical semiconductor, the Fermi wave number is 2.5×10^6 cm^{-1}. The sound velocity in InGaAs is about 4.7×10^5 cm s^{-1}, so that the Bloch–Gruneisen temperature is about 10 K. This lies in the temperature range involved in the experiments discussed in the previous section, so that it is unsafe to do anything other than keeping the full Bose–Einstein distribution. This complicates the evaluation of the F functions in (3.24), but does not prevent their evaluation.

However, it should also be apparent that the power input per electron can be determined from the sample measurements of density and mobility. Then, the power per electron is simply given

$$P(E) = e\mu E^2, \tag{3.27}$$

where here E is the electric field. Hence, knowing the density and sample dimensions allows one to determine the total power entering the sample at a given electric field. The purpose of (3.24) is then to see if the input power can be matched to a given electron temperature. This is shown in figure 3.14, for the Stöger *et al* [23] strained Si quantum well discussed in the last section. Here, the power input is plotted in the main panel as a function of $T_e - T_L$, where the latter is the lattice temperature. The variation with electron temperature is thought to be appropriate for acoustic phonon

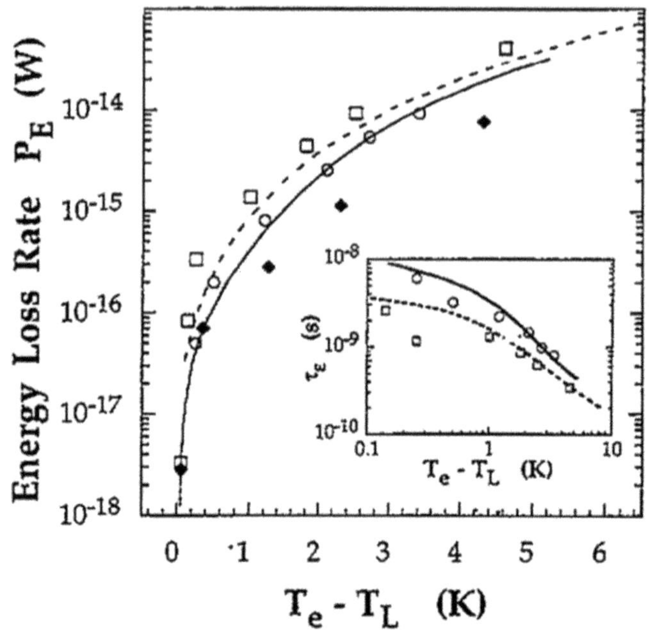

Figure 3.14. Electron loss rate P_E as a function of the electron temperature for a lattice temperature of 1.8 K for sample M1 (circles) and 2.1 K for M2 (squares). The solid and dashed lines are fits for acoustic phonon scattering using the parameters of sample M1. The diamonds come from weak localization measurements. The inset gives the energy relaxation times for these two samples. Reprinted from Stöger *et al* [23], with permission of IOP Publishing.

scattering [34], $p = 2$ in (3.24), in the thin layer of strained Si. In the inset is a plot of the energy relaxation time which is determined from the equation

$$e\mu E^2 = k_B \frac{T_e - T_L}{\tau_E} \tag{3.28}$$

under steady-state conditions. The first obvious point is that the power in the figure is not $p = 1$. While the general form for the energy loss (3.24) has higher powers of the various temperatures, one can always factor out the term $T_e - T_L$, with the remaining temperature terms part of the resulting temperature dependence of the energy relaxation time τ_E. So, in this sense (3.28) is not in conflict with (3.24), but is a general form for the relaxation of a physical quantity, often found in equations such as the Boltzmann equation.

Using the same approach discussed above and in the previous section, where the electron temperature is determined from the Shubnikov–de Haas effect in the InGaAs quantum well device (of figure 3.9), we can determine the energy loss per electron using (3.27). the data for these devices above are plotted as a function of the inferred electron temperature in figure 3.15. The data come from the three values of magnetic field that were used in figure 3.12. It may be seen that the input power per electron all fits on a single curve, although it is difficult to see that this curve fits to $p = 3$ (or any other value). Nevertheless, the slope of the curve at the higher values of

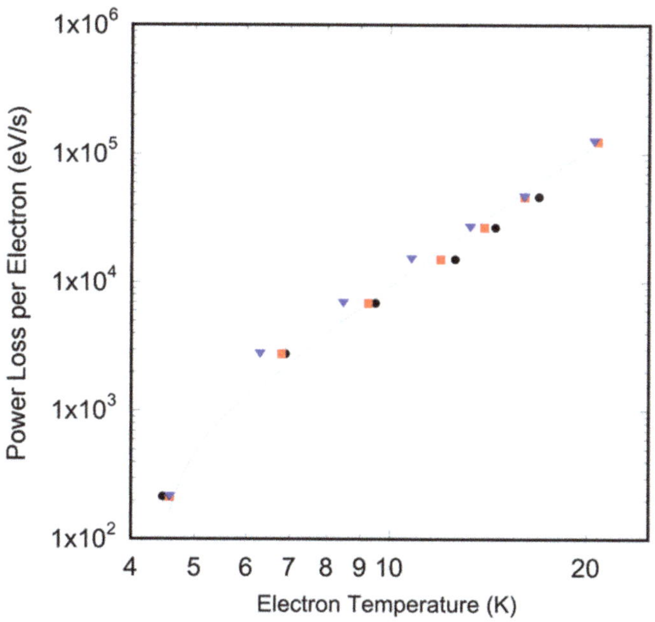

Figure 3.15. The power dissipated per electron for the devices of figures 10.1 and 10.2. The circles are for a magnetic field of 1.6 T, the squares for 2.0 T, and the triangles for 3.2 T.

electron temperature do correspond to this value of the exponent. To see this better, we re-plot the data in figure 3.16 as a function of the temperature expression in (3.24) with $p = 3$. Here, it may be seen that the fit to the power law is very good, although there is a little more scatter in the data. However, it may be seen that slope agrees with each of the three sets of data for different magnetic fields. Thus, it may be inferred that the likely source of the energy relaxation of the excited carriers is due to phonons via the unscreened piezoelectric interaction [33]. In figure 3.17, the energy relaxation time extracted from the data using (3.28) is plotted as a function of the electron temperature.

3.4 Effects in lower dimensions

3.4.1 Quantum dots

Open quantum dots are necessarily coupled to their leads through a pair (at least) of quantum point contacts (QPCs). These dots can also exhibit phase-breaking, but in this case, the phase-breaking generally arises through interactions between electrons within the dot and electrons in the leads (or external environment). The dot considered here is fabricated on a GaAs/AlGaAs heterostructure and lithographically patterned to be 0.8 μm square, with narrow QPC openings to a pair of quasi-one-dimensional leads [35]. These QPCs are patterned to be 150 nm in width at the dot. An optical micrograph is shown in the inset to figure 3.18. The carrier density in the GaAs channel was measured to be 3.8×10^{11} cm^{-2} by Shubnikov–de Haas measurements prior to forming the dot. These measurements also gave a mobility

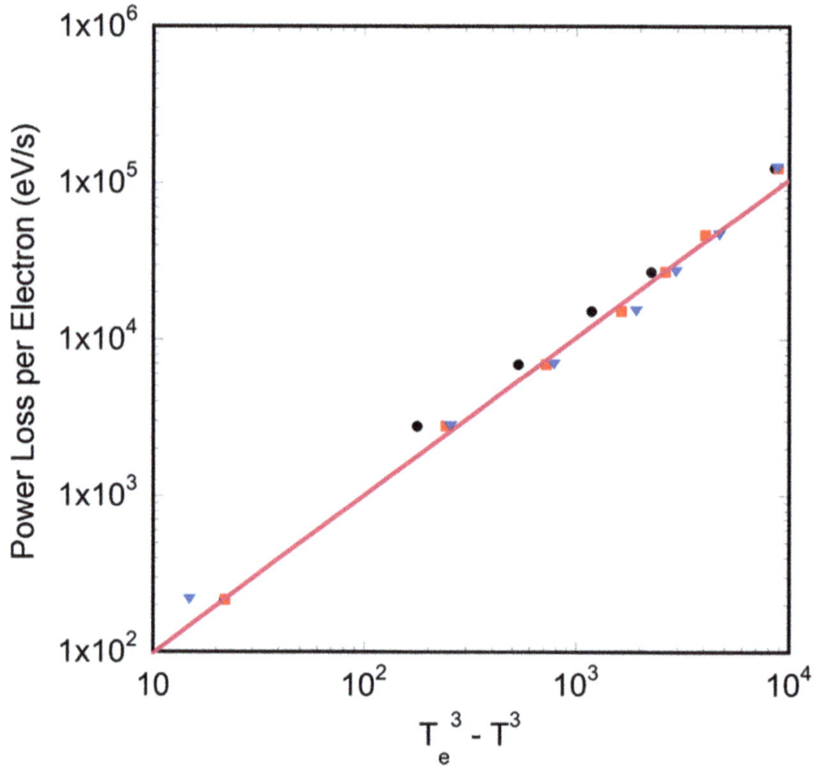

Figure 3.16. The power loss per electron from figure 3.15 re-plotted so that it corresponds to the form of equation (3.24). Reprinted from Ferry [13], with permission of IOP Publishing. All rights reserved.

for the carriers of 3×10^5 cm^2 Vs^{-1} at 50 mK. The dot itself has a self-aligned gate over the top, by which the size of the dot and the QPCs can be modified. The two-terminal resistance of the dot is measured as a function of gate voltage to estimate the size of the dot and QPCs.

These structures exhibit conductance fluctuations for all gate voltages and the correlation functions of these fluctuations are used to estimate the phase-breaking time, as shown in figure 3.18 for three different gate voltages. It will be noticed that the phase-breaking time saturates in all three cases below a certain temperature, which is thought to be related to the mean level separation in the dot as [36]

$$k_B T < \frac{\pi \hbar^2}{m^* A}, \tag{3.29}$$

where A is the area of the dot. Generally, the phase-breaking time for a dot coupled to wires may be written as

$$\frac{1}{\tau_\varphi} = \left.\frac{1}{\tau_\varphi}\right|_{\text{dot}} + \left.\frac{1}{\tau_\varphi}\right|_{\text{wire}}. \tag{3.30}$$

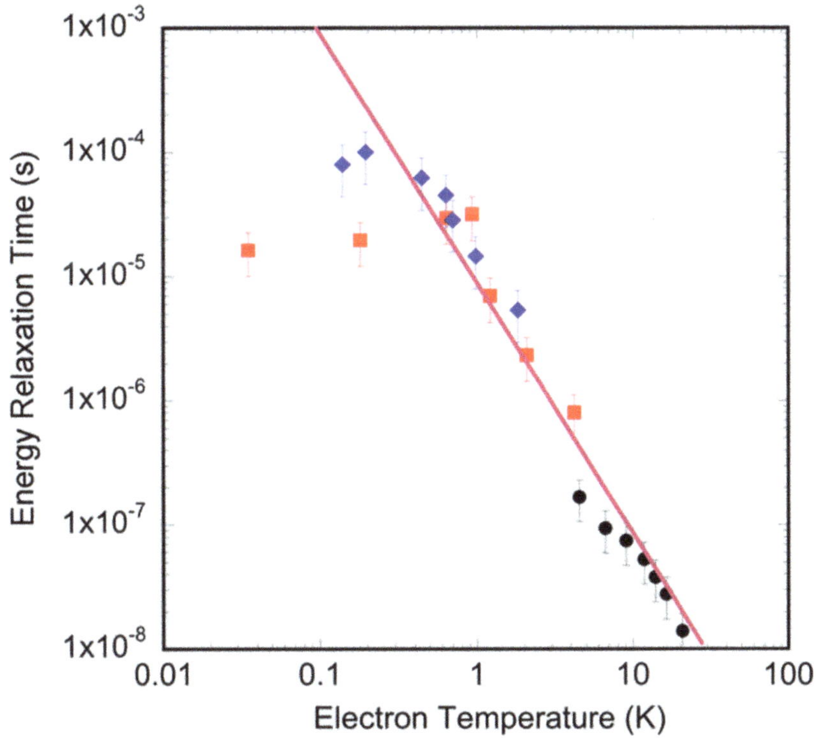

Figure 3.17. The electron energy relaxation time extracted from the power loss data according to (3.24). The colors are for the same magnetic fields as in figure 3.15. The line is a guide to the eye.

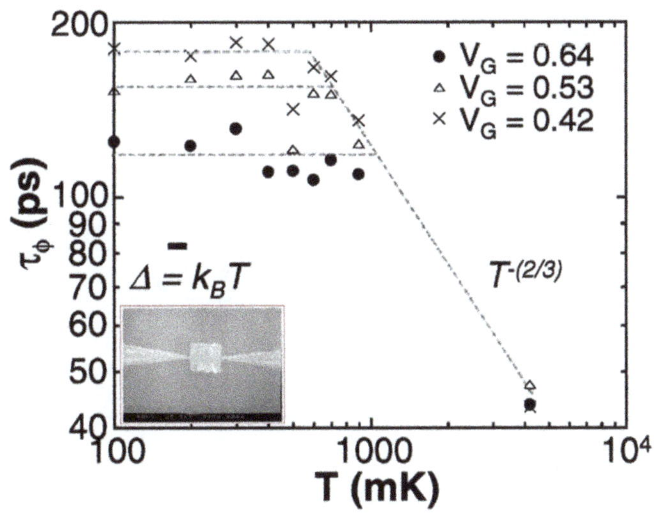

Figure 3.18. The estimated phase-breaking time as a function of the temperature. This is shown for three different gate voltages, which implies three different dot sizes. The mean level separation in the dots is indicated by the short bar on the lower left. The higher gate voltage results in a larger dot and more coupling to the wire, which seems to be important in the low temperature regime. The inset is an optical micrograph of the dot and wire structure. Reprinted from Pivin *et al* [35], copyright (1999) by the American Physical Society.

Here, the decay of the phase-breaking time varies as approximately $T^{-2/3}$, which is characteristic of a one-dimensional wire [26]. Thus, it appears that, at least in this case, the wire–dot interaction is governed by the one-dimensional wires.

In figure 3.19, the phase-breaking time is plotted as a function of the dot size at base temperature (~50 mK) for three different illuminations of the dot (the dot is illuminated once at low temperature to excite the carriers from the impurities, so this corresponds to three different cool-downs of the fridge) [37]. These dots exhibit conductance fluctuations at all temperatures and sizes. But, the data point marked *b* in the figure also exhibits weak localization, whereas the dot marked *a* does not. The fact that these two symbols also indicate two different cool-downs (but with essentially similar densities of electrons) may be important. However, these two markings are for only the dots indicated, and it may be that dot marked *a* has a size smaller than the phase-breaking length l_φ. It is clear that the saturated phase-breaking time at low temperature decreases as the dot size increases, which is consistent with the trend in figure 3.18. This argument can be recast since we know both the dot area and the back-ground carrier concentration. Thus, it is possible to estimate the number of electrons in the dot itself. The data of figure 3.19 is re-plotted in figure 3.20 as a function of the estimated number of electrons within the dot [38].

An array of open quantum dots was created in an AlGaAs/GaAs heterostructure using metal gates to define the array. The samples were fabricated by e-beam lithography and liftoff on a Hall bar measuring 80 μm wide with side arms (see the inset to figure 3.9 for the structure) spaced by 200 μm. The two-dimensional density in the GaAs was 3.7×10^{11} cm^{-2} and displayed a mobility of 2.4×10^5 cm^2 Vs^{-1} at 4.2 K [40]. The lithographic dimensions of the dots were 1.0×0.6 μm, but this size

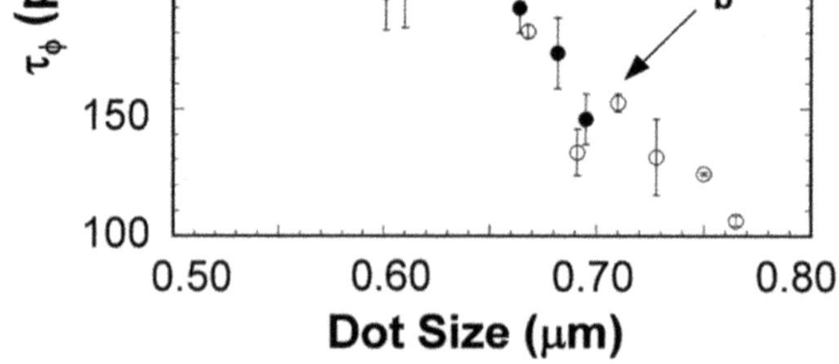

Figure 3.19. Variation of the saturated phase-breaking time with dot size at base temperature (~50 mK) for three different illuminations of the dots. There is a clear trend of decreasing phase-breaking time with increasing dot area. The two symbols *a* and *b* are discussed in the text. Reprinted from Pivin *et al* [37], copyright (1999) with permission from Elsevier.

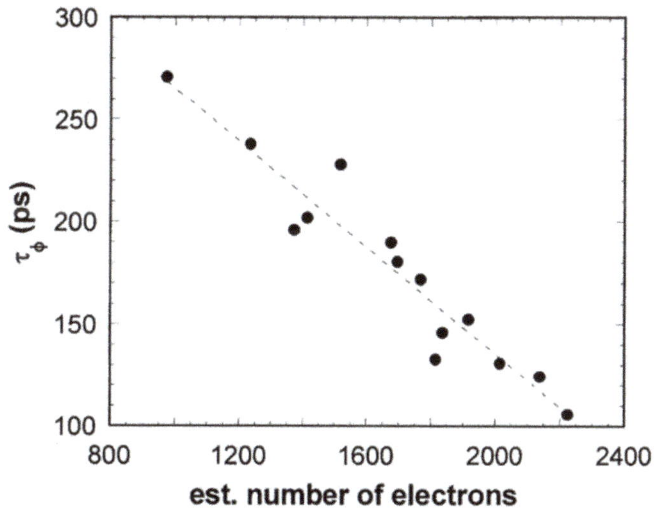

Figure 3.20. The dependence of the saturated phase-breaking time at base temperature (~50 mK) as a function of the estimated number of electrons in the dots. The line is a guide to the eye only. Reprinted from Pivin *et al* [38], copyright (2001) with permission from Elsevier.

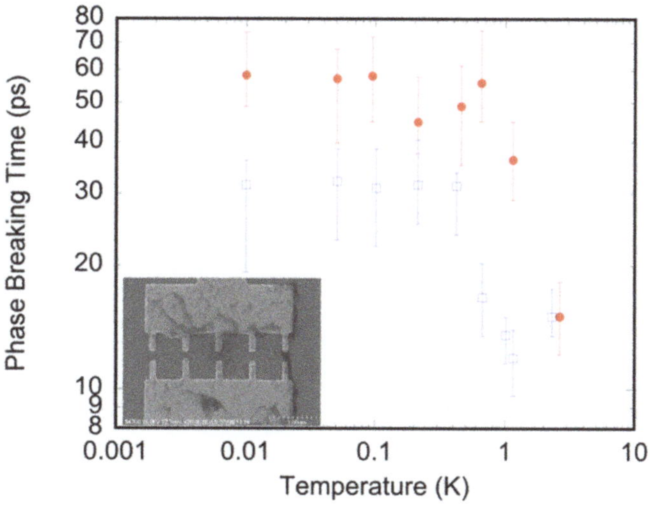

Figure 3.21. The phase-breaking time as a function of the electron temperature for the four dot array shown in the inset. The two curves are for gate voltages of −1.078 V (blue open squares) and −0.55 V (red circles).

was probably reduced by about 0.12 μm at each edge by the fields from the metal gates. The QPCs themselves were lithographically 0.3 μm long. An optical micrograph of the four dot array is shown in the inset to figure 3.21. Measurements were made in a dilution refrigerator over a range of lattice temperatures and current drive. Generally, the measurements were made when the total resistance through the dot array was greater than $2e^2/h$. As this is across five QPCs, the individual QPCs likely had several modes propagating through them. As the dots exhibited conductance fluctuations, the phase-breaking time was determined from the correlation functions as a function of magnetic field. The phase-breaking times for two different bias voltages on the metal gates are shown in the main panel of figure 3.21. As previously, these times show a saturation at low temperature. The temperature decay, however, is dependent upon the dimensionality of the system itself. Here, the temperature decay appears to be somewhere between $T^{-2/3}$ and T^{-1}. The former value is expected for one dimension and the latter value is expected for two dimensions [29], as will be discussed beginning in chapter 6.

Knowledge of the electron temperature now allows us to determine the input power per electron at a given current level and then the energy relaxation time from (3.28). The energy relaxation time determined from this method is plotted in figure 3.22 for the four dot array at the two different gate voltages used earlier. As with the phase-breaking time, a saturation occurs at low temperatures, which is consistent with observations in other systems. This is an expected result for electrons scattering with the plasmons [28, 29], as remarked earlier. The decay with temperature at the higher temperatures appears to very with an exponent of 0.7 ± 0.1, which is consistent with the expectation of an exponent of 2/3 for one-dimensional systems. It is clear that the phase-breaking time shows the transition to saturation at a higher temperature than the energy relaxation time, so these may be due to different physical mechanisms, though that is not assured as the mathematics of energy relaxation according to (3.25) likely has a different energy dependence than the equivalent calculation for the phase-breaking time.

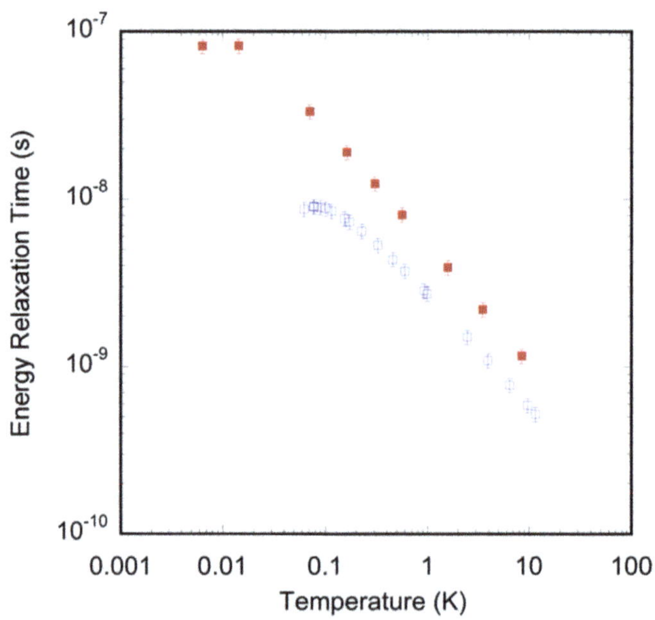

Figure 3.22. The energy relaxation times measured in the four dot array of figure 3.21. The two curves correspond to the two voltages applied to the gates as remarked in the caption to this earlier figure.

3.4.2 Quantum wires

The heterostructure layers of the InAlAs/InGaAs layers that were discussed in the last two sections have been patterned into forming quantum wires within the InGaAs layers. In these structures, the $In_{0.53}Ga_{0.47}As$ quantum well was 25 nm thick. The quantum wires were formed by a wet etch of the layers, and wires of electrical widths of 270, 570 and 770 nm were formed [40, 41]. An optical micrograph of these etched wires is shown in the inset to figure 3.23. All wires were 2.4 μm in length. The electrical widths were determined by the magnetic field necessary for the onset of Shubnikov–de Haas oscillations (the oscillations cannot form until the cyclotron diameter fits inside the wire). These results indicate that there was an almost uniform edge depletion of 125 nm on each side of the wires. As before, the electron temperature was determined from an amplitude analysis of the Shubnikov–de Haas signals once they formed. This allowed the determination of the temperature at each current level for which the wires were excited. In figure 3.23(a), the power input per electron is plotted against the factor $T_e^5 - T^5$ for the two widest wires. The power input per electron for the narrowest wire is plotted in figure 3.23(b) versus the factor $T_e^3 - T^3$. The lines in the two panels are guides to the eye, but have linear slope indicating that the scattering processes in the widest wires are different than in the narrowest wire. The exponent $p = 3$ in (3.24) is thought to arise from unscreened piezoelectric scattering [34], while the exponent $p = 5$ is thought to arise from acoustic deformation potential scattering [34]. It is surprising that such a sharp difference is seen from the narrow to the wide wires.

From the power loss data, one can determine the energy relaxation time from (3.28). This is plotted for the three wire sizes in figure 3.24, with the same color scheme as in figure 3.23. Again, it may be seen that the narrowest has a lower saturation for the energy relaxation time than the wider wires, likely due to the fact that it is more confined and scattering is more backward. But, it was seen earlier with

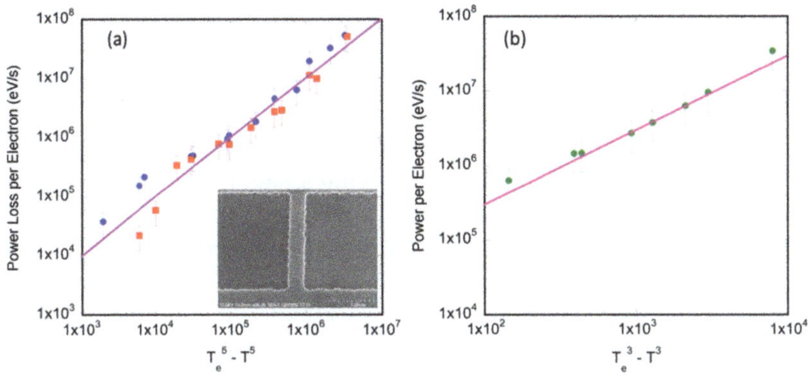

Figure 3.23. (a) The power input for wires of 570 nm (red) and 770 nm (blue) width are plotted as a function of the electron temperature. The inset is a picture of one of the wires. (b) Power loss for a 270 nm wire width as a function of the electron temperature. The lines are guides to the eye but have a linear slope for the two different power laws.

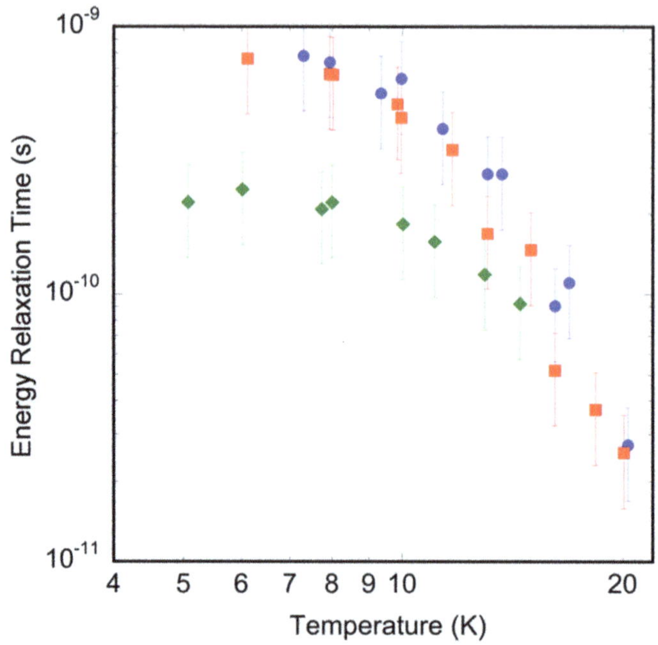

Figure 3.24. The energy relaxation determined from the data of figure 3.21. The colors here correspond to the equivalent colors in this latter figure. It is clear that the narrow wire (green) has a lower value of the saturated energy relaxation time than the two wider wires.

the quantum dots that more confinement generally led to lower values for the phase-breaking time, and this same trend seems to follow for the quantum wires as well.

3.5 Some different systems

3.5.1 GaP

GaP is an indirect band gap material in which the conduction band minimum lies at X with a gap of 2.26 eV. The satellite valleys lie at Γ and L, some 0.52 eV and 0.38 eV, respectively, higher in energy. But, these valleys can be important in a variety of applications. The carrier dynamics can be studied by a variety of techniques, but direct study of the conduction electrons with near band gap photoemission spectroscopy, is realized by having a negative electron affinity surface from which the electrons are emitted [42–44]. Here, a *p*-type GaP crystal, doped to 5×10^{18} cm^{-3} with zinc, was used and a negative electron affinity surface was achieved by annealing in ultra-high vacuum and alternately supplying cesium and oxygen to the surface [43], an approach that is well known [45]. Electrons are photo-excited high into the Γ valley due to the direct transition being more favorable than an indirect one. These carriers subsequently relax in this valley as well as transferring to the X valleys. Two distinct peaks are observed in the emission spectrum, which are presumed to arise from carriers in the X and Γ valleys. By measuring the populations in the various valleys (from the electron emission measurements) one

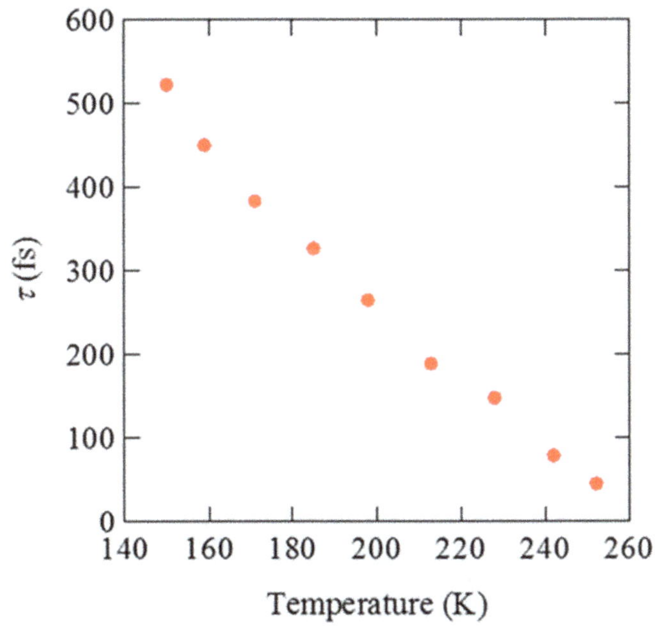

Figure 3.25. Temperature dependence of the relaxation time for carriers moving from the Γ to the X valleys in GaP. Reprinted from Ichihashi *et al* [44] under the Creative Commons license.

can determine both populations and the scattering time from Γ to X. Measurements made as the lattice temperature is varied also allows determination of the temperature dependence of the X valleys, and perhaps also the Γ valley. The scattering rate from Γ to X as a function of temperature is shown in figure 3.25. This relaxation time is determined from the carrier densities and distribution functions in the two valleys.

3.5.2 HgTe topological insulator

Another interesting system is a quantum well in HgTe. Generally, HgTe has what is known as an inverted band gap system, in which one of the normal valence states actually lies above the normal conduction band minimum, which is quite useful in superlattices [46], where the HgTe is a quantum well with CdTe or HgCdTe barriers. The inversion of the bands in HgTe is driven by the intrinsic spin–orbit interaction. As CdTe, or HgCdTe, has the normal band lineup, the two bands that are inverted in the HgTe have to cross at the interfaces between the two materials. This leads to helical edge modes (at the interface) with opposite spin polarization and which propagate in opposite directions, a state known as a topological insulator. Recently, electron heating experiments have been used to study the electron–phonon relaxation processes in these topological edge states [47]. It is felt that (energy) relaxation by the excited carriers appears as relaxation in multiple carrier puddles that appear at the interface due to lateral inhomogeneities, and transport between the puddles appears to be due to tunneling. In this study, the CdTe/HgTe quantum wells were grown by molecular beam epitaxy on (013) surfaces with a quantum well width of

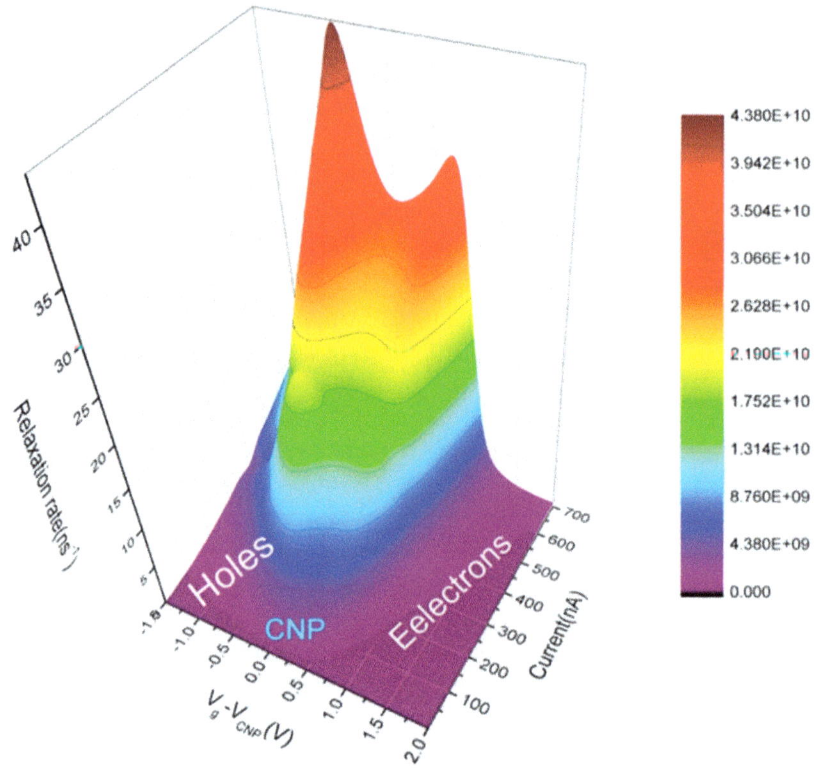

Figure 3.26. 3D plot of the energy relaxation rate showing its evolution with gate voltage and excitation current. The peak at the charge neutral point (CNP) has the most rapid relaxation. Reprinted from Rahim *et al* [47], copyright (2019), with permission from Elsevier.

~8.1 nm. By using a gate on the structure, the density could be varied as ~1.09×10^{11} cm^{-2} V^{-1}. The electron temperature was determined by using the variation of resistance with temperature as a thermometer for carrier heating by drive current. Then from the heating power per electron, the energy relaxation time was determined using (3.28). In figure 3.26, the energy relaxation rate (inverse of the relaxation time) is plotted as a function of the gate voltage and excitation current for both electrons and holes for a lattice temperature of 4.2 K. The peak electron (or hole) temperature is about 27 K.

3.5.3 Group III-nitrides

Incorporating *N* into normal alloy systems can alter the lattice constant and band gap. Thus, it is a useful technique to alter the band gap to one more suitable for optoelectronics devices working at, e.g., 1.3 μm. Recent work in $Ga_{0.68}In_{0.32}N_{0.017}As$ quantum wells, with GaAs barriers, has studied the magnetotransport and carrier heating with annealing temperature [48]. The structures were grown on a (100) GaAs semi-insulating substrate by molecular beam epitaxy. They were then lithographically defined Hall bars with a length of 1.75 mm and width of 0.2 mm. Measurements of the

Shubnikov–de Haas oscillations with lattice temperature and with device current were used to establish the electron temperature at a given current level. The energy relaxation time is shown in figure 3.27 for three different samples, all annealed for different times. No significant difference in the values of the energy relaxation time are observed for these three samples, suggesting that the intrinsic processes leading to energy relaxation are not different for the different annealing. However, the sample annealed the longest (600 s), TN17B seems to lose energy more rapidly at electron temperatures above 25 K. It is thought that this annealing time may make the interaction with acoustic phonons more efficient at this higher temperature.

Recent calculations of the relaxation times in GaN and InN over a range of temperatures, using the ensemble Monte Carlo technique have given some interesting insights [49]. The results are shown in figure 3.28 for both materials; GaN on the left side and InN on the right side (note the different directions of the electric field axes). It is interesting that the energy relaxation rate (inverse of the time) seems to be independent of temperature in InN. On the other hand, the rate is a function of temperature in GaN in the low temperature region, but also not in the high temperature region. In GaN, the relaxation rate rises up to a point where intervalley scattering begins to occur, then drops as the electrons populate the satellite valleys, finally rising once again at the highest fields. The onset of intervalley transfer seems to be a weak function of temperature, and is observed experimentally to happen at about 180 kV cm^{-1} in bulk GaN at room temperature [50].

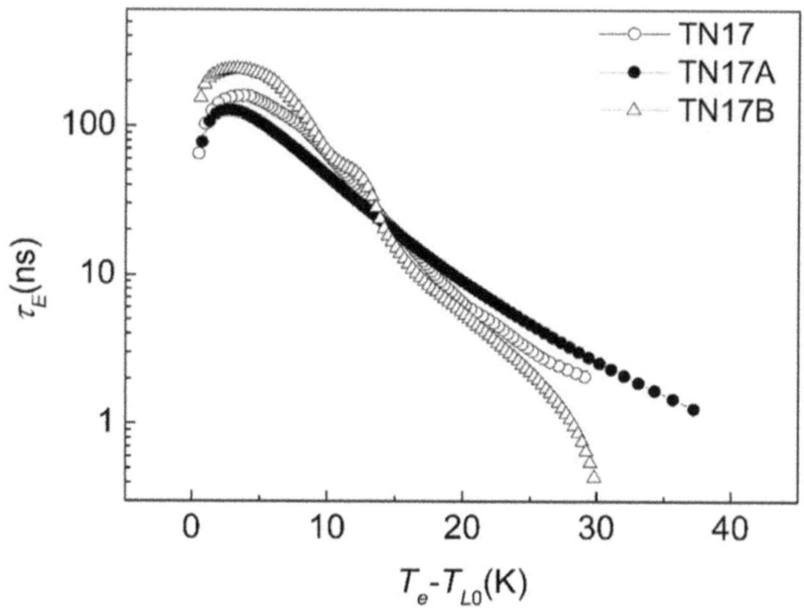

Figure 3.27. Energy relaxation time for the three samples, annealed at different temperatures, as a function of the difference in electron temperature and lattice temperature. Reprinted from Ardali *et al* [48], copyright (2019) with permission from Elsevier.

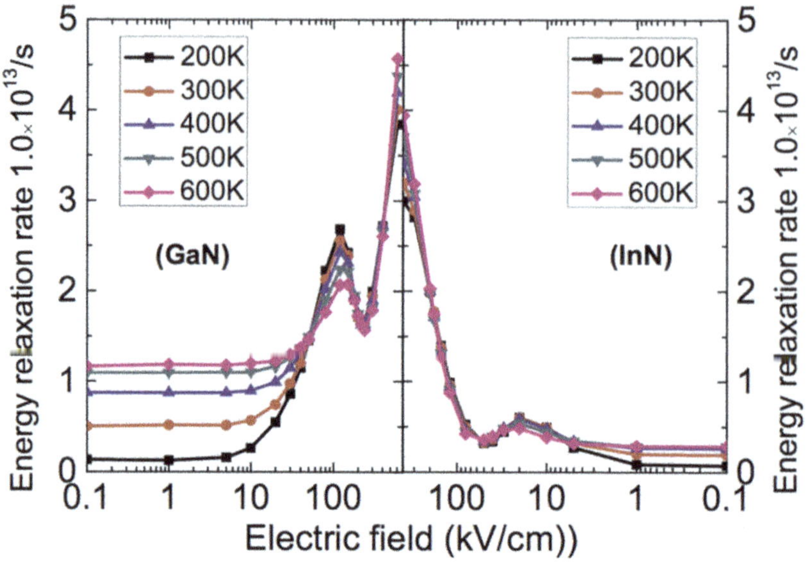

Figure 3.28. The energy relaxation rate in GaN (left) and InN (right) as a function of the applied electric field, with lattice temperature as a parameter. The doping in both materials was 10^{17} cm^{-3}. Reprinted from Wang *et al* [49], copyright (2005) with permission from Elsevier.

3.5.4 Graphene

Graphene, monolayers of graphite, have also been studied at low temperature for their relaxation times. The samples were prepared by exfoliating Kish graphite onto a doped Si substrate upon which 300 nm of SiO_2 were grown [51]. Monolayer and bilayer devices were formed in different samples by this process, with the number of layers being confirmed by optical microscopy and Raman spectroscopy. The samples were contacted with Cr/Au electrodes in a six probe configuration which allowed the electric field to be determined in a manner free from the effect of current contact resistances (as is done in the Hall effect, discussed above for determining the temperature from Shubnikov–de Haas measurements). The samples typically exhibited mobilities in the range 10^3–10^4 cm^2 Vs^{-1} at low temperature. The measurements in the figure are on a bilayer device with a base temperature of 1.8 K and the power input per electron is shown in figure 3.29. In this case, the samples exhibited conductance fluctuations, as also discussed above, and these were used for the thermometer in this situation. As the fluctuations varied with the base temperature and the current passed through the sample, the amplitude of the fluctuations was used to ascertain the electron temperature as a function of the drive current through the device. In the main panel of the figure, the input power per electron is plotted as a function of the temperature according to (3.24) with $p = 3$. Various data sets arising from different gate voltages (bias on the Si substrate), are indicated by the color coding indicated in the figure. These different voltages on the back gate lead to different carrier densities in the graphene layer. The inset shows different values of p for the same data, with each value shown as a dotted line.

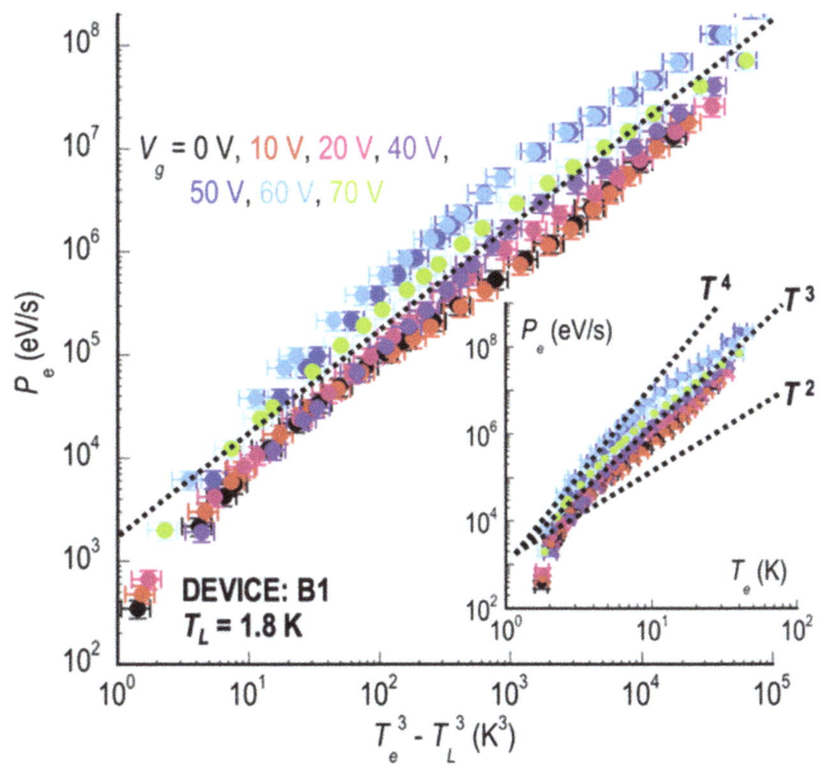

Figure 3.29. The carrier power loss per electron as a function of temperature according to the form of (3.24) at different gate voltages (color coded as indicated). The inset compares the data for slopes of different p values, showing that 3 is the most likely. This is indicated in the main panel as the dotted line. Reprinted with permission from Somphonsone *et al* [51], copyright (2013) American Chemical Society.

The conclusion is that the best fit is to a value of $p = 3$. While this is the same exponent found in the InGaAs heterostructure discussed above, the interpretation of this is much different. First, graphene is not piezoelectric, so the loss process found above does not exist in this material. More importantly, the unique Dirac-like band structure of graphene changes the exponents found for the various scattering mechanisms. For loss via the acoustic phonons, one expects an exponent of 4 in graphene [52–54]. In earlier work, the Oxford group had examined the energy-loss process in chemical-vapor deposition growth of graphene on the Si-terminated face of SiC [55]. In this latter work, the authors claimed a dependence with $p = 4$, although some of the data suggest that it was closer to $p = 3$. Similar results were recently found in another experiment, which also used bilayer graphene as well [56]. The data in figure 3.29, however, clearly do not fit with $p = 4$, and have a much better fit to $p = 3$. Very similar behavior was found for monolayer graphene in several samples. It also has been suggested that a second-order impurity mediated phonon scattering could lead to an exponent of 3 [57]. However, it was found that the strength of this interaction was much too weak to explain the data in the figure.

More recently, it was suggested that plasmon-mediated energy relaxation can explain the energy-loss rate seen in this latter figure [58]. This would be the same mechanism found above in a number of two-dimensional electron systems [28, 29]. Plasmons are excitations of the electron gas itself, so the plasmon loss process actually only redistributes the energy among the electrons. But, in the presence of current, the Fermi sea moves to account for the current flow and is no longer centered about the zero of momentum, as shown in figure 3.30. At high currents, this becomes highly accentuated, with many electrons above the equilibrium Fermi energy. Hence, the plasmon process accounts for scattering of these excited electrons back to states mostly vacated by the shift in the Fermi sea. The process is also well known in the III–V arena [16]. This process reduces the current and the energy input to the electron gas, but ultimate relaxation of the energy must occur by some other process, such as the acoustic phonons. The derivation of the plasmon energy loss follows (3.25), but with the much different energy band structure of graphene [58, 59].

The energy relaxation time determined from the power loss data, with the use of (3.28), is plotted in figure 3.31 for the range of currents studied as a function of the density, and in figure 3.32 as a function of density for various electron temperatures. Here, the energy relaxation time shows a tendency to decrease as the carrier density decreases toward the Dirac point. It also seems to decrease with the current, but this

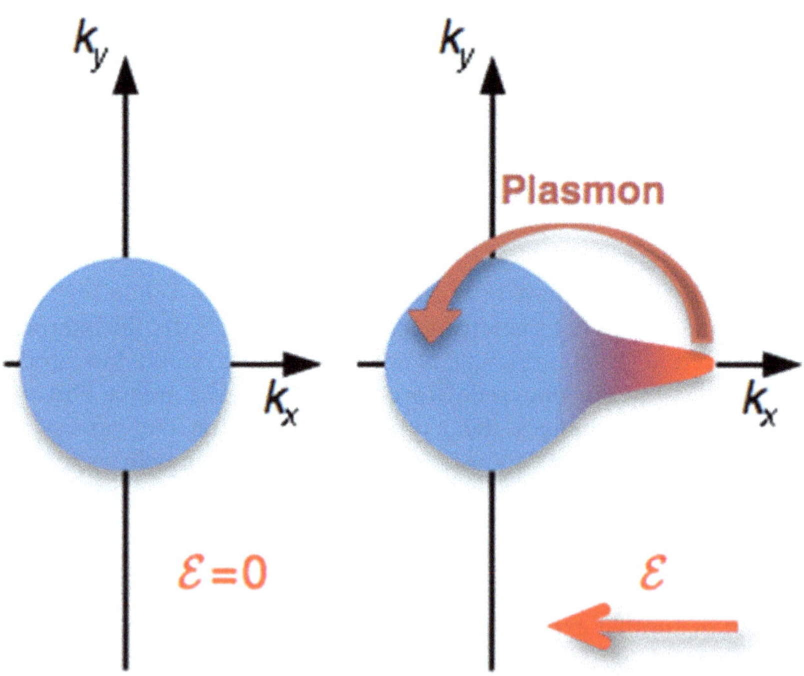

Figure 3.30. (a) A diagrammatic representation of the distribution function in equilibrium. (b) The same distribution function in the presence of a high electric field. The hot carriers occupy the high momentum tail (depicted in red). The script ε is the electric field. Reprinted from Ferry *et al* [59], copyright (2009) with permission of Springer.

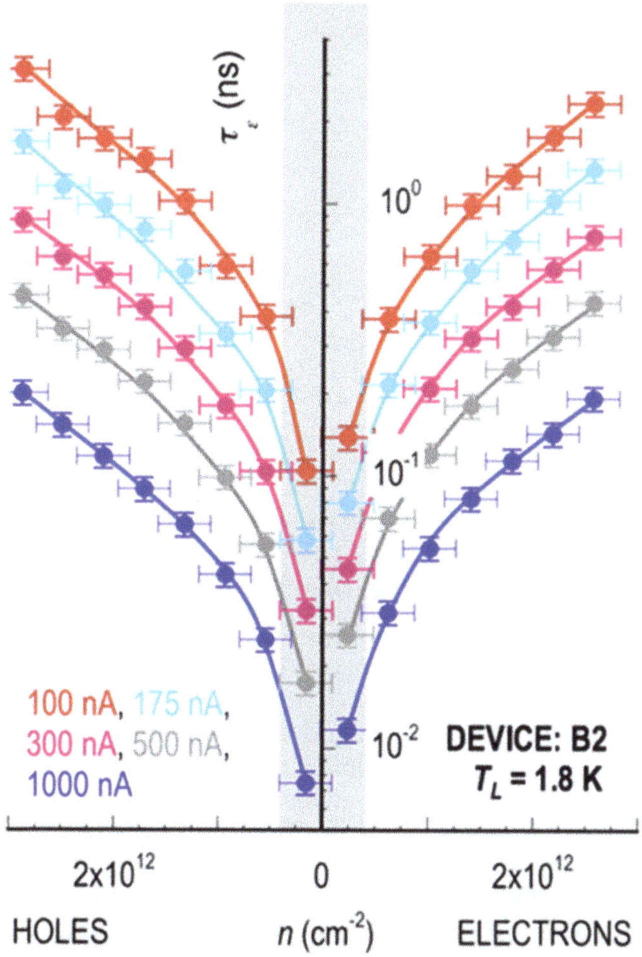

Figure 3.31. Energy relaxation time for a bilayer graphene device determined from the energy loss rate, shown in figure 3.29. This is done at a lattice temperature of 1.8 K and for various currents. Reprinted with permission from Somphonsone *et al* [51], copyright (2013) American Chemical Society.

is expected as higher current leads to a higher carrier temperature. The decrease as the Dirac point is approached is more problematic as the density of states available into which to scatter decreases as one approaches this point. But, nevertheless, the same behavior is expected from the theoretical results, where the energy loss rate is inversely proportional to the momentum scattering time and the square of the Fermi velocity [59]. The energy relaxation time is itself inversely proportional to the power loss rate from (3.28), so the lower loss rate would lead to an increase in the energy relaxation rate at a given carrier temperature.

The plasmon loss mechanism rearranges the energy distribution function, but this energy remains in the electron gas as a whole. Under the excitation current (or electric field), the distribution function is shifted in momentum space to reflect the current, and it spreads as the electron temperature rises. The shift can also distort the

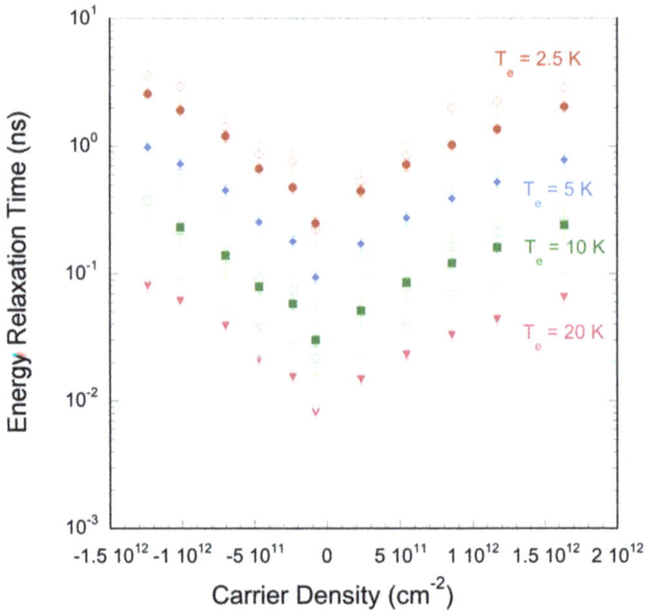

Figure 3.32. Comparison of the theoretical energy relaxation time (solid symbols) with that determined by the experiments of [51] (open symbols). Reprinted from Ferry [59], copyright (2015) with permission of Springer.

distribution to produce an elongated extension in the direction of the field or current, as shown in figure 3.30. The ultimate limit of this elongation is a delta function in the field direction. Between the shift in momentum space, and the elongation, the distribution function has more carriers in the forward direction than the backward direction. The plasmon loss takes carriers from the streaming forward direction and moves them to the backward direction, in the symmetric terms of the distribution function, on the average. This relaxes the streaming distribution, but the ultimate result still remains that another process must pass the energy to the lattice. This can be due to the acoustic phonon process, such as acoustic phonons, or even the super-collision process [57]. However, it is the plasmon interaction that dominates what is seen experimentally as it is this process which moves the current carrying particles out of the forward extension of the distribution function. The other processes which ultimately move the energy to the lattice are thus masked, and this would be even more true with the super-collision process as it has the same temperature dependence (i.e., $p = 3$) as that found here.

3.6 Magnetophonon resonance

The Magnetophonon resonance occurs when the Landau levels in a high magnetic field are separated by an optical phonon energy. This does not have to be adjacent levels, and can occur when any pair of Landau levels are separated by $\hbar\omega_0$. The effect was first predicted by Gurevich and Firsov [60], and seen experimentally by Puri and Geballe [61] and Firsov *et al* [62]. The first significant review appeared already in 1968 [63]. It was first applied to the study of hot electron phenomena in

GaAs in 1968 [64]. Subsequently, studies of hot carriers appeared in CdTe [65], InAs [66], and InSb [67]. In near equilibrium situations, the amplitude of the magnetophonon resonance is much stronger for a transverse magnetic field than for a longitudinal magnetic field. Under warm and hot electron situations, however, is found that the two configurations yield about equal amplitudes [62]. A typical set of curves is shown in figure 3.33 for InP [68]. Notice that it is the second derivative of the magnetoresistance that is plotted, an early method of enhancing the oscillatory component of the signal.

In normal parabolic bands, the Landau levels are equally spaced in magnetic field and the various possible resonances for

$$\hbar\omega_0 = n\hbar\omega_c, \tag{3.31}$$

where ω_c is the cyclotron frequency, all give a set of fields that are linear. Using this linear set of resonances, one can get a good estimate of the effective mass, if the phonon energy is well known, or vice versa. However, in non-parabolic bands, the effective mass changes with the energy of the carriers, usually increasing as the energy increases [16]. This means that the cyclotron frequency depends upon the energy at which the Landau level lies, as the mass changes as one goes up the ladder of levels [64, 69–71]. This was used to measure the temperature variation of the effective mass in several semiconductors [72].

As remarked above, the role of hot carriers on the magnetophonon effect appeared soon after the original studies of the effect itself [64, 67]. It is generally found that the electric field causes a shift in the position of the magnetophonon

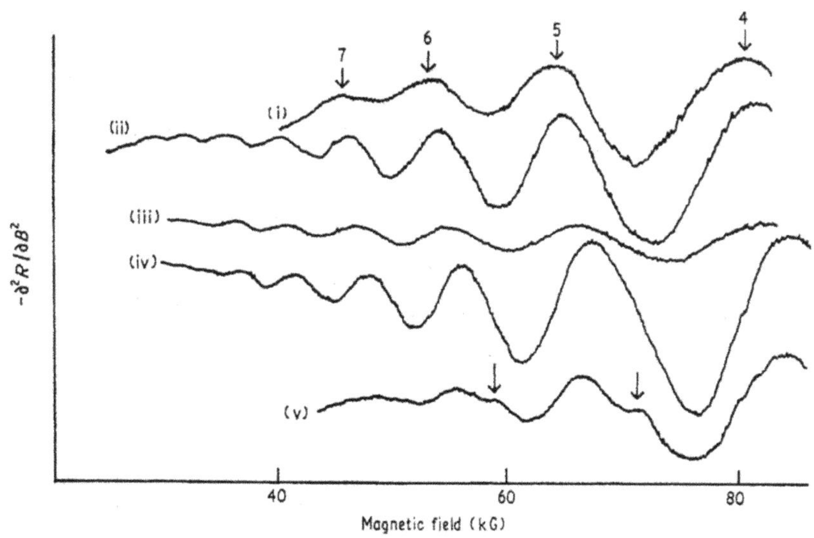

Figure 3.33. Magnetophonon resonances observed in InP for fields up to 90 kG (9 T). Curves (i), (ii), and (iii) are for the magnetic field transverse to the current for temperatures of 290, 175, and 85 K, respectively. Curves (iv) and (v) are for a longitudinal magnetic field and temperatues of 180 and 270 K, respectively. Reprinted from Eaves *et al* [68], with permission from IOP Publishing. All rights reserved.

resonances, presumably due to the change in the effective mass as the average energy of the carriers is raised. Studies in InSb at 77 K indicated that the $n = 1$ increased the resonant magnetic field by as much as 18% in a field of 50 V cm^{-1} [73]. Subsequent studies confirmed the increase in resonant magnetic field caused by the electric field [74, 75]. These studies gave comparable results as summarized by Ferry and Curby [76]. A comparison of the data is shown in figure 3.34 as a function of the electric field for the $n = 2$ and $n = 3$ resonances. The increase is clear and the agreement between the different groups is good. Later theoretical studies also led to the expectation that the resonance would shift quadratically in the field [77], although somewhat smaller exponents were found earlier [78]. This contrasted somewhat with the work of Mori [79], who suggested that there could be splitting in the resonance in which both a spin conserved and a spin–flip line would exist. Most experiments have not shown the spin–flip line.

Hot electron magnetophonon resonance has subsequently been studied in quantum wells at the heterostructure interfaces of GaAs/AlGaAs [80, 81]. In the quantum wells, the theoretical situation may be more difficult to understand, as there are studies which suggest a phase in the signal rather than a splitting [82].

Vaughan *et al* [83] observed magnetophonon resonances in studies of the Landau-level carrier lifetimes in heterostructures. Large changes in the energy relaxation times were observed for carriers within a Landau level, and attributed to oscillatory phonon emission when the resonance condition (3.31) was satisfied. The sample measured was an InAs/GaSb double heterojunction grown on GaAs. First, a 1 μm buffer layer was grown, followed by a thin layer InAs/GaSb superlattice followed by a 40 nm InAs layer. This was all capped with a 20 nm GaSb layer. The sample was cooled to 2 K and illuminated with a far infrared ps pulse from a tunable free

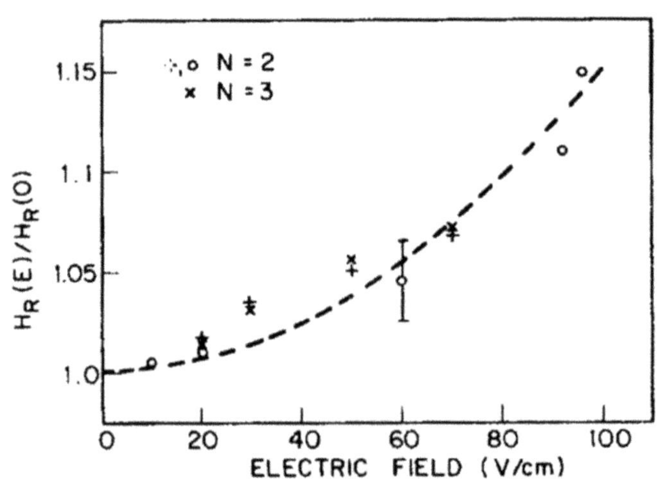

Figure 3.34. Position of the observed magnetophonon resonance as a function of the electric field. Data from [73] appears as circles, while the crosses come from [74, 75]. Typical error bars are shown for the signal at 60 V cm^{-1}. The dashed curve is an approximate quadratic variation. Reprinted from Ferry and Curby [76], copyright (1993) with permission from Elsevier.

electron laser in order to study the optical cyclotron resonance. Lifetimes could be measured thusly by observing the line width and shape and deducing the lifetime for a variety of excitation energies. The excitation would also create hot carriers in the study. It was also found that the energy dependent effective mass could be determined. Earlier studies had seen the oscillatory behavior in the energy relaxation times, but not determined the lifetimes directly [83].

References

[1] Fowler A B, Fang F F, Howard W E and Stiles P J 1966 *Phys. Rev. Lett.* **16** 901
[2] Fang F F and Fowler A B 1970 *Phys. Rev.* **169** 619
[3] Linke H, Bird J P and Cooper J *et al* 1997 *Phys. Rev.* B **56** 14937
[4] Prasad C, Ferry D K and Shailos A *et al* 2000 *Phys. Rev.* B **62** 15356
[5] Prasad C, Ferry D K, Vasileska D and Wieder H H 2003 *J. Vac. Sci. Technol.* B **21** 1936
[6] Bauer G 1974 *Springer Tracts in Modern Physics* vol. 74 (Berlin: Springer)
[7] Ryder E J 1953 *Phys. Rev.* **90** 766
[8] Morgan T N and Kelly C E 1965 *Phys. Rev.* **137** A1573
[9] Seeger K 1963 *Z. Phys.* **172** 68
[10] Hess K and Seeger K 1969 *Z. Phys.* **218** 431
[11] Fang F F and Fowler A B 1970 *J. Appl. Phys.* **41** 1825
[12] Krowne C M and Holm-Kennedy J W 1974 *Surf. Sci.* **46** 197
[13] Ferry D K 2020 *Transport in Semiconductor Mesoscopic Devices* 2nd edn (Bristol: IOP Publishing)
[14] Lee P A and Ramakrishnan T V 1985 *Rev. Mod. Phys.* **57** 287
[15] Newton P J, Mansell R and Holmes S N *et al* 2017 *Appl. Phys. Lett.* **110** 062101
[16] Ferry D K 1991 *Semiconductors* (New York: Macmillan)
[17] Kubo R 1957 *J. Phys. Soc. Jpn.* **12** 570
[18] Chakravarty S and Schmid A 1986 *Phys. Rep.* **140** 193
[19] Beenakker C W J and van Houten H 1988 *Phys. Rev.* B **38** 3232
[20] McCann E, Kechedzhi K and Fal'ko V I *et al* 2006 *Phys. Rev. Lett.* **97** 146805
[21] Hikami S, Larkin A I and Nagaoka Y 1980 *Prog. Theor. Phys.* **63** 707
[22] Sun Y, Zhang M and Dong L *et al* 2018 *AIP Adv.* **8** 045214
[23] Stöger G, Brunthaler G and Bauer G *et al* 1994 *Semicond. Sci. Technol.* **9** 765
[24] Brown C V, Geim A K and Foster T J *et al* 1993 *Phys. Rev.* B **47** 10935
[25] Ferry D K, Edwards G and Yamamoto K *et al* 1995 *Jpn. J. Appl. Phys.* **34** 4338
[26] Fukuyama H and Abrahams E 1983 *Phys. Rev.* B **27** 5976
[27] Altshuler B L, Aronov A G and Khmelnitzky D E 1982 *J. Phys. C: Sol. State Phys.* **15** 7367
[28] Golubev D S and Zaikin A D 1998 *Phys. Rev. Lett.* **81** 1074
[29] Ferry D K, Goodnick S M and Bird J P 2009 *Transport in Nanostructures* 2nd edn (Cambridge: Cambridge University Press)
[30] Linke H, Bird J P and Cooper J *et al* 1997 *Phys. Rev.* B **56** 14937
[31] Prasad C, Ferry D K and Shailos J *et al* 2000 *Phys. Rev.* B **62** 15356
[32] Price P J 1982 *J. Appl. Phys.* **53** 6863
[33] Ma Y, Fletcher R and Zaremba E *et al* 1991 *Phys. Rev.* B **43** 9033
[34] Shinba Y, Nakamura K, Fukuchi M and Sakata M 1982 *J. Phys. Soc. Jpn.* **51** 157
[35] Pivin D P Jr, Andresen A, Bird J P and Ferry D K 1999 *Phys. Rev. Lett.* **82** 4687
[36] Bird J P, Ishibashi K and Ferry D K *et al* 1995 *Phys. Rev.* B **51** 18037

[37] Pivin D P Jr, Andresen A and Bird J P *et al* 2000 *Physica* E **6** 404
[38] Pivin D P Jr, Andresen A, Bird J P and Ferry D K 1999 *Physica* B **272** 72
[39] Prasad C, Ferry D K and Shailos A *et al* 2001 *Phys. Rev.* B **62** 15356
[40] Prasad C, Ferry D K and Wieder H H 2003 *Superlatt. Microstruc.* **34** 475
[41] Prasad C, Ferry D K and Wieder H H 2004 *Semicond. Sci. Technol.* **19** S60
[42] Williams B F and Simon R E 1967 *Phys. Rev. Lett.* **18** 485
[43] Eden R C, Moll J L and Spicer W E 1967 *Phys. Rev. Lett.* **18** 597
[44] Ichihashi F, Tawaguchi T and Dong X *et al* 2017 *AIP Adv.* **7** 115314
[45] Scheer J J and van Laar J 1965 *Sol. State Commun.* **3** 189
[46] Schulman J N and McGill T C 1979 *J. Vac. Sci. Technol.* **16** 1513
[47] Rahim A, Gusev G M and Kvon C D *et al* 2019 *Microelectron. Engr.* **206** 55
[48] Ardali S, Taganov S, Erol A and Tiras E 2021 *Physica E* **125** 114344
[49] Wang S, Wu Z and Haifeng Z *et al* 2019 *Sol. State Commun.* **288** 68
[50] Barker J M, Ferry D K, Koleske D D and Shul R J 2005 *J. Appl. Phys.* **97** 063705
[51] Somphonsane R, Ramamoorphy H and Bohra G *et al* 2013 *Nano Lett.* **13** 4305
[52] Bistritzer R and MacDonald A H 2009 *Phys. Rev. Lett.* **102** 206410
[53] Kubakaddi S 2009 *Phys. Rev.* B **87** 405414
[54] Tse W K and das Sarma S 2009 *Phys. Rev.* B **79** 205404
[55] Baker A M R, Alexander-Webber J A and Altabaeumer T *et al* 2013 *Phys. Rev.* B **87** 045414
[56] Huang J, Alexander-Webber J A and Janssen T J B M *et al* 2015 *J. Phys. Cond. Matt.* **27** 164202
[57] Song J C W, Reizer M and Levitov L S 2012 *Phys. Rev. Lett.* **109** 106602
[58] Ferry D K, Somphonsane R, Ramamoorthy H and Bird J P 2015 *Appl. Phys. Lett.* **107** 262103
[59] Ferry D K, Somphonsane R, Ramamoorthy H and Bird J P 2015 *J. Comp. Electron.* **15** 144
[60] Gurevich V L and Firsov Y A 1961 *Sov. Phys.—Sol. State* **4** 1331
[61] Puri S M and Geballe T H 1963 *Bull. Am. Phys. Soc.* **8** 309
[62] Firsov Y A, Gurevich V L, Parfen'ev R V and Shalyt S S 1964 *Phys. Rev. Lett.* **12** 660
[63] Stradling R A and Wood R A 1968 *J. Phys. C: Sol. State Phys.* **1** 1711
[64] Stradling R A and Wood R A 1968 *Sol. State Commun.* **6** 701
[65] Mears A L, Stradling R A and Inall E K 1968 *J. Phys. C: Sol. State Phys.* **1** 811
[66] Wood R A and Stradling R A 1969 Proc. Int. Conf. High Magn. Fields and Their Appl. (London: Institute of Physics and Physical Society) p 47
[67] Stradling R A and Wood R A 1968 *J. Phys. C: Sol. State Phys.* **3** 2425
[68] Eaves L, Stradling R A and Askenazy S *et al J. Phys. C: Sol. State Phys.* **4** L42
[69] Eaves L, Stradling R A and Askenazy S *et al J. Phys. C: Sol. State Phys.* **5** L19
[70] Kuchar F and Veigl K 1976 *J. Phys. C: Sol. State Phys.* **9** 2797
[71] Kuchar F 1977 *Z. Phys.* **26** 223
[72] Hazama H, Sugimasa T, Imachi T and Hamaguchi C 1986 *J. Phys. Soc. Jpn.* **55** 1282
[73] Curby R C and Ferry D K 1972 Proc. 11th Int. Conf. Phys. Semicond. (Warsaw) (Sci. Publ.) p 312
[74] Ito M, Shirakawa T, Hamaguchi C and Nakai J 1975 *Sol. State Commun.* **17** 717
[75] Shirakawa T, Ito M, Kasai K and Hamaguchi C 1976 *Phys. Lett.* **58A** 123
[76] Ferry D K and Curby R C 1977 *Phys. Lett.* **59A** 471
[77] Suzuki A 1993 *Physica* B **183** 191
[78] Barker J R 1978 *Sol.-State Electron.* **21** 197

[79] Mori N, Nakamura N, Taniguchi T and Hamaguchi C 1988 *J. Phys. Soc. Jpn.* **57** 205
[80] Inoue M, Hida H and Inayama M *et al* 1983 *Physica* **117B/118B** 720
[81] Leadley D R, Brummell M A and Nicholas R J *et al* 1988 *Sol.-State Electron.* **31** 781
[82] Warmenbol P, Peeters F M and Devreese J T 1989 *Sol.-State Electron.* **32** 1545
[83] Vaughan T A, Nicholas R J and Langerak C J G M *et al* 1996 *Phys. Rev.* B **53** 16481
[84] Hawker P, Kent A J and Challis L J *et al* 1989 *J. Phys. Cond. Matt.* **1** 1153

IOP Publishing

Hot Carriers in Semiconductors

David K Ferry

Chapter 4

Optical carrier heating

The properties of photo-excited excess carriers in semiconductors are some of the more interesting aspects of nano-electronics, and indeed of nearly all optically motivated devices today, from the vast and growing field of solar cells to the advanced lasers. The role of the excess carriers is crucial to the operation of such devices. In this chapter, the role of the optical created, nonequilibrium carriers is discussed. The hot carriers created optically are quite different than those created by an electric field and which we studied in chapters 2 and 3. The photoexcited carriers still create a hot distribution. In this sense, the electron system is no longer in thermal equilibrium, especially with the lattice, and one has already made an assumption that a temperature can describe the electron system. However, this distribution is still largely symmetric in momentum space, as the photon momentum is quite small compared to electron momenta. Nevertheless, there may arise some asymmetry to the distribution that arises from the photons themselves.

One approach to the study of optical carrier heating arises from pump–probe experiments in which an intense photon beam excites the carriers from the valence band into the conduction, leaving a dense electron–hole plasma. Then, a second, weaker beam is delayed from the excitation pump and used to explore the properties of the electron–hole system as a function of time after the pump. The photons that excite the electron–hole pairs arrive with an energy $\hbar\omega$, and a momentum toward the semiconductor of $\hbar k = \hbar\omega/c$, where c is the speed of light. That is, the photons arrive at the semiconductor basically with a velocity that is the speed of light, even though their momentum is small. When the photon enters the semiconductor, it is slowed by the dielectric permittivity to $c/\sqrt{\epsilon_\infty}$, where ϵ_∞ is the optical dielectric 'constant.' It is this velocity that leads to the peak of absorption being away from the actual surface upon which the photons impinge. And, this velocity can lead to an asymmetry in the distribution function. This distribution function created by the incident photons has an average energy given by the excess photon energy, and an average velocity away from the surface given by this photon velocity inside the semiconductor. It is not Maxwellian, as carriers which would have a velocity toward the surface are absent in this distribution (one can think about their being reflected at the surface). So, this gives several time scales in the optical problem.

The first important time scale is the time it takes to generate an electron–hole pair by the photon. This is not an instantaneous event. Knowledge of the photo-generated distribution functions of the electrons and holes is inadequate to describe the full range of coherent interactions that are crucial to the temporal behavior of the photon absorption. One reason for this is that information about the inter-band polarization is required. This polarization is created by the coherent interaction of the photon's fields with the wave functions in the conduction and valence band. In fact, the absorption process itself is initiated with the polarization of the electron–hole wave functions, which is then broken up by the final transition of the bound electron from the valence band to the conduction band. This is symbolized by the square magnitude of the optical matrix element that appears in the Fermi golden rule for the absorption process. To describe the polarization, a full quantum mechanical approach is required. There are a variety of such approaches, but most approaches work with the real-time Green's function method [1–3]. We will not dwell upon the approach at this time, but with the physics that occurs.

As was mentioned, the creation of an electron–hole pair by the absorption of a photon is not an instantaneous event. While the photon is envisioned as a particle, it is the electromagnetic wave aspect that is important in creating the electron–hole pair. Whether we are dealing with a semiconductor or an organic molecule the process of creating an electron–hole pair (or an exciton) is much the same. Perhaps the earliest study in this regard was by Geltman, who wished to find out just how long it took to ionize an atom [4]. In ionizing an atom from an incident electromagnetic field (photon), the field first creates a correlation between the initial (ground) and final (excited) states; this in turn may result in Rabi oscillations between these two states [5]. In other words, the unoccupied quantum state in the conduction band becomes entangled with the ground state (electron) in the valence band. This correlation is referred to as the polarization. The first step of the electromagnetic wave of the photon is to create this polarization. The conjugate field then destroys this correlation, resulting in the completed process. The electron might remain in the valence band with the photon going on its way. However, in semiconductors, it usually results in the electron transitioning from the valence band to the conduction band, ultimately creating the electron and hole (or the condensed exciton). Although the electromagnetic field (photon) creates the interaction, in an energy-conserving process it actually is the energy of the absorbed and annihilated photon that is transferred in creating the electron–hole pair. The entire process takes only a few femtoseconds, and is governed by the semiconductor Bloch equations [6–8]. This process is basically coherent, even when the excitation is incoherent such as that from solar radiation. It is important to point out that the ephemeral coherence discussed here in the context of photon absorption, for example in a photovoltaic cell, does not persist once the electron–hole pairs (or excitons) have been created by incoherent and polychromatic sunlight.

While the creation of the electron–hole pair takes a few femtoseconds, the thermalization of the two distribution functions can take a few tens of femtoseconds. This latter is the time scale upon which a so-called 'hot' Maxwellian (or Fermi–Dirac) distribution forms by interactions among the electrons or among the holes, but can also involve electron–hole interactions. However, this thermalization can be interrupted if the excitation energy lies above the edges of the satellite valleys in the conduction band. This is because the inter-valley scattering is also a relatively fast process that occurs in a few tens of femtoseconds (fs) [9]. We can illustrate this with results of very fast optical pump–probe experiments. By the 1990s, the development of lasers with pulse durations as short as 6 fs made it possible to probe very fast nonequilibrium properties of the electron–hole plasma in semiconductors [10]. More progress has been made since that time. But, consider the properties of a 6 fs pulse at 2.0 eV. The frequency of this laser is 4.83×10^{14} Hz, and the period of the oscillation is 0.33 fs. Thus, the 6 fs pulse has less than 20 oscillations of the signal. The 6 ps pulse mentioned was generated by a complicated nonlinear process. The base system uses optical pulses in a colliding pulse mode-locked laser with a set of intracavity prisms [11]. These pulses are amplified in a copper-vapor laser amplifier. The prisms formed a four prism sequence and the signal was then doubled in a KDP crystal to minimize distortion. The value of 6 fs was determined from an interferometric autocorrelation of the pulse with itself. In subsequent work, the signal was split into two, for use in a pump–probe experiment on the transmission of GaAs [12]. By convolving the transmitted signal of the probe with the exciting pulse, and subtracting from the former, the transmittance signal could be identified. Here, it was found that the signal had a fast decay with relaxation time of 33 fs. In distinction from earlier thoughts that this was the cascade time, it was determined that this is the inter-valley transition time for the electrons to scatter from the Γ valley to the L valleys.

While the above is interesting, it is possible to see the inter-valley transfer in a more pictorial manner. Another approach to fs resolution is time-resolved angle-resolved photoemission spectroscopy (TR-ARPES) [13–15], which is the subject of a recent review [16], focused on the electron–phonon interaction. Here, for example, GaAs was probed with ultrafast pulses with photon energies between 1.6 and 2.4 eV, and then probed with TR-ARPES [15]. The electrons injected into the conduction band will equilibrate between themselves in momentum, but not in energy, in a few tens of fs. Relaxation of the carriers is on a much longer time scale and is governed by the interaction of the electrons with a variety of phonons, especially for photo-electron densities less than 5×10^{17} cm^{-3}. The results of experiments on GaAs are shown in figure 4.1. The scattering angle is proportional to the momentum. In panel (a), one sees the initial excitation of the electrons near the Γ point, but already evolving onto the proper band structure in the Γ valley. The solid and dashed white lines are the energy bands in the directions toward L and X. In panel (b), a larger slice of the Brillouin zone is shown with the L valleys, which are already being populated at 20 fs after the pump pulse. Populations, as a function of energy, are shown in panels (c) and (d) at both 20 fs and 70 fs after the pump. Panels (e)–(i) show the temporal distributions for these two valleys at several particular energies (all measured from the conduction

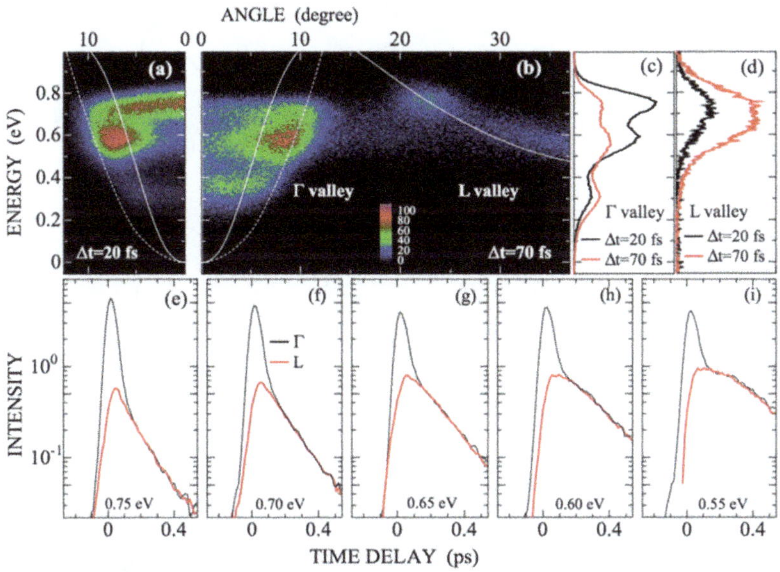

Figure 4.1. (a) Photoemission image of hot electron distribution in the Γ valley mapped in energy and momentum space, 20 fs after excitation by *s*-polarized 2.3 eV photons. (b) Image of the hot electron distribution 70 fs after excitation showing carriers in the L valley as well ($\theta > 13°$). In (a) and (b), the color scale indicates the photoemission intensity, and all energies are referenced to the conduction band minimum. The solid and dashed curves are the energy dispersion along the Γ-L and Γ-X directions, respectively. The angle-integrated, energy resolved distributions, 20 and 70 fs after excitation, are shown in the (c) Γ valley and (d) L valleys. Even at 20 fs, the population of the L valley is not negligible because of finite pulse widths and ultrafast scattering rates. (e)–(i) Time-resolved populations of Γ and L valleys at specific energies in the bands. The populations of the L valleys have been multiplied by a scale constant to bring them to the same order of magnitude as those of the Γ valleys. Reprinted from Tanimura *et al* [15], copyright (2016) by the American Physical Society.

band minimum at Γ). This provides a clear illustration of the scattering to the satellite valleys.

In the following sections, a variety of hot carrier effects that occur in semiconductors will be studied. The oscillatory photoconductive effect will be discussed in the next section. Then intra-valley absorption and higher energy generation of electron–hole pairs will be addressed. The following sections will deal with ultrafast excitation especially in quantum wells and heterojunctions will be addressed.

4.1 Oscillatory photoconductivity

One of the earliest experiments for studying the effect of optically generated carriers at energies above that of the thermal background is known as oscillatory photoconductivity. Apparently, the first report was on low temperature studies of InSb [17, 18]. These early experiments were conducted on doped InSb, and studied with far infrared excitation. By sweeping the photon energy, they observed a series of transitions with the longest wavelength (lowest energy) being supposedly due to the excitation of an electron from the valence band to the acceptor level (Cu, Ag, Au

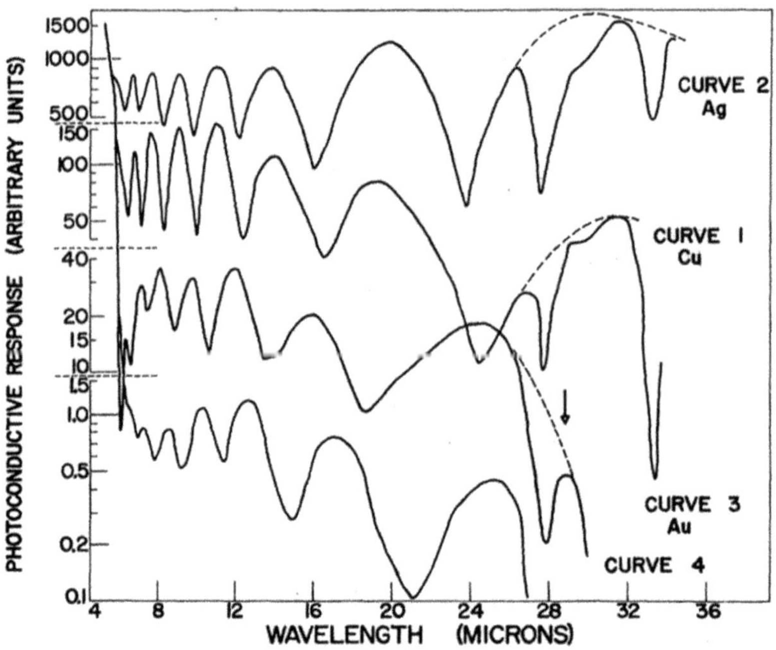

Figure 4.2. Extrinsic photo-response of InSb at low temperature. Reprinted with permission from Engeler *et al* [17], copyright (1961) by the American Physical Society.

acceptors were used), with subsequent resonances corresponding to the transition plus the emission of 1, 2, 3, … phonons. The corresponding photoconductive response is shown in figure 4.2. By measuring the position of the various peaks as a function of the transition number, they were able to associate the oscillation with the known energy of the longitudinal optical phonon at 24.4 meV (at 5 K). In some sense, the oscillatory photoconductivity is the analog of the Hickmott device discussed in section 2.6, where the phonon cascade was observed in tunneling devices.

Direct transitions from the valence band to the conduction were then observed using *p*-type GaSb and InSb [19], as well as in CdS [20]. The semiconductor is excited by an optical source of variable photon energy. As the energy is increased beyond the band gap, the photoconductivity is observed to again oscillate with a period given by the longitudinal optical phonon. It is thought that hot, or at least warm, electrons are involved and the minima in conductivity occur whenever the photo-generated carrier energy reaches the level to emit another optical phonon. Thus, the photo-generated carriers create a phonon cascade as their energy is raised, as indicated in figure 4.3. The green arrow shows the excitation of the electron–hole pair with a photon that has energy $\hbar\omega > E_G$, where E_G is the direct band gap. As many as 15 oscillations have been observed in GaAs at low temperature [21].

It was later found that the oscillatory photoconductivity was essentially the same in either intrinsic or extrinsic InSb [22, 23]. The effect was observed to gradually wash out for fields above ~20 V cm^{-1} at 7 K. It was subsequently seen also for ZnTe

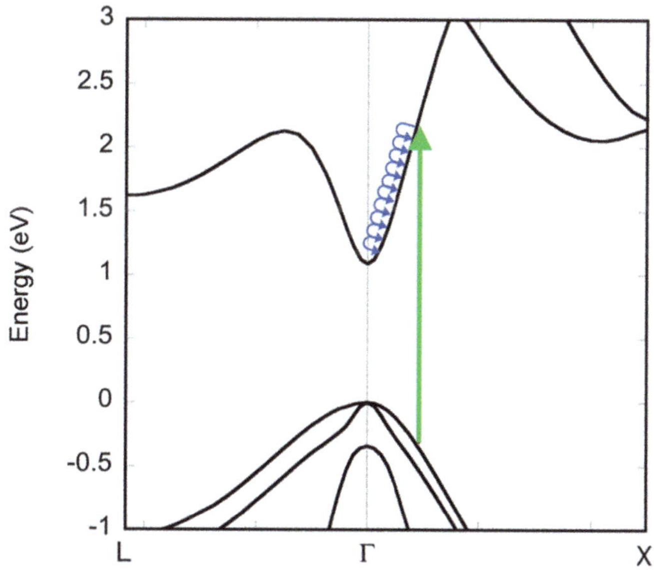

Figure 4.3. Schematic diagram of oscillatory photoconductivity. The photon creates electron–hole pairs, as shown by the green arrow. If, for example, the electron has sufficient energy, it can emit a cascade of phonons (blue arrows). The conductivity will have a local minimum at each energy at which an additional phonon can be emitted, and local maxima between these energies.

[24] and GaAs [25], and it was later shown that the effect would persist even with a relatively high magnetic field [26]. Oscillatory photoconductivity has been reported in the presence of traps in other materials and has also been seen in diamond [27, 28], InP [29], CdTe [30], and BN [31]. Even surface phonons have been observed in ZnO [32] and GaAs [33], and in *p-n* junctions in InSb [34].

Theoretical studies using ensemble Monte Carlo techniques confirmed early on that the response, as expected, arose from the cascade of optical phonons and that the oscillatory photoconductivity should show negative conductivity as well [35, 36]. However, any negative conductivity state is unstable against electric field inhomogeneities that can form, thus masking it from the experiment [37].

In later work, it was shown that using two-photon spectroscopy, where the two photons together had sufficient energy between them to traverse the band gap, also showed the expected oscillatory photoconductivity [38]. Studies in superlattices of GaAs/AlGaAs also exhibited the oscillatory photoconductivity from confined LO phonon modes in the quantum wells at low temperatures [39].

4.2 Free-carrier optics

For incident high-frequency waves of energy insufficient to produce an electron–hole pair, such as microwave radiation, the absorption of the photon can occur either due to the phonons or to free carriers. Here, the absorption of the microwave photon by the carriers is the point of issue. This absorption can only occur in a partially filled band, where there are free electrons or holes capable of gaining energy from the

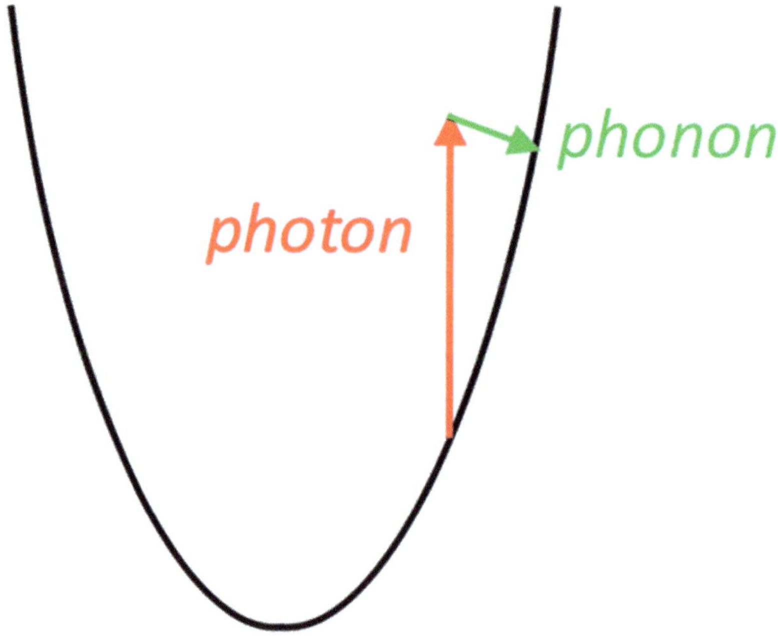

Figure 4.4. Schematic drawing of a free-carrier absorption of a microwave signal. The photon transition is almost vertical, so must be aided by a phonon. The phonon shown here is an emission process, although absorption is also possible. This second-order process is required to conserve both energy and momentum in matching the band structure of the carriers (shown here as electrons).

microwave field, although this absorption also requires scattering processes as well in order to conserve both energy and momentum. Moreover, this is an indirect process, whereby the transition must involve a phonon in order to conserve both energy and momentum, as shown in figure 4.4. This process will lead to the microwave conductivity, although this 'microwave' frequency may well extend into the infrared region, as we have already discussed somewhat in section 2.5. Here, the discussion will go a little deeper.

As was noted already in section 2.5, the idea of studying carriers at microwave frequencies is quite old, and there are two approaches that provide interesting results. First, a small microwave signal is used to monitor the conductance as a function of the high electric field. For example, replacing the microwave frequency with an optical frequency leads to the curve in figure 2.8(b), where the transient velocity in InP appears to peak at a delay of some 300 fs. This would correspond to a peak in the conductance of the material at a frequency of some 6.7 THz. This peak will also transfer into a peak in the frequency spectrum of the conductance of an InP sample measured, for example, with Fourier Transform Infrared Spectroscopy, in order to reach into the THz regime, which is usually hard to measure. Second, one can use very high power microwave, or THz, signals to excite hot carriers in semiconductors and then measuring the properties with small dc fields or with another optical signal. One approach for this latter case is, of course, free-carrier absorption in high microwave fields. It is this latter approach which is of more interest here, as the interest is in photon heating of the carriers.

Early measurements of microwave heating of carriers in solids used pulsed magnetrons for the exciting source and perhaps a small electric field to monitor the conductance. This was done for direct carrier heating [40] and to study the Hall fields in a magnetic field [41], both in GaAs as a method to gain more information about the Gunn effect. Later, Hess *et al* [42] used strictly high power microwaves to study hot carriers in Ge, directly measuring the incident, reflected and transmitted power levels, deducing the sample impedance from the known electric field strength and the absorbed power. Surprisingly, it was found in all these cases that the heating and mobility were determined to be approximately the same as from dc electric fields. Subsequently, very high frequency (890 GHz) from a submillimeter laser was used to probe the conductivity in GaAs and InP in order to determine directly the energy and momentum relaxation times by probing closer to the frequencies corresponding to these relaxation times [43].

As remarked, the primary discussion of microwave heating has already been made in the earlier section. Here, the concept is to turn to the generation and use of THz signals in the study of hot, and warm, carriers in semiconductors. This area has developed since the advent of quantum well and superlattice systems. The idea of using multiple quantum wells and superlattices is actually quite old. Peierls discussed a general distortional instability that would open minigaps in the conduction electron spectrum [44]. The general idea was that if the product of the lattice constant and the Fermi vector were an integer, where twice the Fermi wave vector spanned the Brillouin zone, a structural instability would occur which led to the minigaps, as the distortion would produce a superlattice of period twice the lattice spacing (or more). This structural instability was later sought through the concept of charge density waves, particularly in two-dimensional systems [45]. Later Kazarinov and Suris suggested that in a sufficiently high magnetic field, instabilities would evolve due to the possibility of negative mass in the normal lattice [46], although it was realized that the magnetic field required was beyond that available by orders of magnitude. But, it was recognized that, if one could produce superlattices, with new lattice constants much larger than the normal one, such negative conductance behaviors might be obtainable [47]. The ideas, of course, led to modern growth techniques such as molecular beam epitaxy and metal-organic CVD. Nevertheless, ideas eventually led to studies of lateral surface superlattices [48] and the development of the quantum cascade laser [49]. It should be noticed that the latter devices operate in the far infrared or THz spectrum, so that generation of THz signals can be expected from such multiple quantum wells or superlattices.

A major step in this direction lies in the observation of THz signals from multiple quantum wells [50]. This was accomplished using superlattices grown in the GaAs/AlGaAs system, with dimensions such that at least two sub-bands lay below the threshold for optical phonon emission. Hence, loss of energy to the phonons was greatly inhibited and an applied electric field normal to the planes of the quantum wells would accelerate carriers from the lower sub-band in one well to the second sub-band in the next well, whereupon it would relax by photon emission. In the experiments, the superlattice allowed for electrons from the lowest sub-band in one well to reach the fourth level in the adjacent well, and that would lead to multiple photon emission as the electron relaxed to the lowest level. Signals in the few tens of meV were observed.

4.2.1 Terahertz studies

While the superlattice can produce such THz signals, which may be used for probing other semiconductor properties, one can also produce THz signals by more exotic techniques. Terahertz frequencies are routinely generated by using femtosecond pulses of intense lasers and using nonlinear media to down covert these pulses to the desired frequency range in the time domain. Of course, such frequencies can also be generated by the above mentioned quantum cascade lasers, by synchrotron sources, and so on, but these do not tend to be very short pulses usable for study of dynamic processes. For short-pulse work, one generally uses, for example, a Ti:sapphire laser producing pulses of a few tens of fs to perhaps a few hundred fs. These pulses are split into a variety of beams, one to the pump the sample, and perhaps two others two generate THz signal through nonlinear media that provide the excitation and detection of the time-resolved THz signals [51]. Such approaches have been used to probe bulk semiconductors since at least 2000 [52].

As pointed out already in earlier chapters, field-effect transistors can be used to amplify THz signals, and this has led to the highest performance transistors [53, 54]. In studying the use of FETs for this purpose, there are two important times that characterized the performance. The first is the cutoff frequency f_T given as

$$f_T = \frac{1}{2\pi\tau_t}, \tag{4.1}$$

where τ_t is the transit time through the effective gate length [55]. The second frequency is called the unity gain frequency, or $f_{\max}$, which is a little harder to quantify, as it depends upon the circuit in which the transistor is embedded [56]. The lower of these two frequencies usually determines the highest frequency at which the transistor can provide gain. Optimized design seeks to have $f_{\max} > f_T$. In a sense, exciting THz signals in these FETs is one manner in which semiconductor material can be examined, especially at these frequencies.

Semiconductors which exist as monolayers have become of interest for a variety of applications, including FETs. These materials include graphene, which was the first such material studied for its electronic properties [57]. The transition metal di-chalcogenides (TMDCs) have been studied longer, but the early work was mainly focused on charge (and spin) density waves, as discussed above [45]. However, the lack of a band gap in graphene has led to much more recent interest in the TMDCs for electronic applications [58]. Many of these latter materials have relatively high band gaps of more than 2 eV. Consequently, they can have excitons (correlated electron–hole pairs) even at room temperature. As will be discussed in a later section, these materials can be probed by the normal optical techniques such a fast pulse pump and probe techniques. Here, the excitation pulse and a delayed probe pulse are both in the optical region $\hbar\omega > E_G$. Such techniques can give good information on the excitation dynamics, and the resulting phonon dynamics, but they have trouble dealing with the carrier conduction in either of the two pulses. Here is where time-resolved THz spectroscopy is quite useful, as it monitors the carrier transport properties directly.

An example of the use of THz spectroscopy is given in figure 4.5 [59]. In this study, the sample of MoS_2 was studied with both optical pump–probe techniques and THz spectroscopy. For the former, a Ti:sapphire laser, with a 150 fs pulse, was used to generate the pump and probe signals. A portion of the laser was passed through an optical parametric amplifier with provides *tunable* pump pulses, while a second part was delayed and passed through a sapphire plate, providing a nonlinear medium to produce output over most of the visible spectrum (often called a *white light continuum*). This latter signal is passed through a monochromator to record the excited state spectrum. The THz spectroscopy is obtained by using a second Ti:sapphire laser of somewhat higher power, which is also split into two signals. The first goes through an optical parametric amplifier to produce the tunable pump signals, while the second pulse is focused on a nonlinear medium, here a laser induced air-plasma filament to mix the fundamental and second harmonic, which leads to the THz signal. This latter produces a broad band tunable THz signal in the 0.1–8 THz range. The experiments study the optical transmission through the sample as well as the THz photoconductivity of the sample. We can see the results in figure 4.5, where various responses are shown. The authors assigned a large part of

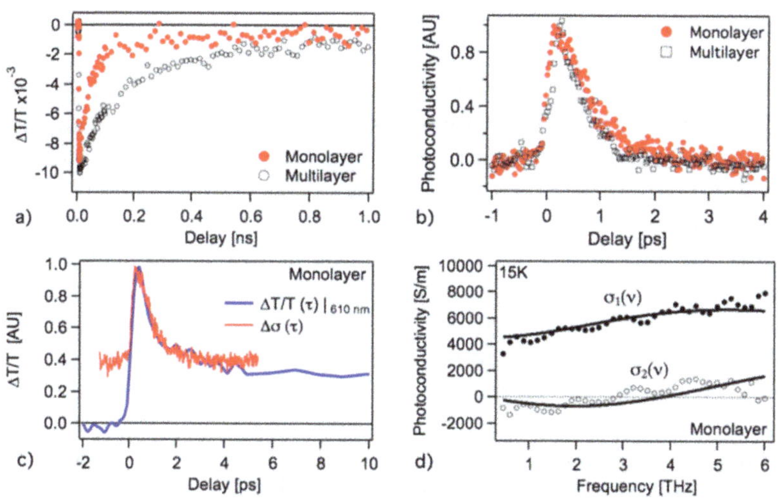

Figure 4.5. (a) Excited state dynamics of multi- (black) and monolayer (red) MoS_2 with a pump at 532 nm and a probe at 490 nm. The plot gives the change in transmission through the sample. (b) The photoconductivity, measured with THz spectroscopy, for the multi- (black) and monolayer (red) material. (c) Comparison of the photoconductivity change and the transmission change. (d) Real (σ_1) and imaginary (σ_2) parts of the photoconductivity as a function of frequency. The solid lines result from a simple one parameter theory of conductivity. Reprinted with permission from Cunningham *et al* [59], copyright (2016) by the American Chemical Society.

the results to charge trapping in localized mid-gap states associated with grain boundaries in the material. However, it is known from other work that chalcogenide vacancies are quite common and lead to both mid-gap states and localization of the carriers due to the random potential produced by the vacancies [60]. So, the states seen in the experiments above may arise from a variety of sources.

THz spectroscopy has also been used to study carriers produced by high energy photon excitation. By high energy photons, it is meant the electron and/or hole are produced high enough in their respective bands to cause additional carriers to be produced by impact ionization [61]. In this work, InN with a band gap of ~0.7 eV was probed with high energy photons over a range of energies. When the femtosecond photon energy was increased to 2.48 eV or higher, carrier multiplication was observed to begin. These carriers were then studied with ultrafast THz signals in the 0–2 THz range, and the temporal response of the conductivity studied. Similar multi-source experiments have examined, for example, graphene [62], WS_2 [63], and ZnTe [64], as well as other semiconductors.

4.2.2 Hot carriers and the quantum Hall effect

The application of high magnetic fields to semiconductors is an invaluable degree of freedom available to the experimentalist in probing the system, and also to the theoretician. Magnetic fields give rise to new fundamental behaviors not observed in bulk-like systems, such as the quantum Hall effect (QHE) [65]. The fundamental quantity characterizing a magnetic field is the magnetic flux density, B, which in the current International System of Units is measured in Tesla (1 Tesla = 10 kG, for older scientists). In the study of semiconductor mesoscopic devices, low fields usually correspond to fields less than 1 T, which is the regime in which low-field magneto-transport experiments such as Hall effect measurements are usually performed. Magnetic intensities of 5–20 T are obtainable in the usual university and commercial research environments, using superconducting alloy coils with high critical magnetic fields (superconductivity can be destroyed by a high magnetic field) immersed in a liquid He dewar.

When free particles are subject to a magnetic field, whether they are free or in a semiconductor quantum well, they experience a Lorentz force which tends to make the particles move in a circular path in the plane normal to the magnetic field. Since the force is always perpendicular to the direction of travel of the particle, the circular motion has an angular frequency given by the cyclotron frequency, given as $\omega_c = eB/m^*$ in terms of the magnetic field and the carrier's effective mass. When the magnetic field is sufficiently high, $\hbar\omega_c > k_B T$ and $\omega_c\tau \gg 1$, the energy corresponding to motion transverse to the magnetic field becomes quantized into Landau levels [66]. This transverse energy is then given by

$$E_n = \left(n + \frac{1}{2}\right)\hbar\omega_c, \quad n = 0, 1, 2 \ldots. \tag{4.2}$$

As the electron is traveling in a closed circular orbit, we can refer to the radius of this orbit as the cyclotron radius

$$r_c = \sqrt{\left(n + \frac{1}{2}\right)\frac{2\hbar}{eB}} = k_{\mathrm{F}}\frac{\hbar}{eB} \equiv k_{\mathrm{F}} l_B^2, \tag{4.3}$$

where k_{F} is the Fermi wave vector ($=\sqrt{2\pi n_s}$) in a quasi-two-dimensional system. The radius is related to the magnetic length through $l_B = \sqrt{\hbar/eB}$. In such a quasi-two-dimensional system, such as a quantum well or a heterostructure, motion in the transverse plane leads to conductivity modulation, or Shubnikov–de Haas oscillations, as shown previously in figure 3.9.

Normally, the QHE is not a hot carrier, or even warm carrier effect, but there are a number of studies which go beyond the standard experiments, such as those of figure 3.10. And these do involve warm, or hot, carriers. Some of these effects will be discussed here.

4.2.2.1 Breakdown of the QHE

Almost as soon as the discovery of the QHE became known throughout the community, it was found that a breakdown could occur. As is well known, in the QHE regime, there is no measurable longitudinal (parallel to the current) electric field, and the field that does exist is the Hall field that is normal to the current (directed across the typical QHE sample). While there is normally no Hall current, the buildup of the Hall field can lead to the breakdown. In studies in Germany, it was found that a breakdown occurred at Hall electric fields of ~80–90 V cm^{-1} [67], above which a longitudinal field would begin to develop. Above breakdown, the longitudinal conductivity is seen to increase with the longitudinal electric field and with temperature. The experiments were performed on a GaAs/AlGaAs heterostructure with a low carrier density (2.2–4.3 × 10^{11} cm^{-2}), so that these carriers were all in the lowest Landau level (n = 0) at 4.7 T. Typical measurements are shown in figure 4.6. The authors discuss the results in terms of an electron temperature model and a energy loss rate. However, the loss rate is zero below the breakdown field, as the electric field is normal to the current. Nevertheless, these d.c. fields, and the onset of breakdown, have many of the characteristics of bulk breakdown discussed in chapter 2.

The breakdown effect was almost immediately confirmed with studies in the U.S. Here, almost similar GaAs/AlGaAs structures were used, but with somewhat higher carrier densities, so that the measurements were carried out in the $\nu = 4$ plateau (B ~ 5.6 T) at 1.1 K [68]. The measurements exhibited a significant jump in longitudinal voltage when the current through the device exceeded ~340 μA. This was accompanied by a drop in the Hall resistance as the longitudinal voltage increased. The measurements were interpreted in terms of an initial localized breakdown which then grew. At the $\nu = 4$ plateau, the Hall resistance is ~6452 ohms, so that the Hall voltage is about 2.2 V, and for the given sample size would lead to a Hall field of ~58 V cm^{-1}. This is lower than that found in the Ebert *et al* measurements, but of the same order of magnitude.

Early theories to explain the breakdown were not very successful, perhaps because few of the involved scientists were acquainted with the early work on breakdown in materials that had started the interest in hot carriers [69, 70]. One

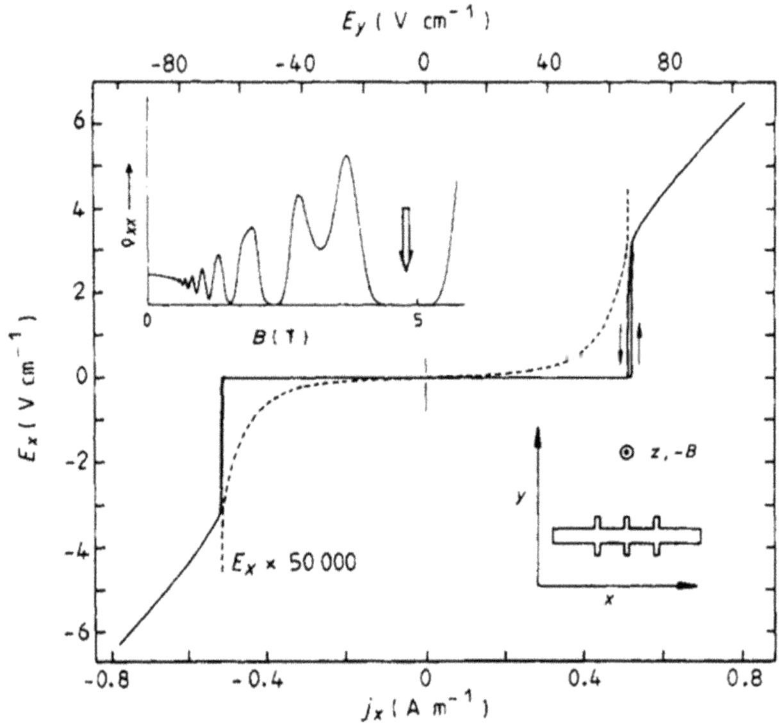

Figure 4.6. Current–voltage characteristics of a $GaAs/Al_{0.28}Ga_{0.72}As$ heterostructure with carrier density of 2.2×10^{11} cm^{-3} at 1.4 K. The device geometry and Shubnikov–de Haas curves are shown in the lower and upper insets, respectively. For the dashed curve, the E_x signal has been multiplied by 5×10^4. The Hall electric field is the top axis and the corresponding current is the lower axis. Reprinted with permission from Ebert *et al* [67], copyright by IOP Publishing. All rights reserved.

approach was a standard Boltzmann equation derivation that failed to fit the experimental details, in particular the sharp onset of breakdown [71]. A second assumed that it occurred due to the onset of acoustic emission whenever the carrier velocity in the edge states exceeded the sound velocity [72]. This was in spite of the fact that the Fermi velocity for a two-dimensional density of 2.2×10^{11} cm^{-2} is about 2×10^7 cm s^{-1} while the sound velocity is of the order of 3–5 $\times 10^5$ cm s^{-1}. This would exclude the QHE in most cases. A third explanation asserted a peculiar S-shaped current–voltage curve that was unique to the QHE [73]. These authors were also apparently unaware of the thermodynamic argument that such S-shaped curves were expected in most breakdown situations [74]. As discussed in chapter 2, the S-shaped curve was likely to arise once the breakdown occurred, but one still needed the source of the breakdown. In the defense of these latter authors, they proceeded to study experimentally the breakdown, and confirmed that the S-shaped curves could be expected for several QHE plateaus, but did not discover the underlying cause [75].

Still, the theories continued to arrive. Pokrovsky *et al* [76] noted that the drift velocity in the edge states were not the Fermi velocity, but that of the $\mathbf{E} \times \mathbf{B}$, where E

is the electric field in the potential at the side of the surface confinement of the sample, and not the Hall field [77]. Nevertheless, these authors assigned the breakdown to phonon-assisted resonant tunneling between Landau levels. Polyakov and Raikh [78], in turn, considered that the breakdown was due to impurity breakdown, an effect known in bulk semiconductors at low temperatures in which the carriers are frozen out of the bands. Dyakonov raised an important point [79]; in the QHE state, the electrons are in the edge states, while the impurities from which the electrons came are more or less uniformly distributed throughout the two-dimensional layer adjacent to the quantum well at the heterostructure interface (this assumes the quantum well is remotely doped). With the electrons moved to the edge states, the impurities are no longer heavily screened. It is normally assumed that the Hall electric field is uniform across the sample, but this can't be the case if there is charge throughout the sample (the charged impurities). Thus, the Hall electric field must be non-uniform, much like the field at a contact, as discussed in chapter 2. Whether or not this is important to the breakdown is not immediately obvious.

Over the intervening years, more experiments clarified some of the observables, such as the bistability at breakdown [80], and the width dependence of the breakdown current [81]. The latter has a corollary that is likely to be important. As the width increases for a given current, the Hall voltage increases as the Hall resistance is constant on a plateau, hence it is likely that the Hall electric field remains at a critical value as the width increases. In fact, these latter authors also suggest that the breakdown occurs due to excitation from a full Landau level (on one side of the sample) to an empty one (on the opposite side of the sample). Breakdown then would likely occur from one side of the sample to the other, in which case the critical Hall electric field would be related to the separation between Landau levels. Optical imaging of the photoluminescence from the Hall bar [82] showed that the initial rise of the breakdown current was associated with response throughout the bulk of the sample. Further work established that the critical current for breakdown is proportional to the sample width, as mentioned, and that the critical electric field for breakdown of a full Landau level is proportional to $B^{3/2}$ [83]. These latter authors supported the idea of inter-Landau level transitions due to the Hall electric field. Cage [84] then found that after breakdown, it was the longitudinal voltage that would become quantized (rather than the longitudinal resistance), and that quasi-elastic inter-Landau level scattering (QUILLS) was indeed the culprit. Further theory brought this together while maintaining the topological nature of the edge states [85]. Interestingly enough, further confirmation of the role of the Hall electric field was found in Corbino disk studies (where there are no edges other than the inner and outer contacts for the current) [86].

Theoretical work then connected the localized states that exist in the bulk of the sample at QHE plateaus with the breakdown itself, suggesting that the field excited carriers out of the localized state to initiate the breakdown process [87]. At the same time, experimental work examined both the integer and fractional QHE and established the gaps in the latter as a process to bring the two together in studies of the breakdown [88], while at the same time providing new evidence for the role of the localized states.

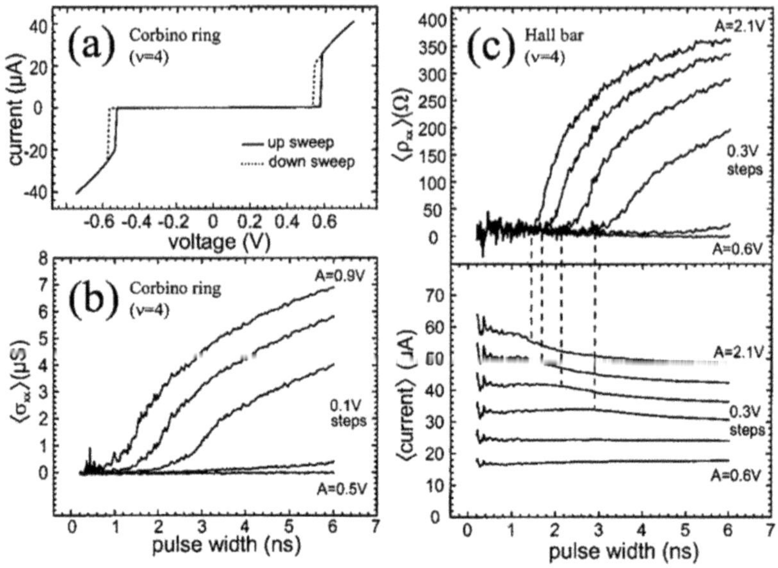

Figure 4.7. (a) *I–V* curves and (b) the longitudinal conductivity of a Corbino disk sample. (c) Average longitudinal resistivity of a Hall bar sample. In (b) and (c) the different curves are for different amplitudes of the voltage pulse applied to the sample, and the data is plotted versus the pulse length. Reprinted with permission from Sagol *et al* [89], copyright (2002) by the American Physical Society.

Short-pulse studies were finally performed on both standard Hall bars and Corbino disks [89], as shown in figure 4.7. Both the Hall bars and the Corbino disks studied had carrier densities in the 2–3 $\times$ 10^{11} cm^{-2} range, so that various plateaus could be studied at relatively low magnetic fields, and experiments were made mostly on the $\nu = 4$ plateau. In the figure, the parameter A is the amplitude of the pulses applied to the current contacts of the sample. This produces a current and a Hall electric field proportional to the amplitude of the pulse. It was found that the breakdown depended upon the pulse length to a large degree, but was independent of the time between the pulses. This meant that it was not a heating issue with the sample. The authors found that the onset of breakdown occurred quicker with a larger value of A. Figure 4.7 shows the results as the amplitude A is increased, for pulses with a duty cycle of 20%. For this filling factor, the breakdown field is 0.73 V, so no increase in sample current is observed for A = 0.6 V. In summary, these authors find that there is a critical time as well as field before breakdown occurs, on the order of nanoseconds, and this time depends upon pulse amplitude, magnetic field, and electron mobility. All of these factors are consistent with impact ionization leading to an increase in carrier density, as discussed in chapter 2, which gives support to the concept of carrier excitation out of localized states (the fields are far too small to be a cross band gap phenomena), and which corresponds to the theories so far developed at the time. Subsequently, contactless measurements using a torque magnetometer, studied the temperature dependence and concluded that the results were consistent with QUILLS model [90].

In general, the community tended to distinguish the breakdown on the basis of the increase in current, with a low field region with what was thought to be a sub-linear increase in current, and then a high current breakdown region with the S-shaped curves. Experiments tended to support the idea that the initial parts of the breakdown, the so-called low current regime was involved with the localized states in the bulk of the sample, and this was reinforced with later measurements on both Hall bars and Corbino disks [91]. It was subsequently found that the breakdown effect occurred also in broken gap InAs/GaSb heterostructures [92], and in a GaAs/AlGaAs superlattice [93]. It has also been seen in graphene [94], which continues to be heavily studied.

4.2.3 Landau level emission

The study of far infrared emission from semiconductors in a magnetic field seems to have begun with the observation of photon emission from impurity levels in InSb in a magnetic field, which may also have contained Landau level transitions [95]. Then, tunable far infrared emission was observed between Landau levels, as well as to the impurity level [96]. In this latter case, magnetic fields up to 2 T were used with the sample cooled to 4.2 K. The radiation was detected by a narrow band Ge/Sb detector with peak response at 9.8 meV arising from the photoionization energy of the Sb impurities in the Ge. Emission spectra from the InSb sample consisted of three peaks that satisfied the requirement $n\hbar\omega_c$, where $n = 1, 2, 3$, and ω_c is the cyclotron frequency for electrons in InSb. The InSb emitter was used early to study Zeeman splitting of impurity levels in GaAs [97]. Clear transitions were observed between the two lowest Landau levels in InSb by the use of photo-excitation of hot carriers with a lattice temperature of 2 K [98].

Emission from the Landau levels in Si MOS inversion layer were also observed [99]. These measurements and recent measurements in InSb, GaAs, and $Hg_{0.8}Cd_{0.2}Te$ were reviewed in a subsequent work [100]. In figure 4.8, the frequencies that can be emitted are plotted for various semiconductors as a function of the magnetic field. Both Landau level transitions and transitions from a Landau level to an impurity are shown.

Emission between the light holes and heavy holes was observed in Ge relatively early [101]. Later, emission between the Landau levels in the light holes of Ge was observed in relatively high electric fields [102, 103]. The emission is sufficiently strong in this system, that it was possible to create lasing action between the Landau levels [104]. This action is interesting in that the non-parabolicity of the light-hole bands leads to a pair of emission frequencies, depending upon which Landau levels are involved [105].

4.2.4 Inter-sub-band emission

As soon as emission from Landau levels had been confirmed, interest turned to seeking the same process from sub-bands in quantum wells or inversion layers. Success was found in a large Si MOSFET fabricated on the *p*-type (100) surface of Si and studied at 4.2 K [106]. The device itself had a square cross-section, 0.25 cm on a

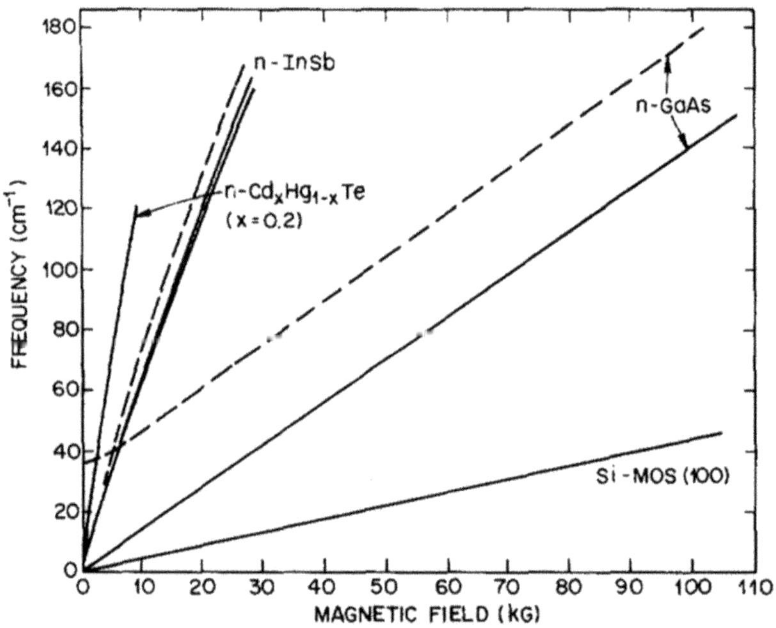

Figure 4.8. Cyclotron emission frequencies for various semiconductors as a function of the magnetic field (1 kG = 0.1 T). The dashed lines indicate the frequency of associated impurity transitions. Reprinted from E Gornik [100], copyright (1979) with permission from Elsevier.

side, and a 0.6 μm thermal oxide. It was excited with pulsed electric fields, with a sufficiently low duty cycle to avoid thermal heating of the device. The inversion layer, sub-bands, and hot carrier distribution are shown in the inset to figure 4.9. The sub-band spacing is determined by the inversion density (which also determines the gate electric field and thus the shape of the quantum well at the interface). Since some of the bulk *p*-type dopants have to be ionized in order to establish a sharp quantum well, gate bias is applied at room temperature and then maintained as the device is cooled, thus locking in the depletion charge in the substrate. In figure 4.9, the measurements of the detector response are shown. The solid curve shows the signal measured with a Ga-doped Ge detector with an As-doped Ge filter to shorter wavelengths, while the dashed curve used a crystal quartz filter. Here, the source–drain field was 20 V cm^{-1} and the inversion layer mobility was 2×10^4 cm^2 Vs^{-1}. No signal was detected if the polarity of the gate voltage was reversed, as was expected. The quartz crystal filter has a sharp absorption peak at 76 μm and the As-doped Ge filter has a sharp absorption peak at 100 μm. These two peaks are indicated in the figure, where the emission has dips. The drop off of signal below $V_G = 12$ V is due to the cutoff of the Ge detector itself. For drain fields below 30 V cm^{-1}, it is estimated that a single emission line with a width of 6 cm^{-1} is present.

Subsequently, inter-sub-band emission was observed from GaAs quantum wells [107]. Still later, it was found that the emission could be observed in quantum wires [108], and in superlattices with the field normal to the hetero-interfaces, so that sequential resonant tunneling normal to the plane of the superlattice was involved

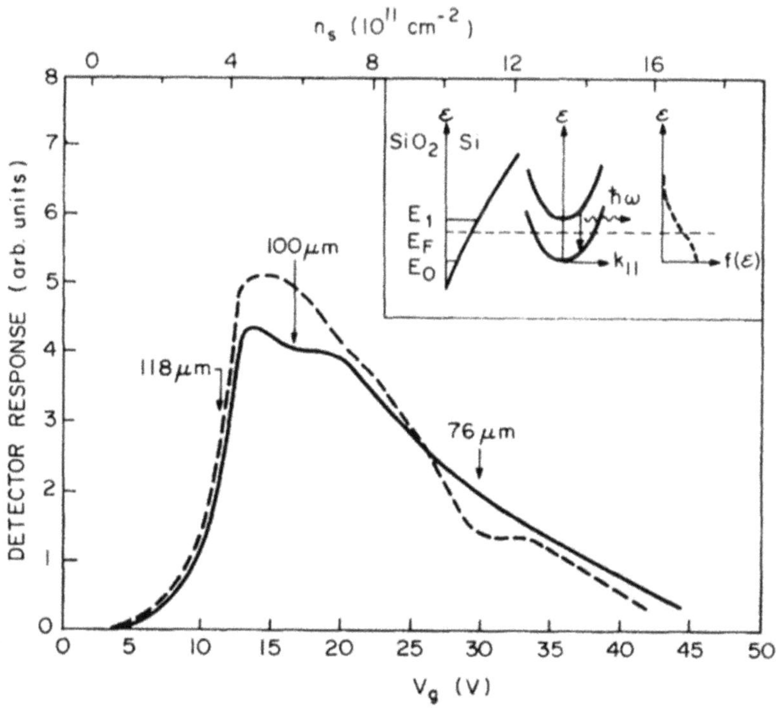

Figure 4.9. The detector response as a function of gate voltage when a channel electric field of 20 V cm^{-1} is applied at 4.2 K. The solid and dashed curves are for different filters in front of the detector (see text). The inset shows the form of the inversion layer and the sub-bands, the inter-sub-band emission, and the hot carrier distribution function. Reprinted with permission from Gornik and Tsui [106], copyright (1976) by the American Physical Society.

[109, 110]. Measurements such as these led to the eventual creation of the quantum cascade laser (QCL) [49]. Once the QCL appeared, it became obvious that it was far superior in generating far infrared signals than either simple inter-sub-band or inter-Landau level emission sources. Thus, interest in the latter waned quite rapidly.

4.3 Optical absorption

At sufficiently high frequencies, where $\hbar\omega > E_G$, the electromagnetic field can induce electrons from the valence band into the conduction band, as is described in the beginning of this chapter. This creates a new electron–hole pair that can contribute to the conduction process. This is the usual optical absorption that provides a method to study the various transitions possible from valence to conduction band, or even between different valence or different conduction bands. The requirement is that the lower energy state must be occupied and the upper energy state empty, to satisfy the Fermi statistics. Of primary interest for the application of hot carriers are the direct transitions, such as between the top of the valence band and the bottom of the conduction band in GaAs at Γ, which is known as the E_0 transition (the direct band gap). But these direct transitions can also occur at the X and L points, although at much higher energies. Direct transitions are strong primarily because of the high density of states in the two involved bands, and the allowed nature of the dipole transition.

The quantity that measures the strength of the optical transition is the absorption coefficient, that describes how much of the optical electromagnetic wave is turned into electron–hole pairs. To find this, first the transition probability for the inter-band transition must be determined. Primarily, the absorption coefficient is measured in units of cm^{-1}, because the researcher is interested in how the electromagnetic power decays with distance as

$$P \sim e^{-\int \alpha dx}, \tag{4.4}$$

where α is the absorption coefficient. The matrix element described at the opening of this chapter is a natural part of this quantity. Here, it will be shown how these matrix elements, needed for the transition probability, arise in a natural manner. Once the transition probability is determined, the number of electrons that make the transition for a given electromagnetic field power can be determined, and this will lead to the absorption coefficient. In figure 4.10, the basic transition for the direct absorption is shown. This transition is vertical, as discussed earlier, because the momentum of the photon is orders of magnitude smaller than the momentum of the electron and/or hole. Note that the transition is made away from $k = 0$. This is because the densities of states for both the valence band and the conduction are zero at $k = 0$, as there are neither full states in the valence band nor empty states in the conduction band exactly at $k = 0$. A transition is shown both for the heavy-hole band and the light-hole band. If the photon has a sufficiently high energy, it can also transition from the spin–split-off valence band. For $\hbar\omega > E_G$, the energy in excess of the gap energy will split as kinetic energy between the valence and conduction band, with the factors for each determined by the effective masses of the two bands. That is, the photon energy splits according to

$$\hbar\omega = E_G + \frac{\hbar^2 k^2}{2}\left(\frac{1}{m_e} + \frac{1}{m_h}\right), \tag{4.5}$$

where m_e and m_h are the electron and hole masses, respectively. The fraction that goes into e.g. the conduction band is

$$\frac{m_h}{m_h + m_e}. \tag{4.6}$$

The first step of creation of an electron–hole pair is the absorption and subsequent annihilation of a photon resulting in the creation of the pair. This step is no different in a semiconductor from what occurs in an atom. Contrary to many beliefs in quantum mechanics, the electron does not just 'jump' into the conduction band via a quantum jump. It transitions through a standard probability function from quantum mechanics in a well-defined process and takes a few femtoseconds to occur, as described at the beginning of this chapter. The field first creates a correlation between the initial state (ground state) wave function and the final state (excited

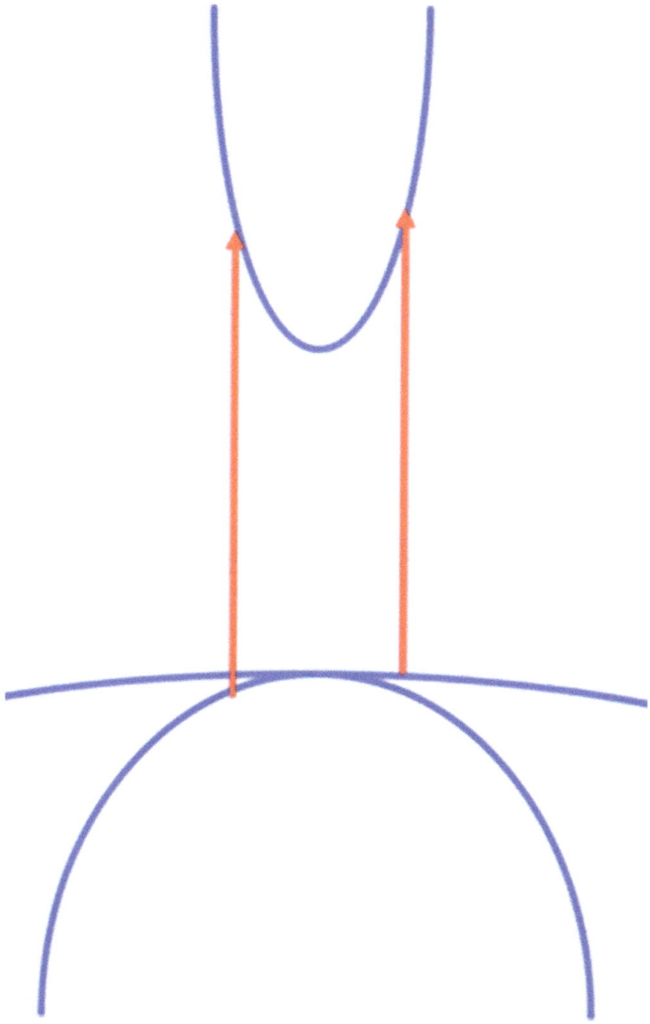

Figure 4.10. Optical transition between the light-hole, and heavy-hole, bands to the conduction band. These transitions are shown as vertical as the photon momentum is quite small compared to those of the electron and hole.

state) wave function. This correlation is referred to as the polarization within the semiconductor. The conjugate field then destroys this polarization (the field and its conjugate appear in the magnitude squared of the matrix element). Once the polarization is destroyed, the process is complete. But, the electron may remain in the valence band with the photon proceeding on its merry way, or the electron may wind up in the conduction band with the photon being destroyed. The fraction of these processes that lead to actual absorption of the photon is known as the quantum efficiency. Although it is the electromagnetic field (photon) that creates the interaction, conservation of the energy leads to the photon energy being divided between the energy gap and the kinetic energy of the electron and hole as described in (4.5). The entire process takes only a few femtoseconds and is governed by the semiconductor optical Bloch equations [6–8]. They will be discussed in chapter 8.

The process by which the above transition is determined is just standard time-dependent perturbation theory in which, the transition coefficient may be written as

$$t_{12} = -\frac{H_{12}}{\hbar}\left[\frac{e^{i(\omega_{21}+\omega)t} - 1}{\omega_{21} + \omega} - \frac{e^{i(\omega_{21}-\omega)t} - 1}{\omega_{21} - \omega}\right], \tag{4.7}$$

where

$$\hbar\omega_{21} = E_c(k_2) - E_v(k_1)$$

and

$$H_{12} = \langle 2 \mid H_F \mid 1 \rangle = \int \psi_v^{\dagger}(k_1) H_F \psi_c(k_2) d^3\boldsymbol{r} \tag{4.8}$$

is the matrix element for the transition, which as usual is the spatial integral of the perturbing term weighted by the initial and final wave functions. As is standard in the Fermi golden rule, if the long time limit is taken (here, this limit is defined as a time large compared to the few femtoseconds over which the polarization is created and destroyed), the square brackets reduce to the energy-conserving delta function $2\pi\delta(\hbar\omega_{21} \pm \hbar\omega)$. Note that the two signs give us either absorption or emission. Because we are interested only in the absorption, we will use the negative sign in the following. Again, in the standard Fermi golden rule approach, the two momenta are equal by conservation of this quantity and we may write the probability as

$$c_{21} = \frac{\mid t_{12} \mid^2}{t} = \frac{4 \mid H_{12} \mid^2}{\hbar^2} \frac{\sin^2[(\omega_{21} - \omega)t/2]}{(\omega_{21} - \omega)^2 t^2}. \tag{4.9}$$

The full derivation can be found in many textbooks [9], but the key part of the electromagnetic wave perturbation term appearing in the Schrödinger equation leads to the leading term in the matrix element to be

$$H_{\text{pert.}} = \frac{i\hbar e}{m_c}\boldsymbol{A} \cdot \nabla, \tag{4.10}$$

where $\boldsymbol{A}$ is the vector potential. Then, the magnitude squared of the matrix element leads to

$$c_{12} = \frac{e^2 A_0^2 t}{m_0^2 \hbar^2} \mid \boldsymbol{a} \cdot \boldsymbol{P}_{12} \mid^2 \frac{\sin^2[(\omega_{21} - \omega)t/2]}{(\omega_{21} - \omega)^2 t^2}, \tag{4.11}$$

where

$$\boldsymbol{a} \cdot \boldsymbol{P}_{12} = ih \int d^3\boldsymbol{r} u_c^{\dagger}(\boldsymbol{r}) \boldsymbol{a} \cdot \nabla u_v(\boldsymbol{r}) \tag{4.12}$$

is the dipole matrix element between the conduction and valence bands. In direct bands, such as those in GaAs, the conduction band minima is primarily composed of cation s-functions, while the valence band maxima are primarily composed of anion p-functions [111]. This parity difference between two bands is required for the dipole matrix element. A standard development from (4.11) [9], leads to the absorption coefficient

$$\alpha = \frac{e^2(2m_R)^{3/2}f_{\text{osc}}}{\pi m_0 \hbar^2 G_\omega}(\hbar\omega - E_G)^{1/2}, \tag{4.13}$$

where the reduced mass m_R is the reciprocal of the term in parentheses in (4.5), and the oscillator strength is the standard optical strength

$$f_{\text{osc}} = \frac{2\langle P_{12}^2\rangle}{m_0\hbar\omega}, \tag{4.14}$$

and the expectation value of the dipole is found easily from (4.12).

The square-root variation in (4.13) is just that expected from the density of final states to be found in the conduction band. The numerator mass involved is the reduced mass accounting for both the valence and conduction states. The denominator mass, however, arose from the Hamiltonian, and is prior to the effective mass approximation, as it still contains the full potential of the crystal, and thus should be the free electron mass. Excellent tabulations of the properties of materials, including the optical absorption coefficient (as a function of energy), can be found at [112, 113]. In figure 4.11, the absorption coefficient is shown for $Ga_{0.47}In_{0.53}As$ at room temperature. The form is typical for direct gap semiconductors, and only needs to be shifted in the energy axis as the band gap changes, although the exact form will change as other, higher energy, transitions can occur.

In the above discussion of the direct gap material, it was clear that the optical transition must be vertical, as the momentum of the photon is much smaller than the electron or hole momentum. In indirect band-gap semiconductors, however, this would lead to very large energy photons being required, unless a phonon becomes involved in the momentum conservation. For these indirect gap materials, we can write the momentum conservation condition becomes

$$\boldsymbol{k}_1 + \boldsymbol{k}_0 - \boldsymbol{k}_2 \pm \boldsymbol{q} = \boldsymbol{0}, \tag{4.15}$$

where $\boldsymbol{k}_0$ is the photon momentum, $\boldsymbol{q}$ is the phonon momentum, and the other two momenta have the same description as above. Energy conservation requires

$$E_c(k_2) - E_v(k_1) = \hbar\omega \pm \hbar\omega_q. \tag{4.16}$$

These transitions are illustrated in figure 4.12 for an indirect band gap. Both phonon absorption and emission processes are allowed, although it is the absorption of a phonon that requires the lower energy photon of the two processes.

An astute reader will immediately comment that there are no band structure states at the point where the photon connects with the phonon. That is, when the band structure appears as in figure 4.12, there are no allowed states at this energy and momentum site. But, the band structure is generally computed for those wave functions which have propagating behavior; e.g., allowed waves could exist with real

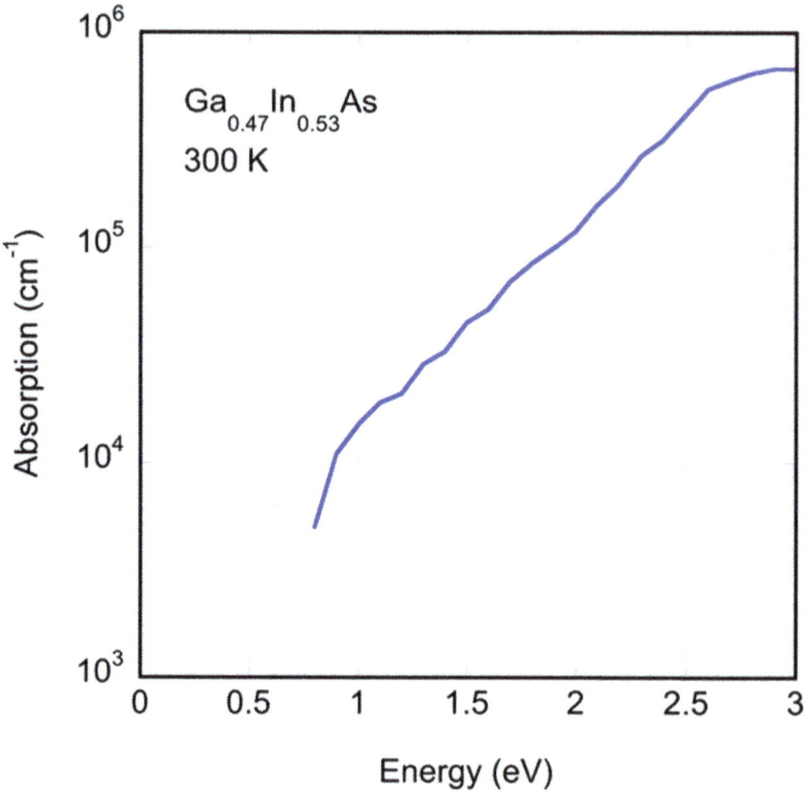

Figure 4.11. The absorption coefficient for the alloy $Ga_{0.47}In_{0.53}As$ at room temperature. This alloy has a band gap of about 0.75 eV.

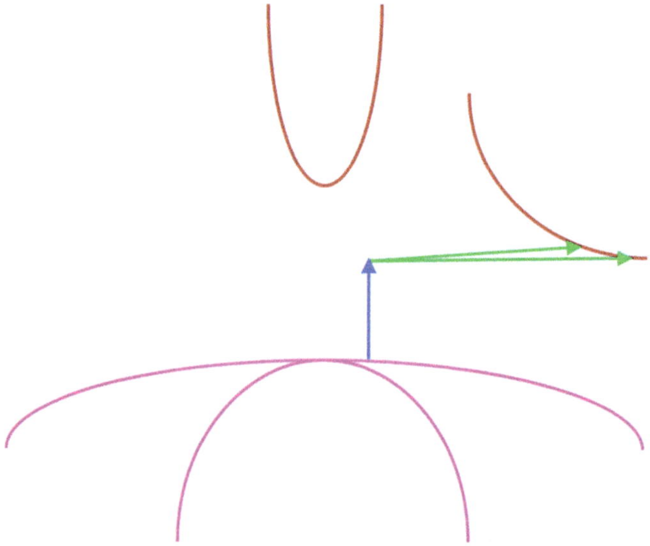

Figure 4.12. The phonon-assisted indirect optical transition proceeds through an evanescent state. Reproduced from [9].

momentum. It is well known that impurities and defects can introduce states in the gap regions, but there are other states that exist in this region as well. These states actually come from the band structure, but *are not propagating* states. They are the so-called evanescent states that have imaginary momentum. Consider a simple two-band model, which here is the hyperbolic bands from $k \cdot p$ theory written as

$$E(E - E_G) - k^2P^2 = 0. \tag{4.17}$$

If the momentum is real, $k^2 > 0$, this leads to the two mirror image bands

$$E = \frac{E_G}{2}\left[1 \pm \sqrt{1 + \frac{4k^2P^2}{E_G^2}}\right], \tag{4.18}$$

that are the so-called hyperbolic bands. The zero of energy lies at the top of the valance band. It should be understood that there is no physical reason to use only $k^2 > 0$. Of course, this gives propagating waves. But, it would be equally correct to choose $k^2 < 0$. In this case, (4.17) leads to the energies

$$k = \pm iP\sqrt{E(E_G - E)}, \tag{4.19}$$

which gives us the so-called *complex* band structure. In this case, k is imaginary, so that the state decays away exponentially from its site in real space. This imaginary k is plotted in figure 4.13 as a negative k (red curve), while the real in (4.18) is plotted in blue. The complex band structure is much more complicated than the real band structure, because every pair of bands, such as the simple bands of (4.18), is connected by the imaginary k curves, and this leads to enormous complexity in crossing bands. These are so-called evanescent states, which are highly localized and are certainly not propagating waves. Yet, they are important as they provide the localized states throughout the gap that allow the transitions shown in figure 4.12. A photon can pump an electron into one of these states, where the electron can subsequently make a transition to the conduction band through phonon scattering. And this is the process of the indirect transition—it is a two-step process.

The perturbation treatment must be expanded to account for the two quite distinctly different processes that compose the indirect transition. This is second-order perturbation theory and has one perturbation from the photon and one for the phonon. The second-order process here involves both an optical matrix element and an electron–phonon matrix element. Then, the transition probability involves the magnitude squared of each of these two matrix elements. This must now be integrated over the initial and final states of the hole and the electron and have the phonon properties inserted. The latter, of course, involve the Bose–Einstein distribution. In general, all the unknown sins are incorporated in the magic oscillator strength. Here, only the final results are presented, which are expressed in terms of the energy dependence of the absorption coefficient [114], which gives

$$\alpha \sim \frac{x}{e^x - 1}\left(\hbar\omega - E_G \pm \hbar\omega_q\right)^2, \quad x = \frac{\hbar\omega_q}{k_B T}. \tag{4.20}$$

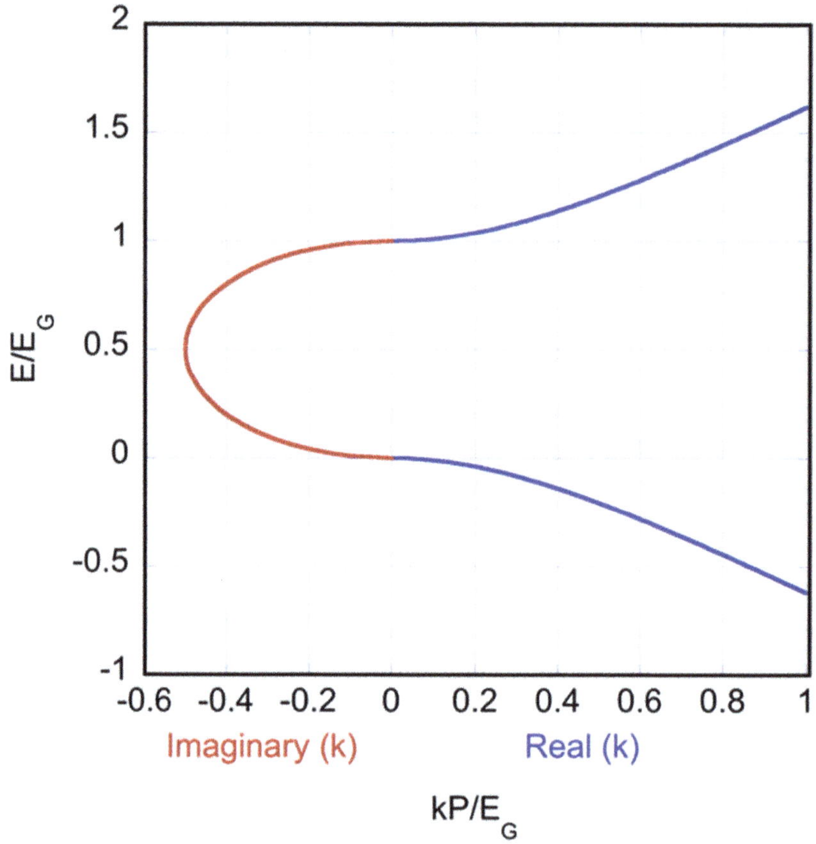

Figure 4.13. Complex band structure for a simple $\boldsymbol{k} \cdot \boldsymbol{p}$ two-band model. The blue curves are for real k, while the red curve is for imaginary k. The energy gap is actually $2E_G$. Reproduced from [9].

The indirect transition has a very significant temperature dependence which arises from the phonon contribution. In addition, there is a much stronger dependence upon photon energy, although the pre-factor is considerably smaller. In general, the second-order process is usually at least an order of magnitude weaker than the first-order process. The absorption coefficient for Si is shown in figure 4.14. It may be seen that near the threshold, the absorption is more than an order of magnitude smaller than for the direct gap III–Vs. There is also an increase in the absorption near 4 eV that arises from direct transitions into a secondary conduction band at Γ.

4.4 Ultrafast excitation studies

Physicists have used lasers to study semiconductors for a very long time. The earliest studies were attempts to determine band gaps and other material properties, but things like oscillatory photoconductivity (discussed above) focused their attention on the interesting properties of these materials. The use of short-pulse lasers began in the early 1970s, primarily with studies of Ge. Early studies of Ge by ultrafast pump–probe experiments looked primarily at the transmission of the signal through the

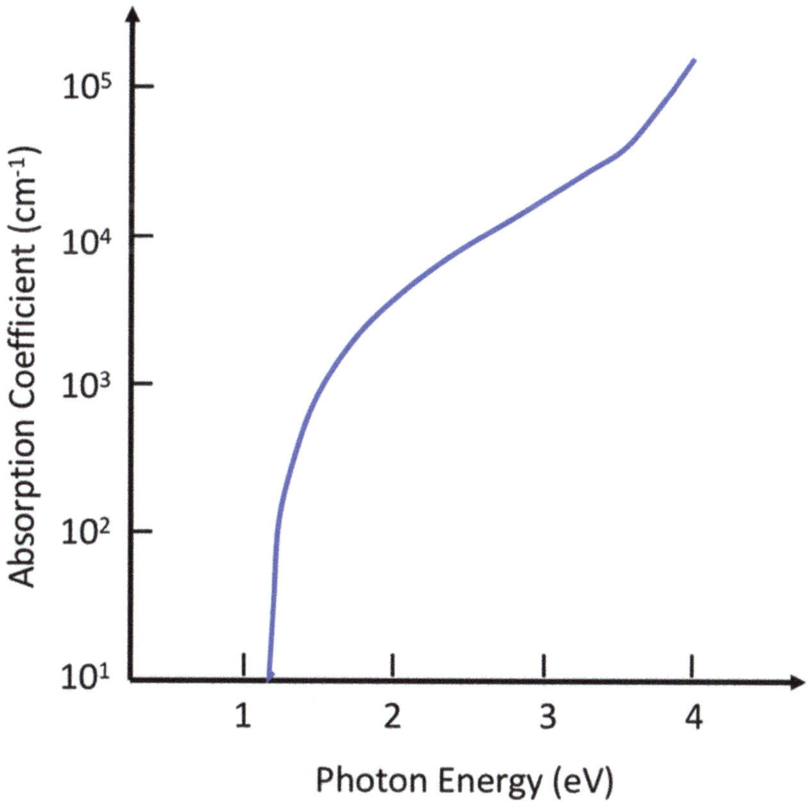

Figure 4.14. Absorption coefficient for Si at 300 K. The rise at ~4 eV is due to direct transitions to the second conduction band at Γ. Reproduced from [9].

wafer [115, 116]. These studies indicated that the energy relaxation of the photo-excited carriers was less than 5 ps, and that a saturation of the absorption could occur in the Γ valley. While Ge is an indirect gap semiconductor, with a gap of 0.66 eV, and with the conduction band minima at the L points in the Brillouin zone, the Γ valley lies only 0.8 eV above the valence band maximum. Hence, the photons in these experiments were primarily absorbed into the Γ valley, although they could then relax to the L valleys by inter-valley phonon emission. The observation of the saturation in the absorption was likely due to state-filling in the Γ valley, due to the low density of states, although this would be mitigated in some sense by the band gap narrowing arising from the electron–electron interactions and the resulting many body effects [117]. The observation that the saturation recovered rapidly, within the pump pulse duration, likely is a result of the many body effects, although there certainly could be other explanations [116].

Actually, slightly before the Ge work, initial studies had begun on GaAs, due to its known direct gap. The laser creates electron–hole pairs with considerable excess energy, as discussed above, particularly for the electrons in direct-gap semiconductors. The carriers then cool by the emission of LO phonons, the polar LO modes in compound semiconductors [118]. It was found that exciting a semiconductor with intense laser radiation could lead to nonequilibrium populations of the longitudinal optical (LO) modes in polar materials [119, 120], especially at room temperature and with temporally short pulses of this radiation. If sufficient energy is transferred to the phonon system, from the relaxing carriers, the phonon distribution can be driven out of equilibrium. This population can be studied by Raman scattering [120], among other methods, and the population estimated by the relative strengths of the Stokes and anti-Stokes transitions. The nonequilibrium phonons will be discussed in chapter 5.

By the late 1980s, the development of lasers with pulse durations as short as 6 fs made it possible to probe very fast nonequilibrium properties of the electron–hole plasma in semiconductors [10]. More progress has been made since that time. But, consider the properties of a 6 fs pulse at 2.0 eV, as discussed in the introduction to this chapter. The frequency of this laser is 4.83×10^{14} Hz, and the period of the oscillation is 0.33 fs. Thus, the 6 fs pulse has less than 20 oscillations of the signal. This is still large. The 6 ps pulse was generated by a complicated nonlinear process. The base system uses optical pulses in a colliding pulse mode-locked laser with a set of intracavity prisms [11]. These pulses are amplified in a copper-vapor laser amplifier. The prisms formed a four prism sequence and the signal was then doubled in a KDP crystal to minimize distortion. The value of 6 fs was determined from an interferometric autocorrelation signal. In subsequent work, the signal was split into two, for use in a pump–probe experiment on the transmission of GaAs [12]. By convolving the signal transmitted through the sample with the exciting pulse, and subtracting it from the latter, the transmittance signal could be identified. Here, it was found that the signal had a fast decay with relaxation time of 33 fs. In distinction from earlier thoughts that this was the cascade time, it was determined that this was the inter-valley transition time from the Γ valley to the L valleys, showing that the scattering time to the satellite valleys was only a few tens of fs.

When carriers are highly excited into states well beyond their normal equilibrium states, whether through the application of an electric field or by photo-excitation, they relax their excess energy by phonon emission, either before or after spending some time in the L valleys. When they are in the Γ valley, such as for direct gap III–Vs, this usually involves the polar LO modes. This was illustrated in figure 4.3, where the situation for optical excitation was shown. If the laser photon has energy below that needed for transfer to the satellite valleys, the cascade begins immediately. In the figure, only excitation from the heavy-hole band is shown, and roughly 90% of the excess energy goes into the photo-electrons. In GaAs, the dominant LO mode has an energy of ~36 meV. Still, the average photo-electron will emit something like 15–20 polar LO phonons. The average time required to emit a phonon like the LO phonon is about 160–200 fs, so that the cascade will take just over 3 ps.

While the average time is relatively long, it remains an *average* time. It is possible that the phonons will begin to be emitted somewhat earlier. This can be illustrated with ensemble Monte Carlo simulations for photo-excitation of GaAs [121]. Photoexcited carriers from a 200 fs, 1.64 eV laser pulse are excited in the material,

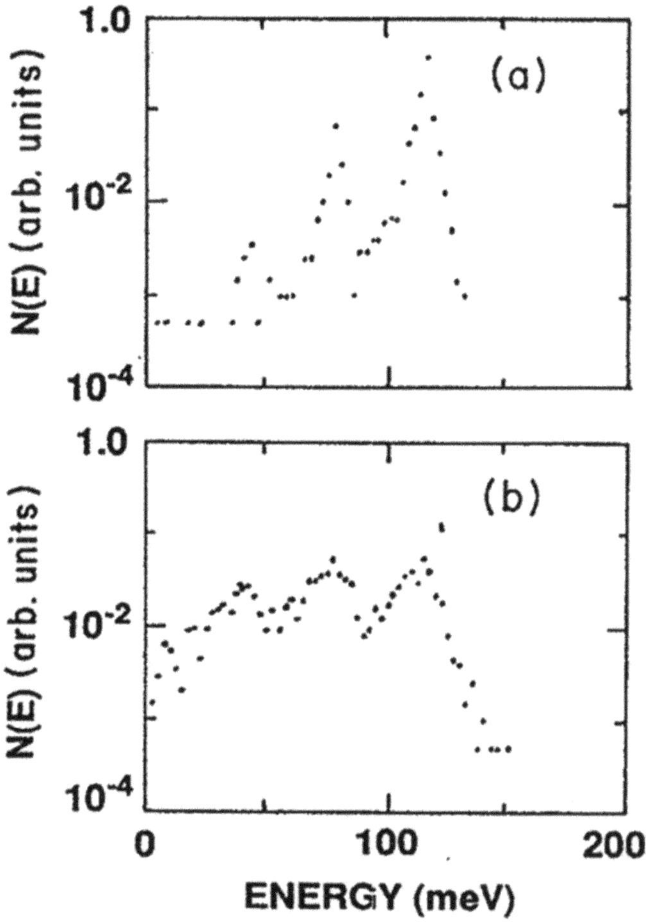

Figure 4.15. The electron energy distribution after excitation by a 200 fs pump pulse from a 1.64 eV laser. This generated an electron–hole density of 5×10^{16} cm^{-3}. (a) Distribution 100 fs after the pump. (b) 500 fs after the pump. Reprinted with permission from Osman and Ferry [121], copyright (1987) by the American Physical Society.

which puts the photo-electrons below the satellite valley minima. The role of the various scattering processes on the phonon cascade can be studied in this case. The excited electron–hole density is 5×10^{16} cm^{-3}. In figure 4.15, the electron population is shown at two different times after the start of the laser pulse. At 100 fs, it is clear that at least two phonon replicas of the excitation distribution are seen, so that even at this early time, at least two steps of the phonon cascade are observed. By 500 fs, however, these replicas are beginning to broaden substantially, due to the electron–phonon, electron–electron, and electron–hole scattering. If the excited density is increased to 10^{18} cm^{-3}, the entire spectrum is washed out by the electron–electron scattering already at 100 fs. In both cases, the peaks are broadened more to the low energy side, so that many electrons have already lost some energy due to other scattering processes prior to emitting the first phonons.

4.4.1 Hot carrier luminescence

Perhaps the earliest study of hot carrier photoluminescence was the seminal work from Shah on photo-excited carriers in GaAs at 2 K [122]. Several of the luminescence spectra are shown in figure 4.16. The samples were epitaxial GaAs with a background doping of 10^{15} cm^{-3} (both *n*- and *p*-type). Excitation was provided by a high power argon laser focused to a spot size of some 100 μm diameter. It was observed that the electrons and holes equilibrated among themselves quickly and the luminescence exhibited a thermal behavior with a temperature that increased with excitation level. At low excitation levels, the emission consisted of a number of sharp lines which were different for *n*- and *p*-samples. As the intensity increased, the emission spread into a broad thermal emission, and was the same for both types of samples. Only a few of the measured spectra are shown for both types of samples. At these higher excitation levels, it was

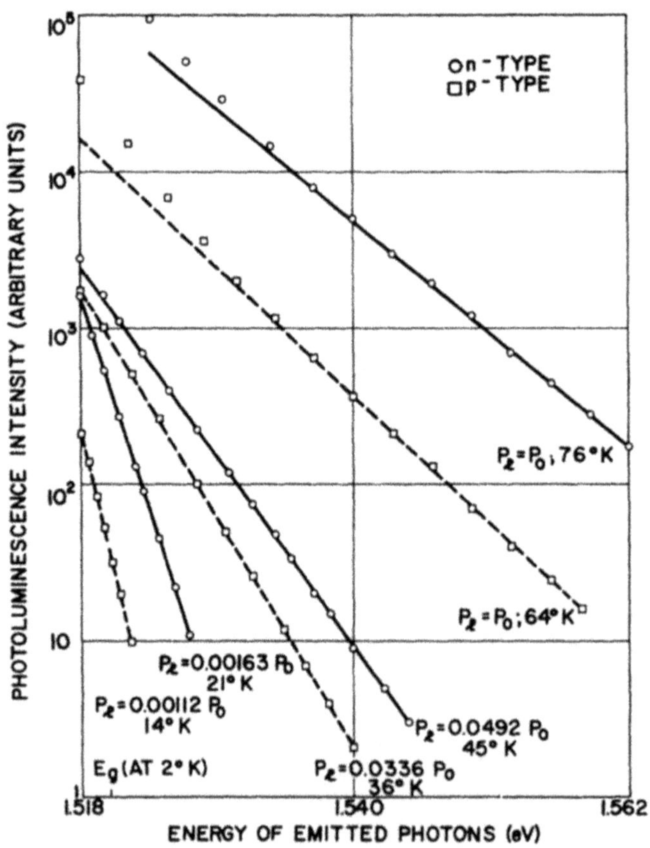

Figure 4.16. The high energy tails of the photoluminescence spectra of *n*- and *p*-type GaAs at 2 K. The data are for three different excitation levels, of the total of eight such spectra. The points are taken from a continuous emission spectra. The lines define a characteristic carrier temperature T_c for each curve. P_0 is the laser power at maximum excitation level. Reprinted with permission from Shah and Leite [122], copyright (1969) by the American Physical Society.

found that the emission followed an exponential decay, varying as $\exp(-T_0/T_c)$, where T_0 corresponded to an energy of approximately 33 meV, which is close to the LO phonon energy near Γ. In subsequent work, it was shown with Raman scattering that it is the LO phonons that account for the power lost by the carriers, and that these phonons are driven out of equilibrium [120].

Shah pointed out that the hot photo-excited electrons can decay by at least two mechanisms: the emission of LO phonons or by the interaction directly with the electron gas [118]. The relevant rates for energy loss were given by [123–125]

$$\left(\frac{dE}{dt}\right)\bigg|_{e-e} = -\frac{ne^2}{4\pi\epsilon\sqrt{2m^*E}}, \tag{4.21}$$

and

$$\left(\frac{dE}{dt}\right)\bigg|_{\mathrm{LO}} = \frac{m^*e^2}{2\pi\gamma\epsilon_0\sqrt{2m^*E}}\left[N_q \sinh^{-1}\left(\frac{E}{\hbar\omega_0}\right) - \left(N_q + 1\right)\sinh^{-1}\left(\frac{E}{\hbar\omega_0} - 1\right)\right], \tag{4.22}$$

where γ is given by the Fröhlich interaction [126, 127] as

$$\frac{1}{\gamma} = \omega_0^2\left(\frac{1}{\epsilon_{r,\infty}} - \frac{1}{\epsilon_{r,0}}\right). \tag{4.23}$$

The two relative dielectric constants are the optical and static values. The interaction with the electromagnetic waves and the lattice polarization splits the LO and TO modes at Γ. The difference between the optical and static dielectric constants is connected to this LO–TO splitting through the Lyddane–Sachs–Teller relation [128]. Because of the lattice polarization, the LO mode is commonly referred to as the polar LO mode. Importantly, the polar LO mode can scatter electrons and holes through the resulting Fröhlich interaction (4.23).

It is clear from the above equations that carrier–carrier interactions will play a larger role with higher excited carrier densities, as was already apparent from the discussion around figure 4.15. Neither of the two scattering processes ever goes away. But, the dominant one will change as excitation increases. The form of energy relaxation (4.22) does not include any screening of the interaction. However, since it is a Coulombic interaction, screening will play a role [9], even though screening by the carriers is reduced as the carrier temperature rises.

Studies of quantum wells, especially those of GaAs/AlGaAs, also focus primarily upon the luminescence and its decay with time [129, 130]. In 1985, Shah *et al* measured the actual carrier loss rates after excitation with a fast pump pulse [131]. The samples were both *n*- and *p*-type modulation-doped GaAs/AlGaAs quantum wells, measured at 1.8 K. These samples were excited with relatively low power (<10 mW cm^{-2}) from a dye laser at 770 nm. A pulsed electric field (>40 V cm^{-1}) was applied and the luminescence was measured only during the pulse. Various width wells and barriers were measured, but the results seems to be relatively independent of these parameters, as may be seen in figure 4.17, where the reciprocal carrier temperature is shown as a function of the power loss of the carriers (in the figure, d_1, d_2, and d_3 are the quantum well width, the doped AlGaAs width, and the AlGaAs spacer layer, respectively, and η is the number of periods for the multiple quantum wells). From the inset to this figure, it may be seen that the electrons and holes clearly have separate temperatures, and it is clear that the holes have a factor of 25 larger power loss than the electrons at any temperature below ~100 K. This lower power loss in the case of the electrons is attributed to the creation of nonequilibrium LO phonons with which primarily the electrons interact. As has been noted above, the holes tend to emit TO phonons more than the LO modes. While the energy loss

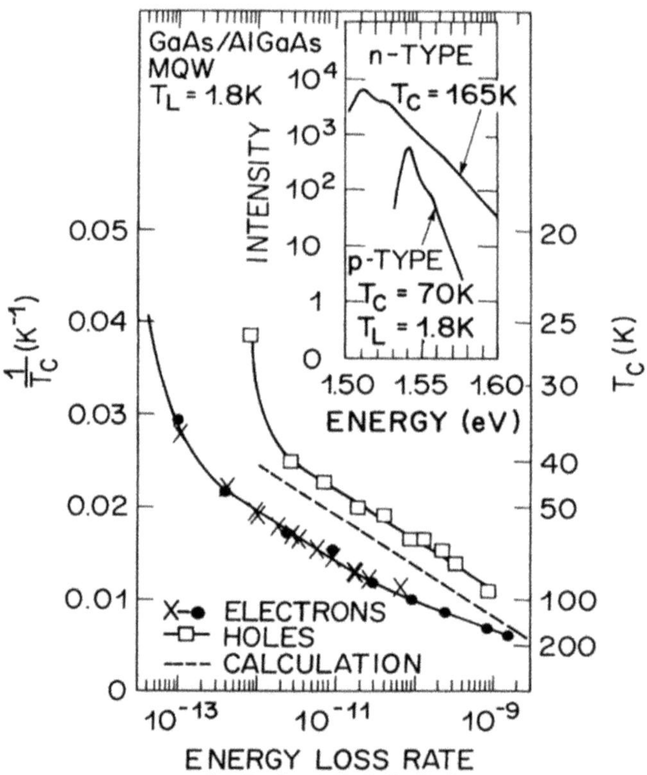

Figure 4.17. Inverse of carrier temperature as a function of the energy loss rate to the lattice for one p-type and two n-type modulation-doped multiple quantum wells in GaAs/AlGaAs. The dimensions for these wells are for Sample 1 (crosses): $n = 7 \times 10^{11}$ cm^{-2}, $x = 0.2$, $d_1 = 26.2$ nm, $d_2 = 31.7$ nm, $d_3 = 16.3$ nm, $\mu = 6.3 \times 10^4$ cm^2 Vs^{-1}, $\eta = 20$; for sample 2 (circles): $n = 3.9 \times 10^{11}$ cm^{-2}, $x = 0.23$, $d_1 = 25.8$ nm, $d_2 = 28.4$ nm, $d_3 = 11.8$ nm, $\mu = 7.9 \times 10^4$ cm^2 Vs^{-1}, $\eta = 15$; and for sample 3: $p = 3.5 \times 10^{11}$ cm^{-2}, $x = 0.45$, $d_1 = 9.4$ nm, $d_2 = 4.3$ nm, $d_3 = 25.6$ nm, $\mu = 3.6 \times 10^4$ cm^2 Vs^{-1}, $\eta = 15$. The solid curves are merely guides to the eye. The inset shows typical luminescence curves for samples 1 and 3 at 1.8 K, and 750 and 1000 V cm^{-1}, respectively. The values of carrier temperature are deduced from such luminescence curves. Reprinted with permission from Shah *et al* [131], copyright (1985) by the American Physical Society.

rates seem to agree with simple calculations, the energy loss rates are anomalously low, and this led to the conclusion that the LO modes are out of equilibrium, thus slowing the electron loss rates.

Subsequently, detailed ensemble Monte Carlo simulations, that included a careful analysis of the nonequilibrium LO modes, confirmed that the low electron loss rates were indeed due to nonequilibrium phonons [132]. A carrier generation rate of width 0.6 ps, peaking at 1 ps was used to excite *n*-type samples with quantum well widths of 5, 15, and 25 nm with a background doping of 2.5×10^{11} cm^{-2}. An excited carrier density of 5×10^{11} cm^{-2} was assumed in the simulation. In figure 4.18, the electron distribution as a function of energy is shown at several different times. It may be seen that the generation remains at a specific energy during the pulse, but rapidly decays into a thermal distribution in less a ps. By 10 ps, the distribution has cooled considerably. However, it is not clear if any of the distributions are properly thermal in nature, particularly as the second sub-band appears to still be occupied at 10 ps. The nonequilibrium LO population is shown in figure 4.19 at several different times in the simulation. Also the time variation of the population at two different wave numbers is shown in the inset to the figure.

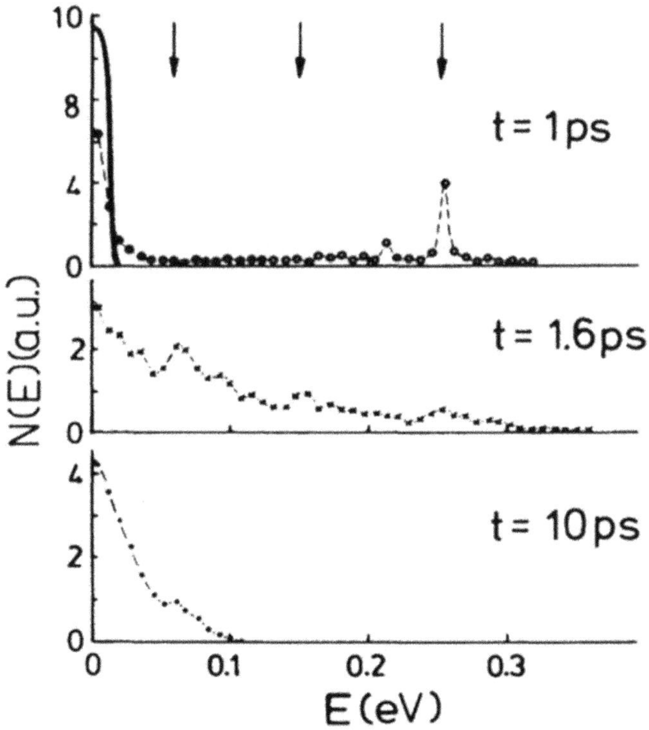

Figure 4.18. The electron distribution function for several times: before (solid curve in top plot), 1 ps, 1.6 ps, and 10 ps after initiation of the laser carrier generation in the 15 nm quantum well. The arrows indicate the position of the higher sub-bands in the quantum well relative to the ground state. Reprinted with permission from Lugli and Goodnick [132], copyright (1987) by the American Physical Society.

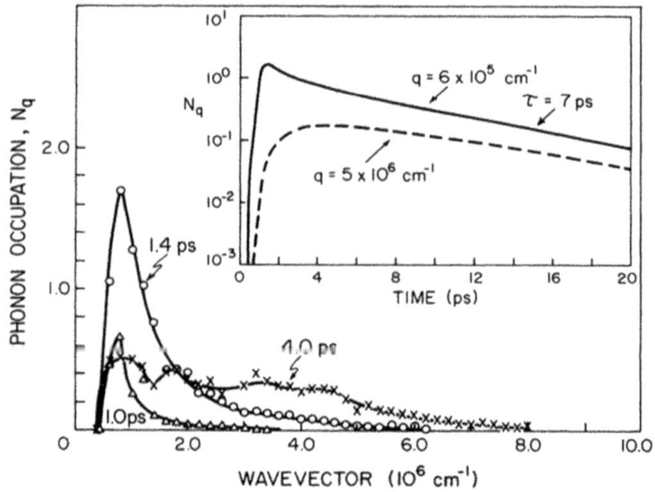

Figure 4.19. Calculated population of the LO phonons during the simulation of laser excitation at several different times. The inset shows the temporal behavior of the population at two different wave numbers. Reprinted with permission from Lugli and Goodnick [132], copyright (1987) by the American Physical Society.

Not surprisingly, it was found that the carrier cooling was much slower in the quantum wells than in the bulk. From such studies, it was learned that the electrons and holes relaxed differently and did not have a common temperature [133], while in many cases the carrier populations were not thermal at times below ~0.2 ps [134]. Nevertheless, the cooling in the quantum wells also appeared to slow down as the excitation was raised, suggesting the presence of a bottleneck, that will be discussed in chapter 5.

Theoretical studies of the photo-excited carriers in such quantum wells showed that, on the picosecond time scale for which temperatures could be defined, electrons, light holes, and heavy holes all exhibited different temperatures, only becoming the same at times on the order of 100 ps [135]. The holes mainly interacted through the TO modes rather than the LO modes, and this led to both the LO and TO modes being driven out of equilibrium. Interestingly enough, the light holes efficiently reabsorbed the phonons, so that these phonons were sufficiently out of equilibrium, the light holes could be heated to temperatures higher than that of the electrons (and much higher than that of the heavy holes). At long times, recombination and degeneracy of the various carrier concentrations also played a role.

The idea of hot carrier solar cells (HCSC) was introduced by Ross and Nozik [136], and we have previously discussed the concepts in chapter 1. The important ideas were to delay the cascade of phonons emitted by the relaxing electrons and to extract only the hot carriers. Later, it was definitively shown that using quantum wells rather than bulk GaAs would lead to a much slower cooling of the photo-excited carriers, as shown in figure 4.20 [137]. In this latter study, two multi-quantum well samples were studied, one with 20 periods of 20 nm GaAs wells and 8 nm AlGaAs barriers ($x = 0.52$), and a second with 14 periods of 13.5 nm wells and 40 nm barriers ($x = 0.48$). The samples were excited with 10 ps pulses from a dye

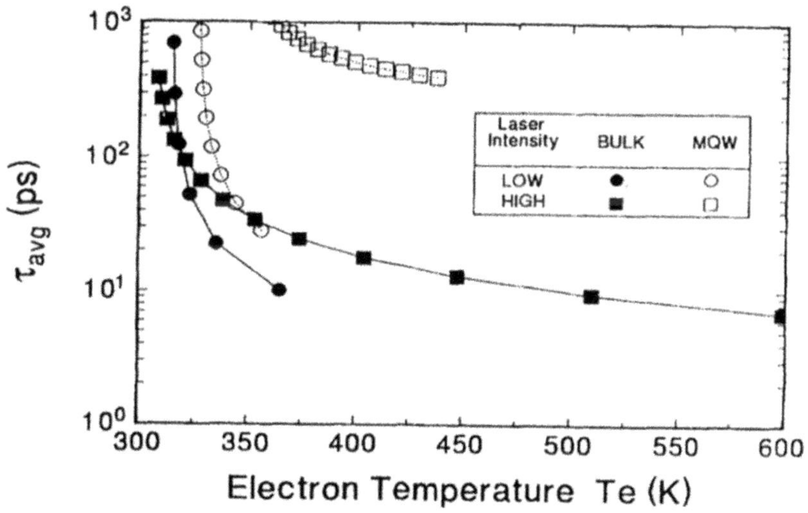

Figure 4.20. Time constant for electron cooling as a function of the deduced electron temperature for bulk GaAs and for a multi-quantum well (sample 2) at the high and low excitation densities. Reprinted with permission from Rosenwaks *et al* [137], copyright (1993) by the American Physical Society.

laser operating at 600 nm. The samples were excited at two densities of 2–4 × 10^{18} cm^{-3} (noted as 'high' in the figure) and 3–5 × 10^{17} cm^{-3} (noted as 'low' in the figure). Detailed measurements of the photoluminescence were then made to extract temperatures and decay times. It is projected that the slower decay in the quantum wells is due to the LO modes being driven out of equilibrium and providing a bottleneck as first suggested by Shah *et al* [131]. It was suggested that this slowing could be enhanced by special consideration of materials that would provide a hindrance to the LO mode decay [138, 139].

In the search for the HCSC, there has been extensive interest and research, beyond these initial studies. The hot carrier electroluminescence has been measured from GaAs/InGaAs and GaAs/AlGaAs *p-i-n* diodes [140]. In the former case, electroluminescence from both materials, which have direct gaps, was observed. The luminescence from each material displayed somewhat different carrier temperatures as may be expected. In the GaAs/AlGaAs material, the luminescence had a break in the response as the indirect luminescence occurred at a higher energy. The AlGaAs luminescence had a distinctly lower temperature than that from the GaAs. In both cases, the luminescence increased as the bias was raised.

Quantum well HCSCs with InGaAs wells were examined for their photoluminescence behavior [141]. These were single quantum wells 10.7 nm thick of $In_{0.16}Ga_{0.84}As$ embedded in GaAs. The devices were studied at 10 K, using laser illumination at 1.52 eV. The photoluminescence was seen to have an increasing intensity as the laser intensity was increased and exhibited increasing electron temperature with increasing laser intensity, as expected. What may be the first report

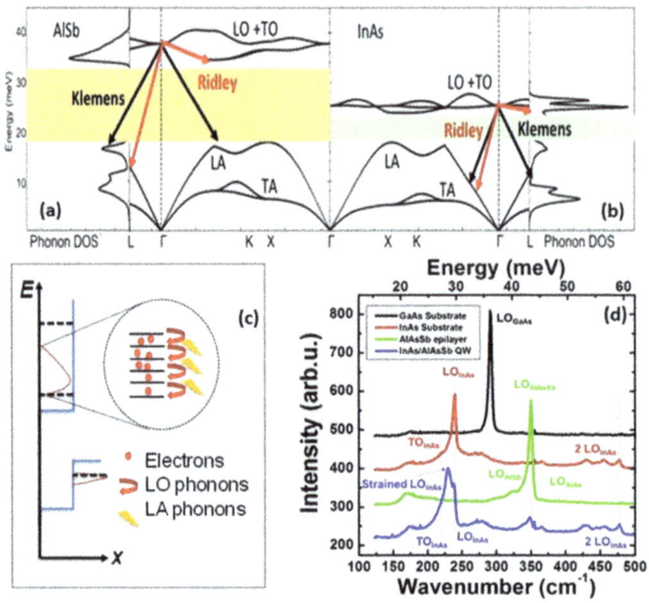

Figure 4.21. Phonon dispersion and density of states for (a) AlSb and (b) InAs. The yellow and green shaded regions are the phonon gaps. Also shown are examples of Klemens and Ridley decay paths (see chapter 5). (c) A schematic of the hot electron distribution (light red curve) and phonon decays for the electrons and holes. (d) Raman spectra for the sample. Reprinted from Esmaielpour *et al* [144], under the creative commons open access license.

of the use of an energy selective barrier to extract only the hot carriers use resonant tunneling from a GaAs absorber, through a double barrier quantum well into an AlGaAs collector [142]. The observation of photo-current from an operating quantum well HCSC was also seen at this time [141].

Quantum wells at room temperature were studied with both electroluminescence and photoluminescence to evaluate the hot carrier contributions to the response [143]. Here, 7.4 nm wells of $In_{0.78}Ga_{0.22}As_{0.81}P_{0.19}$ were embedded in a wider 150 nm quantum well of $In_{0.8}Ga_{0.2}As_{0.435}P_{0.565}$. These cells were examined with intense laser illumination at 980 nm, and the electron temperature was observed to rise dramatically as the fluence was increased beyond 1 kW cm^{-2}. Temperatures above 1000 K were easily reached.

Enhanced electron lifetimes were observed in quantum wells designed to inhibit the phonon relaxation process [144]. A multi-quantum well sample with InAs wells embedded in $AlAs_{0.16}Sb_{0.84}$ barriers, of 2.4 nm and 10 nm, respectively, used 30 periods, all grown on a semi-insulating GaAs substrate. A 2 mm InAs strain relaxation layer was grown prior to the quantum well structure. The quantum well structure has a quasi-type I barrier at low temperatures, which transitions to a modified type-II structure in which the holes are relatively free at high temperature. The ideas behind this cell are shown in figure 4.21. In panels (a) and (b) one sees the phonon spectrum for the binaries AlSb and InAs, with the possible decay routes (discussed extensively in chapter 5). The quantum structures are shown in panel (c) with the phonon emission routes. Panel (d) shows the Raman scattering from the structure. There are substrate phonons (red and black) as well as those in the quantum wells and the barriers (blue and green). The idea is that the separation of the electrons (in the wells) and holes (in the barriers), as well as the structure will slow the carrier cooling. The samples are studied by pump and probe with a laser near-infrared probe at 1.03 eV and a time-resolved THz probe, as shown in the inset to figure 4.22(a). This figure also shows the THz measurement of the transient conductivity following the laser probe excitation. The normalized differential THz transmission through the sample is shown in panel (a). Note that it is separated into

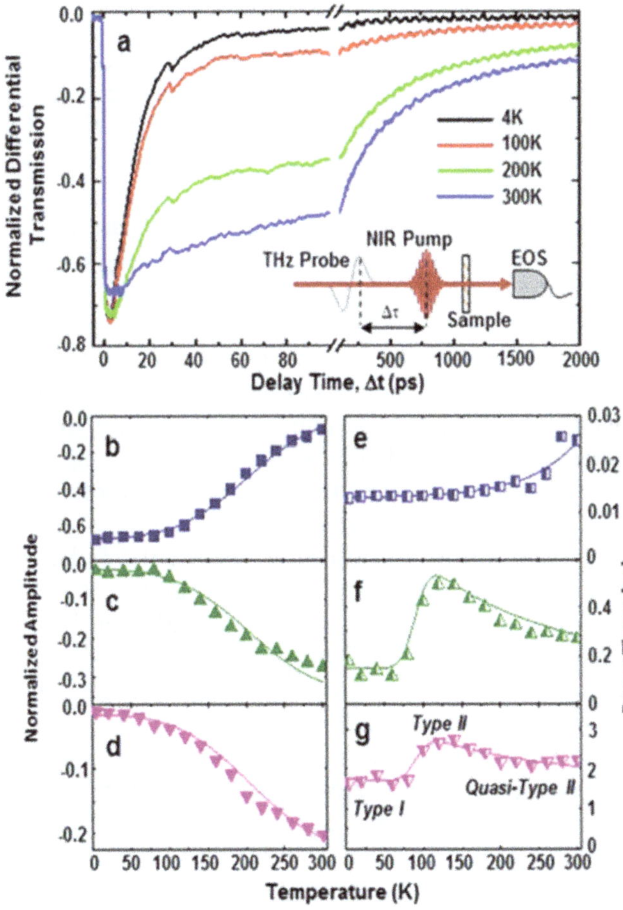

Figure 4.22. Carrier dynamics of the multi-quantum well system. (a) Normalized differential THz transmission for a range of temperatures. The experimental pump–probe arrangement is shown in the inset. The amplitude (b–d) and decay time (e–g) of the fast, intermediate, and slow responses of the signal. The transition from quasi-type I structure to quasi-type-II structure is denoted in (g). Reprinted from Esmaielpour *et al* [144], under the creative commons open access license.

two different time regimes, to account for both fast and slow response of the carriers. The other panels show the temperature dependence of the amplitude (b–d) and decay times (e–g) of what the authors call the fast, intermediate, and slow decay processes at work in the THz signal. Note the different delay time scales in panels (e–g). The authors conclude that the signals are representative of robust hot electrons in the multi-quantum well structure at temperatures above 100 K, which are the result of the type-II structure and inhibited LO phonon dissipation from the active region. It is thought that this leads to an acoustic phonon bottleneck and re-absorption of the nonequilibrium LO phonons by the electrons.

In later work, it was demonstrated in InAs/InAlAs multi-quantum well structures that delayed cooling could also be achieved through transfer of the electrons to the satellite valleys of the InAs [145]. As noted earlier in this chapter, inter-valley transfer represents a storage of the electrons and keeps them from much of the phonon cascade. It remains to be seen if they can be extracted from these upper valleys in a sufficiently timely manner to provide a true HCSC.

At this point, it is probably important to talk about the time scales involved in these various studies. In the introduction to this chapter, the time it takes to form an electron–hole pair was discussed, including the detailed multi-step process this follows (including the polarization). It was also pointed out that the photon is traveling at a relatively high velocity. For example, consider the InGaAs whose absorption coefficient is shown in figure 4.11. Before entering the semiconductor, the photon is traveling at the speed of light, 3×10^{10} cm s^{-1}. Once the photon enters the InGaAs, it is slowed by the optical dielectric constant to about 8.8×10^{9} cm s^{-1}. From figure 4.11, it may be seen that the absorption coefficient for a 2 eV photon is about 1.5×10^{5} cm^{-1}. Then, the average photon will travel about 0.67×10^{-5} cm, or about 67 nm. Given the velocity of the photon, it will cover this distance in about ~0.8 fs. This distance may be regarded as giving the mean range of the 2 eV photon upon entering the semiconductor. The time is somewhat more important, because it gives the average flight time *before initiating* the electron–hole generation. Using the same logic, a 3 eV photon will still travel for about 0.2 fs before beginning the absorption process. The important point is that the 2 and 3 eV photons will be some 67 and 15 nm, respectively, deep into the semiconductor before generation of electron–hole pairs begins. There is, of course, a caveat, and that is: curves such as figure 4.11 are really for bulk material, not for thin films! It has been assumed that this curve holds, even near the surface. There is no assurance that this is the case. The numbers above assume that the so-called range (inverse of the absorption coefficient) into the semiconductor decreases with the energy of the photon. If ions are injected into the semiconductor, their range increases with energy, as their stopping power arises from the Coulombic interaction with the electrons around the atoms. While the photons have a stopping power much different, there is no real reason to assume that the delay time before initiating electron–hole pair generation decreases with energy as was found above. Suppose it were constant. Then, the absorption curve near the surface would be much different than that of figure 4.11. The point of all of this is that not much is really known of the details of the photon absorption on the fs time scale, particularly in confined geometries.

In contrary to the above, the fact that the momentum normal to the plane of a quantum well or thin film is no longer a good quantum number and the energy structures changes into a set of quasi-two-dimensional sub-bands. Thus, the nature of the absorption coefficient changes from that of a bulk material in that it is no longer a continuous function. As a consequence, when light absorption is required, multi-quantum well systems are utilized [146, 147]. When the device is intended for solar cell applications, then other techniques such as light trapping are used to increase the efficiency of the quantum well solar cell [148, 149].

In general, however, the overall time scale for the optical pump creation of electron–hole pairs can be described fairly well. The photons are absorbed and electron–hole pairs are formed on time scales of the order of one to a few fs. Interactions among the electrons, the holes and each other, begin on times of the order of 10–30 fs, depending upon the density of the electron-pairs. Similarly, the emission of optical phonons, dominantly the LO modes, begins on the few 10s of fs. If the energy of the electrons is above the minimum of the satellite valleys, then inter-valley scattering begins on this same scale. In this latter case, the electrons may be stored in the satellite valleys on the few ps time scale, as the reverse scattering back to the Γ valley is a slow process due to the differences in the density of states in the various valleys. Either on the few 10s of fs, in the case of low energy electrons, or on the few ps after inter-valley transferred carriers return, the phonon cascade continues until the electrons are thermalized near the bottom of the Γ valley. This thermalized distribution has a high temperature which gradually decays on a longer time scale.

4.4.2 Measuring the electron distribution

Raman scattering is usually associated with measuring the nonequilibrium phonons that are created by energy relaxation of the hot carriers, as we shall discuss in chapter 5. However, Raman scattering is not limited to measuring the optical phonons. It can be used to measure acoustic phonons, which is normally referred as Brillouin scattering. Additionally, it can measure free particle scattering from an electron or hole gas. This free particle scattering is useful to measure parts of the hot carrier distribution [150–152]. Generally, one uses a *p-i-n* structure to create a sizable electric field in the intrinsic region, which is adjustable by an applied bias. Light is sent through the device, parallel to the applied field. The Raman configuration is usually prepared to sense only the single spin scattering [152]. As with the shift due to the wave number of the phonons, the back scattering signal from the sample is shifted by the wave momentum of the electrons themselves. Normally, the scattered signal is primarily from the Γ valley electrons. Using sub-ps laser pulses also allows one to estimate the short-time overshoot velocity of the electrons as well [153], as will be discussed in the next section.

In figure 4.23, a typical sample configuration is shown for the study of GaAs hot carriers (several more examples have already appeared as figures 2.6–2.9). The sample is grown by molecular beam epitaxy on a (001) GaAs substrate, and subsequently patterned into the shape shown in the figure. Be-doping produces

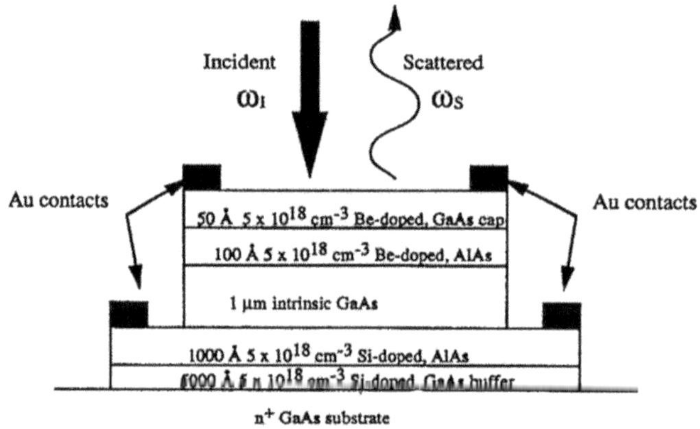

Figure 4.23. Schematic cross-section of the *p-i-n* sample used in the study of carrier distributions. Reprinted with permission from Grann *et al* [152], copyright (1995) by the American Physical Society.

the *p*-type layers, while Si-doping typically produces the *n*-type layers. Au contacts allow the application of a voltage to produce the high electric field. An area of 0.25 mm^2 was opened in the top Au contact in order to allow penetration of the photons. Measurements were made at 80 K. A cavity-dumped, tunable DCM dye laser synchronously pumped by a mode-locked cw Nd:YAG laser. The dye laser gave 3 ps pulses at 7.6 MHz. This excited electrons from all three valence bands, as photons at 1.92 and 1.951 eV were used. The single-particle Raman experiments were carried out in the $Z(X,Y)Z$ configuration, where $Z = (001)$, $X = (100)$, and $Y = (010)$, so that only the single-particle signal associated with the spin-density fluctuations would be measured. The back-scattered signal was detected with a standard Raman photon-counting system. An important aspect of this is that the Raman cross-section is inversely proportional to square of the effective mass. Thus, the scattering is primarily from electrons in the Γ valley where the mass is quite small ($0.067m_0$ in GaAs), and we can study primarily the hot carriers in this valley, without worrying about inter-valley transfer.

Figure 4.24 shows the single-particle scattering for a photon energy of 1.951 eV and a photo-excited carrier density of 10^{17} cm^{-3}, for various electric fields. Asymmetry in the distribution is already seen at $E = 0$, because of the built-in electric fields in the *p-i* junction through which many of the photo-excited carriers have to traverse. Thus, the field indicated is the added field produced by the applied voltage. At densities higher than that used here, the spectrum broadens and shifts toward the Stokes side. We illustrate this in figure 4.25, where the velocity distribution along the electric field is shown for an $In_{0.53}Ga_{0.47}As$ sample (at 300 K) for an electric field of 5 kV cm^{-1} [154]. Two different excited densities are shown. For a density of 5×10^{16} cm^{-3}, it can be inferred that the distribution is very non-spherical and well-peaked along the electric field direction. On the other hand, at a carrier density of 2×10^{18} cm^{-3}, the distribution appears to have become more spherical around the shifted momentum. This move toward symmetrization is usually attributed

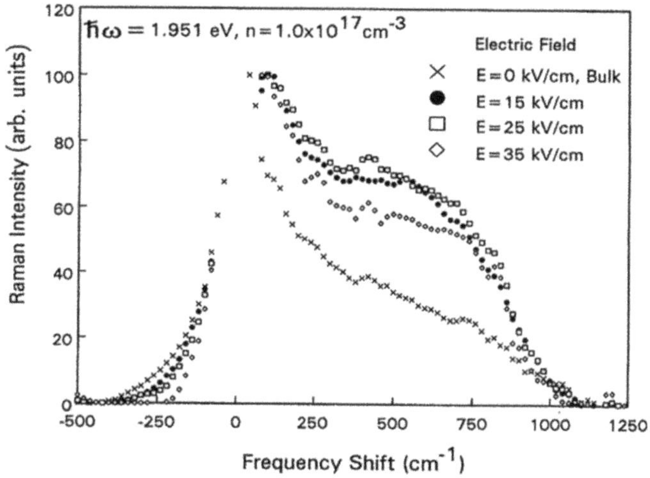

Figure 4.24. Single-particle scattering spectra for $\hbar\omega$ = 1.951 eV for four different electric fields. The background luminescence from the E_0 band gap has been removed from the spectrum. The resulting spectrum is then normalize to its peak intensity. Reprinted with permission from Grann *et al* [152], copyright (1995) by the American Physical Society.

to electron–electron scattering, and already sets in for densities above 10^{17} cm^{-3}, where the distribution function begins to become degenerate, especially in this material. The velocity is converted from the momentum k by the usual non-parabolic band approximation.

4.4.3 Overshoot velocity

The ideas of velocity overshoot have been discussed already in section 2.2. When the electron distribution is excited by an electric field, it initially rises with the equilibrium momentum relaxation time. As the distribution heats over a time scale given by the energy relaxation, the momentum relaxation time decreases and this can lead to a velocity overshoot, otherwise often referred to as ballistic transport. This can also be probed by fast sub-picosecond optical techniques, particularly pump and probe single-particle Raman scattering, discussed in the previous section. Electro-optic sampling of pulsed photoconductivity was used early to study transient transport, and the results were interpreted as being due to velocity overshoot in the device [155]. However, once one has the carrier population as a function of the velocity in the electric field direction, as is given in figure 4.26, a simple integration over the population gives an estimate of the average velocity, and can exhibit directly the velocity overshoot [156], as shown in figure 4.27. In these latter studies, the sample was a GaAs/AlGaAs *p-i-n* structure such as figure 4.23. In the near-intrinsic region, the background electron concentration was about 5×10^{14} cm^{-3}, while the photo-excited density was about 10^{17} cm^{-3}. The sample was excited with the same pulsed laser system described in the previous section. When the excitation

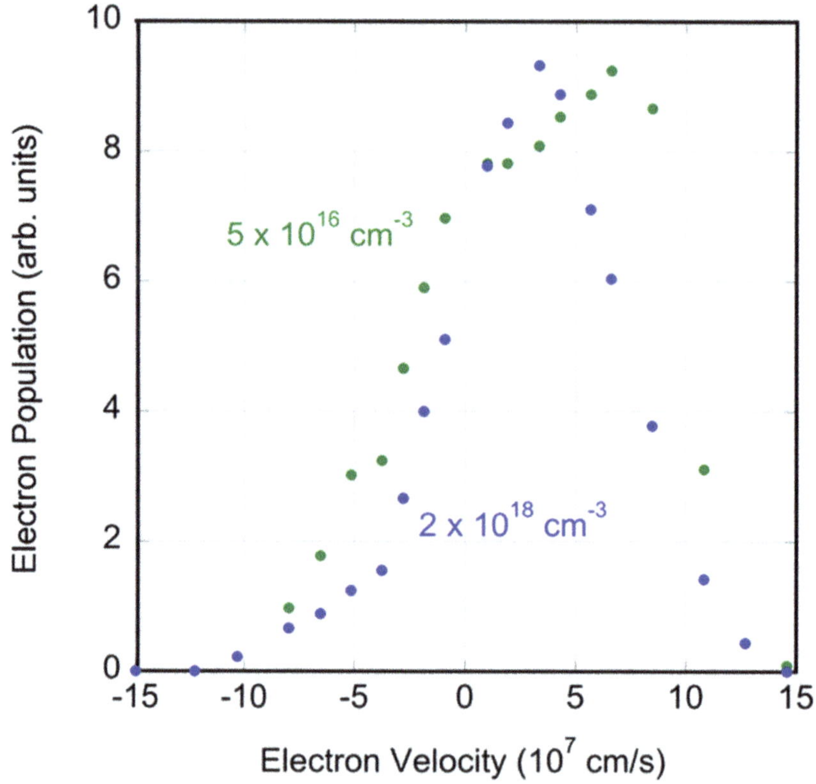

Figure 4.25. Electron velocity distribution along the electric field in an $In_{0.53}Ga_{0.47}As$ *p-i-n* structure for an electric field of 5 kV cm^{-1}, and two different photo-excited carrier densities. Data is taken from Tsen *et al* [154].

pulse was 600 fs, the inferred velocity at 20 kV cm^{-1} was 4.5×10^7 cm s^{-1}, and at 25 kV cm^{-1} it was 5.8×10^7 cm s^{-1}. If the pulse was widened to 3 ps, the peak velocity at both fields dropped to 2.5×10^7 cm s^{-1}. Clearly, the data exhibited the properties of an overshoot velocity. Again, it should be remarked that the Raman signal samples primarily the carriers in the Γ valley due to its variation with the inverse of the square of the effective mass. If the excited carrier density is increased by an order of magnitude, the measured velocity for the 600 fs pulse decreased by more than a factor of 2, presumably due to the much increased carrier–carrier scattering, which rapidly reduces the streaming in the field direction [147]. Studies in InP, with a photo-excited carrier density of 10^{18} cm^{-3}, showed a velocity of 4.7×10^7 cm s^{-1} for the 600 fs pulse and 2.9×10^7 cm s^{-1} for the 3 ps pulse [157], both at 15 kV cm^{-1}. The overshoot velocity in InP is considerably large than that of GaAs for an equivalent situation. In figure 4.26, the carrier populations in GaAs, InP, and $In_{0.53}Ga_{0.47}As$ for a field of 15 kV cm^{-1} for a density of 5×10^{17} cm^{-3} at 300 K for 600 fs pulsed excitation at 1.43 eV [158]. The inferred velocities for these three

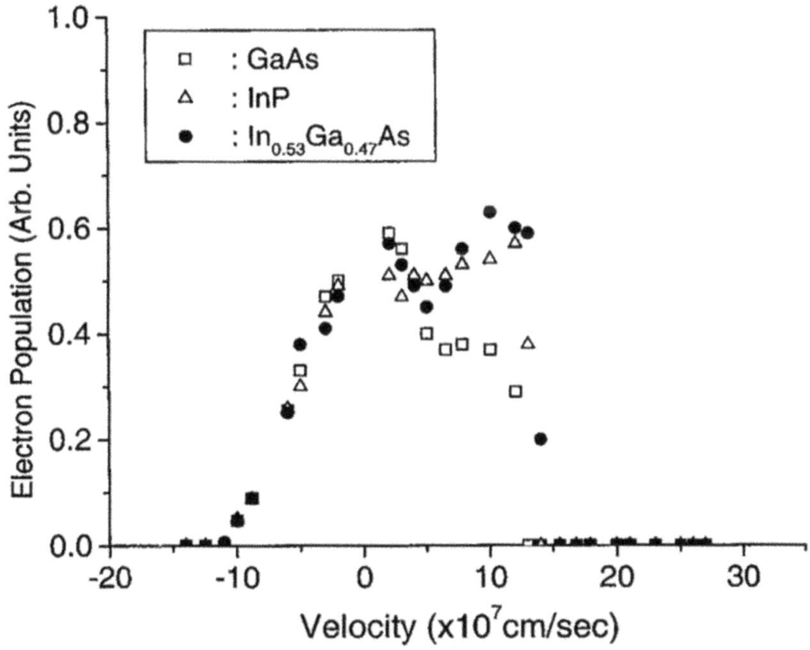

Figure 4.26. The electron distributions for three different semiconductors taken under the same experimental conditions. Reprinted from Liang *et al* [158], with permission from IOP Publishing. All rights reserved.

materials and conditions were 3.2×10^7 cm s^{-1}, 4.0×10^7 cm s^{-1}, and 5.4×10^7 cm s^{-1}, respectively.

Further studies of $In_{0.53}Ga_{0.47}As$ were made with a higher laser energy, 1.92 eV with an 800 fs pulse width, for a wide range of photo-excited carrier concentrations [158]. With this higher photon energy, there is more scattering to satellite valleys and quicker decay of the ballistic velocity peak. The results for an electric field of 5 kV cm^{-1} are shown in figure 4.27. It may be noted that the velocity decreases slightly as the carrier density increases. This may be a result of the carrier screening of the electron–LO phonon interaction or of the carrier–carrier interactions themselves.

4.5 Real-space transfer

In chapter 2, the idea of real-space transfer of carriers under hot electron conditions was discussed. Optical studies of this effect are in fact quite interesting as well. One of the earliest studies was of the photoluminescence coming from a GaAs/AlGaAs heterostructure during real-space transfer, with the electric field parallel to the interface at 4.2 K [159]. The photoluminescence signal was used to infer the hot carrier temperature, which would drop as soon as transfer began to occur. Here, it was found that fields as low as 0.4 kV cm^{-1} were sufficient to induce the transfer. Electroluminescence was studied in a similar structure, but with a strained 10 nm $In_{0.2}Ga_{0.8}As$ quantum well embedded within the normal GaAs well, with 50 nm of GaAs on either side of the InGaAs well. AlGaAs barriers were placed outside this combined well [160]. This structure also allowed holes to be trapped, and light-

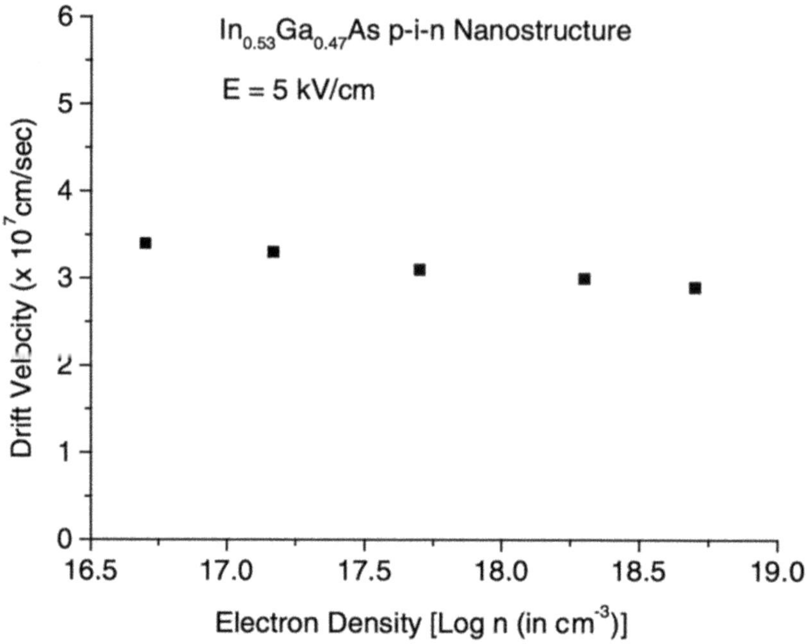

Figure 4.27. The electron drift velocity as a function of the photo-excited carrier density for an $In_{0.53}Ga_{0.47}As$ *p-i-n* structure at 300 K. Reprinted from Tsen *et al* [156], with permission of IOP Publishing.

emission was observed from the structure, with the peak luminescence around 0.9 μm. Strong infrared radiation, and possible hole population inversion was observed from multi-quantum well of InGaAs/GaAs [161]. Here the quantum wells had indium composition $0.1 < x < 0.2$ and thickness of 5–10 nm. The GaAs was uniformly 20 nm thick and 20 such quantum wells were used. Emission was observed at 4.2 K for electric fields above 0.5 kV cm^{-1}.

A more interesting study used lateral transport in a GaAs layer above several AlGaAs layers, with the structure shown in figure 4.28(a) [162]. The two-dimensional electron gas lies deep in the structure at the top of the GaAs buffer, shown by the blue dashed line in the figure. Carriers come from modulation doping in the second layer up, which is Si doped. The contacts allow the field to be applied along the surface. The observed observations observed in the current in panel (b) arise from the Gunn effect, and the creation of traveling domains (see section 2.3.2), where the field can rise to hundreds of kV cm^{-1}. It is believed that impact ionization in these domains begins at voltages of 50–60 V, as this is where the luminescence begins to appear. This luminescence is thought to occur from recombination of electrons and holes created by the impact ionization. While only the lower energy luminescence is shown in panels (c) and (d) of the figure, luminescence is observed up to 2 eV at the higher biases shown in panel (b). A second peak in this is observed at 1.84 eV, which is believed to arise from carriers that have transferred in real space to the AlGaAs.

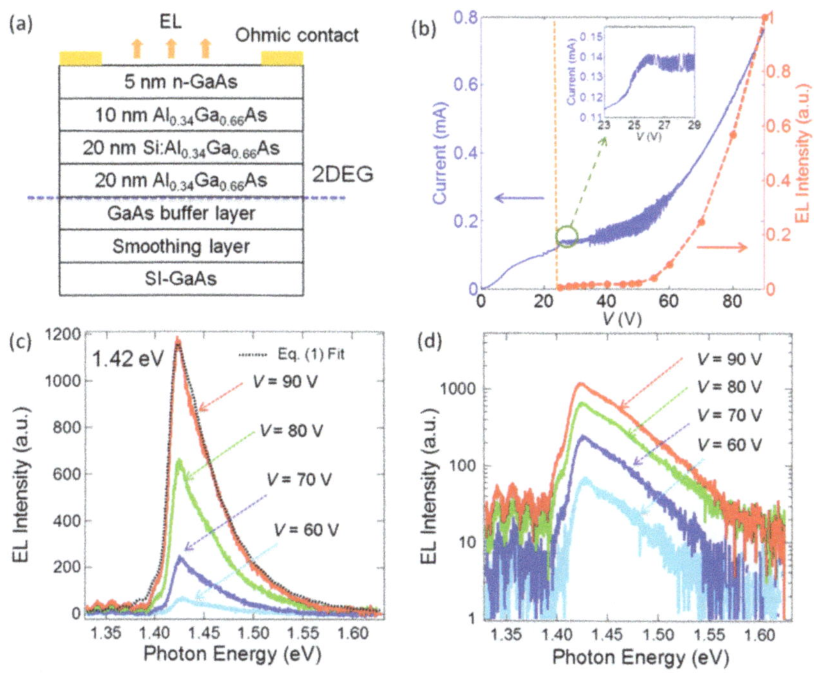

Figure 4.28. (a) Vertical structure of the $GaAs/Al_{0.34}Ga_{0.66}As$ heterostructure studied. The blue dashed line indicates the position of the electron gas. (b) The bias dependence of the current (blue) and normalized, integrated luminescence (red). The spectrum of the luminescence on a linear (c) and logarithmic (d) scale for various applied voltages. The black line is a fit to a thermal distribution. Reprinted from Gao *et al* [162] under the creative commons open access license.

References

[1] Haug H 1985 *J. Luminesc.* **30** 171
[2] Schmitt-Rink S, Ell C and Haug H 1986 *Phys. Rev.* B **33** 1183
[3] Kuznetsov A V 1991 *Phys. Rev.* B **44** 8721
[4] Geltman S 1977 *J. Phys.* B **10** 831
[5] Haan S L and Geltman S 1982 *J. Phys.* B **15** 1229
[6] Lindberg M and Koch S W 1988 *Phys. Rev.* B **38** 3342
[7] Haug H 1988 *Optical Nonlinearities and Instabilities in Semiconductors* ed H Haug (San Diego, CA: Academic)
[8] Kuhn T and Rossi F 1992 *Phys. Rev.* B **46** 7496
[9] Ferry D K 2020 *Semiconductor Bonds and Bands* 2nd edn (Bristol: IOP Publishing)
[10] Shah J 1989 *Sol.-State Electron.* **32** 1051
[11] Fork R L, Brito-Cruz C H, Becker P C and Shank C V 1987 *Opt. Lett.* **12** 483
[12] Becker P C, Fragnito H L and Brito-Cruz C H *et al* 1988 *Appl. Phys. Lett.* **53** 2089
[13] Papalazarou E, Faure J and Mauchain J *et al* 2012 *Phys. Rev. Lett.* **108** 256808
[14] Kanasaki J, Tanimura H and Tanimura K 2014 *Phys. Rev. Lett.* **113** 237401
[15] Tanimura H, Kanasaki J and Tanimura K *et al* 2016 *Phys. Rev.* B **93** 161203
[16] Sjakste J, Tanimura K and Barbarino G *et al* 2018 *J. Phys. Cond. Matt.* **30** 353001
[17] Engeler W, Levinstein H and Stannard C Jr. 1961 *Phys. Rev. Lett.* **7** 62

[18] Engeler W, Levinstein H and Stannard C Jr. 1961 *J. Phys. Chem. Sol.* **22** 249
[19] Habegger M A and Fan H Y 1964 *Phys. Rev. Lett.* **12** 99
[20] Park Y S and Langer D W 1964 *Phys. Rev. Lett.* **13** 392
[21] Shaw R W *Phys. Rev.* B **3** 3283
[22] Stocker H J, Levinstein H and Stannard C R Jr. 1966 *Phys. Rev.* **150** 613
[23] Stocker H J and Kaplan H 1966 *Phys. Rev.* **150** 1966
[24] Nahory R E and Fan H Y 1966 *Phys. Rev. Lett.* **17** 251
[25] Nahory R E 1969 *Phys. Rev.* **178** 1293
[26] Barbarie A and Fortin E 1974 *Sol. State Commun.* **14** 267
[27] Collins A T, Lightowlers E C and Dean P J 1969 *Phys. Rev.* **183** 725
[28] Nesládek M, Meykens K and Haenen K *et al* 1999 *Phys. Rev.* B **59** 14852
[29] Instone T and Eaves L 1979 *J. Phys. C: Sol. State Phys.* **12** L345
[30] Nicholas R, Carter A C and Fung S *et al* 1980 *J. Phys. C: Sol. State Phys.* **13** 5215
[31] Remes Z, Nesladek M and Haenen K *et al* 2005 *Phys. Stat. Solidi (a)* **202** 2229
[32] Lüth H 1972 *Phys. Rev. Lett.* **29** 1377
[33] Morawski A and Lagowski J 1974 *Phys. Lett.* **47A** 219
[34] Cavenett B C 1972 *Phys. Rev.* B **5** 3049
[35] Barker J R and Hearn C J 1969 *Phys. Lett.* **29A** 215
[36] Barker J R and Hearn C J 1969 *J. Phys. C: Sol. State Phys.* **2** 2128
[37] Barker J R and Hearn C J 1970 *Phys. Lett.* **31A** 51
[38] Seiler D G, Littler C L and Heiman D 1984 *J. Appl. Phys.* **57** 2191
[39] Moore K J, Duggan G and Dawson P *et al* 1989 *Phys. Rev.* B **39** 1219
[40] Hamaguchi C, Kono T and Inuishi Y 1967 *Phys. Lett.* **24A** 500
[41] Acket G A 1967 *Phys. Lett.* **25A** 374
[42] Hess K, Nimtz G and Seeger K 1969 *Sol. State Electron.* **12** 79
[43] Philipp A, Kuchar F and Seeger K 1977 *Phys. Stat. Solidi (b)* **79** 115
[44] Peierls R E 1955 *Quantum Theory of Solids* (Oxford: Oxford Univeristy Press) p 108
[45] Grüner G 1988 *Rev. Mod. Phys.* **60** 1129
[46] Kazarinov R F and Suris R A 1972 *Sov. Phys. Semicond.* **6** 120
[47] Esaki L and Tsu R 1970 *IBM J. Res. Develop.* **14** 61
[48] Ferry D K 1992 *Prog. Quantum Electron.* **16** 251
[49] Faist J, Capasso F and Sivco D L *et al* 1994 *Science* **264** 553
[50] Helm M, England P and Colas E *et al* 1990 *Surf. Sci.* **229** 199
[51] Ulbricht R, Hendry E and Shan J *et al* 2011 *Rev. Mod. Phys.* **83** 543
[52] Beard M C, Turner G M and Schmuttenmaer C A 2000 *Phys. Rev.* B **62** 15764
[53] Radisic V, Leong K M K H and Mei X *et al* 2012 *IEEE Trans. Microwave Theory Techniq.* **60** 724
[54] Mei X, Yoshida W and Lange M *et al* 2015 *IEEE Electron. Dev. Lett.* **36** 327
[55] Akis R, Ayubi-Moak J and Faralli N *et al* 2008 *IEEE Electron. Dev. Lett.* **29** 306
[56] Akis R, Faralli N and Goodnick S M *et al* 2010 *Phys. Stat. Solidi (c)* **7** 2502
[57] Neto H C, Guinea F and Peres N M R *et al* 2009 *Rev. Mod. Phys.* **81** 109
[58] Kaasbjerg K, Thygesen K S and Jacobsen K W 2012 *Phys. Rev.* B **85** 115317
[59] Cunningham P D, McCreary K M and Hanbicki A T *et al* 2016 *J. Phys. Chem.* C **120** 5819
[60] Zhu W, Low T and Lee Y-H *et al* 2013 *Nature Commun.* **5** 3087
[61] Jensen S A, Versluis J and Cánovas E *et al* 2012 *Appl. Phys. Lett.* **101** 222113
[62] Hafez H A, Chai X and Sekine Y *et al* 2017 *Phys. Rev.* B **95** 165428

[63] Gustafson J K, Cunningham P D and McCreary K M *et al* 2019 *J. Phys. Chem.* C **123** 30676
[64] Deng H, Chen L and Qu W *et al* 2020 *Opt. Engr.* **59** 056101
[65] von Klitzing K, Dorda G and Pepper M 1980 *Phys. Rev. Lett.* **45** 494
[66] Landau L 1930 *Z. Phys.* **64** 629
[67] Ebert G, von Klitzing K, Ploog K and Weimann G 1983 *J. Phys. C: Sol. State Phys.* **16** 5441
[68] Cage M E, Dziuba R F and Field B F *et al* 1983 *Phys. Rev. Lett.* **51** 1374
[69] von Hippel E 1933 *Zeitschr. Elekrochem.* **39** 506
[70] Fröhlich H 1937 *Proc. R. Soc.* A **160** 230
[71] Cai C, Lei X L and Ting C S 1985 *Phys. Rev.* B **31** 4070
[72] Smrcka L 1985 *J. Phys. C; Sol. State Phys.* **18** 2897
[73] Komiyama S, Takamatsu T, Hiyamizu S and Sasa S 1985 *Sol. State Commun.* **54** 479
[74] Ridley B K 1963 *Proc. Phys. Soc.* **82** 954
[75] Takamatsu T, Komiyama S, Hiyamizu S and Sasa S 1986 *Surf. Sci.* **170** 202
[76] Pokrovsky V L, Pryadko L P and Talapov A L 1990 *J. Phys. C: Cond. Matt.* **2** 1583
[77] Chklovskii D B, Shklovskii B and Glazman L I 1992 *Phys. Rev.* B **46** 4026
[78] Polyakov D G and Raikh M E 1990 *Sol. State Commun.* **74** 1209
[79] Dyakonov M 1991 *Sol. State Commun.* **78** 817
[80] Ahlers F J, Hein G and Scherer H *et al* 1993 *Semicond. Sci. Technol.* **8** 2062
[81] Kawaji S, Hirakawa K and Nagata M *et al* 1994 *J. Phys. Soc. Jpn.* **6** 2303
[82] Shashkin A A, Kent A J and Harrison P A *et al* 1994 *Phys. Rev.* B **49** 5379
[83] Okuno T, Kawaji S and Ohrui T *et al* 1995 *J. Phys. Soc. Jpn.* **64** 1881
[84] Cage M E 1996 *J. Res. NIST* **101** 175
[85] Ishikawa K, Maeda N and Tadaki K 1996 *Phys. Rev.* B **54** 17819
[86] Yokoi M, Okomoto T and Kawaji S *et al* 1998 *Physica* B **249-51** 93
[87] Shizuya K 2000 *Physica* E **6** 148
[88] Matthews A J, Watts J P and Zhu M *et al* 2000 *Physica* E **6** 140
[89] Sagol B E, Nachtwei G and von Klitzing K *et al* 2002 *Phys. Rev.* B **66** 075305
[90] Matthews A J, Kavokin K V and Usher A *et al* 2004 *Phys. Rev.* B **70** 075317
[91] Oto K, Katagiri T and Tsubota T 2005 *AIP Conf. Proc.* **772** 523
[92] Takashima K, Nicholas R J and Kardynal B *et al* 2006 *Semicond. Sci. Technol.* **21** 1758
[93] Bykov A A, Marchinshin I V and Bakarov A K *et al* 2007 *JETP Lett.* **85** 63
[94] Singh V and Deshmuhk M M 2009 *Phys. Rev.* B **80** 081404
[95] Gornik E 1972 *Phys. Rev. Lett.* **29** 595
[96] Kobayashi K L I, Komatsubara K F and Otsuka E 1973 *Phys. Rev. Lett.* **30** 702
[97] Gornik E, Müller W and Kohl F 1974 *IEEE Trans. Microwave Theory Technq.* **22** 991
[98] Gornik E, Nguyen V T and Damen T C 1976 *Appl. Phys. Lett.* **29** 169
[99] Gornik E and Tsui D 1978 *Surf. Sci.* **73** 217
[100] Gornik E 1979 *J. Magn. Magn. Mat.* **11** 39
[101] Komiyama S 1982 *Adv. Phys.* **31** 255
[102] Vasil'ev Y B and Ivanov Y L 1984 *Sov. Tech. Phys. Lett.* **10** 398
[103] Bespalov A V, Gol'tsman G N, Semenov A D and Renk K F 1991 *Sol. State Commun.* **80** 503
[104] Unterrainer K, Kremser C and Gornik E *et al* 1990 *Phys. Rev. Lett.* **64** 2277
[105] Heiss W, Unterrainer K and Gornik E 1994 *IEEE J. Quantum Electron.* **30** 2778

[106] Gornik E and Tsui D 1976 *Phys. Rev. Lett.* **37** 1425

[107] Gornik E, Schawarz R and Tsui D *et al* 1981 *Sol. State Commun.* **38** 541

[108] Briggs S, Jovanovic D and Leburton J P 1989 *Sol.-State Electron.* **32** 1657

[109] Helm M, England P and Colas C *et al* 1989 *Phys. Rev. Lett.* **63** 74

[110] Allen S J, Brozak G and Colas E *et al* 1992 *Semicond. Sci. Technol.* **7** B1

[111] Slater J C and Koster G F 1954 *Phys. Rev.* **94** 1498

[112] Madelung O 2004 *Semiconductors: Data Handbook* 3rd edn (Berlin: Springer)

[113] http://ioffe.ru/SVA/NSM/Semicond/

[114] Smith R A 1969 *Wave Mechanics of Crystalline Solids* (London: Chapman and Hall)

[115] Kennedy C J, Matter J C and Smirl A L *et al* 1974 *Phys. Rev. Lett.* **32** 419

[116] Auston D H and Shank C V 1974 *Phys. Rev. Lett.* **32** 1120

[117] Ferry D K 1978 *Phys. Rev.* B **18** 7033

[118] Shah J 1978 *Sol.-State Electron.* **21** 43

[119] Vella-Coleiro G P 1969 *Phys. Rev. Lett.* **23** 697

[120] Shah J, Leite R C C and Scott J F 1970 *Sol. State Commun.* **8** 1089

[121] Osman M and Ferry D K 1987 *Phys. Rev.* B **36** 6018

[122] Shah J and Leite R C C 1969 *Phys. Rev. Lett.* **22** 1304

[123] Conwell E 1967 *Solid State Physics* ed F Seitz, D Turnbull and H Ehrenreich (New York: Academic)

[124] Stratton R 1958 *Proc. R. Soc. (London)* **A246** 406

[125] Hearn C J 1965 *Proc. Phys. Soc.* **86** 881

[126] Fröhlich H 1937 *Proc. R. Soc.* A **160** 280

[127] Fröhlich H 1954 *Adv. Phys.* **3** 325

[128] Lyddane R H, Sachs R G and Teller E 1941 *Phys. Rev.* **59** 673

[129] Shank C V, Fork R L and Yen R *et al* 1979 *Sol. State Commun.* **47** 981

[130] Ryan J F 1984 *Physica* B+C **127B** 343

[131] Shah J, Pinczuk A, Gossard A C and Wiegmann W 1985 *Phys. Rev. Lett.* **54** 2045

[132] Lugli P and Goodnick S M 1987 *Phys. Rev. Lett.* **59** 716

[133] Höpfel R A, Shah J and Gossard A C 1986 *Phys. Rev. Lett.* **56** 765

[134] Knox W H, Hirliman C and Miller D A B *et al* 1986 *Phys. Rev. Lett.* **56** 1191

[135] Joshi R P and Ferry D K 1989 *Phys. Rev.* B **39** 1180

[136] Ross R T and Nozik A J 1982 *J. Appl. Phys.* **53** 3813

[137] Rosenwaks Y, Hanna M C and Levi D H *et al* 1993 *Phys. Rev.* B **48** 14675

[138] Conibeer G J, König D, Green M A and Guillemoles J F 2008 *Thin Sol. Films* **516** 6948

[139] Conibeer G J, Guillemoles J F, Yu F and Levard H 2014 *Advanced Concepts in Photovoltaics* ed A J Nozik, G J Conibeer and M C Beard (London: Royal Society of Chemistry) ch 14

[140] Nelson J, Barnes J and Ekins-Daukes N *et al* 1997 *Appl. Phys. Lett.* **82** 6240

[141] Hirst L C, Walters R J, Führer M F and Ekins-Daukes N 2014 *Appl. Phys. Lett.* **104** 231115

[142] Dimmock J A R, Day S and Kauer M *et al* 2014 *Prog. Photovol.: Res. Appl.* **22** 151

[143] Nguyen D-T, Lombez L and Gibelli F *et al* 2017 *Nat. Energy* **3** 236

[144] Esmaielpour H, Whiteside V R and Piyathilaka H P *et al* 2018 *Sci. Repts.* **8** 12473

[145] Whiteside V R, Esmaielpour H and Mishima T D *et al* 2019 *Semicond. Sci. Technol.* **34** 094001

[146] Barnham K W J 1990 *J. Appl. Phys.* **67** 3490

[147] Elkins-Daukes N J, Barnham K W J and Connolly J P 1999 *Appl. Phys. Lett.* **75** 4195
[148] Campbell P and Green M A 1987 *J. Appl. Phys.* **62** 243
[149] Li K H, Li P C and Hu D C *et al* 2014 *Appl. Phys. Lett.* **115** 044303
[150] Mooradian A and McWhorter A L 1970 Proc. of the 10th Int. Conf. on Physics of Semiconductors S P Keller, J C Hansel and F Stern (Oak Ridge, TN: US Atomic Energy Commission) p 380
[151] Ralph S E and Wolga G J 1990 *Phys. Rev.* B **42** 11353
[152] Grann E D, Sheih S J and Tsen K T *et al* 1995 *Phys. Rev.* B **51** 1631
[153] Grann E D, Tsen K T and Ferry D K 1996 *Phys. Rev.* B **53** 9838
[154] Tsen K T, Kiang J G and Ferry D K 2006 *J. Phys. Cond. Matter* **18** L585
[155] Meyer K, Pessot M and Mourou G *et al* 1988 *Appl. Phys. Lett.* **53** 2254
[156] Grann E D, Tsen K T and Sankey O F *et al* 1995 *Appl. Phys. Lett.* **67** 1760
[157] Tsen K T, Ferry D K and Wang J-S *et al* 1996 *Appl. Phys. Lett.* **69** 3575
[158] Liang W, Tsen K T and Ferry D K *et al* 2004 *Semicond. Sci. Technol.* **19** S23
[159] Makiyama M, Kasai K, Ohori T and Komeno J 1992 *Semicond. Sci. Technol.* **7** B248
[160] Yeh Y-H, Lai J-T and Lee J Y-M 1997 *J. Appl. Phys.* **81** 3607
[161] Aleshkin V Ya, Andronov A A and Antonov A V *et al* 1998 *Physica* B **249-51** 971
[162] Gao W, Wang X and Chen R *et al* 2015 *ACS Photon.* **2** 1155

IOP Publishing

Hot Carriers in Semiconductors

David K Ferry

Chapter 5

Nonequilibrium phonons

Interest in nonequilibrium phonons apparently became of interest in studies of spin relaxation in metals at very low temperatures [1]. The observation in modeling the spin relaxation was that the phonons could not remain at the lattice temperature in the presence of the energy transfer arising from the spin relaxation process. Subsequent work in the rare-earth salts showed that the nonequilibrium phonons might even create a bottleneck that would slow the relaxation of the nonequilibrium phonons (which has subsequently become of some interest, particularly in hot carrier solar cells) [2], and subsequent work found the presence of a bottleneck arising from the near equality of emission and absorption processes [3]. In semiconductors, it was found that the spin-lattice relaxation for the spins on dangling bonds in Si could lead to a bottleneck [4]. In all of these early studies, it was the acoustic phonon modes which were driven out of equilibrium, as optical modes were not excited at these low temperatures. And the controversial views over the presence, or not, of a phonon bottleneck have reappeared in later work.

The earliest interesting aspect of the excitation of the acoustic modes was the observation that one could achieve spatially inhomogeneous amplification of the acoustic modes from the carriers in a high electric field, and the propagation of high-field domains were observed in CdS [5]. This was attributed to the process of stimulated emission of acoustic phonons when the carrier's drift velocity was faster than the velocity of sound. The interaction was considered to be the piezoelectric interaction between the applied field and the acoustic modes. In this case, the microwave sound waves were introduced to the sample through transducers in a manner in which they propagated along the electric field [6]. It was subsequently observed that this would lead to a kink in the current versus electric field and a sub-linear overall effect [7]. It should be noticed that this work was contemporary with the observations of the Gunn effect (chapter 2), and led to some confusion over the interpretation of the latter. This was especially true with the observation that the propagating acoustoelectric domains could lead to current oscillations [8], even without the introduction of external sound waves.

doi:10.1088/978-0-7503-3947-6ch5

Nevertheless, the observations of the interactions between the acoustic modes and the electrons led to the field of microwave acoustics [9]. In this regard, acoustic spectroscopy became a field in which the acoustic phonons could be used to study properties of the semiconductors, such as defects, or even carrier motion itself [10, 11]. It was subsequently found that a MOSFET at 4.2 K, when cooled to 4.2 K, would emit a pulse of acoustic phonons when excited with a pulsed electric field [12]. It was already known at this time that hot carriers at low temperatures could lose their energy to the acoustic modes, but not at this strength.

Some years after the onset of study of the acoustic modes, it was found that exciting a semiconductor with intense laser radiation could lead to nonequilibrium populations of the longitudinal optical (LO) modes in polar materials [13, 14], especially at low lattice temperature and with temporally short pulses of this radiation. The laser creates electron–hole pairs with considerable excess energy, particularly for the electrons in direct gap semiconductors. The carriers then cool by the emission of LO phonons, particularly the polar LO modes in compound semiconductors [15]. If sufficient energy is transferred to the phonon system, from the relaxing carriers, the phonon distribution can be driven out of equilibrium. This population can be studied by Raman scattering [15], among other methods, and the population estimated by the relative strengths of the Stokes and anti-Stokes transitions. In many studies, it is found that the LO modes are coupled to the free-carrier plasma modes and can appear as hybrid coupled modes [16, 17].

A bottleneck of these polar LO phonons have been found in low dimensional structures such as quantum wells and quantum wires in the compound semiconductors [18–21]. What is meant by a 'bottleneck' is that the population of optical phonons has reached a value so that $N_q \sim N_q + 1$, and the phonon emission rate is comparable to the absorption rate. This leads to a slowing in both the carrier energy relaxation process and the phonon relaxation as well, which depend upon the difference in the emission and absorption processes. The existence of the bottleneck was confirmed theoretically through simulations utilizing ensemble Monte Carlo techniques [22]. It was also shown to be present by direct use of Raman scattering, as discussed above [23].

Separate from the optical studies, nonequilibrium phonons and possible bottlenecks have been found in semiconductor devices and studied for some time. Hot phonon populations for the optical phonons and acoustic phonons have been studied for Si MOSFETs [24], GaN HEMTs [25–28], as well as carbon nanotubes [29]. Generally, the decay of the optical phonon modes into lower energy modes and the thermal conductivity both involve three-phonon processes involving primarily the third-order elastic constants [30]. The important fact is that the decay of the optical modes, and the possibility for a bottleneck is the same whether the dynamic carriers are excited by photons or by an electric field in the semiconductor device. The physics of the phonon decay and lifetime will be the same.

In this chapter, the nonequilibrium system of acoustic and optical modes will be discussed, as they are connected to the system of hot carriers. In the next section, the nature of the problem will be discussed. Then, various experimental approaches to the study of these hot phonons will be treated. The last two sections will treat the development of the current understanding of the acoustic and optical phonons in their nonequilibrium situations.

5.1 The nature of the problem

In order to clarify the processes that are of interest, consider figure 5.1, which is a block diagram of the flow of energy from the source into the lattice. Energetic electrons are created at high energies in the conduction band (or in the valence band for holes) by either a high electric field or by the absorption of photons. Even with the photons, the carriers have an energy that arises from the fact that the photon energy is usually considerably larger than the band gap of the material. Consider, for example, carriers excited by a 2 eV photon in GaAs, where the band gap is 1.4 eV. These carriers go into the Γ valleys as GaAs is a direct gap material. The excess energy (0.6 eV) is split between the created electron and hole, with most going to the electron, due to its much smaller effective mass.

If the electrons are created above the minima of the satellite valleys, the electrons subsequently will be transferred to the satellite valleys. Transfer to the satellite valleys is a rapid process, when compared to phonon emission. For our example of GaAs, the L valleys lie 290 meV above the bottom of the conduction band. Intervalley transfer is known to occur in nearly all direct gap semiconductors as well as in Ge. It can even be present in Si when strain or quantum confinement is present or even when a high electric field is oriented in a direction other than one of the three principal axes in the Brillouin zone [31–33]. It is a property of nearly all modern semiconductor devices, as was discussed in chapter 2.

In the situation here, the carriers are stored in the satellite valleys for a few picoseconds, and then return to the Γ valley, and begin to lose their energy via the emission of several LO phonons, or even directly by the emission of acoustic modes,

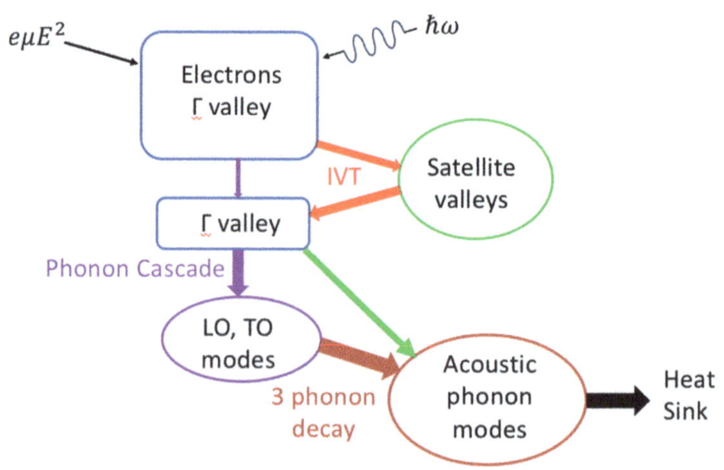

Figure 5.1. Schematic view of the flow of energy from the carriers, whether excited by photons or an electric field, to the phonons and thus to the heat sink.

especially at low temperature. This is the phonon cascade, which can lead to a nonequilibrium distribution function for the phonons if a sufficient number are emitted during this cascade. As the phonon population builds up, it is reduced by relaxing to lower energy phonons through a third-order elastic constant [30], or in some cases through a second-order piezoelectric interaction [34]. These three-phonon processes will be discussed in the next section. But, this points to an additional consideration on inter-valley transfer, as this is known to be important in very-high frequency HEMTs for raising the breakdown voltage and reducing the role of the phonon cascade [35]. Similarly, using the satellite valleys has also been suggested as a method of reducing the phonon cascade [36]. Once the energy reaches the acoustic phonons, it can be transferred to the heat sink, but still is characterized by a rise in the temperature of the device.

5.1.1 The phonon spectrum

In diatomic lattices, such as the III–V and II–VI semiconductors, the polar nature of the LO and TO modes arise from the fact that there are two dissimilar atoms per primitive unit cell [37]. These two atoms form the basis and have a different number of outer shell electrons which contribute to the bonding. With the tetrahedral bonding, each atom has, on average, four electrons which are shared with its nearest neighbors to fill the shells. For a binary material, such as GaAs, the two atoms of the basis will each have a small, but opposite charge, called the *effective charge* e^*. This leads to a dipole force between the two atoms, or between any one of the two charges and its four neighbors by symmetry. This dipolar field can interact with external electromagnetic waves, which leads to a polarization of the lattice, and this contributes to the dielectric function and provides a different permittivity depending upon whether the frequency of the waves lies above or below the characteristic frequency of the lattice polarization [38]. These two values of the permittivity are referred to as the optical and static permittivities (often called dielectric constants). Importantly though, it is this dipolar field which scatters the electrons through a Coulombic interaction.

The interaction with the electromagnetic waves and the lattice polarization splits the LO and TO modes at Γ. The difference between the optical and static permittivity is connected to this LO–TO splitting through the Lyddane–Sachs–Teller relation [39]. Because of the lattice polarization, the LO mode is commonly referred to as the polar LO mode. Importantly, this polar LO mode can scatter electrons and hole through the resulting Fröhlich interaction [38, 40] with an effective interaction parameter

$$\frac{1}{\gamma} = \omega_{\mathrm{LO}}^2\left(\frac{1}{\epsilon_\infty} - \frac{1}{\epsilon(0)}\right), \tag{5.1}$$

where ϵ_∞ and $\epsilon(0)$ are the optical and static permittivity, respectively. The field and potential that arise from the charge dipole of the two atomic charges lead to a long-range scattering potential of Coulomb nature. Hence, the scattering rate, which depends upon the squared magnitude of the matrix element in perturbation theory, has a variation with scattering wave momentum of the order

$$\frac{1}{\tau} \sim \frac{1}{\left(q^2 + q_0^2\right)^2}, \tag{5.2}$$

where q_0 is the screening wave vector. It is important to note that this scattering process falls off very rapidly with wave number q. While the conduction band (or valence band) may have multiple valleys, this scattering process is important really only for intra-valley scattering due to this fact. There are much stronger scattering mechanisms that can account for inter-valley scattering.

As an example, the phonon spectrum of bulk GaAs is shown in figure 5.2, although this can be considered as any generic direct gap III–V. This spectrum was calculated using a valence force field model [37] with the 'band structure lab' tool, available at NanoHub.org. The expected splitting in the LO and TO modes is clearly evident near the Γ point. Note that the LO and TO modes cross between Γ and the zone edges. Experimentally, the crossing are a little closer to Γ than shown here [41]. In a simple group IV semiconductor such as Ge or Si, or as we will see in the TMDCs, LO and TO are degenerate at Γ, as there is no dipolar field. Also, the LO and LA modes are degenerate at X. This latter degeneracy is broken by the fact that the two atoms in the III–V materials have different masses. The LA branch describes the motion of the heavier of these two atoms while the LO branch describes the lighter atom motion. This splitting in GaAs is found to be only about 1.8 meV as determined by neutron spectrometry [41]. But, the splitting is also a reflection of the dipolar nature of the diatomic lattice.

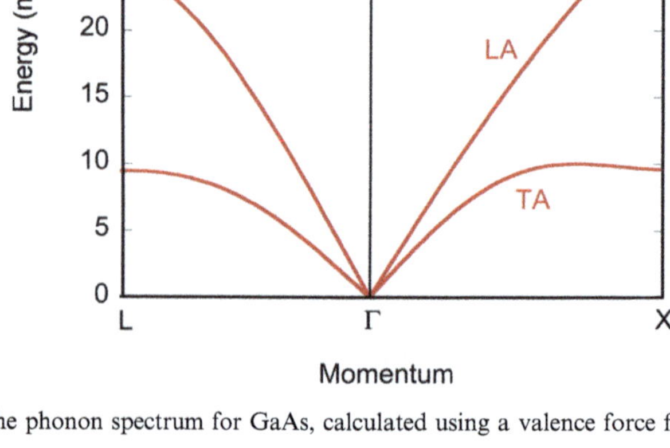

Figure 5.2. The phonon spectrum for GaAs, calculated using a valence force field model [37].

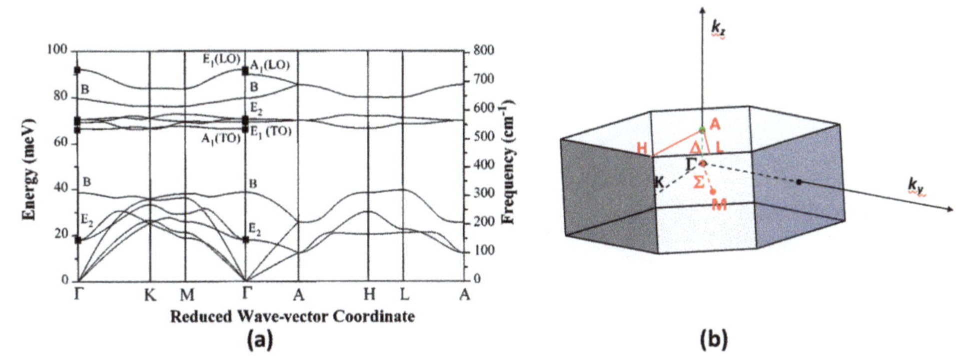

Figure 5.3. (a) Phonon spectrum for GaN in the wurtzite phase. Doubling of the lattice in the z-direction gives four atoms per unit cell and hence 12 phonon modes. Reprinted with permission from Siegle *et al* [43], copyright (1997) by the American Physical Society. (b) The Brillouin zone for GaN.

The wurtzite structure for the III–Vs is often considered as a relaxation of the zinc-blende lattice that occurs when the effective charge, and resulting ionicity, becomes too large [42]. This relaxation doubles the unit cell in the z-direction, normal to the hexagonal planes. This doubling produces four atoms per unit cell and 12 phonon modes, as shown in figure 5.3(a) [43]. The various points in the dispersion curve for GaN in panel (a) can be identified from the Brillouin zone in panel (b). The phonons may be divided into modes that lie in the basal plane of the material and modes with polarizations normal to the basal plane. The A_1 mode represents the two N atoms moving in one direction and the two Ga atoms moving oppositely with the motion along the (0001) axis (the c-axis). The E_1 mode is comparable motion but with the motion in the (x, y) plane. The E_2 modes are also in the latter plane, but with one the atoms (e.g., Ga) fixed and the other atoms moving relative to them. The B modes are similar to the E_2 modes, but with the motion along the c-axis. As one can see from figure 5.33(a), there are three pairs of LO–TO modes, which have a wide variety of orientations. Normally, the low energy B and E_2(low) are often dark modes, which means that they don't couple to light. Each of the LO modes can lead to scattering of the electrons and this contributes to nonequilibrium phonons in a wurtzite material.

5.1.2 Layered compounds

The best known layered semiconductor is graphene, which is a single layer of carbon atoms. In the single layer, the atoms are arranged in a hexagonal structure with two atoms per unit cell. Instead of the tetrahedral sp^3 bonding of the zinc-blende and wurtzite structures, the covalent bond is relaxed to sp^2 with the hybrids in the planar layer. The p_z orbital is directed normal to the layer and provides the electrons for the conduction and valence bands.

The phonons may be divided into modes that lie in the plane of the material and modes with polarizations normal to the layer. The latter modes usually play little role in the transport, but are certainly present in the rippling of the monolayer that is often observed. In figure 5.4(a), the graphene phonon modes are shown [44], while the corresponding Brillouin zone is depicted in figure 5.4(b). As graphene is

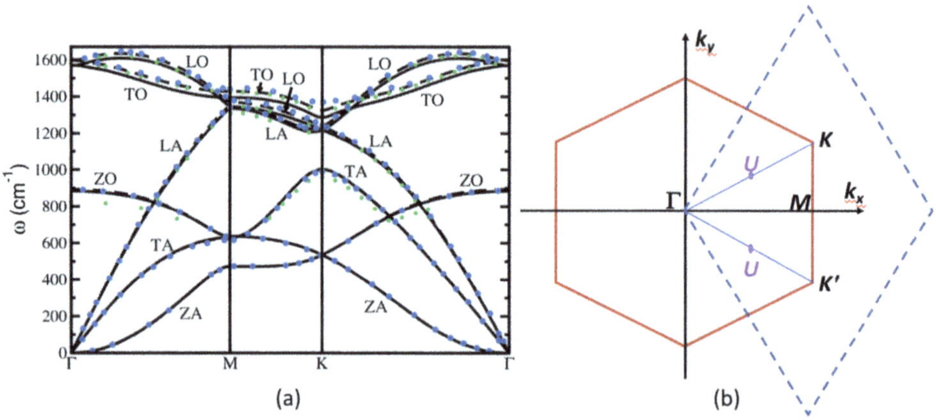

Figure 5.4. (a) Several different *ab initio* calculations of the phonon dispersion curves for graphene are shown. Reprinted from Wirtz and Rubio [44], copyright (2004) with permission from Elsevier. (b) The Brillouin zone for graphene, identifying critical points that appear in panel (a). The primitive unit cell of the Brillouin zone is shown by the dashed diamond shape. The ZA and ZO phonons are modes with the vibration normal to the graphene plane.

composed of a single type of atom, the LO and TO modes at Γ are degenerate and there is a single permittivity that is the same at static and optical frequencies.

If we consider tiling the plane with the hexagonal cells, then we can describe the primitive cell of the reciprocal lattice. This cell is an elongated diamond shape with the long axis passing through the K and K′ points and reaching to the Γ points in the hexagonal cells above and below these two points, as shown in figure 5.4(b). The minor axis runs from Γ through M to the Γ point in the next hexagon to the right along the k_y axis. This diamond cell may be sliced and diced into four parts, which may be relocated by reciprocal lattice vectors into the hexagonal shape, which is the one usually used to describe the properties of graphene [44].

The graphene primitive cell in real space has two C atoms in it, which are denoted the A and B atoms. The primitive cell in the reciprocal lattice is the diamond discussed above, that contains two structurally equivalent K points. These are denoted the K and K′ points in figure 5.4(b), and the wave functions at these two points have a phase difference. The conduction and valence bands both have their primary extrema at the K and K′ points in the zone, and the bands are degenerate at these points with a zero band gap. As there are two minima (maxima) for each band, one usually assigns a valley index as a pseudo-spin parameter, and the wave functions have a corresponding spinor component [37]. As a result of the zero gap, the bands around these points are characterized as linear in momentum k. This gives graphene some of its better properties as well as some unwelcome properties, such as in devices where a band gap is essential. It is important to note that the M points are points of high symmetry, and have a gap of ~4.5 eV, with the satellite minima of the conduction band lying ~2 eV above the zero gap point at K [45].

The transition metal dichalcogenides (TMDCs) can be prepared as almost monolayer compounds. The word ‘almost’ is used as all the atoms in a single layer

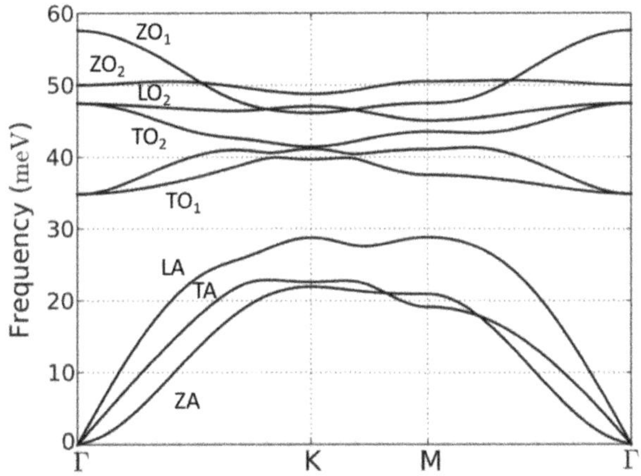

Figure 5.5. Phonon spectra of MoS_2, with the various modes labeled according to the common interpretation. Note that the two LO–TO pairs are degenerate at Γ, so that there is no Fröhlich interaction. Adapted with permission from Kaasbjerg *et al* [48], copyright (2012) by the American Physical Society.

are not in a single plane. They have a unit cell in both real and reciprocal space equivalent to graphene. There are two 'atoms' in each unit cell in real space—a metal atom and a pseudo-atom composed of two chalcogenide atoms, with one located above the metal plane and one below this plane, with the two atoms sitting above one another, in the 2H phase. Hence, the reciprocal lattice is the same as figure 5.4(b) above, and the discussion of the Brillouin zone in graphene may be carried over to the the TMDCs. Generally, the *s-p* bonding leaves a large band gap, so that the conduction and valence bands are primarily derived from the metal *d* states. These latter states give conduction band minima and valence band maxima at the K and K′ points *in the monolayer*. The monolayer does not have an inversion point, similar to the III–V compounds. However, if a second layer is created, it will be rotated by π around the z-axis, and restore the proper symmetries. The conduction band minima in the bulk material is located at the U points[1] indicated in figure 5.4(b). Like graphene, there are also satellite valleys at the M points, so it is quite likely that hot electrons (or holes) will occupy these satellite U and M valleys through intervalley scattering in high electric fields [46, 47].

The phonon spectrum for MoS_2 is shown in figure 5.5 [48]. Because there are now three atoms per unit cell, the phonon spectra has two LO–TO pairs for vibrations in the monolayer plane. Each of these two pairs is degenerate at Γ as there is no lattice polarizability and the optical and static permittivity are the same. Hence, according to the Lyddane–Sachs–Teller relation, there is no Fröhlich (polar) interaction. As for graphene, the ZO and ZA modes are out of the monolayer plane. The degenerate LO_2–TO_2 modes have motion in the plane with the pair of chalcogenide atoms moving together and opposite of the metal motion. The degenerate LO_1–TO_1 modes have the metal atoms relatively fixed with the two chalcogenide atoms moving opposite to one another in the monolayer plane.

[1] These points are often denoted as T, Q, or Λ in the literature.

5.1.3 Alloys

Ternary alloys, such as $Ga_{1-x}Al_xAs$ provide a challenge to our understanding of the physics behind such alloys, especially the disorder-induced effects. While the random alloy theory would suggest that this material should have a single LO mode that arises from the average mass contributions from Ga and Al, this is not what is found. It is well known that this material is a two-mode alloy (an effect found in most alloys), in which LO modes arise from both GaAs-like phonons and from AlAs-like phonons [49, 50], as shown in figure 5.6. In a true random alloy, one would expect a single mode whose frequency varied from the AlAs mode to the GaAs mode smoothly with x. However, what is normally found is that there are these two modes extending from either end of the alloy composition, but moving directly to the other extreme, following a sometimes linear curve. In figure 5.6, the experimentally determined phonon modes for GaAlAs are shown, with samples prepared by different growth methods [51]. Measurements were made with Raman scattering over a wide range of temperatures. Here, there are LO and TO modes of both GaAs-like behavior and AlAs-like behavior. The AlAs-type modes for the LO and TO phonons start at the AlAs LO and TO frequencies [51]. Then, as the alloy concentration is reduced they both *decrease* in

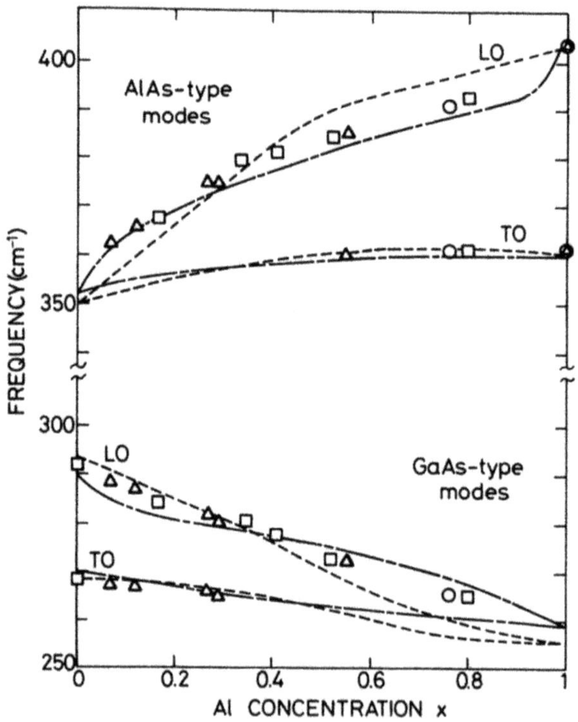

Figure 5.6. Frequency variations of the GaAs-like modes and the AlAs-like modes in the AlGaAs alloy as a function of composition. Samples grown by MBE (triangles), LPE (squares), and VPE (circles) are represented. The curves are theoretical values, and are discussed in the text. Reprinted with permission from Jusserand and Sapriel [51], copyright (1981) by the American Physical Society.

frequency with decreasing concentration until they merge (become degenerate) at $x = 0$. This merging point lies well above the GaAs LO mode frequency, so they do not merge into one of the pure GaAs modes. On the other hand, the GaAs-type modes begin at $x = 0$ with the pure GaAs LO and TO modes. Then, as the concentration is increased, both modes *decrease* in frequency, rather than approaching the AlAs modes. Again, the two modes merge at $x = 1$, and thus no longer contribute polarization. Experiments show that the data deviate slightly from this linear behavior in the figure and are compared with theoretical curves from Bonneville [52] and Tsu *et al* [53].

With such two-mode behavior, it becomes necessary to establish an 'effective dielectric constant' midway between the static and optical values if the Frohlich interaction (5.1) is to be evaluated for these two separate modes. In some cases, it becomes rather difficult and subject to variations with the electron density. For example, we show the case for AlGaAs in figure 5.7. In panel (a), the blue data dots are taken from Kash *et al* [54], who found that the strength of the AlAs-like mode increased monotonically with Al composition. In panel (b), the blue data dots are from Tsen *et al* [55], who found that the AlAs-like mode strength increased monotonically only up to $x = 0.2$ and then began to decrease. This apparent contradiction was determined to be a function of the carrier density and the manner in which the Raman signal was interpreted. The red data in the figure is from a full Monte Carlo simulation taking into account the Raman signal from the two modes, and the nonequilibrium populations of the two modes [56]. It was found that there are two major contributors to the differences in the two panels. First, the Raman probe does not measure the main part of the disturbed phonon distribution with either of the two phonons. Secondly, the spread of the distribution through continuous absorption and emission of phonons leads to a complicated temporal behavior of the phonon Raman signal, which causes the integrated behavior to differ from the simple peak behavior. In addition, the carrier density plays a significant

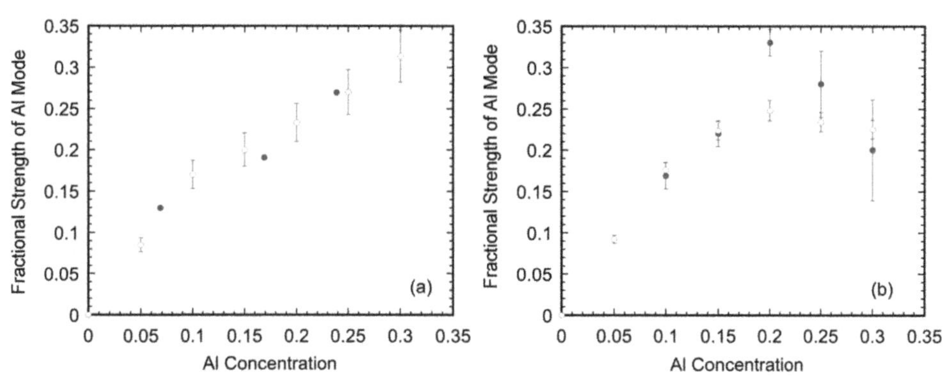

Figure 5.7. The compositional dependence of the AlAs-like mode strength. (a) The data of Kash *et al* [54] (blue dots) is compared to Monte Carlo simulation results (red) for a Raman signal measured at the *peak* of the scattering intensity. (b) The data (blue) of Tsen *et al* [55] is compared with Monte Carlo simulation results (red) for a Raman signal measured as the time *integrated* phonon signal. The differences in these two results arises from a complicated temporal behavior of the phonon Raman signal and not a simple temporal decay. Reprinted from Shifren *et al* [56], copyright (1999) with permission from Elsevier.

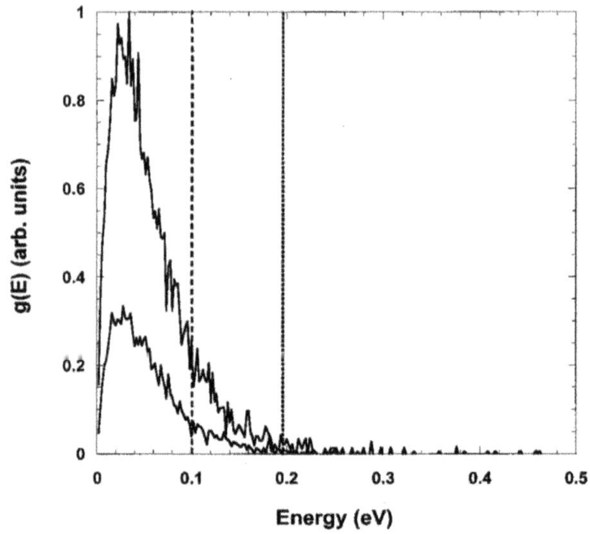

Figure 5.8. The electron distribution for two different carrier densities: $n = 2.5 \times 10^{16}$ cm^{-3} (top curve) and $n = 7.5 \times 10^{16}$ cm^{-3} (bottom curve). The dashed line corresponds to the minimum energy of the carriers for which they can emit a GaAs-like LO phonon at the Raman wave vector. The dotted line corresponds to the minimum energy of the carriers for which they can emit a AlAs-like LO phonon at the Raman wave vector. It is clear that the strength of the two modes will vary depending upon the density. Reprinted with permission from Shifren *et al* [57], copyright (2000) by the American Physical Society.

role in this second factor [57]. This is clearly shown in figure 5.8. At the lower density in the figure, the AlAs-like mode has few electrons with which to interact, and this leads to the apparent non-monotonic behavior in figure 5.7(b).

More complicated behavior has been seen in $In_xGa_{1-x}As$, such as one mode following the smooth transition and linear in x, and one at lower frequencies [58]. In general though, the strength of each mode varies with the composition of the alloy. Here, the GaAs-type modes also *decrease* with composition just as in $Ga_{1-x}Al_xAs$. But, here, both modes merge at the InAs LO mode, as may be seen in figure 5.9. Also shown in the figure are the results from Kuan and Su [59]. With the InAs-type modes, the LO mode frequency *decreases* as the concentration is reduced, while the TO mode *increases* in frequency until they merge (at $x = 0$) at a frequency well below the GaAs TO mode. As this merge point does not correspond to either of the GaAs mode frequencies, it has been referred to as merging at a ‘gap mode’, which must be evanescent. More importantly, experiments show that only the highest energy mode has true LO behavior and only the lowest energy mode has true TO behavior [58]. The other LO-like mode shows mixed mode behavior, likely due to disorder in the crystal, and is not effective in scattering the electrons. The lines in the figures come from reflectivity data, which covers more of the phonon range [60].

One can also alloy on the group V site, such as the case of $GaAs_{0.51}Sb_{0.49}$, which not only has important device applications, but provides a challenge to our understanding of the physics behind such alloys, especially the disorder-induced

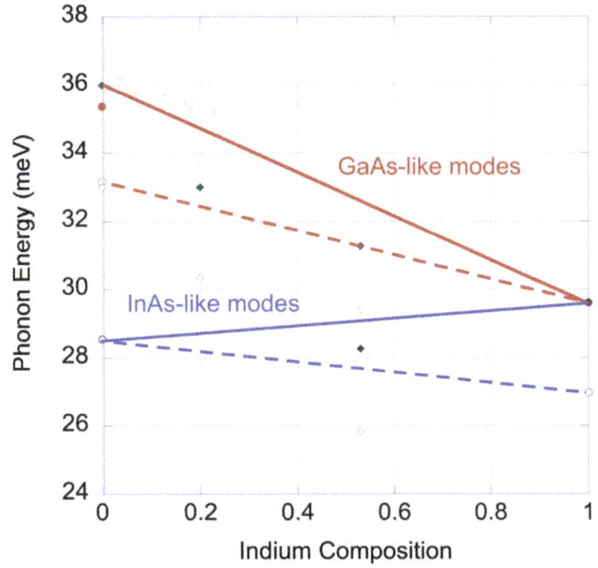

Figure 5.9. The optical phonon modes in InGaAs. The diamonds are data from Pearsall *et al* [58], while the crosses are data from Kuan and Su [59]. The lines are linearly extrapolation from the two endpoints.

effects. It is well known that this material is a two-mode alloy (just as discussed above for the alloys on the group III site), so that here there are LO modes that arise from both GaAs-like phonons and from GaSb-like phonons [61, 62]. In general though, the strength of each mode varies with the composition of the alloy. Here, the GaAs-type modes *decrease* with increasing composition of GaSb. While the data in this alloy is insufficient to be sure, normally both of these modes would merge at the GaSb LO mode. At the same time, the GaSb-like LO mode also *decreases* as the concentration is reduced, while the TO mode *increases* in frequency until they merge (at $x = 0$) at a frequency well below the GaAs TO mode. The optical phonons in GaAsSb have been measured by Raman scattering for x in the range 0–0.6 and in pure GaSb [61]. The results support the above discussion and are shown in figure 5.10. It is clear that the GaSb-like LO mode is very weak and appears in the Raman signal only at $x = 1$. Similarly, while the GaSb-like TO mode is present, it too is quite weak. By heating the lattice to several hundred degrees centigrade, the GaSb-like LO mode can be observed in Raman scattering, but it remains quite weak [62].

5.2 Acoustic spectroscopy

The idea of using sound waves as a spectroscopic tool is probably as old as the study of the elastic properties of materials like semiconductors. The normal elastic constants are contained, for cubic material, in a 6×6 matrix that addresses both the simple strain fields for compression and tension, but also for the shear waves that can exist in the solid [37]. Measuring the velocity of sound in various crystallographic directions will lead to being able to determine the three independent strain

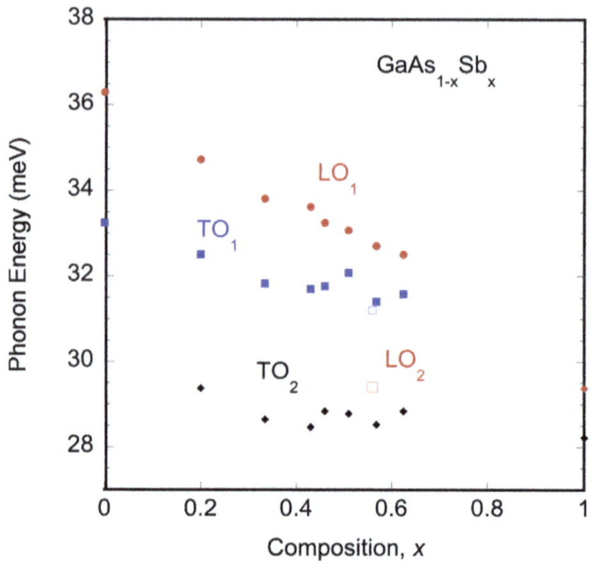

Figure 5.10. The dual mode nature of the optical phonons in GaAsSb are shown here. The solid symbols come from [61], while the open symbols come from [62]. Subscript 1 refers to the GaAs-like modes while the subscript 2 refers to the GaSb-like modes.

constants for cubic material: C_{11}, C_{12}, and C_{44}. Even an estimate of the third-order elastic constants (required for the decay of the nonequilibrium modes and for the thermal conductivity) can be made by adding pressure to the crystal while measuring the sound velocities to determine their nonlinear behavior under this pressure [63, 64]. Once it is realized that even the acoustic phonons are quantized, it is apparent that these can be used to probe structures within the semiconductor that have small energy level separations, such as impurities or defects [11].

The use of phonon spectroscopy seems to have begun in the early 1960s [11], when low frequency sound was used to probe the impurities and defects mentioned above, but expanding to the use of high frequency sound to probe even electrons in the thin surface layers extent in MOSFETs and other two-dimensional transport situations. The sound waves can be generated by transducers over a wide range of frequencies or even by heat pulses applied locally to part of the material structure. In either case, the acoustic phonons are certainly driven out of equilibrium and then used to probe the spectral properties of other parts of the solid. For example, a microwave transducer can use the piezoelectric properties of the III–Vs to launch a longitudinal sound wave parallel to an electric field. Thermal excitation, on the other hand, launches a broad spectrum of sound waves, but again with a specific direction.

We have already mentioned the use of such sound waves in CdS to study the amplification of these waves by the drifting electrons [5]. The general study of amplification of these sound waves continued for many years. The use of microwave acoustics to study properties of Si were reported already in 1964 [9]. It was even discovered that optical Brillouin scattering could be used to study the piezoelectric domains in GaAs [65]. In a relatively early study, the emission of acoustic phonons from warm and hot carriers in a quasi-two-dimensional electron gas in a Si MOSFET, under an applied drain bias, was carried out using an array of CdS transducers placed on the underlying crystal [12]. Studies were also carried out with laser generated nonequilibrium phonons on the luminescence properties of multi-quantum wells in GaAs/AlGaAs at low temperature [66].

The MOSFET at low temperatures, and in a magnetic field, could also be used to detect the phonon waves created by a heat pulse [67]. In this case, the Si MOSFET was fabricated on a 5 mm thick Si (100) wafer and had an area of 1×1 mm^2, with a 4.2 K mobility of 4500 cm^2 Vs^{-1}. A Constantan heater was placed on the back surface of the wafer, opposite to the MOSFET. Electrical pulses of 0.1–0.2 μs applied to the heater generated the acoustic phonon beam. In figure 5.11, the MOSFET signal detected is shown for two different heater powers as a function of the gate voltage at a magnetic field of 7 T. Also shown for comparison is the Shubnikov–de Haas signal as function of gate voltage (the oscillations arise as a function of the Fermi energy in the device, which is proportional to the density and thus the gate voltage). It is clear that peaks in phonon detection occur near the peaks in the source–drain resistance of the device arising from the Landau levels moving through the Fermi energy. The authors felt that a very small fraction of the ballistic phonons available were actually absorbed by the electron gas in the MOSFET, but that this was still significant enough to produce a noticeable rise in the electron temperature due to the very small specific heat of the electrons. This in turn was enough to see a change in the source–drain resistance in the magnetic field. Moreover, they demonstrated that a Si MOSFET can be used as a frequency selective phonon detector, the selectivity being dependent upon the Landau level width, which was about 300 GHz in this particular case. This work was followed by using a larger CdS array on the backside of the Si crystal to image the emission of phonons by the hot electrons in the Si MOSFET at low temperature [68]. Between these

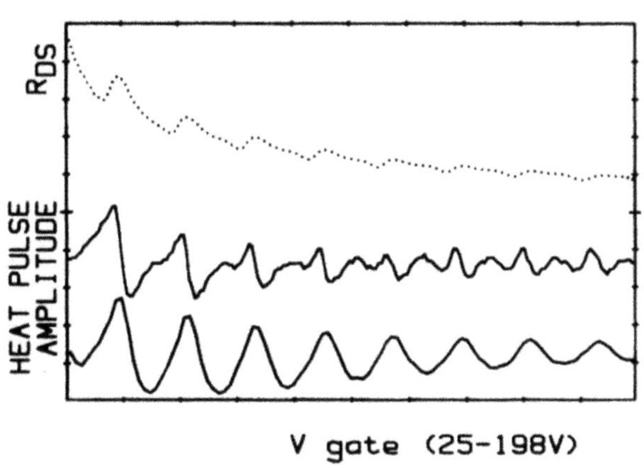

Figure 5.11. Variation of the detected signal amplitude with gate voltage variation at $B = 7$ T. Two heater powers are shown: upper trace, 5 W mm^{-2} and lower trace, 0.5 W mm^{-2}. Also shown for comparison is the Shubnikov–de Haas signal in the source–drain resistance (top, dotted line). Reprinted with permission from Kent *et al* [67], copyright (1988) by the American Physical Society.

two studies, it was concluded that such experiments provided a probe of the electron–phonon interactions within the electron gas.

In chapter 4, the excitation of semiconductors by intense laser pulses was discussed, and in section 5.1 the resulting cascade of LO phonons was described. However, the LO modes decay into acoustic modes, as will be discussed below. But, the cascade of acoustic phonons, and their nonequilibrium behavior has seldom been discussed. But, this can be measured. The ultrafast response and study of acoustic modes in a GaAs quantum well has been studied by Baumberg *et al* [69]. Here, the GaAs quantum well was of several different widths, as shown in figure 5.12(a), and was clad with normal AlGaAs barriers. Infrared 70 fs pulses were obtained from a Ti:sapphire laser and directed onto the quantum well to excite electrons and holes. A part of the laser pulse was doubled and used to monitor the phonons on the back surface of the device, as also shown in figure 5.12(a). The electron–hole pairs are generated with low excess energy in the first sub-band of the quantum well in a manner to avoid inter-sub-band scattering. A phonon wave packet is generated in the quantum well and propagates away from it in both directions. The doubled photon signal is used to probe the surface deformation caused by the arrival of the phonon wave packet. This signal is deflected by the packet and a slight change in the refractive index within the ~16 nm absorption depth of the probe beam, and this deflection is detected by a balanced photo-diode pair. The detected signals are shown in figure 5.12(b). Confirmation of the deflected detection is confirmed by translating the probe spot to the other side of the pump focus, which reverses the signal (compare traces *a* and *b*). If the undoubled infrared probe is used, a much stronger signal is found, as shown in trace *c*. The initial rise in the top two traces comes from a return signal from the GaAs cap layer, which cannot be ignored in the process.

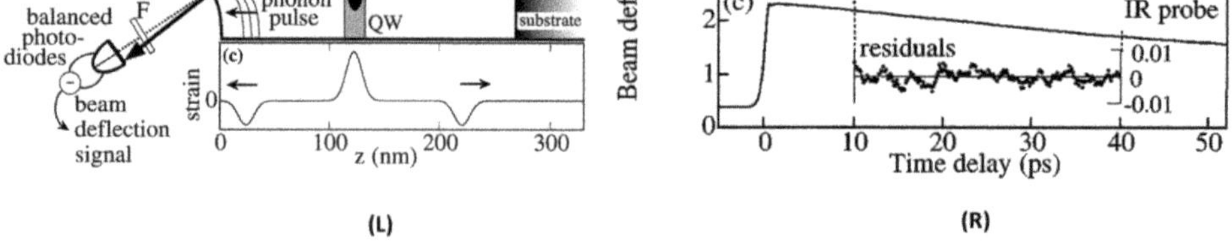

Figure 5.12. *Left panel*: (a) Vertical structure of the three samples studied. (b) Schematic of the experimental arrangement, with filter (F) and beam splitter (BS). (c) Calculated strain pulse. *Right panel:* Time-resolved deflection signal from sample A. (a) Probe beam on one side of focus spot. (b) Probe beam on other side of focus spot, as discussed in the text. (c) Signal using an infrared probe. Nothing is seen, even in the residual, after subtracting the dominant relaxation. Reprinted with permission from Baumberg *et al* [69], copyright (1997) by the American Physical Society.

The oscillations may be from different detection mechanisms or from interference between direct and indirect paths taken by the ballistic phonons. It is clear that ultrafast phonon wave packets produced in single quantum wells can be detected and resolved as they arrive at the surface. Several different phonon paths are thought to be found in the signals.

The above effects have been used since then for evaluation of nonequilibrium acoustic phonons with quantum devices. It has been found that the use of heat-generated nonequilibrium acoustic phonons can be sensed in the THz range in the luminescence of CdTe/ZnTe quantum wells [70]. Then, such nonequilibrium phonons have been used to study the spatial and frequency distribution of hot photo-excited carriers in Si, Ge, and GaAs [71].

In more recent work, coherent phonons generated by a surface acoustic wave have been used to probe the quantum levels of single-electron tunneling double quantum dots [72]. In this work, the double quantum dot (DQD) was fabricated in a GaAs/AlGaAs heterostructure with the two-dimensional electron gas 100 nm below the surface. The surface waves were excited with a standard interdigital transducer (IDT) excited at 1.92 GHz. The IDT is fabricated with Ti/Au metallization. Generally, two such transducers are used. In figure 5.13, panel (a) shows an image

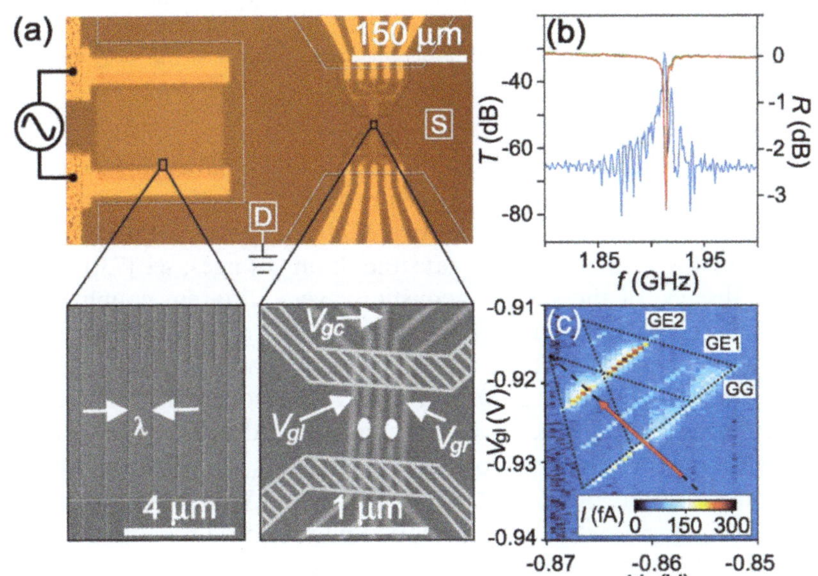

Figure 5.13. (a) Picture of the device with an IDT (left) and DQD (right) visible. Contacts for applying bias to the DQD are shown as source (S) and drain (D), as well as for the gates visible in the picture. The enlarged views below show the interdigitated fingers of the IDT and the location of the DQD. The separation of the odd fingers gives the surface wavelength $\lambda = 1.4$ μm. (b) Transmission (blue) and reflection (red) between two transducers is shown, with the peak/dip occurring at the IDT resonance at 1.92 GHz. (c) Color plot of the DQD current spectrum in the absence of any surface wave, and as a function of the gates V_{gl} and V_{gr} for a source–drain bias of 500 μV. The conductance triangles are enhanced by the dotted lines, and three resonant tunneling peaks (GG, GE1, and GE2 are shown). Reprinted with permission from Naber *et al* [72], copyright (2006) by the American Physical Society.

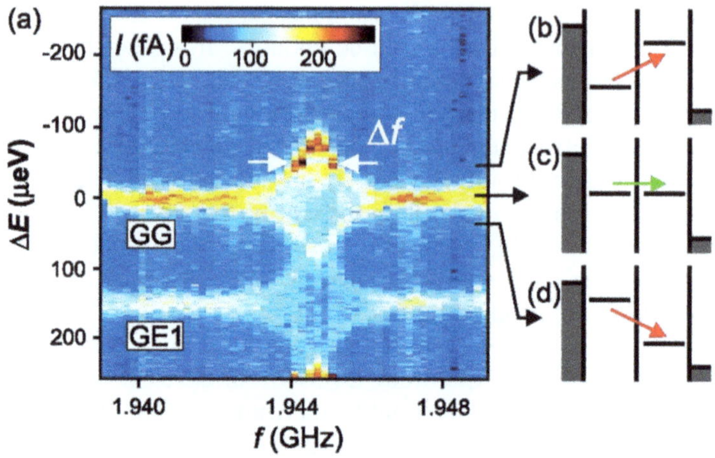

Figure 5.14. (a) Color scale plot of the DQD single-electron current as a function of the level space ΔE and microwave frequency applied to the IDT. The two gate voltages are swept along the red arrow in figure 5.13(b). The currents at $\Delta E = 0$ and 150 μV correspond to resonant tunneling through the ground state (GG) and through the left ground state and right excited state (GE1), respectively. A clear resonance is observed at 1.9446 GHz corresponding to the IDT resonant frequency. The inelastic current is due to the emission and absorption of the SAW acoustic phonons depicted in the energy diagrams (b) and (d). Diagram (c) corresponds to elastic tunneling. Reprinted with permission from Naber *et al* [72], copyright (2006) by the American Physical Society.

of the transducer and double quantum dot is shown. In panel (b), the transmission and reflection between two transducers is shown, and a clear peak in transmission is seen at the resonance frequency of the two IDTs. In panel (c), the single-electron current spectrum (as a function of the gate and drain voltages; see [73]) is seen for the DQD in the absence of the surface acoustic waves. Different coupling conditions and current lines are labeled for later reference. In figure 5.14, the excited current of the DQD is seen when the acoustic wave is present. There is a clear increase in the current as a function of frequency when the wave is resonant with the separation of DQD energy levels. The surface wave creates a time-dependent potential through the piezoelectric interaction of the form of $V_{ac}\cos(2\pi f_{SAW}t)$ within the DQD [72]. As the separation d of the two dots is 220 nm, and the SAW wavelength is 1.4 μm, the local potential V_{ac} is a fraction η of the actual piezoelectric potential, where $\eta = \sin(\pi d/\lambda) \sim 0.47$. The time-dependent level spacing between the two dots is then $\Delta E + \eta V_{pe}\cos(2\pi f_{SAW}t)$. In panel (a) of figure 5.14, the peak splitting at the resonance frequency can be explained as the action of the SAW on the DQD absorption. Panels (b)–(d) show different conditions for signals at different resonant tunneling lines given in figure 5.13, which should occur at $\Delta E + \eta V_{pe}$. Hence, the presence of the SAW leads to inelastic tunneling in a DQD two-level system. The DQD is a sensitive detector for surface acoustic waves and should have interesting uses for studying the electron–phonon interaction.

One can go beyond the above experiment to consider that one dot system can emit phonons which are then detected by a second dot system [74, 75]. The generation of these phonons in such quantum dots can be visualized using ultrafast electron diffraction (discussed more in the next section) [76], as the phonons themselves lead to a strain in the crystal that can be imaged through this diffraction.

5.3 Measuring the nonequilibrium phonons

From the earliest measurements of nonequilibrium polar LO modes in CdS [13] and GaAs [14], it is clear that Raman scattering primarily is used to gather an estimate of the hot phonon distribution function. It is also known that this measured distribution is not going to be comparable to the Bose–Einstein distribution that describes the phonons in equilibrium, it remains the fact that the estimate can be a good measure of the degree to which the phonons are driven out of equilibrium. Indeed, in the case of GaAs, it was established that the estimate of the non-equilibrium phonon population followed an increase that was linear in the laser power, with a deviation from linearity only when it was obvious the substrate was being heated as well [14]. For example, the measured phonon population in InP determined from a Monte Carlo simulation for laser excitation at 1.96 eV, with a 0.6 ps pulse, is shown in figure 5.15 as a function of the phonon wave number. To begin with, the wave numbers here are quite small, the largest being at most about 1/20 of the zone edge; that is, they are all quite close to the Γ point. Second, the LO phonon branch is quite flat near the Γ point, so that the phonon energy is assumed to be constant, independent of q (see figure 5.2). For that situation, the Bose–Einstein distribution is independent of q. Even with the small variation that can be seen in figure 5.2, this would not offset the very rapid drop in population beyond the peak

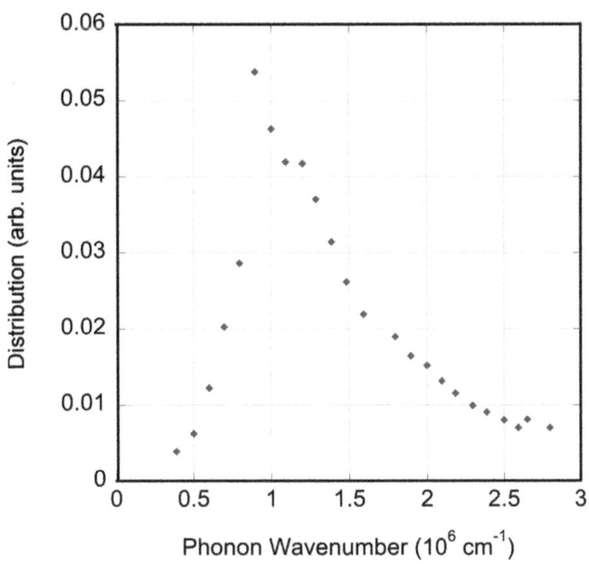

Figure 5.15. The MC simulated phonon distribution in InP for excitation by a 1.96 eV, 0.6 ps laser pulse and measured 2 ps after the pulse. This distribution is very non-thermal in nature.

shown in figure 5.15. Hence, it must be concluded that this is a non-thermal distribution. We will return to this point below, in connection with figure 5.27, and illustrate it in a more convincing manner.

In this section, the various approaches to measure the nonequilibrium phonons will be discussed, beginning with the Raman scattering approach. This will be followed by some new technologies including that of ultrafast diffraction which has been especially useful for studying the acoustic modes.

5.3.1 Raman scattering

For the early work on GaAs, the same laser was used to generate the hot carrier distribution and to perform the Raman scattering experiments [14]. In later work, a traditional pulse and probe approach was utilized. In such an experiment, the hot carriers are produced by the initial intense laser pulse. A fraction of this pulse is created by a beam splitter and then delayed relative to the initial pulse. This delay can be varied so that one can examine the time-dependent Raman signal [23, 77, 78]. The advantage of this delay approach lies in being able to study cases where the peak in the Raman signal occurs several picoseconds after the initial pulse. Such a pulse and probe apparatus is shown in figure 5.16, although it has to be said that there are a great many different approaches to the experimental setups used for these measurements. Here, the laser beam is split into two beams, one of which is directed normal to the sample surface (the break in the path is to indicate that it must be delayed to achieve the same path length as the second pulse at minimum delay). The second beam, the probe, passes through box A, which may be a parametric oscillator to shift the frequency or just a pass through or even another laser or an amplifier. This beam is then delayed by the movable reflector to adjust the time at which it passes through the sample. To delay an optical pulsed by 1 fs requires an additional path length of 300 nm (1 ps thus requires 300 μm). The movable reflector is moved just one half of this distance. So path lengths must be carefully adjusted throughout the experiment. This probe beam is often brought in at an angle to the pulse beam to avoid interference from the initial pulse beam. Or, the polarizations of the two beams can be adjusted to achieve this non-interference. Either the transmitted or reflected probe beam (depending upon the transmittivity of the sample) is then detected, along with the scattered beams.

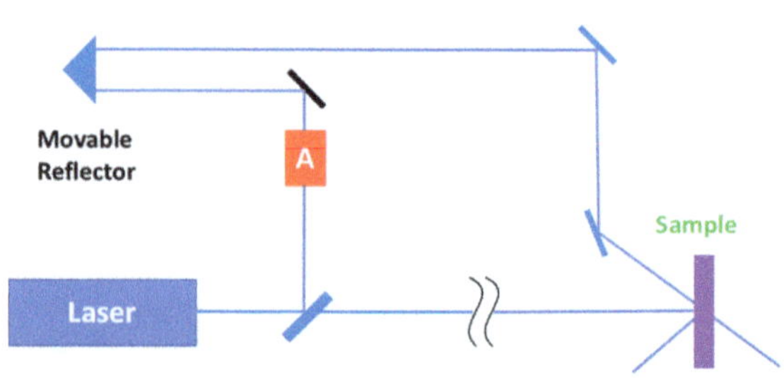

Figure 5.16. A generic pulse and probe laser experiment. Details are discussed in the text.

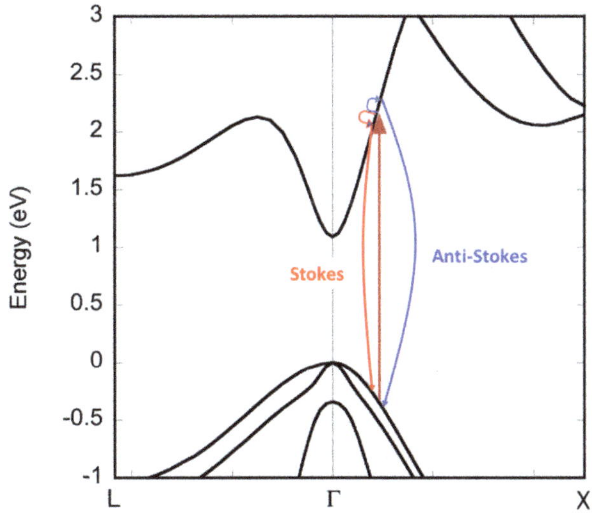

Figure 5.17. Raman scattering in semiconductors. The Stokes line arises from emission of a phonon by the electron–hole pair created by the incoming photon. The anti-Stokes line arises from the absorption of a phonon by the electron.

Raman scattering is in essence the scattering of the input optical signal by an excitation within the target material, which in the study of phonons is the scattering from the polar LO phonons. In this case, the input laser photon is given by the vertical deep red absorption arrow in figure 5.17. There are two possibilities. In one case the created electron–hole pair can emit a phonon and recombine giving the scattered photon, shown in red in the figure (the scattered photon is red-shifted). This is the Stokes signal. In the second case, one of the created particles can absorb a phonon prior to recombining to yield the scattered photon. This is the case in blue in the figure (the scattered photon is blue-shifted), and yields the anti-Stokes signal. The two possible scattering photons will be proportional to the phonon population through a variation as

$$\begin{aligned} I_S &\sim N_{LO} + 1, \\ I_{AS} &\sim N_{LO}, \end{aligned} \tag{5.3}$$

for emission and absorption of phonons, respectively. This can be rearranged to give

$$N_{LO} = \left(\frac{I_S}{I_{AS}} - 1 \right)^{-1}. \tag{5.4}$$

Thus, the amplitude of the phonon distribution function can be estimated from the ratio of the two Raman signals.

The Raman probe excites a particular wave number that is excited by the incident photon. For a back-scattered Raman photon, the wave number is reversed from the incident one, so that the scattering wave number is

$$q_{Raman} = \frac{2\omega_{photon}}{c}\sqrt{\epsilon_\infty}. \tag{5.5}$$

The index of refraction appears since the scattering actually occurs inside the semiconductor. To monitor the nonequilibrium phonon distribution, this wave number must be in the range of wave numbers for which the hot carriers can emit or absorb phonons, which will be discussed further below. Generally, photon wave numbers are much smaller than electron wave numbers, so it is the minimum q that is important. In general, then,

$$q_{\text{Raman}} > \frac{m^* \omega_{\text{LO}}}{\hbar k} \tag{5.6}$$

is required. In the last equation, $\hbar k$ derives from the excess energy of the photo-excited particle, which is assumed to be much larger than the phonon energy. But, there have been situations, such as the dual-mode alloy $\text{Ga}_{1-x}\text{Al}_x\text{As}$, where the Raman signal has actually almost missed one of the two LO modes in a study of the relative scattering strengths of the two modes [57]. This means that it is necessary to be sure that the various wave numbers are really compatible for the desired experiment.

5.3.2 Ultrafast experiments

Years ago, the use of the phrase 'ultrafast' would mean anything faster than ps response. We have already been exposed to this in the 200 fs pulse discussed earlier. These days, even this is seldom termed ultrafast. Although the electron scattering time for the emission of a phonon has been cited earlier as about 200 fs, phonons can be emitted much faster, as the scattering time is an average over the electron distribution. Scattering was occurring earlier than even 50 fs in figures 4.1 and 4.12. Yet, there are processes which occur even faster. For example, the time required to create an electron–hole pair was discussed in the early parts of chapter 4. It is probably not unexpected that the time required to actually emit or absorb a phonon occurs on a comparable time scale. An approach using true many-body non-equilibrium Green's functions to study this duration identified a proper energy-dependent collision duration [79]. As expected, it takes a few fs to complete a collision, just as to create the electron–hole pair. This and the excitation of the electron–hole pair are just two of the processes for which ultrafast sources would be useful for understanding the processes.

By the 1990s, the development of lasers with pulse durations as short as 6 fs made it possible to probe very fast nonequilibrium properties of the electron–hole plasma in semiconductors [80]. More progress has been made since that time. But, consider the properties of a 6 fs pulse at 2.0 eV. The frequency of this laser is 4.83×10^{14} Hz, and the period of the oscillation is 0.33 fs. Thus, the 6 fs pulse has less than 20 oscillations of the signal. This is still large. The 6 ps pulse was generated by a complicated nonlinear process. The base system uses optical pulses in a colliding pulse mode-locked laser with a set of intracavity prisms [81]. These pulses are amplified in a copper-vapor laser amplifier. The prisms formed a four prism sequence and the signal was then doubled in a KDP crystal to minimize distortion. The value of 6 fs was determined from an interferometric autocorrelation signal. In subsequent work, the signal was split into two, for use in a pump–probe experiment on the transmission of GaAs [82]. By convolving the transmitted signal of the probe with the exciting pulse, and subtracting from the former, the transmittance signal could be identified, and used to estimate the time required for the onset of inter-valley scattering.

While the above describes the ultimate at that time, time has advanced, and the shortest pulse at the present time appears to be the 0.043 fs pulse generated in an x-ray laser [83]. Having said that, most work still appears at longer times, and most ultrafast studies still utilize e.g. a Ti:Sapphire mode-locked laser, typically with a pulse repetition rate on the order of 80 MHz and a 780 nm center wavelength, to generate both the excitation pulse and the probe. This system usually can do pulse widths in the range 35–500 fs. However, in some advances, the probe pulse is used to generate a THz temporal signal [84], or to generate a white-light continuum signal (generated, for example, in a micro-structured fiber) [85].

The THz generation is created, for example, by the use of a strained InGaAs/InAlAs short period superlattice, excited by pulses from a Ti:Sapphire laser, which leads to the emission of a THz signal [86]. The rapid decay of the photo-excited carriers (~1.7 ps) leads to a THz signal with less than a full oscillation, and hence a broad band of frequencies, ranging up to about 3 THz. The white light continuum is similarly generated typically with the Ti:Sapphire laser by focusing the pulse onto a micro-structured fiber [86] or by focusing the laser on a sapphire plate [87]. This can produce a rather broad spectrum; that from the fiber has been shown to stretch across the 1.2–1.6 μm wavelength range. These techniques are often used to study transient reflection and/or transient absorption in a given sample.

In fact, both THz and white-light continuum sources have their origin in the Ti: sapphire laser pulses, such as the 70 fs pulses mentioned above [69]. Such a pulse is, in actuality, an amplitude modulated pulse, and the 70 fs width corresponds to 1/2 of the period of the modulation signal. Hence, the 70 fs pulse width corresponds to a modulation frequency of 1.14 THz. Thus, if we pass this pulse through a nonlinear system, the modulation part can be extracted as a stand-alone signal. This is shown in figure 5.18 from the review by Ulbricht *et al* [88]. Here, the incident laser (not shown) beam is split into three parts: one to excite the sample as the pump pulse and one to generate the THz signal with which the sample is probed. The third beam will be explained below. Passing the generation signal through a nonlinear medium such as an air plasma [89] or a strained crystal such as ZnTe or GaP, or even a photo-switch, will cause the desired signal to be generated. This signal is then passed to the sample as the probe signal. Note that both the probe signal and the third signal are both delayed to provide a coincidence of these two signals and to provide the time response of the probe itself. The THz signal could, of course, be directly detected with a diode, which would give the amplitude of the signal passing through the sample. However, heterodyne detection is more sensitive, and for this a signal with the THz modulation is needed; hence, the third beam from the laser. This is provided by mixing the third beam with the THz signal to provide the detected signal. By the use of the two balanced detectors, and the polarizers shown in the figure, the polarization of the detected THz pulse can be determined.

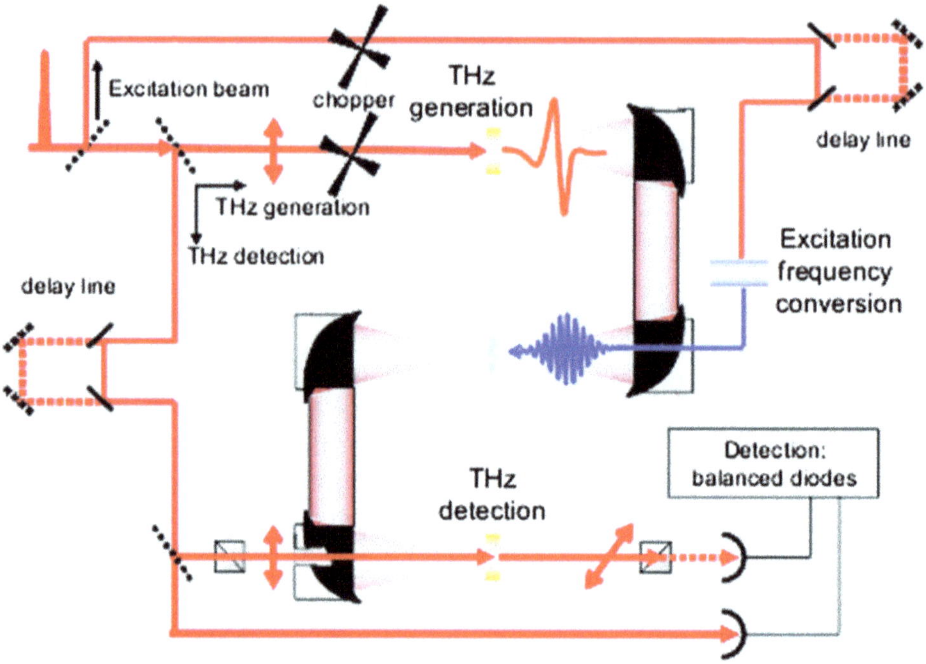

Figure 5.18. Schematic representation of an optical pump-THz probe system. A train of ultrashort laser pulses enters from the upper left, and is split into three beams for sample pump, THz generation, and THz detection. Mechanical choppers are used for synchronous detection of the THz signal. In the balanced detector scheme shown, the polarization of the detection pulse can be determined (in consort with the polarizers shown). Reprinted with permission from Ulbricht *et al* [88], copyright (2011) by the American Physical Society.

For the white light continuum, the THz generation loop is generally replaced by a micro-structured fiber [86]. The laser pulse can be considered to be made up of a wide range of frequencies in order to generate a single wave packet. A nonlinear medium such as a micro-structured fiber is dispersive; that is, different frequencies travel at different velocities. Hence, sending the wave packet through the fiber means that different frequencies in the wave packet arrive at different times at the output of the fiber. This separates the packet into a broad excitation that can actually cover the visible range. This broad spectrum is referred to as the white-light continuum. The experimental setup is essentially the same as for the THz generation, except that the THz generation segment is replaced by the continuum generation material, such as the fiber. Other materials can be used, such as a nonlinear CaF_2 crystal.

Another fast approach is time-resolved angle-resolved photoemission spectroscopy (TR-ARPES) [90–92], which is the subject of a recent review [93], focused on the electron–phonon interaction. Here, for example, GaAs was probed with ultrafast pulses with photon energies between 1.6 and 2.4 eV, as shown in figure 4.1 of the previous chapter, and then probed with TR-ARPES [92]. The electrons injected into the conduction band would equilibrate between themselves in momentum, but not in energy, in a few tens of fs. Relaxation of the carriers was on a much longer time scale and was governed by the interaction of the electrons with a variety of phonons, especially for

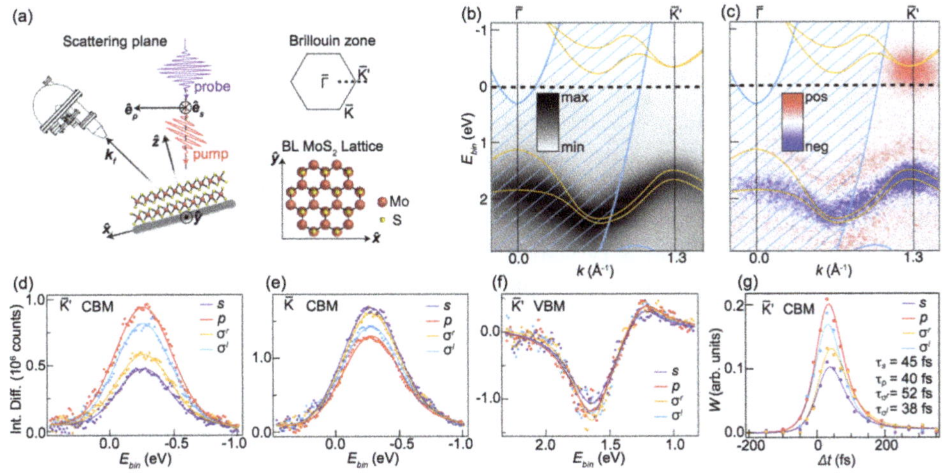

Figure 5.19. (a) TR-ARPES setup: Geometry of the scattering plane and coordinate system, photo-electron wave vectors and polarization directions. The bilayer lattice and Brillouin zone for the measurement geometry are shown in the right panels. The dashed line in the Brillouin zone shows the measurement direction. (b) ARPES intensity along the Γ-K′ high-symmetry axis before arrival of the pump pulse. (c) Intensity difference between the spectrum in (b) and that existing at the peak of the pump pulse 40 fs into the excitation. The dispersion of the bilayer MoS_2 has been overlaid (orange curves) as has the bulk continuum of Ag (111) (blue spectrum). (d–f) Energy distribution curves for the intensity difference for different pump polarizations, fitted with Gaussians (lines), for the conduction band minimum at K′ (d) and K (e), as well as the valence band maximum at K′ (f). (g) Time dependence of the spectral weight in (d), fitted with an exponential rise and a single exponential decay given by the indicated time constants. Reprinted with permission from Volckaert *et al* [94], copyright (2019) by the American Physical Society.

photo-electron densities less than 5×10^{17} cm^{-3}. Another TR-ARPES experiment is shown in figure 5.19, where linear dichroism in MoS_2 is studied [94]. Here, a bilayer of MoS_2 is grown on a Ag(111) surface. The signal is collected with a 32.5 eV probe pulse that is linearly polarized parallel to the scattering plane. The time resolution is 40 fs (width of the probe pulse) at 300 K. The energy distribution curves obtained from panel (c) are averaged over a momentum range from −0.02 to 0.02 nm^{-1} around K′ and are fit with Gaussians for the *s*- and *p*-polarized pump pulses. Note that the sign of the energy is reversed from what one normally assumes in band diagrams.

5.3.3 Ultrafast diffraction

When strain is introduced into a crystal, the change in the unit cell can be measured by diffraction techniques. The use of short-pulse diffraction seems to have begun around 1970 in the study of shocks to various materials [95–97]. This approach progressed to laser-induced shocks and sub-nanosecond time resolution some time later [98], in which the x-rays were obtained by focusing a second pulsed laser onto a solid target. This leads to the approach named ultrashort pulse x-ray diffraction (UXD). Measurements of the acoustic phonons in Si were made during laser-induced melting/regrowth with short (sub-ns) pulses of x-rays [99]. Shorter pulses were subsequently achieved by the laser interaction with a plasma [100] and by transversal of a laser pulse across a focused electron beam [101]. Electrons also have wave properties and can be utilized for diffraction studies. In the present interest, this is important as electrons have scattering cross-sections many orders of magnitude larger than x-rays. The first sub-picosecond electron source used a Ti:sapphire laser, with sub-50 fs pulses, to excite a photocathode, for which the resultant electron-beam emisions could be focused and raster scanned [102], to create ultrafast electron diffraction (UED). These electron beams typically act as the probe beam in usual pump–probe style measurements. By moving to rf photogeneration and MeV accelerating voltages, much shorter pulses with higher spatial resolution can be achieved [103, 104]. A typical source might use a dc biased photocathode followed by an rf resonator to bunch the electrons, so as to shorten their temporal spread [105]. Since the diffraction largely measures the change in the crystal structure, it is largely limited to studying the acoustic modes of the phonons, as the optical modes generally do not lead to changes in the crystal structure. Thus, as will be seen below, most of the studies using UXD and UED have been focused upon the acoustic modes of the lattice.

An example of ultrafast x-ray diffraction is shown in figure 5.20 [106]. The experiment is examining the ultrafast changes in charge density in a thin film of diamond following a two-color x-ray pump (7.8 eV) and probe (11.7 eV) experiment. With the intense x-ray pump, it is found that bonds are broken and electrons become localized around the atoms within 5 fs after the peak of the pump pulse (normally, in covalent bonds, the bonding electrons tend to sit between the tetrahedrally coordinated atoms).

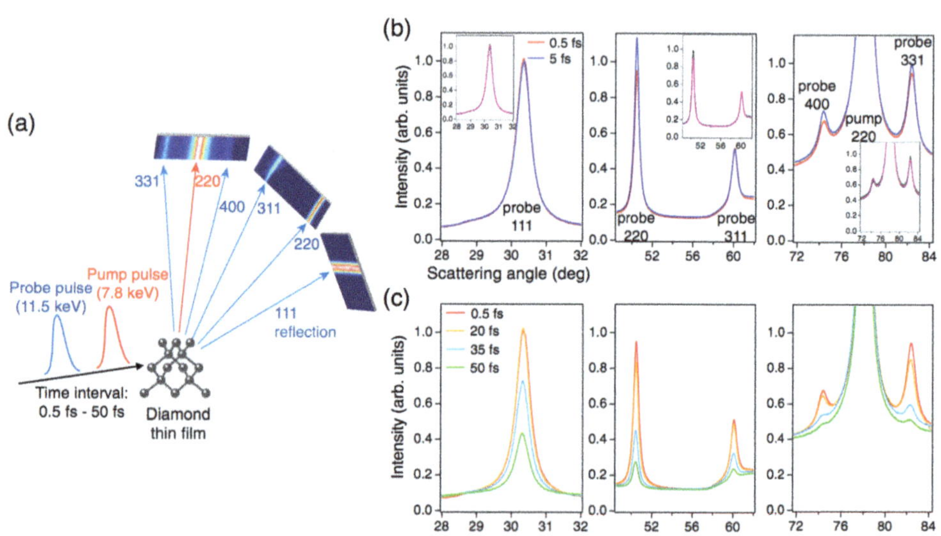

Figure 5.20. (a) A schematic illustration of the experiment. The focused two-color pulses arise from an x-ray free-electron laser and are used for pump and probe of a diamond film. The diffraction beams were measured with multi-port CCD detectors. (b) Diffraction intensity profiles for delay times of 0.5 and 5.0 fs, which are normalized to the (111) intensity. (c) Diffraction intensities for 0.5, 20, 35, 50 fs. The lines are lined up with the same lines in (b). Reprinted with permission from Inoue *et al* [106], copyright (2021) by the American Physical Society.

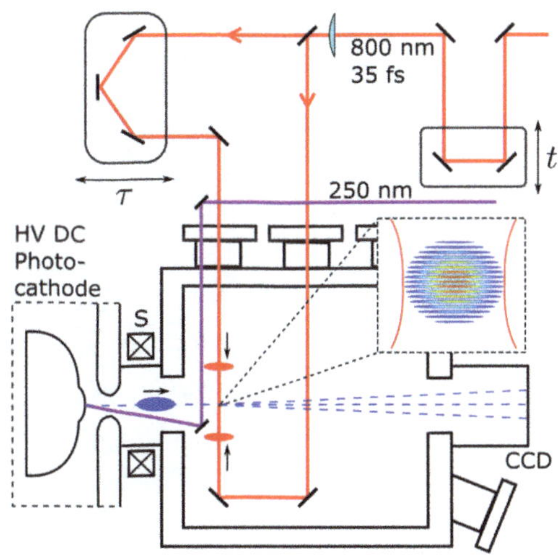

Figure 5.21. A schematic of the electron diffraction system using a magnetic lens (S). The standing wave created by the two colliding pulses within the sample is shown in the dotted box, with amplitude increasing from blue to red. Reprinted from Morrison *et al* [109], under the open access creative commons license.

A typical electron diffraction source might use a dc biased photocathode, in which the electrons are photo-emitted from the cathode, followed by an rf resonator to bunch the electrons, so as to shorten their temporal spread, as mentioned above [105]. Modern systems, of course, have limits on the temporal and spatial resolution [107]. Nevertheless, the modern approach utilizes commercial TEMs which are adapted for fast pulsing, and that is essentially what is shown in figure 5.21. The laser source produces 35 fs pulses at 800 nm and a power of 3 mJ pulse^{-1}. The laser is split into two beams, one to pump the sample and the second sent to e.g. a LaB_6 photocathode in TEM [108]. In figure 5.21, we show a typical electron diffraction approach [109]. The pump pulse is split into two beams and brought onto the sample from opposite directions. The pump is delayed to allow synchronization of the three signals (electron pulse and the two laser pump pulses). The two laser beams brought onto the sample from opposite directions create a standing wave within the sample. The purpose of the double probe pulse is to increase the power delivered from the laser to the sample. A third beam from the laser is used to pump an optical parametric amplifier to produce the 250 nm photons used to excite the photocathode. The electron-beam system is also phased-locked to the pulsed laser used to pump the sample.

The concept of ultrafast electron diffraction is illustrated in figure 5.22 [110] (which could also be applied to ultrafast x-ray diffraction if the electron pulse is replaced by an x-ray pulse). Since there is also diffuse scattering of the electrons, this approach has the ability to provide both a wave-vector resolved view of the interactions with the phonons as well as the coupling between the electrons and the lattice. Since the electrons have a relatively large cross section for scattering and the Ewald sphere is relatively flat, there exists the possibility of seeing both the average lattice structure from the Bragg scattering and the lattice dynamics from the diffuse scattering. This requires a relatively thin sample, such as a monolayer of a

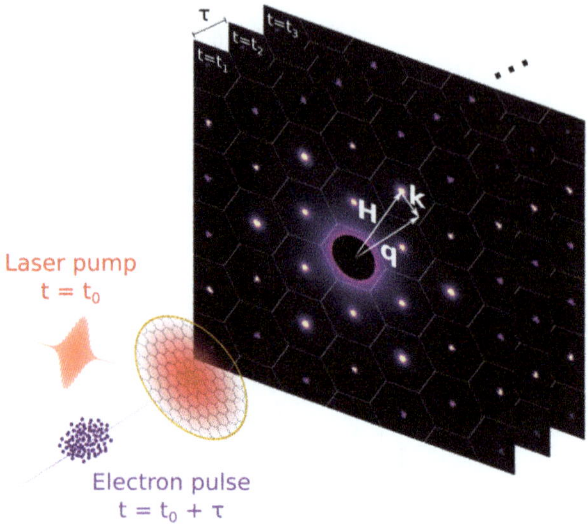

Figure 5.22. Schematic view of ultrafast diffuse electron scattering experiments. Samples are excited with ultrashort pulses of light at the initial time (35 fs pulses in this case). After a delay τ, an ultrashort electron pulse scatters through the sample. Elastic and inelastic signal are collected after transmission through the sample. By scanning the time delay, a stroboscopic movie of the dynamics can be assembled. This figure also shows the example of a scattering vector **q** and its associated reduced wave vector **k**, which are related by the nearest Bragg reflection **H**. The hexagons are related to the Brillouin zone of two-dimensional materials such as graphene or MoS_2. Reprinted with permission from René de Cotret *et al* [110], copyright (2019) by the American Physical Society.

two-dimensional material. This allows for both time-resolved measurements and momentum-dependent information on the phonon population and dynamics.

5.4 Rise and fall of the phonons

When carriers are highly excited into energy states well beyond their normal equilibrium states, whether through the application of an electric field or by photo-excitation, they relax their excess energy by phonon emission. When they are in the Γ valley, such as for direct gap III–Vs, this usually involves the polar LO modes, as discussed previously. In figure 5.23, we show the situation for optical excitation. Here, we show the central part of the band structure for $In_{0.52}Al_{0.48}As$ [111], although this is generic for any of the direct gap materials of interest. In this material, the direct gap is 1.5 eV and photons of 2.5 eV are assumed in the figure. Most of the excess energy goes into the electrons exciting them into higher energy states in the conduction band, as discussed in chapter 4. These carriers then relax by the emission of the LO modes as shown by the blue cascade in the figure. Only excitation from the heavy-hole band is shown, and roughly 90% of the excess energy goes into the photo-electrons. The dominant LO mode has an energy of ~44 meV. Hence, the average photo-electron will emit something like 20 polar LO phonons. The average time required to emit an LO phonon is about 160 fs, so that the cascade will take just over 3 ps. While the

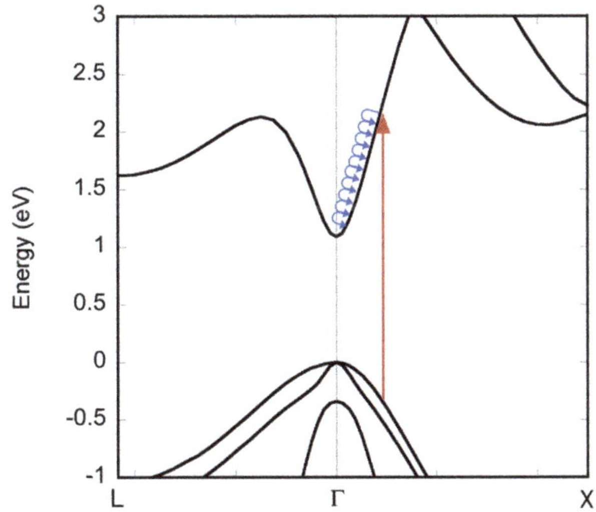

Figure 5.23. Highly excited electrons, such as those shown here being photo-excited from the valence band to the conduction band (red), can relax by emitting a cascade of optical phonons (blue). The large number of emitting phonons can drive the phonon distribution out of equilibrium.

average time is relatively long, it remains an *average* time. It is possible that the phonons will begin to be emitted somewhat earlier, as was observed in figure 4.1.

It is possible to see the cascade by a variety of methods, and it is quite likely that the resulting phonon distribution, normally the Bose–Einstein distribution, will be driven out of equilibrium for those wave numbers that can be excited by the electrons. It is important to point out that this nonequilibrium distribution does not have the traditional Bose–Einstein shape, as there are preferred wave numbers that are found in the scattering process. That is, there is a minimum and a maximum value of phonon momentum q that can be excited by the emission by the electrons, which was shown in figure 5.8. This leads to a very narrow distribution of nonequilibrium phonons. A typical phonon distribution was shown in figure 5.15. The change in the phonon population at a particular value of wave number can be written as

$$\frac{dN(q)}{dt} = G(q) - \left.\frac{dN(q)}{dt}\right|_{sc}, \tag{5.7}$$

where

$$G(q) = \Gamma_{\text{em}} - \Gamma_{\text{abs}} \tag{5.8}$$

is the net rate of phonon emission by the electrons—the difference in phonon emission and phonon absorption processes, given in more detail in later chapters.

The most common input to the generation rate $G(q)$ is from the emission of phonons by the hot electrons. In figure 5.24, the scattering rates for $GaAs_{0.51}Sb_{0.49}$, for scattering by the LO phonons while in the Γ valley, are plotted as a function of the kinetic energy of the electrons. This curve can be considered generic as it doesn't

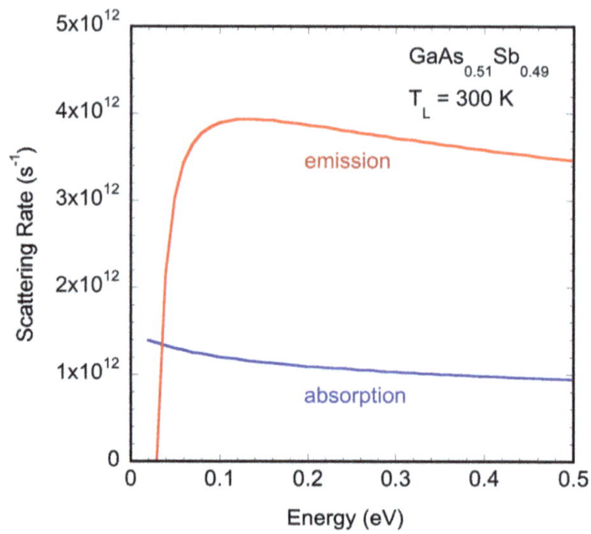

Figure 5.24. Scattering rates for electrons, by the emission or absorption of an LO phonon in the Γ valley, in GaAsSb as a function of energy for the case in which the phonons are in equilibrium at 300 K.

vary much from that for GaAs or many other III–V semiconductors. It is important to note however that the phonons are considered to be in equilibrium at 300 K in this plot. Since these scattering rates depend upon the phonon distribution, they will change as the phonons go out of equilibrium, which we discuss further below in connection with the bottleneck. The emission curve does not start at $E = 0$, because the electron cannot emit a phonon unless its energy is $E > \hbar\omega_{LO}$.

5.4.1 The phonon cascade

Conceptually, and from simulations, it is clear that one should observe a phonon cascade arising from hot carriers, as illustrated in figure 5.23. This cascade was already apparent in oscillatory photoconductivity, discussed in section 4.1. It is even apparent in photoluminescence under the right conditions. Peterson *et al* [112] were able to show that oscillations in electroluminescence energy arose from a GaAlAs/GaAs *n-i-p* structure. In this structure, the *n*-layer and *i*-layer were GaAlAs, while the *p*-layer was GaAs. Then, with bias applied to the structure, the GaAlAs layer injected hot electrons into the GaAs, where they decayed by the phonon cascade. As these hot electrons recombined with the holes in the *p*-type GaAs, the luminescence signal would have peaks at the energies between the allowed phonon emissions, leading to oscillations in the luminescence signal as a function of energy.

From the scattering rates in figure 5.24, it can be see that the scattering rate for the emission of a phonon is about 3.5×10^{12} s^{-1}, so that it will take about 4 ps to emit 13–15 phonons by the hot carriers. In this time, the carrier may reabsorb only about 2–3 phonons, so that there is a net loss to the phonon population. As the phonons are heated, these statistics will change. The two processes depend upon the population as [37]

$$\begin{aligned} \Gamma_{em} &= \frac{me^2}{4\pi\gamma\hbar^2\omega_{LO}}(N_{LO}+1)W_{em}(k) \\ \Gamma_{abs} &= \frac{me^2}{4\pi\gamma\hbar^2\omega_{LO}}N_{LO}W_{abs}(k) \end{aligned}, \tag{5.9}$$

where γ is given by (5.1) and the W terms are given as

$$\begin{aligned} W = \ln&\left(\frac{q_{max}^2+q_0^2}{q_{min}^2+q_0^2}\right) \\ &-\frac{1}{2}\frac{4q_0^2\sqrt{k^2 \mp \frac{2m\omega_{LO}}{\hbar}}}{\left(\frac{2m\omega_{LO}}{\hbar}\mp q_0^2\right)^2+4k^2q_0^2} \end{aligned}, \tag{5.10}$$

with the minimum and maximum wave numbers given by

$$\begin{aligned} q_{min} &= k - \sqrt{k^2 - \frac{2m\omega_{LO}}{\hbar}} \\ q_{max} &= k + \sqrt{k^2 - \frac{2m\omega_{LO}}{\hbar}} \end{aligned} \tag{5.11}$$

for emission and

$$\begin{aligned} q_{min} &= \sqrt{k^2 + \frac{2m\omega_{LO}}{\hbar}} - k \\ q_{max} &= \sqrt{k^2 + \frac{2m\omega_{LO}}{\hbar}} + k \end{aligned} \tag{5.12}$$

for absorption. We will derive these rates more formally in chapter 6. In (5.10), the upper sign is for emission and the lower sign for absorption. The important point here is the dependence upon the phonon occupation.

In the presence of inter-valley scattering, the phonon cascade is interrupted and charge is transferred to the satellite valleys where it is stored for some time. Scattering of carriers from the Γ valley to the L or X valleys does not arise from the polar LO scattering, but from phonons near the Brillouin zone boundary that have sufficient momentum for the valley change, and this is a very rapid process. The scattering to and from the satellite valleys can follow a variety of paths. In the above case, the electrons are scattered from Γ to X, and from here they can either return to Γ, or scatter to an L valley or another X valley. Within the satellite valleys, the dominant scattering is the inter-valley scattering between equivalent valleys. But, in any case, the electrons will end up back in Γ, after some period of time. The major point is that inter-valley scattering delays the ultimate decay of the nonequilibrium phonons and replaces some of the polar LO population with energy transfer to zone-edge modes. But, there can still be significant numbers of the polar LO mode excited, unless other steps are taken.

5.4.2 Phonon bottleneck

Discussion of the possible existence of a phonon bottleneck dates from the earliest discussions of nonequilibrium phonons [2, 3]. At that time, it was associated with equality in the emission and absorption processes. Generally, a so-called bottleneck occurs when the emission and absorption processes for the decay of polar LO phonons become comparable. Then, it is observed that the energy relaxation process for the excited 'hot' carriers is slowed. From (5.8), it can be inferred that this will happen when the two Γ's are comparable. These scattering rates depend upon the phonon population as shown in (5.9) as

$$\begin{aligned} \Gamma_{\text{em}} &\sim N_{\text{LO}} + 1, \\ \Gamma_{\text{abs}} &\sim N_{\text{LO}}, \end{aligned} \tag{5.13}$$

where N_{LO} is the phonon mode population (here, it has been indicated as the LO mode, but it can be any phonon mode). Normally, at 300 K, the value of this population, for a 36 meV phonon, is about 0.33 for, e.g., the LO modes in GaAs, so emission is about four times larger than absorption (but, this is obviously dependent also upon the carrier energy, when $E > \hbar\omega_{\text{LO}}$). If, however, the population is increased significantly, this ratio drops dramatically, approaching unity for large phonon populations. It is in this case where emission and absorption become comparable and the bottleneck occurs.

But, the bottleneck only indicates that the relaxation process is slowed, not eliminated. The factor of unity in (5.13) never goes away, so phonon emission always will be larger than absorption so long as the system is out of equilibrium. The only time the two are equal is when detailed balance occurs for an equilibrium situation with no electrical or optical excitation. Detailed balance is a situation when the total number of emitted phonons equals the total number of absorbed phonons. To achieve this, the rates in (5.9) have to be averaged over the total electron (or other carrier) distribution, so that detailed balance can only be achieved when the electron temperature is the same as the lattice temperature; e.g., see the discussion around equation (1.4). In the hot carrier world, detailed balance is extremely difficult to even approach.

Nevertheless, there is considerable evidence for a bottleneck occurring. Early measurements of optical excitation in Ge with photons of 1.06 eV with 5 ps pulses, showed exceptionally long decay times [113]. The long relaxation times are attributed to a long-lived excitation of the phonons which lead to a bottleneck for electron cooling. Using two-phonon Raman scattering, Tsen *et al* then showed that the TA modes were also driven out of equilibrium, thus slowing the decay of the LO modes in GaAs [114]. (We will see below that this evidence of the TA mode being out of equilibrium in Si speaks to the decay process that can exist.) Further measurements solidified the view of a bottleneck on the cooling of the carriers.

The study of the bottleneck became more interesting with experiments on quantum wells, especially early in GaAs/AlGaAs heterostructures [19, 20].

Experimentally, it was found that an unusually large difference was found in the cooling rates of electrons and holes, and this was ascribed to a phonon bottleneck that slowed the cooling of the electrons [19]. Simulations certainly supported this conclusion [22, 115]. However, continued studies in GaAs quantum wells showed that it was important to consider intervalley transfer as well, as this slowed the cascade—a simple cascade model was inadequate [116]. The bottleneck effect was confirmed by measurements in quantum wells of different sizes, so that the inter-sub-band transitions had energies both above and below the phonon energy [21, 117]. That is, by adjusting the size of the quantum wells and the barriers, it is possible to create a situation where the sub-band width in energy is less than an optical phonon, and the gap between the maximum energy of one sub-band and the minimum energy of the next sub-band is larger than an optical phonon energy. Hence, the phonons are prohibited from being scattered at all!

More recently, studies have returned to investigating multiple quantum wells to further understand the bottleneck mechanism and the adjustment of sub-bands and gaps to suppress phonon emission, not least because of the use of these structures for hot carrier solar cells [118, 119]. Indeed, the use of multiple quantum wells seems to slow the cooling of the carriers as expected, and leads to an enhanced bottleneck effect at lower illumination intensities. This also might arise from the restriction on carrier movement normal to the interface, which leads to further confinement of the phonons, although this repetitive well/barrier structure will lead to zone folding of the phonons, which can defeat the idea of the blockade. Such was evident in the use of inelastic x-ray scattering to study $In_{0.17}Ga_{0.83}As/GaAs_{0.8}P_{0.2}$ multiple quantum wells, which has given results suggesting that zone folding of the phonons in the superlattice leads to *an enhancement* in electron–phonon coupling [120]. That is, from the zone folding effect, there are a great many more modes with which the electrons (or holes) can interact. This could certainly change the nature of the phonon emission and bottleneck effect.

5.4.3 Anharmonic decay

With all of these excess phonons, and the fact that the polar LO phonons may be driven out of equilibrium by the emission of these modes by the electrons and holes, it is time to discuss how to get rid of these excess phonons. In most situations, the decay of these resulting nonequilibrium phonons can be described by a lifetime, which will be addressed below. The decay process arises primarily from a three-phonon interaction in which e.g. the polar LO phonon is annihilated and two (often acoustic) phonons are created in a manner that conserves both the total phonon energy and the momentum. This interaction can occur through the anharmonic terms of the lattice potential energy, similarly as for nonpolar modes such as the acoustic phonons mentioned in the introduction [121]. In addition to the anharmonic decay, it is also possible for the polar mode to decay through coupling of the dipole polarization to a pair of acoustic modes through the second-order piezoelectric effect [34], which will be discussed in the next section.

The polar LO modes do not effectively transfer their energy to the surface of a semiconductor because of the extremely small group velocity of these modes. Hence, the transfer of energy from these LO modes to the heat sink is generally accomplished through the anharmonic interaction sending the energy to acoustic modes which can propagate to the device boundaries. Phenomenological treatment of the decay process has been discussed by Weinrich [122], Cowley [123], and by Klemens [124]. The third-order interaction can be added to the Hamiltonian through a cubic term in the strain in the crystal potential via a term of the form

$$V_3 \sim \sum_{q,q',q''} \boldsymbol{u}_q \boldsymbol{u}_{q'} \boldsymbol{u}_{q''} \vdots \boldsymbol{A} \tag{5.14}$$

where the $\boldsymbol{u}$'s are the Fourier transformed mode amplitudes and $\boldsymbol{A}$ is the third-rank generalized force tensor. This tensor has 27 elements, so there are a variety of pathways that can be found in this interaction, some of which will be described below. There remains, even today, very little data available on the third-order elastic constants, especially for the polar mode. Consequently, one usually relies upon a fit to the measured experimental data on the lifetime to adjust the effective coupling constant. In considering the decay of the LO mode to lower energy modes, energy and momentum must be conserved, so that this requires

$$\omega_{\mathrm{LO}} = \omega_1 + \omega_2, \tag{5.15}$$

and

$$q_{\mathrm{LO}} = q_1 + q_2. \tag{5.16}$$

In these expressions, 1 and 2 refer to the two phonons into which the LO mode decays. Since q_{LO} is quite small, this means that the two wave numbers on the right of (5.16) must almost be equal and opposite in direction.

There have been several calculations of the LO phonon lifetime, and these have considered different possible phonons combinations. The first process, known as the Klemens channel, the decay route was LO → 2LA, which is the easiest to conceive [123, 124]. This channel is illustrated by the green arrows in figure 5.25(a). It may be seen that the two momenta have opposite signs as needed by (5.16). Ridley and Gupta also considered the Klemens channel, but also suggested the decay route LO → TO + TA (the TA is shown here, but it could also be a LA) [125]. This process is shown in green in figure 5.25(b). While they considered the general decay into an optical and acoustic phonon, the channel LO → LO + TA can just be made to satisfy energy and momentum conservation in GaAs, as the TO mode is quite flat. Then, Vallée and Bogani considered this latter channel plus a LO → LA + TA near the X points [126]. This is shown in figure 5.25(b) in magenta. Later, Barman and Srivastava looked anew at all these various channels and suggested some new ones for wurtzite material (discussed later). In fact, it is possible to separate these decay mechanisms in studies of the nonequilibrium LO modes, which will be discussed in a following section.

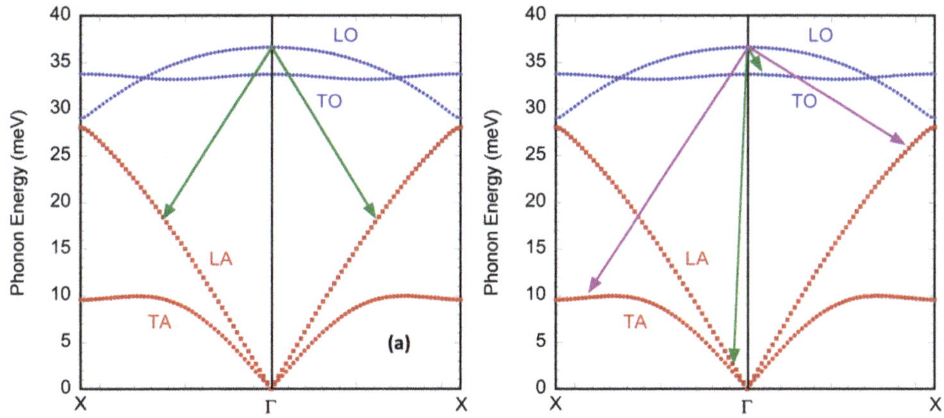

Figure 5.25. Schematic of the popular decay channels, shown on the phonon spectrum of GaAs. (a) The Klemens channel. (b) The Ridley–Gupta channel (green) and the Vallée–Bogani channel (magenta).

In any of these channels, the matrix element for the decay will follow from (5.14). The **u**'s are standard harmonic oscillator amplitudes, so that the matrix elements become [127]

$$|M|^2 = \frac{\hbar^2 A^2}{8MN^3\omega_{\rm LO}} \sum_{q_1 q_2} \left(\frac{1}{\omega_1 \omega_2} \right)^{3/2} N_{\rm LO}(N_1 + 1)(N_2 + 1). \tag{5.17}$$

A summation over one of the wave vectors has absorbed the conservation of momentum relation (5.16), but energy must be conserved in using this matrix element in the calculation of the lifetime. The two secondary phonon populations, and their frequencies, will depend upon which decay mechanism is considered in computing the lifetime. The N and M in the prefactor denote the number of atoms and the atomic mass (or an average mass in a compound), respectively. Using standard perturbation theory, the lifetime can be found to be [127]

$$\Gamma_{\rm LO} = \frac{\hbar A^2}{8\pi\rho_M \omega_{\rm LO} v_1 v_2} N_{\rm LO}(N_1 + 1)(N_2 + 1), \tag{5.18}$$

where the velocities enter from $\omega_1 = q_1 v_1$ and $\omega_2 = q_2 v_2$, and ρ_M is the mass density. The prefactor differs from an equivalent derivation from Weinrich [122] only by some numerical factors. For InSb, a value of 1.2×10^{10} s^{-1} is found for the prefactor in (5.18) using the data of Drabble and Brammer for the third-order elastic constant [63]. But, this number uses the low energy velocity of sound, which is far too big for the Klemens channel, and so what will boost the value considerably is we take the actual slopes of the dispersion at the energies of the two LA modes. Hence, a scattering rate on the order of 10^{12} s^{-1} is not unexpected. The important aspect is the dependence upon the three-phonon populations.

5.4.4 Piezoelectric decay

In addition to the anharmonic decay, it is also possible for the polar mode to decay through coupling of the dipole polarization to a pair of acoustic modes through the piezoelectric effect [34]. The potential in this case couples the polarization of the LO mode to a pair of secondary electrons via the second-order piezoelectric effect. That is, the polarization produces an electric field and this couples to the phonons by this piezoelectric interaction. The potential in this case can be written in analogy to (5.14) as

$$V_3 \sim \sum_{q,q',q''} \boldsymbol{E}_q \boldsymbol{u}_{q'} \boldsymbol{u}_{q''} : \boldsymbol{Q}, \tag{5.19}$$

where, as before, the $\boldsymbol{u}$'s are the Fourier transformed mode amplitudes and $\boldsymbol{Q}$ is the coefficient of the second-order piezoelectric interaction, while the electric field arises from the polarization and is given by [37]

$$\boldsymbol{E}_q \sim \epsilon_0 \boldsymbol{P}_q \sqrt{\frac{\hbar}{2\gamma\omega_{\mathrm{LO}}}}\, \boldsymbol{e}_q \left(a_{\boldsymbol{q}_1} e^{i\boldsymbol{q}_1 \cdot \boldsymbol{r}} + a_{\boldsymbol{q}_1}^{\dagger} e^{-i\boldsymbol{q}_1 \cdot \boldsymbol{r}} \right), \tag{5.20}$$

and γ is given by (5.1). This gives the matrix element to be

$$|M|^2 = \frac{\hbar^3 Q^2}{2M^2 \gamma N \omega_{\mathrm{LO}}} \sum_{q_1 q_2} \left(\frac{1}{\omega_1 \omega_2} \right)^{3/2} N_{\mathrm{LO}}(N_1 + 1)(N_2 + 1), \tag{5.21}$$

which differs only in details of the prefactor from (5.17). From this, the decay rate for the optical modes through the piezoelectric interaction may be found to be

$$\Gamma_{\mathrm{LO}} = \frac{\hbar Q^2}{2\rho_M M \gamma \omega_{\mathrm{LO}} v_1 v_2} N_{\mathrm{LO}}(N_1 + 1)(N_2 + 1). \tag{5.22}$$

There is very little data on the second-order piezoelectric coefficient, especially for InSb, which was used as the model in the previous section. However, we can estimate the relative strengths from calculations by Flytzanis [128], on two-phonon Raman scattering. This would suggest that the phonon decay time for piezoelectric scattering is somewhat smaller than that by the three-phonon elastic term. However, this may be true for the zinc-blende and diamond materials, but the III–N materials, with their wurtzite structure, may be quite different. It is known that they have large piezoelectric coefficients, which together with the strain in a heterostructure, lead to the ability to induce inversions densities at an interface without any dopants being present [129]. This would also suggest that the second-order piezoelectric coefficient is large and phonon decay by this process is quite likely.

5.4.5 Phonon lifetime

As we have discussed above, it is clear that the polar LO phonons may be driven out of equilibrium by the emission of these modes by the electrons and holes. In most situations, the decay of these resulting nonequilibrium phonons can be described by a lifetime. The decay process arises from either a three-phonon lattice strain interaction, or by a second-order piezoelectric interaction. In either case, the polar LO phonon is annihilated and two (often acoustic, see figure 5.25) phonons are created in a manner that conserves both the total phonon energy and the momentum. The interaction through the anharmonic terms of the lattice potential energy also occurs for nonpolar modes such as the acoustic phonons mentioned in the introduction [121].

The lifetime of the excess polar optical phonons or even the nonpolar optical phonons, that are emitted by the electrons, is related to the rate at which these modes can decay into the acoustic modes. Thus, one can write a continuity equation for the optical phonons as given in (5.7) and (5.8). The last term in (5.7) is the one of interest here, and it can be written as [37]

$$\begin{aligned}\frac{dN_{\text{LO}}}{dt} &= -P[N_{\text{LO}}(N_{\text{LA}}+1)(N_{\text{LO-LA}}+1)-(N_{\text{LO}}+1)N_{\text{LA}}N_{\text{LO-LA}}] \\ &= -PN_{\text{LO}}\left[1+N_{\text{LA}}+N_{\text{LO-LA}}-\frac{N_{\text{LA}}N_{\text{LO-LA}}}{N_{\text{LO}}}\right]\end{aligned} . \tag{5.23}$$

The prefactor P is the leading terms in the various interactions discussed above, and the term in square brackets will vanish in equilibrium. Thus, if we write $N_{\text{LO}} = N_{\text{LO}} + n_{\text{LO}}$, where the last term is the deviation from equilibrium, then the lifetime can be written as

$$\frac{1}{\tau_{\text{LO}}} = P(1 + N_{\text{LA}} + N_{\text{LO-LA}}). \tag{5.24}$$

All of these populations are temperature dependent, especially if the secondary ones remain in equilibrium. Thus, by doing temperature dependent studies of the lifetime, one can separate the different processes. This is illustrated in figure 5.26 for measurements on wurtzite GaN [130]. Here, the measured data is given by the solid circles. The orange curve is for the Klemens decay LO → 2LA while the blue curve depicts the Ridley decay LO → TO + TA. From this data, it appears that, at least in this material, the primary decay channel is the Ridley–Gupta type. While these curves are for the A_1(LO) mode, similar behavior is found for the E_1(LO) mode [131].

5.4.6 Nature of the nonequilibrium population

Earlier, in figure 5.15, a plot of the nonequilibrium LO phonon population was given, and it was remarked that it was certainly not a thermal distribution. At this point, it is worth examining this comment in somewhat more detail. The Bose–Einstein distribution is given by

$$N_{\text{BE}} = \frac{1}{\exp\left(\frac{\hbar\omega}{k_B T}\right) - 1}. \tag{5.25}$$

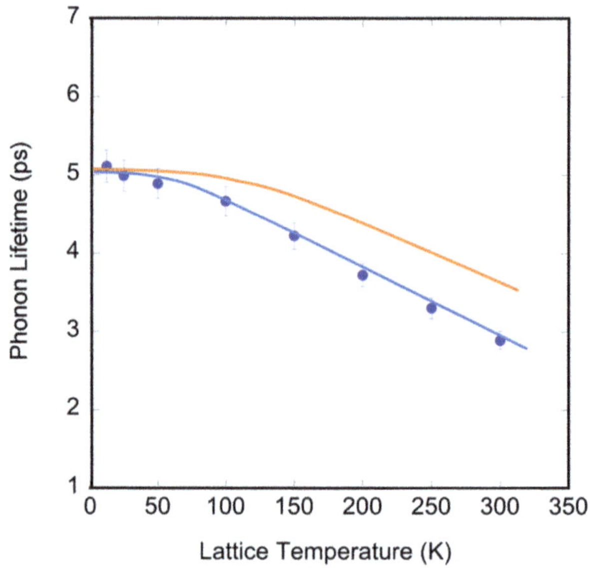

Figure 5.26. The phonon lifetime, as deduced from Raman scattering, as given as a function of temperature for wurtzite GaN. The orange curve is the Klemens [124] channel while the blue curve is the Ridley–Gupta [125] channel. The data is adapted from Tsen *et al* [130].

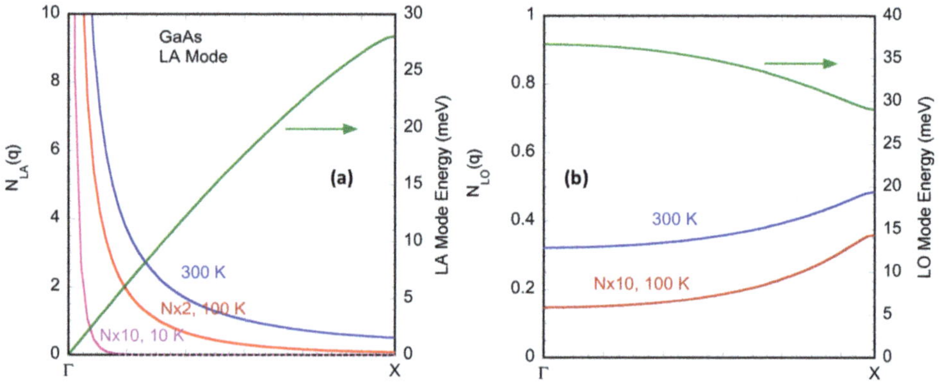

Figure 5.27. (a) Plot of the LA mode dispersion and the Bose–Einstein distribution for several temperatures. The 10 and 100 K curves have been multiplied by a factor of 10 and 2, respectively. (b) Plot of the LO mode dispersion and the Bose–Einstein distribution for two temperatures. The 100 K curve has been multiplied by a factor of 10 for ease of viewing. The 10 K is orders of magnitude smaller.

In figure 5.27, this distribution is plotted as a function of wave number in the Brillouin zone. The phonon dispersion is also shown. In panel (a), the distribution for the LA modes is plotted at temperatures of 10, 100, and 300 K. The curves for the two lower temperatures have been multiplied by factors of 10 and 2, respectively, to bring them close to the values for the higher temperature. In this case, the exponential decay is clearly evident, due to the dependence of the phonon energy upon the wave number. In panel (b), the same curves are shown for the LO modes, although the value for 10 K is just too small to plot, and the curve for 100 K has been multiplied by a factor of 10 to bring it close to that for 300 K. It may be noted in this latter case, that the energy of the phonon decreases with wave number so that the Bose–Einstein distribution clearly increases as one moves across the zone!

The plot of figure 5.15 is actually the population of the LO modes, so that it represents the product of the distribution function (5.25) and the number of states at that energy, or wave number. It may be noted that the relevant wave numbers in this figure that go up to 3×10^6 phonon wave numbers are less than 3% of the zone boundary. In this case, they are located near Γ, and we can approximate the number of states by a normal spherical coordinate transformation. In this case, the density of states rises, in three dimensions, proportional to the square of the wave number. The rise in figure 5.15 is less rapid than this quadratic behavior, although it is more rapid than a linear variation. But, the fall of the population beyond 10^6 cm^{-1} is certainly not due to the Bose–Einstein distribution. Instead, the entire shape of the phonon population is likely due to the distribution of the electrons that are emitting the phonons.

The population in figure 5.15 might be characterized by its width, so that (5.11) could be used to say that

$$q_{\max} - q_{\min} \sim 2\sqrt{k^2 - \frac{2m^*\omega_{\mathrm{LO}}}{\hbar}} \sim 106\ \mathrm{cm}^{-1}. \tag{5.26}$$

This means that the energy at the peak of the population seems to be only a few meV above the LO phonon energy. It is likely that the rise in figure 5.15 occurs primarily due to the onset of the electrons having sufficient energy to emit a phonon, and the decay occurs because of the decay of the electron population above this energy. But, this figure is not unique as very similar shapes and wave numbers are found in the study of AlGaAs [57]. However, it should be remarked that both figure 5.15 and the results on AlGaAs are for photon initiated hot carriers. As a result, the hot carrier distribution function does not show the streaming behavior found when electric fields are present.

The above is not the only possible result. Some studies of, e.g., GaN in a high electric field do show populations at wave numbers well above those of figure 5.15 [25]. Another look at (5.11) shows that, for very large values of the electron momentum, e.g., when $E \gg \hbar\omega_{\mathrm{LO}}$, the phonon wave number can vary from near zero to almost $2k$, where k is the electron wave number. That is, the phonon wave number can lie almost anywhere on the constant energy surface $E - \hbar\omega_{\mathrm{LO}}$. These large wave numbers in the phonon population for GaN are shown in figure 5.28. Here, the range of the wave numbers is more than an order of magnitude larger than in figure 5.15. In fact, the maximum wave number shown is that for an electron above the minimum of the satellite valleys at M or L in the wurtzite structure (figure 5.3) or X in the cubic form. Even when plotted on a semi-log plot, the distribution does not decay very rapidly, and is quite likely also determined by the electron population, as discussed above.

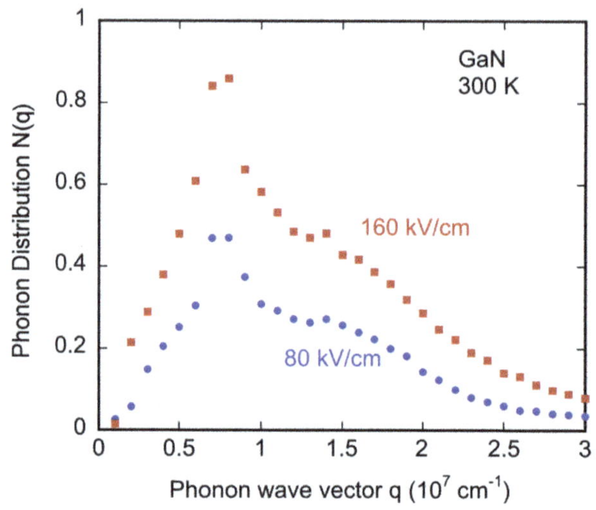

Figure 5.28. Phonon population at two different electric fields in GaN at room temperatures. These wave numbers are more than an order of magnitude larger than those of figure 5.15, and are more indicative of streaming in the electron distribution.

5.5 Measuring the lifetime

While much early work concentrated on the III–Vs, the study of simpler materials such as Si and Ge came later. In fact, much less is known for these group IV materials than the others. Early studies of Ge by ultrafast pulse-probe experiments investigated the transmission of the signal through the wafer [132]. These studies probed the optical properties by using intense picosecond laser pulses of sufficient energy to create electron–hole pairs by direct absorption, for example into the central Γ valley in the conduction band [133]. The nonlinear optical properties have been probed, not only by the absorption of the intense pulse, but also by delayed probe transmission with a second pulse [132, 134], and by reflectivity measurements [15]. These studies indicated that the energy relaxation of the photo-excited carriers was less than 5 ps, and that a saturation of the absorption could occur in the Γ valley. Somewhat later, ultrafast x-ray pulses from a free-electron laser were used to study the phonons in Ge [135]. These authors showed that they could measure large portions of the phonon spectrum via the diffuse scattering of these x-rays. They were able to study acoustic modes on the fast time scale and observed oscillations in the scattering intensity in the THz regime. It was suggested that these oscillations were in the coherence of the phonon scattering and arose from the strain induced by the x-rays. This approach was extended to observe the oscillations as a function of the infrared probe delay [136]. It was noted that the x-ray probe excites a pair of phonons with opposite momenta, and the oscillations likely were the result of interference between these TA modes.

Ultrafast experiments on Ge with a near infrared pump and an ultrafast XUV probe studied the transient carrier dynamics [137]. The XUV excites carriers from the 3d core levels of Ge and allows a separation into separate signals for the electrons and holes. The led to the observation of state blocking (band filling) and a red shift in the bands themselves. The latter occurred over some ~10–15 fs after the excitation pulse of 8×10^{20} cm^{-3} carriers. Clear electron and hole relaxation in under 200 fs was observed, and there were suggestions of the possibility to measure the phonon interactions.

Si phonons were studied by UXD by Kishimura *et al* [138], where it was observed that the laser generated electron–hole pairs recombined through the emission of LO phonons, which themselves decayed into LA modes, although no lifetime was deduced from these experiments. Si has also been studied by UED, but primarily for the acoustic modes with no lifetime being reported [139, 140]. Si nanowires were studied using third harmonic photons at 4.66 eV as the excitation pulse and near infrared signals as the probes [141]. These nanowires had diameter >10 nm, so that the bands were basically those of bulk material. Probes at 570 and 1045 nm could only be absorbed by phonon-assisted indirect transitions, while a probe at 1350 nm allowed absorption through a free-carrier transition. With an excited carrier density of 3.2×10^{19} cm^{-3} carriers, the initial decrease in transmission of the 570 nm signal follows excitation of photo-electrons into the L valleys, with subsequent decay into the X valleys on the ~5–25 ps time scale, followed by a longer energy relaxation of 350–400 ps. With the 1350 nm excitation, the initial decay time increased continuously for wire diameters up to 100 nm, where the decay time was ~75 ps. No direct measurements of the phonons was reported. So far, only the measurements of the phonon lifetimes and nonequilibrium properties remain those of Tsen *et al* [115], in which the 2TA signal was monitored.

Electron–phonon interactions in Ge/SiGe heterostructures have been studied with the use of pump–probe spectroscopy [142]. Effectively, the study probed the phonon dynamics of two-dimensional Ge, where the dynamics of the deformation-coupled optical phonons is more difficult to determine. The work is significant because of the determination of a value for that deformation potential. Of interest is the differences between a sample that had 21 nm $Si_{0.2}Ge_{0.8}$ barriers and a well of 24 nm, and a second sample with the same barriers but with a well of total width of 24.5 nm. This second sample had a step in the well of 13 nm of $Si_{0.03}Ge_{0.97}$. These samples were labels S1 and S2, respectively. Excitation and probing of the samples used a 5 ps pulse from a free electron laser, which was tuned to the energy separation of the two lowest L levels in the Ge quantum wells. However, the inter-sub-band absorption was monitored by FTIR. In figure 5.29, the results for these two structures are shown. The work uses the predicted deformation potentials from Jacoboni *et al* [143], from Monte Carlo simulations of the behavior of high-field transport. The latter authors assumed two effective phonon energies and deformation potentials for the inter-sub-band, (f phonons) and the intra-sub-band (g phonons), respectively, as discussed in section 2.3.1. These values were 3×10^8 eV cm^{-1} and 5.5×10^8 eV cm^{-1}, respectively. The experiments show that the elastic scattering is important, but that the intra-valley optical phonons need a weaker deformation potential, near to 1.4×10^8 eV cm^{-1}. Finally, the oscillations in the pump–probe data of figure 2.29(f) seem to have a period of some 20 ps, which might correlate with the oscillations in energy in figure 2.29(d) that are a few meV. The nature of these oscillations has not been discussed.

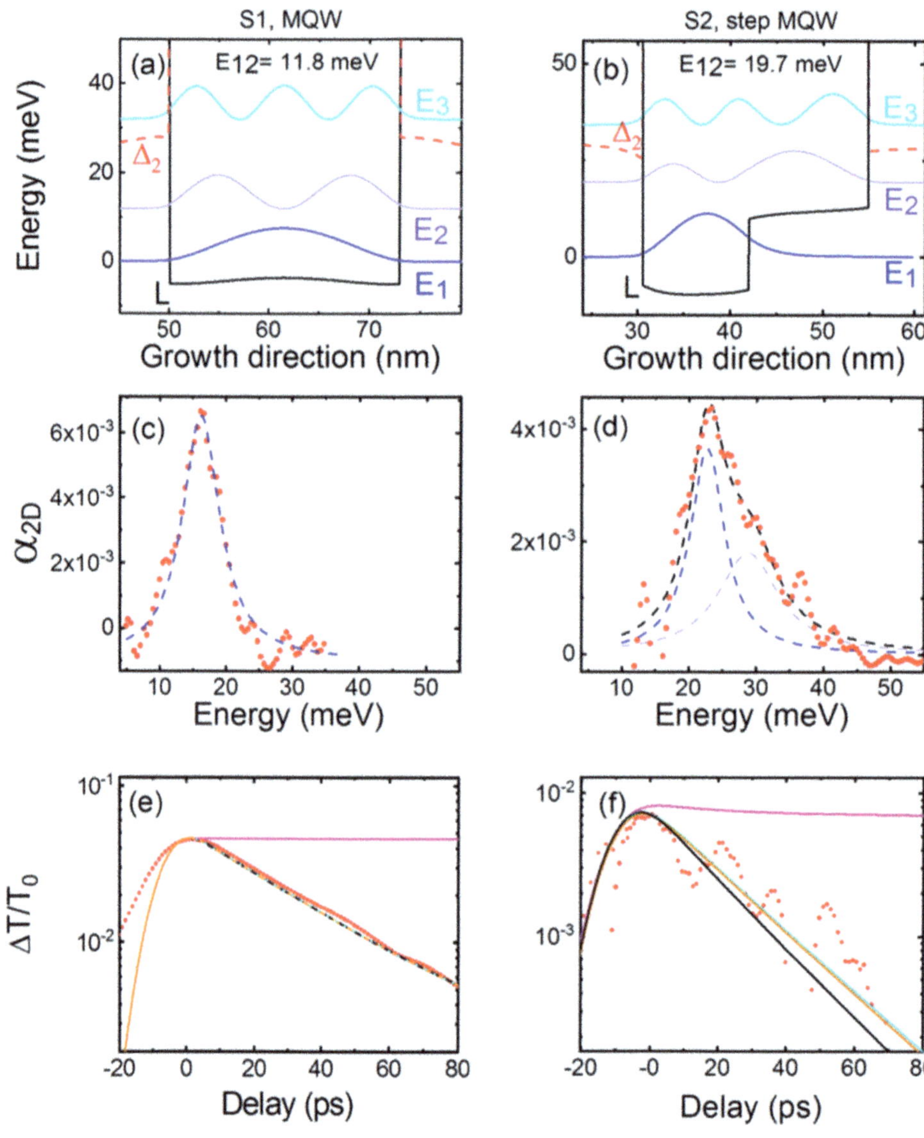

Figure 5.29. (a,b) Band-edge profiles of samples S1 and S2, respectively. Satellite valleys are shown as Δ. Theoretical values for the transition between the two lowest sub-bands are also shown. (c,d) The corresponding absorption spectra for transitions between the sub-bands for the two samples. The dashed violet curve in (d) indicates some E_{13} absorption. (e,f) The pump–probe signals for the two sub-bands at $T = 4$ K. The various curves are theoretical simulations: (black) using deformation potentials from Jacoboni *et al* [143], (magenta) turning off elastic scattering and using only the deformation potentials from Jacoboni *et al* [143], (cyan) using *only* elastic scattering, (green) using all scattering but with a new valley of deformation potential (discussed in the text) for the optical phonons. Reprinted from Ciano *et al* [142] copyright (2020) by the American Physical Society.

5.5.1 Zinc-blende materials

The zinc-blende materials have a crystal structure that is mainly cubic as discussed above in section 5.1.1. Here, the dominant optical phonon is the polar LO mode, although the acoustic modes play a role in the presence of a bottleneck and the ultimate decay of the carrier energy to the lattice and environment. Each material, for which important studies have been taken will be treated separately.

5.5.1.1 GaAs

The nonequilibrium optical phonons in GaAs were some of the earliest studies in the field [14, 15]. In these early experiments, a range of photon energies were used as were a range of laser powers. But, a phonon lifetime was estimated to be ~5 ps at a carrier density of ~10^{15} cm^{-3}. This was felt to be in good agreement with an earlier value of 3 ps that had been deduced from a study of the Raman linewidth [144]. In subsequent work at low temperature, a slowing in the carrier cooling was observed [145], which was thought to be due to a bottleneck. Subsequent Monte Carlo studies supported this conclusion and showed the phonons were definitely out of equilibrium with sufficient strength to create a partial bottleneck [22].

Early study of GaAs by photoluminescence (PL) at low temperature showed that electrons could be captured by donors, with the emission of multiple LO phonons [146]. Other work demonstrated that the study of such PL would give information on the hot carrier distribution [147]. It was then demonstrated how the PL line shape and intensity would vary with an electric field [148]. The latter group also demonstrated how measuring the PL at low temperature (77 K) would show a kink in the spectrum at the LO phonon energy [149]. With the appearance of such kinks, the distribution could no longer be considered to be a Maxwellian with a well-defined temperature; one then had to use only the tail of the distribution to define such a temperature.

In later work, Von der Linde *et al* used a mode-locked rhodamine-6G dye laser at 2.16 eV to excite ~10^{17} electron–hole pairs at 77 K [150]. From time-resolved Raman scattering, they deduced a lifetime of 6.3 ps from the decay of the anti-Stokes signal, which depends directly upon the phonon population, as shown in (5.3). Kash *et al* [151] used a similar laser to produce a carrier density of the order of ~10^{17} electron–hole pairs at 300 K, and deduced a phonon lifetime of 4 ps. Tsen and Morkoc [152] used a dye laser at 1.92 eV to photoexcite a density of ~10^{17} electron–hole pairs at 80 K, and measured a phonon decay time of 6 ps. Zhou *et al* [153] were able to show a slowing in the carrier cooling due to the presence of nonequilibrium phonons and deduced a bottleneck effect, using 1.72 eV photons that induced ~10^{18} cm^{-3} electron–hole pairs in As-doped GaAs. These latter authors concentrated more on the carrier–phonon interactions rather than a phonon lifetime. It seems that many of these measurements may have been deduced from the carrier cooling, and these may have been affected by intervalley transfer. Nevertheless, some of these studies looked at the variation of the phonon lifetime as a function of temperature (5–215 K), and suggested that the decay channel for the optical phonons was through a TA phonon and an LO phonon of opposite wave vector, both near the L point in the Brillouin zone [127]. This likely is thought to be the Vallée–Bogani channel, but it could also be ascribed to a variant of the Ridley–Gupta channel.

Ryan *et al* [154] used PL to study hot carrier luminescence in GaAs/AlGaAs heterostructures. Laser photons at 2.03 eV were used to excite the structure, and it was found that the carrier cooling time was a function of excited electron–hole plasma, as shown in figure 5.30. The authors found that the electron–phonon interaction appeared to be weakened in this structure, which may have been due to a bottleneck having formed or the fact that the reciprocal of the wave number for coupling to the LO modes is smaller than the thickness of the quantum wells. On the other hand, it was shown that carrier cooling in such GaAs quantum wells was still dominated by the LO phonon [155]. By plotting the inverse of the electron temperature, it was clear that, above 40 K, the slope was that of the LO phonon mode.

Studies of GaAs quantum wells at room temperature with PL in the presence of a high electric field showed that the carriers reached higher effective temperatures [156]. These authors identified two methods of computing the carrier temperature: (1) fitting the decay of the PL distribution gave an effective temperature which represented the combined effects of the electrons and holes, and (2) removing the hole temperature with an effective mass argument led to an electron temperature. In further PL studies of GaAs/AlAs quantum wells of various sizes, particularly the barrier width, it was learned from the observation of phonon emission peaks that the GaAs modes and the AlAs modes would begin to hybridize if the barrier was too thin [157], as shown in figure 5.31. For the sample with larger barrier widths, the phonon mode approached that of the AlAs phonons, while at the thinner widths the phonon line dropped to 320 cm^{-1} which is above the GaAs mode, but is well below the AlAs mode. The behavior of this mode could be explained by the alloy variation

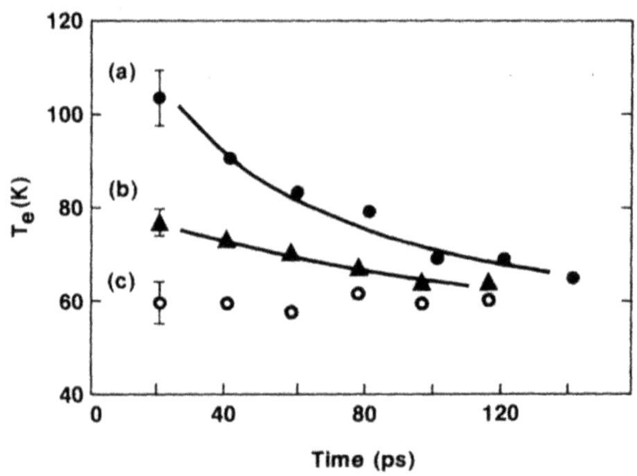

Figure 5.30. Time dependence of carrier temperatures in a GaAs–AlGaAs multi-quantum well structure for excited carrier densities of: (a) 5×10^{18} cm^{-3}, (b) 6×10^{16} cm^{-3}, and (c) 9×10^{14} cm^{-3}. The solid lines are obtained from a simulation assuming a relaxation time of 7 ps and a specific heat of nk_B. Reprinted with permission from Ryan *et al* [154], copyright (1984) by the American Physical Society.

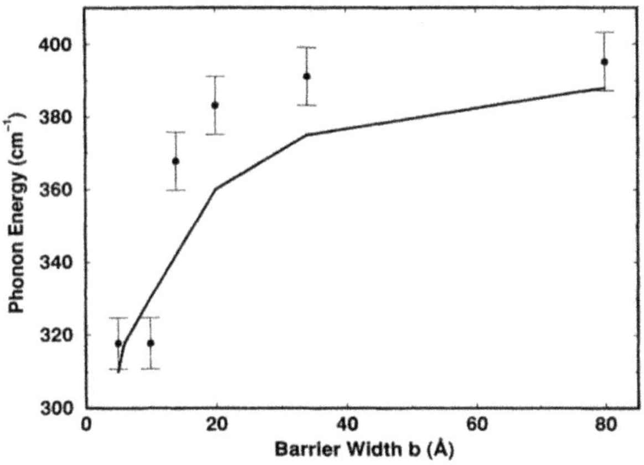

Figure 5.31. Measured energy difference of the 1LO emission line in the hot LO spectra of GaAs/AlAs multi-quantum wells as a function of the barrier thickness. The solid line is the calculated effective phonon energy based upon the interaction of the GaAs and AlAs modes in the structure. Reprinted with permission from Sapega *et al* [157], copyright (1995) by the American Physical Society.

as shown in figure 5.6 for the thicker barriers, but this is certainly not the case for the thinner barriers.

Langot *et al* [158] studied hot electron and hole cooling in GaAs at room temperature for a variety of densities, using 1.53 eV excitation and a single- and double-wavelength absorption saturation technique. The 50 fs laser pulse is fed into a single mode optical fiber which broadens the original spectrum by 14–130 nm, and this signal is split into two for the experiment. The pump energy creates electrons below the threshold for intervalley scattering. From their measurements of the electron and hole relaxation, they found a phonon lifetime of 2.1 ps.

Earlier, Tsen *et al* [114] had also studied the 2TA spectra in Raman scattering in GaAs as well as Si, and showed that these peaks could be measured easily. It was then found that these phonons could be driven out of equilibrium by the decay of the LO modes, and this would lead to a bottleneck in its own right. It was found that the TA modes could be excited either by the LO mode decay at any zone edge or by a TO–TA pathway, which means that both the Klemens and the Ridley pathways were possibly active.

In a different approach, which does not create extra electron–hole pairs, intra-band (free carrier) optical absorption, requires the additional emission or absorption of an optical phonon, as discussed in section 4.2. In these experiments [159], a pair of optical pump pulses are used to create a diffraction grating, and a subsequent delayed probe pulse is scattered off this grating (the setup is much like that of figure 5.21, except that infrared optical pumps are used). Here, they infer a lifetime similar to that given by infrared Raman scattering of 2 ps [160].

In the quantum wells, the electrons and holes relaxed differently and did not have a common temperature [161], while in some cases the carrier populations could not be described with a thermal distribution for times below ~0.2 ps [162]. Actual measurements of phonon lifetimes in quantum wells were made later using 5 nm quantum wells (with 10 nm barriers) with two independently tunable dye lasers operating at 1.86 eV pump and 1.68 eV probe [23]. This yielded a measured LO lifetime of 8 ps at 10 K. Later studies of quantum wells established that the amplitude of the nonequilibrium phonons depended upon the well thickness and that there were interface modes that could be excited [163]. Finally, it was determined that the average electron–LO phonon scattering time at 77 K was about 170 fs [164].

The time-dependent response of the LO phonons in these quantum wells was subsequently measured by anti-Stokes Raman scattering [165, 166]. It was observed that the LO phonons decayed with a lifetime of about 5 ps at 77 K, which is faster than the bulk GaAs at this temperature. It was later confirmed that the cooling rate for the photo-excited carriers was affected by the presence of nonequilibrium phonons [153].

The situation became much more complicated in quantum wires, due to the complex folding of the various phonon modes. Several LO modes could be identified in the Raman spectrum of such quantum wires [167]. Ultrafast measurements of arrays of GaAs core–shell nanowires, were probed with sub-10 fs broad band (480–950 nm) pulses [168]. It was found that the nanowires had carrier cooling on the sub-0.5 ps time scale, and that the cladding had little effect on this cooling. The authors conclude that hot optical phonons, and the subsequent decay into acoustic phonons was important in the cooling process. In studies of tunneling in GaAs/AlGaAs double quantum well structures, the interaction of the tunneling carriers with the acoustic modes was observed [169]. Both Stokes and anti-Stokes lines were seen to appear in the tunneling current as phonon-assisted tunneling behavior. Laser excitation of GaAs has also been observed to launch a ballistic beam of acoustic phonons that can be observed propagating through the sample [71]. More recently, Vanacore *et al* [76] have used ultrafast electron diffraction to study the emission of high energy acoustic phonons from GaAs quantum dots. An ultrafast laser excitation at 800 nm and 150 fs pulses, is nearly resonant with the bound states of the dot, producing excitons which themselves couple to high energy acoustic phonons through the deformation potential interaction. This creates a local stress in the dots, shifting the atomic positions. The time to excite the acoustic modes is about 1.3 ps experimentally and their decay seems to be about 112 ps.

5.5.1.2 AlGaAs

GaAlAs is a common ternary alloy material used in quantum well and heterostructures as an 'insulating' layer on top of GaAs for a wide range of devices. This alloy has phonons which show a two-mode system in that there are GaAs-like LO modes and AlAs-like LO modes, as discussed in section 5.1.3 above. The relative strength of the AlAs-like mode increases with respect to the GaAs-like mode as the AlAs composition increases up to 0.25, although for higher compositions, this increase depends critically upon the excited carrier density, and how the phonon peaks are measured. This is a result of the wave number of the Raman signal which lies off the central part of the peak of nonequilibrium phonons created by the hot carriers, so that the measured phonon amplitude is a function of density. Studies at

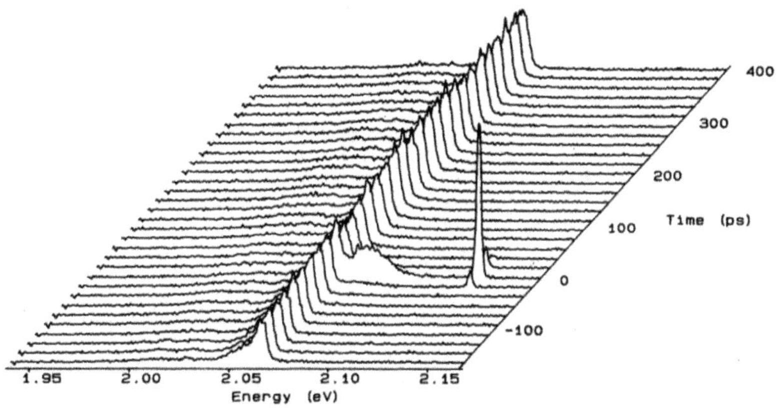

Figure 5.32. Three dimensional plot of the temporal behavior of the PL in AlGaAs with $x = 0.44$. The major long-lived peak at 2.06 eV is due to the electrons in the X valleys. This lifetime is longer than the repetition period of the laser excitation! The flash signal at 2.08 eV, and the pulse at higher energy, arise from electrons in the Γ valley. Reprinted with permission from Rühle *et al* [172], copyright (1989) by the American Physical Society.

low temperature showed that the lifetime for both modes at 10 K was 7 ± 1 ps [170]. In thin quantum wells of GaAs–AlAs, the AlAs LO mode could still interact with the carriers in the GaAs wells, although this interaction decreased rapidly with well thickness [171], and the effective phonon energy varied with the barrier thickness, as shown in figure 5.31 above.

PL studies of $Al_xGa_{1-x}As$ with values of x up to 0.44 have shown additional information [172]. For compositions below 0.41, the material has a direct band gap with the conduction band minimum at Γ. However, for $x = 0.44$, the material is indirect with the conduction band minima at X. Laser energies in the range of 1.55–1.97 eV were used to study several compositions. For the direct transitions with $x < 0.44$, the behavior was much like that of GaAs, with a phonon cascade in the Γ valley, with cooling curves similar to figure 5.30, but with the excess energy rather than the density being varied. When $x = 0.44$, however, the PL becomes long-lived as the photo-excited electrons transfer quickly to the X valleys, where they have a very long recombination time due to the indirect nature of this latter quantity. This is shown in figure 5.32, where the long-lived peak at 2.06 eV arises from electrons in the X valleys. The lifetime of this peak is longer than the repetition period of the laser excitation, so cannot be determined to any degree of accuracy. However, a faster behavior at 2.08 eV comes from electrons in the Γ valley that have not transferred to the X valleys. Another important result here is the measurement of the acoustic phonon deformation potential for scattering by holes in this material, as shown in figure 5.33. It was assumed that the variation in the scattering of the holes as the alloy composition was changed was due to a variation in the deformation potential, which is a reasonable assumption.

5.5.1.3 GaP

The LO phonons in GaP were studied by measuring the infrared coherent anti-Stokes Raman line [160]. Here, two incident laser signals, whose frequency

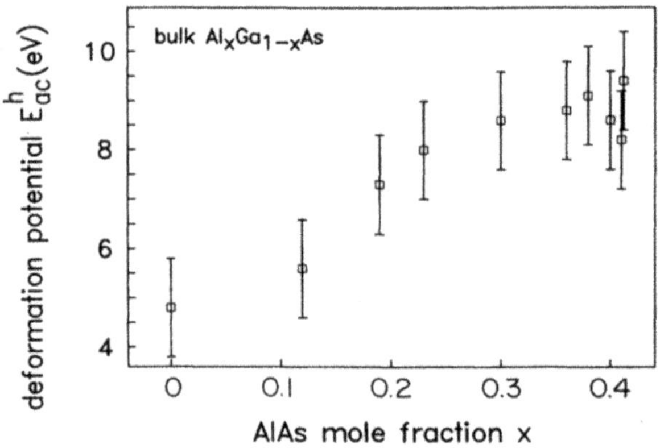

Figure 5.33. Dependence of the deformation potential of the holes on the composition of the alloy. This assumes that the variation in the scattering arises from a change in the deformation potential. Reprinted with permission from Rühle *et al* [172], copyright (1989) by the American Physical Society.

difference corresponds to the phonon frequency, are used for the pump and another delayed laser pulse is used to measure the dephasing time of the phonons. This dephasing time is related to the lifetime of the phonon population. For delay times longer than 5 ps, definite exponential decay was observed. A value of the lifetime of 6.8 ± 0.6 ps was obtained at 300 K, and it was observed to get longer as the temperature was reduced. Using a visible version of the coherent anti-Stokes Raman scattering, it was found to have a lifetime of ~7 ps at 300 K, and ~22 ps at 5 K [173, 174]. Moreover, these latter authors suggest that, due to the phonon dispersion in GaP, only the Klemens channel is available for phonon decay into acoustic modes (see figure 5.25). At higher temperature, in other work they have shown that there is a definite contribution from the Ridley–Gupta channel as well [125].

5.5.1.4 GaSb

Some of the first measurments of GaSb were studies of hot carrier PL, in which the samples were illuminated with 3 ps pulses at 2.04 eV, from a rhodamine-6G dye laser [175]. The signal was upconverted and detected with photon counting. The PL signal is shown in figure 5.34 for several different lattice temperatures. The luminescence seems to follow a Maxwellian decay with energy, so that a temperature can be determined, and this is shown in figure 5.35 for both GaSb and a layer of InGaAs, given for reference purposes (this material is discussed below). In studies to try to simulate the hot carrier relaxation, the authors assumed the same LO phonon lifetime as for GaAs, which was about 7 ps. Both bulk GaSb bulk and GaSb/AlSb quantum well structures were studied by the same authors using standard optical pump and probe techniques at ~2 eV excitation to study coherent optical phonon signals [176]. GaSb is an interesting material, in that the satellite L valleys are only a few tens of meV above the conduction band minimum, and intervalley scattering is a very fast process [177, 178]. These studies indicate a phonon decay on the 9 ps scale at 12 K. Somewhat different behavior is observed at room temperature, where an initial fast decay on the 2 ps scale is followed by a much slower relaxation at later

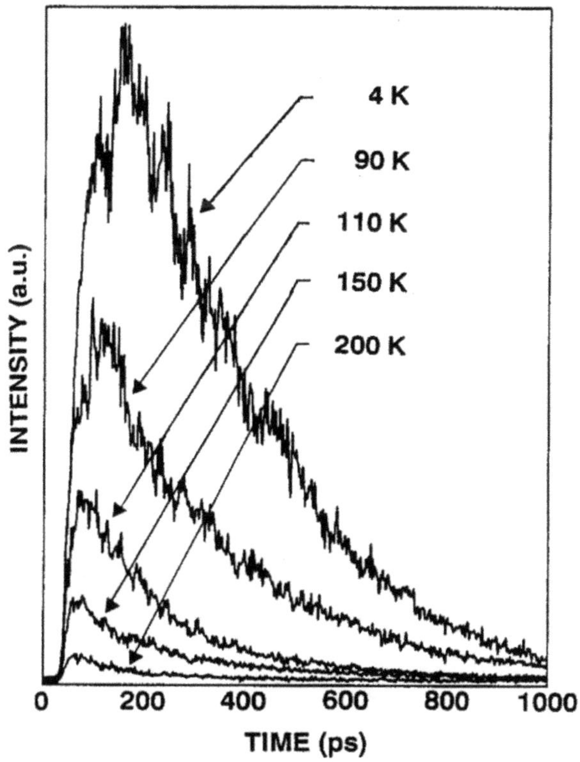

Figure 5.34. Time dependence of the spectrally integrated PL for undoped GaSb for several lattice temperatures. Reprinted from Snow *et al* [175], copyright (1989) with permission from Elsevier.

time which is indicative of a bottleneck in the phonon decay [179]. In quantum wells, scattering from both the LO modes in the wells and in the barriers was observed, when probed with a lower photon energy of 1.55 eV [179]. From these signals, one could determine the decay of the coherent signals. These latter authors found that the decay times were of the order of 1.1 ps for the bulk modes and longer in the quantum wells due to the folding of the phonon modes in the superlattice. They also found that the amplitude increased with the laser power, so that both modes were in a nonequilibrium state.

5.5.1.5 AlAs

The AlAs-like mode in a GaAlAs layer was mentioned earlier to have a lifetime of approximately 6 ps at low temperature [170]. Later studies of AlAs on its own, using pump and probe Raman measurements, gave a similar result of about 9.7 ps at low temperature [180]. These latter authors measured the linewidth of the Raman signal to estimate the lifetime and the data seem to indicate a lifetime of ~2.1 ps at room temperature.

5.5.1.6 InAs

InAs is a typical narrow band gap (0.4 eV) material with a relatively high carrier mobility. The presence of nonequilibrium LO phonons at 70 K was detected by an

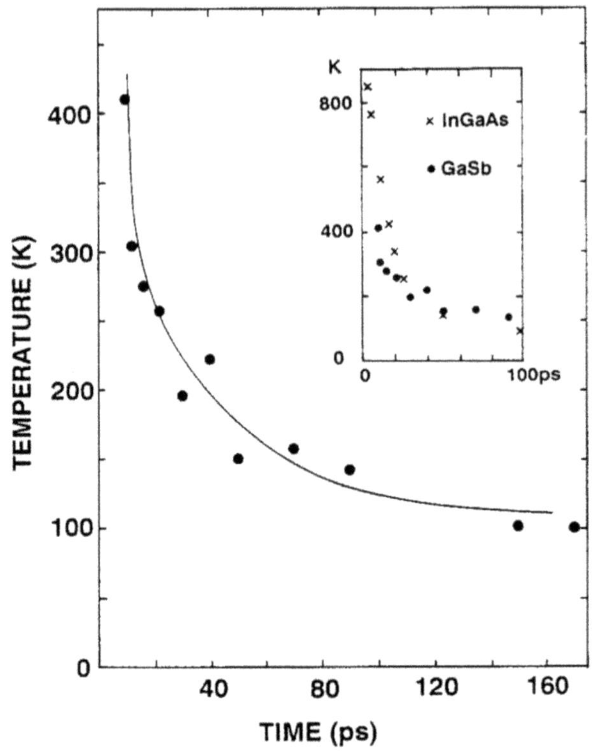

Figure 5.35. Experimental carrier cooling curve for undoped GaSb. The inset shows the curve for InGaAs under similar conditions. Reprinted from Snow *et al* [175], copyright (1989) with permission from Elsevier.

increase in free-carrier absorption in bulk InAs [181], using photons of 0.19 eV. With this low energy, no two-photon absorption can occur, so only heating of the residual carriers (usually electrons) can occur through free-carrier absorption. However, the LO phonon lifetime was masked by the cooling rate of the excess carriers, which decayed within 70 ps due to the fast recombination in this material. In this case, the InAs was heavily doped to about 10^{18} cm^{-3}, which made the material very heavily degenerate. The density of absorbed photons was about 5×10^{17} cm^{-3}.

The use of sub-picosecond optical pump and probe pulses to study Raman scattering in InAs was reported by Grann *et al* [182]. These measurements at room temperature give an LO phonon lifetime of 1.4 ps. Here, excitation was carried out with a dye laser at 1.95 eV, with the power adjusted to create ~3×10^{15} cm^{-3}. The results are shown in figure 5.36 along with simulations using a range of parameters for the lifetime of the phonons and the deformation potential for intervalley scattering. The experimental data is shown as the blue dots, while the ensemble Monte Carlo simulation is the other symbols, which show the variation in parameters used. Curve 1 is for a phonon lifetime of 1.8 ps and a Γ-L phonon coupling constant of 1.4×10^8 eV cm^{-1}, curve 2 is for a lifetime of 1.8 ps and a coupling constant of 1.4×10^9 eV cm^{-1}, and curve 3 is for a lifetime of 4 ps and a coupling constant of 1.4×10^9 eV cm^{-1}. It appears that curve 2 gives the best fit to

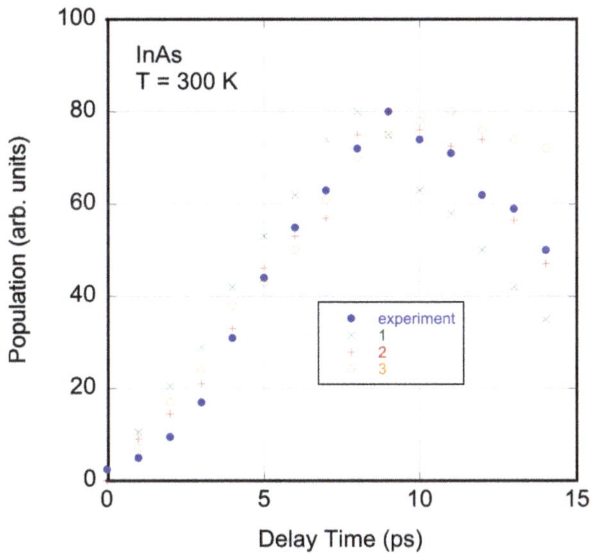

Figure 5.36. Rise and fall of the anti-Stokes line from Raman scattering from the polar LO phonons in InAs at 300 K. The experimental data is the blue dots, while theoretical curves are discussed in the text. The data is taken from Grann *et al* [182].

the data. The fact that the amplitude peak occurs at 9 ps after the excitation pulse illustrates the effectiveness of satellite valley storage of the electrons. Also, using the simulation allows one to determine a number of the important parameters of the total interaction. The decay of the L electrons back to Γ is a simple exponential with the decay time of ~2.6 ps. In InAs/AlSb quantum wells, a strong bottleneck was observed through the much longer lifetime, when excitation was provided by sub-THz rf signals [183]. These quantum wells were grown by MBE and etched into Hall bar shapes. At 10 K, the sheet density of the two-dimensional electron gas was about 2×10^{12} cm^{-2}. The dc conductivity at 6 K was measured with short electric field pulses, and the changes induced by illumination with 0.66 THz signals was monitored. Although they could monitor the incident power per electron, they did not find values for the LO phonon lifetimes. The presence of the bottleneck led to an increase of the lifetime to several ps at low temperature.

Similar effects have been found recently in InAs/InAlSb multi-quantum well systems [184]. This heterostructure has a staggered band lineup with electron levels in the InAs and hole levels in the InAlSb. Relatively narrow quantum wells were used to give a large ground to first excited state separation (~0.7 eV). PL was measured at several temperatures with the sample excited by a laser pump, as shown in figure 5.37. Panel (a) shows the PL and its fit at three temperatures, where band-to-band recombination dominates. Panel (b) shows the shift in the PL peak as a function of the lattice temperature. The approximate temperature rise of the hot carriers is shown in panel (c) for the three curves of (a) as a function of the excitation power. Dynamics of the carriers and the phonons were studied by time-resolved THz probe signals. The carriers in the quantum wells remain long-lived due to the

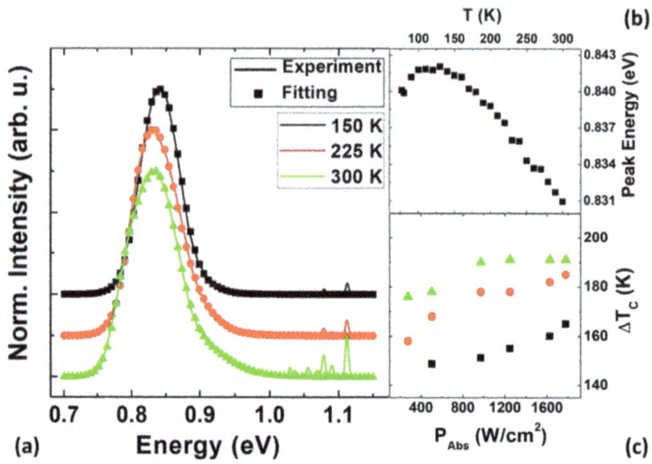

Figure 5.37. (a) Normalized PL from the InAs/InAlSb multi-quantum well system at three temperatures. (b) The peak PL energy as a function of lattice temperature. (c) The increase in the carrier temperature as a function of the incident power for the curves of (a). Reprinted from Esmaelpour *et al* [184] under the creative commons 3.0 open access license.

presence of a phonon bottleneck. Decay of the InAs LO modes is thought to be through the Klemens channel into a pair of LA modes. The multi-quantum well structure inhibits the motion of the LA modes through to the substrate, so that these modes are quite likely out of equilibrium as well.

5.5.1.7 InGaAs

$In_{0.53}Ga_{0.47}As$ was studied as early as 1983, where the two-mode phonon structure was studied [58]. While one expects two-mode behavior, these authors found that the Raman scattering indicated only a single LO mode at 270.5 cm^{-1}. It was also found that studies using magnetophonon resonance indicated that the LO modes reached a temperature of 385 K, while the lattice was at 130 K.

Studies of the nonequilibrium phonons in InGaAs quantum wells were done as early as 1988 [185]. Samples were probed with 3 ps pulses from a rhodamine-6G dye laser at 607 nm. PL was measured at several delay times after the pump pulse, showing that the distribution was almost thermal already at 20 ps. These studies focused on the cooling of the electrons without careful study of the phonon lifetime. More recently, ultrafast pumping of an InGaAs/InAlAs strained quantum well system was used to show rapid generation of THz signals in this structure [86]. However, the time delays of the experiment were such that the optical phonon decays were not determined. More recently, the use of InGaAs of various concentrations has become of extreme interest for hot carrier solar cell application, as has been discussed.

5.5.1.8 InP

Early studies of InP with infrared absorption gave considerable information on the phonon spectrum of the material [186]. Subsequently, PL was studied in InP/AlInAs

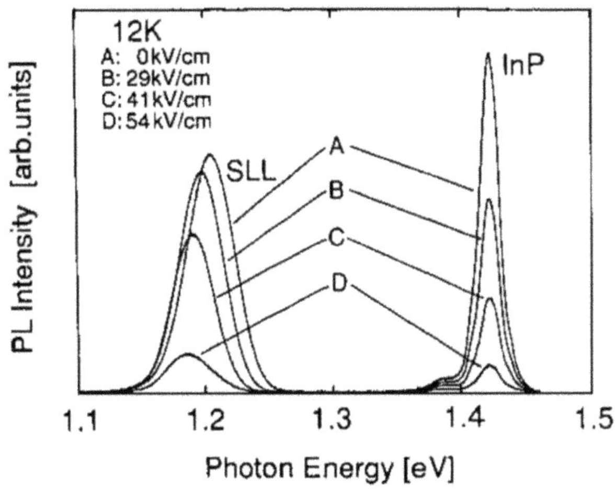

Figure 5.38. PL signal for various gate voltages (indicated by the electric field). The two signals arise from the InP directly and from electrons in the InP accumulation layer with holes in the AlInAs at the interface (SLL). Reprinted from Sakamoto *et al* [187], copyright IOP Publishing. All rights reserved.

heterostructures [187]. The layers were modulation-doped heterostructures with a 5 nm AlInAs spacer layer between the undoped InP and the Si-doped AlInAs upper layer. PL was observed to have primarily two lines: one from the electron–hole recombination in the InP itself, and a lower energy line from recombination between electrons in the InP accumulation layer and holes in the quantum well in AlInAs at the interface, as shown in figure 5.38, where the PL is shown for varying gate voltages that change the electron density in the InP accumulation layer. The PL line from InP was essentially independent of the gate bias, while the second line showed a weak reduction in energy at large positive gate voltage.

Raman studies of InP with a 1.952 eV pump excitation showed that intervalley scattering to the L valleys was prevalent, and this led to some difficulty in directly measuring the phonon lifetime [182]. However, fitting the observed carrier distributions and the phonon decay with EMC simulations showed that a good fit to the experiment could be obtained with a lifetime of 2.3 ps at room temperature and 5 ps at 77 K. Earlier results obtained by monitoring just the coherent anti-Stokes Raman scattering suggest that the lifetime may be longer at 7.6 ps at low temperature [160], although a lifetime inferred from the Raman scattering linewidth suggests a lifetime of 2.4 ps [188], closer to that of the direct Raman study discussed above.

InP self-assembled quantum dots were grown on a GaAs substrate, between two layers of GaAsP [189]. PL was observed from the dots, which was centered around 1.7 eV at 2 K and showed clear signals from the bulk InP LO modes.

In recent years, there have been suggestions that the Klemens decay channel for the LO modes would be blocked by a phononic gap in InP, and that this would lead to a longer lifetime than in GaAs [190]. But, this is incorrect. The LO phonon frequency at Γ is 10.35 THz, while the LA(X) frequency is 5.8 THz [191]. The two numbers are actually closer together as found with recent neutron scattering values [192]. Hence, the Klemens channel is open, although the LA modes involved must be

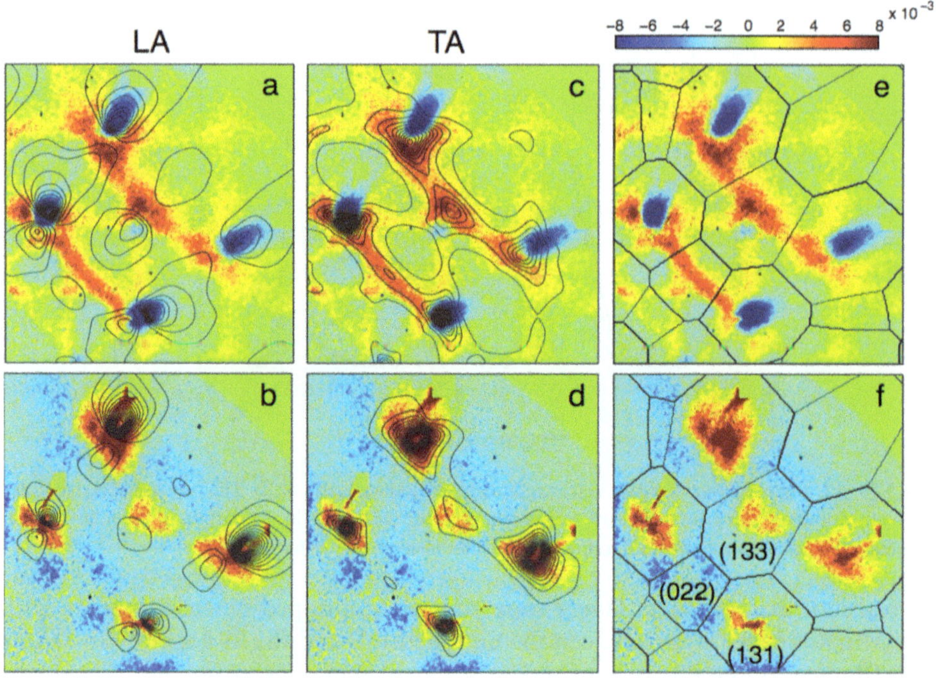

Figure 5.39. Contributions to the thermal diffuse scattering of the x-rays from the TA and LA modses in InP and InSb. The second left singular value plot is shown in (a) for InP and (b) for InSb. The contours show the values for the LA modes. The TA modes are shown in (c) and (d). (e) and (f) show the Brillouin zone boundaries for the two materials, and some Miller indices are shown in (f). Reprinted with permission from Trigo *et al* [193], copyright (2010) by the American Physical Society.

quite near the zone edge. Indeed, studies show that the Klemens channel is involved with the LO decay, but must be supplemented by the Ridley–Gupta channel in order to fit the measurements as a function of temperature [126].

InP has also been studied with UXD using a laser at a wavelength of 680 nm to create the diffraction pattern and 100 ps x-ray pulses to probe the system [193]. Here, it was observed that TA phonons continued to be excited for as long as 1 ns after the laser pulses, although the phonon lifetimes were a few ps, but not determinable due to the long x-ray pulse. The diffraction image is shown below, in figure 5.39, along with that of InSb. However, it was found that the dominant TA signals were along the (111) and (010) directions near the zone edges.

5.5.1.9 InSb

The phonon relaxation has been probed along with the carrier relaxation with THz pump and probe spectroscopy [194]. The THz signal was varied in the range 0.2–1.6 THz, and was generated by 800 nm laser pulses. The effort was directed toward the impact ionization of InSb at both 80 and 200 K, although little data on generation rates or ionization coefficients was obtained. While they presume an overly long phonon relaxation time [127], they establish that the LO modes relax through the LA modes (the Klemens mechanism).

Most other studies of InSb have utilized UXD. Lindenberg *et al* appear to have been the first to utilize this approach, although they were limited to frequencies below 0.1 THz, which held them to study only low energy acoustic modes [195]. The sample was excited with 800 nm laser pulses, synchronized to the x-rays pulses from a storage ring. For low laser fluences, the authors observed oscillations generated by coherent acoustic phonons. Feuer *et al* generated their x-rays from laser excitation of a Si crystal (generating an electron–hole plasma) with the dominant x-rays at ~0.665 nm (He_α) and ~0.712 nm (K_α) lines [196]. This was combined with laser excitation of the sample at 796 nm. However, they also were able to study only low energy acoustic modes, as their effort was concentrated upon characterizing the x-ray source. This trend continued with the work of Larsson *et al* who excited coherent acoustic phonons and were able to determine the sound velocity in the (111) direction [197]. In this latter work, the sample was irradiated with 800 nm photons and the x-rays came from a synchrotron storage ring. The extracted sound velocity was 4×10^5 cm s^{-1}.

Trigo *et al* also used UXB to study the acoustic modes throughout the zone [193]. While they do not determine a lifetime, they do find that the LA modes do not seem to show much time dependence, while the TA modes show relatively homogeneous excitation throughout the Brillouin zone which suggests the Ridley decay mode is dominant. To assist in their evaluation, they produced a singular value decomposition of the data, which gave them a left singular value of the scattering data, which is plotted in figure 5.39.

5.5.1.10 CdTe

Raman scattering of both the phonons and the electrons, by single-particle scattering, has been studied for CdTe, for a range of carrier densities [198]. These densities ranged from a few times 10^{15} cm^{-3} to a few times 10^{17} cm^{-3}, and the phonon population increased almost linearly with the density. The LO phonon lifetime was deduced to be ~0.75 ps at room temperature.

Phonon spectroscopy was used to study ballistic phonons in CdTe/CdMnTe semimagnetic quantum wells [199]. The nonequilibrium phonons were able to be seen in the PL data, and the latter was suggested as a way of detecting the sub-THz phonons.

5.5.1.11 ZnSe

Coherent anti-Stokes Raman scattering was used to study ZnSe [174]. These measurements gave a phonon lifetime of the order of 1.2 ps at 300 K, but the excitation density was not reported. Study of the temperature decay of the LO phonons indicated that the decay channel at room temperature was likely a Ridley–Gupta channel with the combination of a TA and a LA phonon with a $\boldsymbol{q} \sim 0.85X$. It was thought that the decay at low temperatures could be due to the Klemens channel. It was also noted that the LO phonon decay slowed at longer times, which likely indicated the presence of a blockade.

Subsequently, pump and probe Raman experiments using a 514.5 nm argon-ion laser over a range of temperatures were made [200]. From the phonon dispersion curves, there are several available channels for both the LO and TO modes to decay, including both the Klemens and Ridley mechanisms. The authors were able to determine lifetimes to be 1.89 ps for the LO modes and 2.26 ps for the TO modes at room temperature. These lifetimes are a factor of 2–3 larger at low temperatures. From the temperature dependence of the decay, these authors found that their data clearly fit only a Klemens channel with two LA phonons along the (100) direction in the crystal.

Similar Raman pump and probe experiments were conducted with a cavity dumped dye laser on ZnSe grown on GaAs [201]. They reported that the Raman signal varied with the square of the excited density, and that the decay of the phonons had a lifetime of about 4.5 ps, at 20 K, but it was not clear if this was dominated by the hot carrier relaxation or not.

ZnSe grown on GaAs was also studied by Mehendale *et al* [202]. Here, differential reflectivity of the 400 nm pump and probe pulses was used to study the carrier cooling with varying excitation intensity, and consequently with varying carrier density. They observed a definite slowing in the carrier cooling rate at a density of $\sim 3 \times 10^{17}$ cm^{-3}, which they interpreted as the onset of a phonon blockade. The cooling observed was consistent with an electron–LO phonon emission time of 45 fs.

5.5.2 Wurtzite material

The wurtzite materials have a great many more phonon modes than the zinc-blende materials, because they have twice as many atoms per unit cell. In essence, this is a relaxation of the zinc-blende structure for materials that are more ionic than those of the previous section. The presence of more phonon modes means that they have more channels for the Klemens decay mechanism due to the additional lower energy B and E_2 modes in figure 5.3(a). But, the higher energy B and E_2 modes in this latter figure also provide more channels for the Ridley decay mechanism, as well as the Vallée–Bognani mechanism. It will be seen in the following that these materials, particularly the group III–N materials, tend mostly to follow the Ridley–Gupta channel for their LO phonon decay process, although this is not universal. This was already illustrated above in figure 5.26.

5.5.2.1 AlN

The phonon modes in AlN and GaN have been investigated by micro-Raman scattering using the 514.5 nm line of an argon-ion laser, and studying the linewidth at room temperature [203]. In general, the lifetimes of the various phonons fall into two categories, one in which the E_2^1 mode (upper E_2 mode in figure 5.3(a)) has a relatively long lifetime and a second group of the E_2^2 (lower E_2 mode in figure 5.3(a)), E_1(TO), A_1(TO), and A_1(LO) that have much shorter lifetimes. The observed lifetimes for these five modes in AlN were found to be 4.4, 0.83, 0.91, 0.76, and 0.45 ps, respectively. While these were measured in crystallites, similar values for the E_2^2 mode were found in thin films of AlN.

In a similar approach, the full temperature dependence of the various phonon modes were subsequently studied by micro-Raman scattering using the 488 nm line of the argon-ion laser [204]. Again, the lifetimes were extracted from the Raman linewidths. A value of 0.57 ps was measured for the A_1(LO) line at room temperature, and the measurements from 10 to 1250 K showed that this phonon decays predominantly into two phonons with equal energy (the Klemens channel). The E_2^1 mode exhibited a lifetime of about 1.7 ps at room temperature, and the temperature variation suggested that this phonon decays into two different phonons of frequency of 176 cm^{-1} and 483 cm^{-1}, which might be an acoustic mode and the E_2^2 mode (the Ridley–Gupta channel).

5.5.2.2 CdS

The earliest work on CdS appears to be that of PL at 77 K [205]. A pulsed N laser at 3.68 eV was used as the excitation source. PL peaks were observed to shift to lower energy as the excitation increased, and were thought to arise both to carrier correlation and to nonequilibrium LO phonons. While the phonon lifetime was not found, the effective temperature of the phonons reached almost 1800 K at the highest excitation levels.

Later studies of CdS used both a pump and probe approach and a transient grating approach [206]. A Nd:YAG laser was used with the pump at the third harmonic at 355 nm and the probe at the second harmonic with the sample at 5 K. The PL showed an exponential decay over ~300 ps for a range of excitation power. This long decay is assumed to be longer than the lifetime of the phonons, so that the latter could not be obtained.

Similar pump and probe measurements as well as luminescence measurements were undertaken using the third harmonic of the Nd:YAG laser at 3.5 eV [207]. It was found here that the carrier relaxation exhibited two regions of decay which they suggested reflected the anharmonic decay of the nonequilibrium LO phonon system, but a second generation of phonon creation, which is more likely to be the onset of a phonon blockade due to the nonequilibrium of the secondary phonons. The lifetime of the LO phonons appears to be about 2 ps.

Femtosecond dynamics of the electron–hole plasma have been observed by the luminescence up-conversion technique (a process by which the normal luminescence is shifted upward by the absorption of a probe pulse) [208]. Here, both normal LO phonon decay and its slowing by a bottleneck effect were reported. At low densities, they observe normal LO decay into two equal LA phonons, appropriate for the Klemens channel, but at higher density, there seems to be a LO mode amplification, which is likely due to a bottleneck effect, with a lifetime of ~0.6 ps. However, the authors feel that this bottleneck arises from the build-up of the acoustic modes that slows the LO phonon decay. The disturbance of the acoustic phonons occurs on the sub-ps level, likely due to a fast decay of the LO modes, and provides a reduction of the energy loss rate of the carriers due to a slowing of the LO modes.

5.5.2.3 CdSe

CdSe seems to have been studied first by Shank *et al* [209], using traditional pump and probe techniques. Here, the pump was at 307.5 nm from a dye laser, and the probe was provided by the second harmonic of this laser. The probe was used to measure the reflectivity of the sample. This change in reflectivity is due to the electron–hole plasma generated by the pump. A very fast decay of this plasma was observed, where the carriers lost about 0.8 eV each in the first 2 ps. It was felt that this was due to the emission of optical and acoustic phonons.

A similar approach used a Nd:glass laser at a wavelength of 1060 nm [210]. This signal was doubled to produce single-photon excitation of the electron–hole plasma. The decay of the luminescence was about 230 ps for all excitation levels, with the peak luminescence occurring some 20–30 ps after the pump pulse. Their results lead to an estimate of the phonon lifetime of ~5 ps at 12 K.

Studies using a pump pulse at 2.12 eV and a white light continuum for the probe, with the sample at 4.2 K, were carried out by Masumoto and Sasaki [211]. This allowed an ultrafast measurement of the absorption of the crystal, with both pump and probe pulses being in the fs range. This enabled determination of the phonon emission time by electrons of ~120 fs.

Studies of the hot carrier loss rates at 8 K [212], suggested a phonon lifetime of 6–9 ps, depending upon the laser excitation. Here, excitation with 1.8 ps pulses from a rhodamine-6G dye laser was used to study the material with luminescence up-conversion. The probe pulse generates a dense electron–hole plasma with dominant electron–hole scattering. As a result, they deduce that the cooling is due to TO phonon emission from the holes, and not by LO scattering as this latter is more or less blocked by the nonequilibrium generation of the phonons. Hence, they estimate a slow 6–9 ps decay time for the LO modes. Later measurements pushed this down to 5 ps at 8 K [213]. However, these authors also found a clear density dependence of the LO decay, with the lifetime increasing with density. They speculate that the density dependence actually arises from an increase in the acoustic phonon population, which leads to the blockade mentioned earlier.

5.5.2.4 GaN

Over the past few decades, wurtzite GaN has become of great interest in both electronic and optical devices. Nonequilibrium phonons have been studied not only in the bulk material, but also in the heterostructures important for field-effect transistors. The apparent earliest measurements were made using a R6G dye laser at 2.18 eV, which is below the band gap [214], so that two-photon absorption was required with the sample at 25 K. The phonon distribution was found to increase with the square of the laser power, indicating that the hot carriers were produced by this two-photon absorption. It was determined that only the A_1(LO) mode and the E_1(LO) mode were driven out of equilibrium. Phonon lifetimes could not be observed because of the 3 ps pulse width. Subsequent time-resolved measurements with the photon energy at 4.18 eV were able to detect the phonon lifetime of the A_1(LO) mode to be ~3.5 ± 0.8 ps at 300 K [215]. Further measurements refined the value to 3.0 ± 0.1 ps at 300 K and 5 ps at 25 K [132].

As mentioned above, the phonon modes in AlN and GaN have been investigated by micro-Raman scattering using the 514.5 nm line of an argon-ion laser, and studying the linewidth at room temperature [203]. In general, the lifetimes of the various phonons for GaN fall into two categories, one in which the E_2^1 mode (upper E_2 mode in figure 5.3(a)) has a relatively long lifetime and a second group of the E_2^2 (lower E_2 mode in figure 5.3(a)), E_1(TO), and A_1(TO) that have much shorter lifetimes. The observed lifetimes for these four modes in GaN were found to be 10.1, 1.4, 0.95, and 0.46 ps, respectively. The A_1(LO) was not observed presumably because of plasmon damping.

At about this time, a problem was discovered. In studies of GaN/AlGaN HEMTs, it was found that the experimental properties of these devices could be explained only if a much shorter lifetime, 0.35 ps, was assumed for the hot phonons [216]. A second, more careful comparison of experiments and simulation of such HEMTs concluded that the phonon lifetime could not be more than 1 ps [217]. Studies of the nonequilibrium phonons were then made over a range of carrier densities [218], using the third harmonic of a Ti:sapphire laser to excite the sample, and pump and probe measurements were made as a function of the photo-excited carrier density. It was found that a lifetime of 2.5 ps was found at a density of 10^{16} cm^{-3}, but decreased to 0.35 ps at 2×10^{19} cm^{-3} for the A_1(LO) mode. The density dependence of the A_1(LO) modes for both GaN and InN are shown in figure 5.40. Other measurements, at fewer densities, confirmed the dependence and found a similar behavior for the E_1(LO) mode [219]. These latter studies also showed that the phonons in the higher energy E_2, the A_1(TO), and the E_1(LO) modes were all driven out of equilibrium, whether directly by the electron–phonon interaction or the A_1(LO) mode decay. The reason for the density dependence is still an open question.

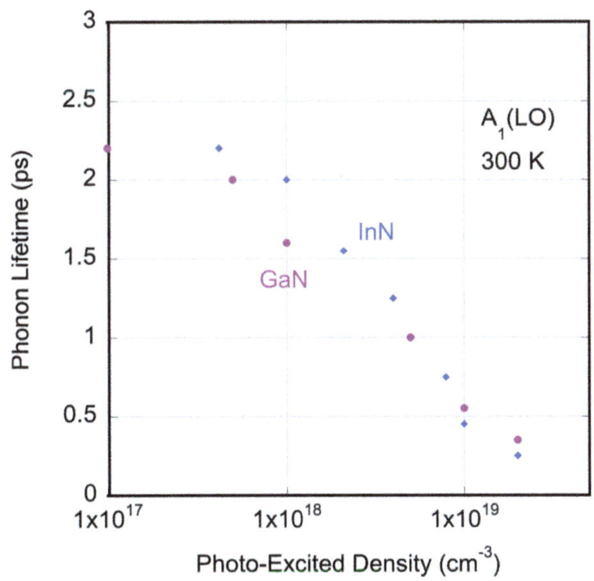

Figure 5.40. Density dependence of the LO phonon lifetime as a function of the photo-excited carrier density. Similar density dependence is found for the E_1(LO) mode as well. Data is taken from Tsen *et al* [218] for GaN and from Tsen *et al* [225] for InN.

PL from a moderately dope layer of GaN grown by MBE was observed at room temperature, with excitation from a He:Cd laser [220]. With a carrier density of 10^{18} cm^{-3}, the data could only be fit with a phonon lifetime in the range of 3–4 ps. This estimate seems to be high by about a factor of 2, according to the data of figure 5.40, which could be the case if a blockade of the LO modes was not considered. Measurements in a GaN/AlGaN HEMT have shown even longer lifetimes [221]. In these devices, the two-dimensional electron density was about 2.5×10^{13} cm^{-2}. The devices were excited at 208 nm (fourth harmonic of the 832 nm laser). Measurements of the PL now gave an LO phonon lifetime of 4.2 ps.

5.5.2.5 InN

While scientists have long sought to find materials in which one or more of the LO phonon decay paths are blocked, InN may be one of the few materials for which this may actually be possible; that is, there is a gap between the optical and acoustic branches that precludes the Klemens mechanism from occurring. The LO phonons at Γ are given by the A_1 and E_1 modes at 586 cm^{-1} and 593 cm^{-1}, respectively [222]. The low side of these high energy modes at Γ are the A_1(TO) and E_1(TO) modes at 447 cm^{-1} and 476 cm^{-1}, respectively. The top of the low energy modes at Γ is the B mode at 220 cm^{-1}. This latter mode, away from Γ, would be the top candidate for the Klemens or the Vallée–Bognani pathways, but clearly it is too low in energy for these processes. But, there certainly are many possibilities for the Ridley pathway, that will still allow for the decay of the LO modes.

When InN was first studied, it was thought that this material was a wide band gap material, much like GaN. But, Raman studies, and the careful analysis of their possible routes of excitation in terms of the electron momentum required to absorb or emit an LO phonon, contributed to the eventual understanding that the gap was more like InGaAs in the infrared [223]. Today, the gap is thought to be closer to 0.7 eV. While, of course, others were coming to this same conclusion, the latter work certainly provided some convincing data for the argument.

Early measurements of the phonon lifetime were obtained from temperature dependent linewidths of the Raman signals [224]. The Raman signal was studied over a temperature range of 80–300 K, using a diode laser at 785 nm as the pump source. The Raman signal for the A_1(LO) and the E_2(high) modes are found at 597 cm^{-1} and 494 cm^{-1}, respectively, at 80 K. The modes tended to shift downward in energy as the temperature was raised. These suggest a lifetime of ~0.7 ps and ~1.75 ps for the A_1(LO) and E_2(high) modes, respectively, at 300 K. It was thought that the A_1 mode decays through the Ridley channel, with the secondary phonons at 102 cm^{-1} and 495 cm^{-1}. But, the E_2(high) mode could not be explained with such a simple mechanism, and it was suggested this mode decayed into three lower energy phonons, all of equal energy at 165 cm^{-1}. That is, this decay occurs through a four-phonon process.

Later, time-resolved, Raman measurements suggested lifetimes of 0.75 ps for both A_1(LO) and the E_1(LO) modes, and decay through the Ridley channel for both modes [131]. These pump and probe measurements used a Ti:sapphire laser at 442 nm, with a 100 fs pulse width. The decay of the A_1(LO) mode is thought to involve the same two phonons as above, while the E_1(LO) is also found to follow the Ridley channel with a large TO mode and a smaller TA or LA mode being involved. Later work, using photons at 532 nm, showed that the lifetime was very density dependent, as in GaN, and values varied from 2.2 ps down to 0.25 ps for both modes at room temperature [225, 226], as already shown in figure 5.40. This latter work also demonstrated that the decay was due to the Ridley channel, using temperature curves such as those in this figure.

Hot carrier PL was studied in InN, for an MOCVD layer grown on a GaN nucleation layer on sapphire [227]. The PL was obtained by excitation with an argon-ion laser at 514.5 nm. From a fit to the density dependent electron temperature, it was found that the LO phonon lifetime at room temperature is about 0.89 ps, a value assumed to be due to the presence of hot LO modes.

Later PL studies of InN grown on GaN indicated that both the LO modes and acoustic modes were involved in the energy relaxation of the hot electrons [228]. The PL was obtained by excitation with an argon-ion laser at 514.5 nm, as above. It was found that the energy loss was mainly due to the acoustic modes for $T < 50$, and by a mix of the two phonons, but predominantly the LO mode, for higher temperatures.

The use of InN for a hot carrier solar cell has also been studied [229]. Time-resolved PL was studied using a Ti:sapphire laser at 1.59 eV for excitation. The conclusions of the effort were that the hot carriers cooled to the band edges too rapidly to offer any possible efficiency increase from the use of InN.

5.5.3 Layered compounds

In the previous two sections, the materials were measured for their bulk properties. The materials which may be obtained in near monolayer form have also been measured for their nonequilibrium phonon behavior, but these are quite often described as surface properties as there just isn't much bulk in monolayer materials.

5.5.3.1 Graphene

Monolayer graphene, the monolayer form of graphite, was isolated early in this century. Prior to this, there was considerable effort on the study of the phonons and the nonequilibrium state of these phonons in graphite. Even so, the early studies were long after those in the III–V materials. Studies of the phonons in graphene began almost as soon as it was isolated. The phonon dispersion was studied as soon as materials were available, both theoretically and experimentally. The dispersion curve is shown in figure 5.4 above and was obtained by a detailed analysis and careful fitting to the experimental data [44].

Few layer graphite, or multi-layer graphene depending upon one's view, was studied by time-resolved THz spectroscopy. The hot carrier distributions and the interaction with the LO–TO modes were studied [230]. The experiment used a Ti:sapphire laser and the generation of 8–20 THz signals by difference frequency mixing in GaSe crystal. The latter signals were delayed so as to provide a time-resolved THz probe for spectroscopy to study the phonon dynamics. The electrons and holes can interact with Γ phonons within a single K minimum or with K phonons for scattering between the two K minima. These measurements suggested that the phonon lifetime was about 7 ps. These authors found that most of the excitation energy given to the hot carriers was provided to the optical phonons within 500 fs, and the latter decayed with the lifetime mentioned above.

Graphene exfoliated onto an oxidized Si substrate was used to study the phonons [231]. Pump–probe measurements with the pump at 800 nm and the probe at 1300 nm were used. The pump was a standard Ti:sapphire laser, which also drove a secondary oscillator to generate the 1300 nm probe signal. The measurements gave an initial fast decay of the excited carriers with a decay constant in the range of 0.1–0.3 ps, followed by a slow decay with a decay constant an order of magnitude larger. It is the slower decay that is thought to arise from the interaction with the phonons, and this delay time increased with the number of graphene layers in the sample.

Breusing *et al* [232] studied layers of graphite in the range 20–30 nm thick using 7 fs pump and probe pulses to study the differential transmission. They find a rapid intra-band carrier equilibration on the 30 fs time scale, with subsequent intra-band cooling due to phonons on the 100 fs time scale. Subsequent measurements on graphite by Yan *et al* [233] used a Ti:sapphire pump at 800 nm and a doubled probe at 400 nm to study the nonequilibrium phonons. These studies used the decay of the anti-Stokes Raman line, and suggested that the zone center optical mode was the primary carrier scattering center. These were driven out of equilibrium and subsequently decayed with a phonon lifetime of ~2.2 ps.

Few layer graphene was studied with ultrafast pump and probe techniques using a Ti:sapphire laser at 780 nm for both signals [234]. The graphene samples were grown on the C face of SiC, or on Ni with subsequent transfer to quartz. Time-resolved differential transmission was studied with a probe pulse with a width of ~100 fs. These studies found phonon lifetimes of 2.5 ps. Later studies used ultrafast absorption spectroscopy to study multi-layer graphene grown on the Si face of SiC [235]. These latter studies use a Nd:YAG laser for micro-Raman studies and a Ti:sapphire laser for the differential absorption measurements. In general, they found a fast decay of the hot carriers in less than 0.2 ps (limited by the instrument resolution), and a slower decay attributable to the hot phonons, with a lifetime near ~2 ps, depending upon the pump intensity.

Time-resolved anti-Stokes Raman scattering was used to study both graphite and monolayer graphene [236]. Graphene was exfoliated onto oxidized Si wafers. A Ti:sapphire laser was used to produce both pump and probe beams. The time-resolved anti-Stokes signal of the G band Raman signal is shown in figure 5.41. Here, plots are shown for both graphite and graphene. Also shown is the probe autocorrelation signal to illustrate the time resolution available. From these plots, one gathers the phonon lifetime to be 2.4 and 1.2 ps for graphite and graphene, respectively.

Subsequently, single-layer graphene exfoliated onto a mica substrate was studied with ultrafast 30 fs pump photons at 1.5 eV from a Ti:sapphire laser [237]. Under these pulses, the graphene exhibited light emission from the visible into the near ultraviolet. The ultrafast PL from the hot carriers was studied, where it was determined that the phonon decay was about 1 ps. Subsequently, time-resolved anti-Stokes Raman scattering was used to study single-layer graphene [238].

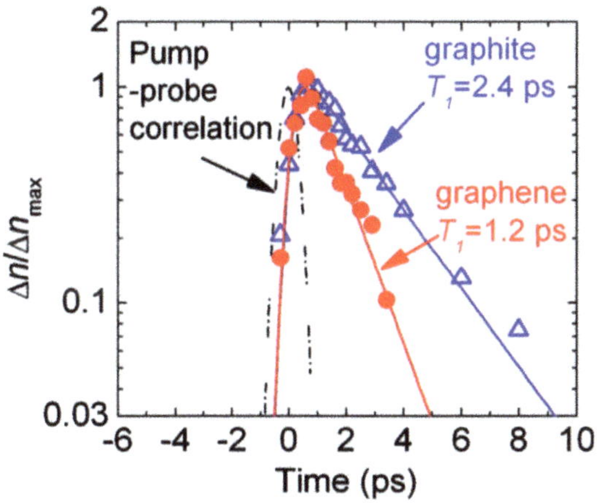

Figure 5.41. Normalized intensity of the anti-Stokes Raman signal from graphite and graphene. The probe autocorrelation function is also shown to illustrate the temporal resolution. The extracted phonon lifetimes are shown for each curve. Reprinted with permission from Kang *et al* [236], copyright (2010) by the American Physical Society.

Free-standing, supported, and gated monolayers of graphene were studied. These studies showed that when as few as three layers of graphene were present, the phonon lifetime had become essentially that of graphite. But, for a single layer, the lifetime was found to be 2 ps.

More recently, ultrafast electron diffraction (UED) was used to probe the dynamical aspects of the phonon dynamics in graphite [239]. Here, the pump is provided by 35 fs, 800 nm photons, while the probe uses the bunched electrons to diffract from the sample. These studies identified the principle phonon for inter-valley scattering is A_1' phonon (the highest energy TO mode in figure 5.3) and the intra-valley scattering arises from the E_{2g} mode (the degenerate LO+TO mode at Γ in figure 5.4). No information on the decay of the nonequilibrium LO modes was given, although the overall phonon system appeared to have a decay time of 6.5 ps, indicating that the acoustic modes were likely out of equilibrium as well as the optical modes.

Also using UED, it was found that decaying optical phonons in single-crystal graphite led to the population of acoustic phonon modes along the Γ–K and Γ–M directions in less than 2 ps [240]. Here, the same Ti:sapphire laser as above was used for the pump, with the electrons serving as the probe. It was thus felt that the actual high energy modes had a lifetime that was smaller than this time scale.

Ultrafast pulses of ~45 ns duration, obtained from a Ti:sapphire laser at 780 nm were used to study graphene grown on BN [241]. These authors used differential reflection spectroscopy to study the nonequilibrium phonons. They obtained phonon lifetimes in the 200–500 fs, depending upon the laser fluence (and thus on the excited carrier densities), which are significantly faster than those measured on SiO_2 substrates. They attribute this speed-up to elimination of any phonon bottleneck with the higher thermal conductivity of the BN substrate.

Graphene grown on foils and subsequently transferred to low stress SiN membranes was studied by micro-Raman scattering from a 532 nm laser pulse [242]. Both first and second-order Raman spots were used to show that the optical and acoustic modes were both out of equilibrium at different, but close, temperatures. The ZA phonons were found to be important in dissipating the heat.

The electron–phonon interaction in graphene has been probed by superconducting tunnel spectroscopy [243]. CVD grown graphene is transferred to an oxidized Si wafer and patterned by normal electron-beam lithography. A hexagonal BN layer was deposited on top of the graphene to serve as the tunnel barrier. The tunneling served to characterize the hot electron distribution function and study the screened electron–phonon interactions, as well as the phonon cooling. At these low temperatures, carrier cooling was predominantly by the acoustic phonons.

5.5.3.2 TMDCs

Much of the early interest in materials such as graphene arose for their electronic applications, but its lack of a band gap causes certain limitations. More recently, interest has shifted to the transition metal dichalcogenides from the fact that they have sizable band gaps. While a fair amount of work has been done on their electrical properties [244], not so much attention has focused on their lattice dynamics. However, with the development of the ultrafast diffraction approaches, the monolayer materials are a natural for use with these tools. Thus, it is UED that mostly has been used to study the dynamics of the phonons in these materials. Since these techniques measure strain in the crystal under pump excitation, the focus is often on the acoustic modes and not upon the optical lifetimes. Nevertheless, the existence of nonequilibrium modes has been confirmed in many cases.

Picosecond x-rays have also been used to study the boundary heat conduction for TMDC films [245]. Here, several different TMDC materials were studied using an optical pump at 515 nm, and 60 fs x-ray pulses from a synchrotron source. The various samples were on sapphire substrates during measurement. They then studied the heat flow across the samples, but focused significantly on flow normal to the layer.

The excitons in MoS_2 as in other TMDCs have a relatively large condensation energy due to the large band gap, and the formation of these excitons involves the phonons as well. Both conduction and valence bands are spin-split due to the strong spin–orbit interaction, and this leads to two excitation energies for excitons of opposite spin. The A excitation pumps from the upper valence band to the lower conduction band, while the B excitation pumps from the lower valence band to the upper conduction band. These excitons cannot be assigned to specific spin transitions, as the ordering of the bands in the K′ valleys is exactly opposite to that in the K valleys. This is a result of mixing of the pseudo-spin valley index and spin index, to create spin–valley interactions [246]. There will be exciton studies mentioned in the following.

In one of the early studies, femtosecond optical excitation at 400 nm, above the band gap of MoS_2, created the photo-excited electrons and holes [247]. UED was then used to follow the atomic displacements and as an atomic-scale probe. The

MoS_2 was prepared on sapphire substrates. An increase in the diffracted intensity followed 1.7 ± 0.3 ps after the excitation pulse, and this could be studied with a temporal resolution of ~0.4 ps. It was found that the excitation led to large amplitude in-plane atomic displacements and strain and out-of-plane rippling of the monolayer. Many of the changes are explained in figure 5.42. The onset of the

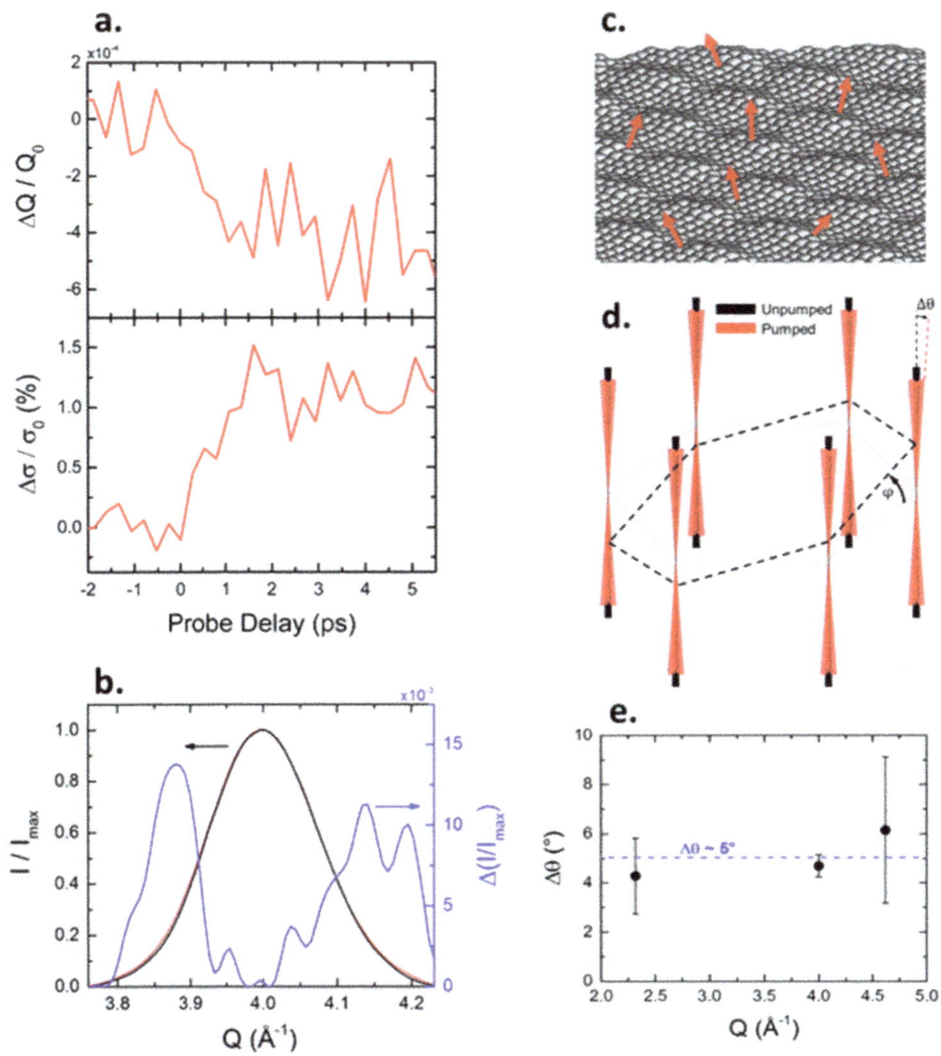

Figure 5.42. (a) Time-dependent shift (top) and broadening (bottom) for the $\{11\bar{2}0\}$ diffraction peak showing rippling of the monolayer on the few ps time scale. (b) Normalized data for this reflection with the shift removed, so that the broadening can be seen between the initial reflection (black) and the reflection after the pulse (red). The blue curve is the differential broadening from these curves. (c,d) Diffraction intensities from MoS_2 are given by sharp rods in reciprocal space parallel to the surface normal. If the monolayer is rippled or buckled, the diffraction intensity is given by the sum over many such rods. Optical pumping increases the rippling by an angle $\Delta\theta$, which broadens the cone. This angle is shown in (e). Reprinted with permission from Mannebach *et al* [247], copyright (2015) by the American Chemical Society.

rippling is likely due to the fact that the clamping of the sample hinders its expansion on the time scale of the experiment. Nevertheless, there were large-scale in-plane atomic displacements that accompanied the rippling. It was felt that the temporal rise in diffracted intensity represented the electron–phonon coupling time, but was not indicative of phonon lifetimes. The authors felt that such approaches will open new possibilities for the study of dynamic fluctuations through the inelastic scattering evident in the diffusively scattered electrons observed in this experiment.

MoS_2 was studied with both optical pump–probe techniques to study transient absorption and THz spectroscopy [88] in MoS_2. For the former, a Ti:sapphire laser, with a 150 fs pulse, was used to feed an optical parametric amplifier to generate tunable pump signals, while part of the original signal was used to generate a white light continuum from a sapphire plate with which to probe the transient absorption. For the latter, the second pulse was passed through an air-plasma to generate THz signals in the 0.1–8 THz region. They observe an ultrafast photoconductivity response that matches a corresponding linewidth response and a decay of the ground state bleaching (by the incident laser pump pulse). They assign this response to mid-gap states associated with grain boundaries or surface defects. However, it is now known that the TMDCs suffer from chalcogenide vacancies which lead to a band of mid-gap states as well as localization of the band edges and a mobility edge [248].

Thin, but bulk-like, WSe_2 has been studied with resonant photo-excitation—photons of 1.55 eV were used, which had a significant spectral weight at the A-exciton at 1.59 eV, and then studied with UED [249]. Focus was on the energy transfer between the electrons and the phonons. It was observed that the excited electrons would transfer to the Σ valleys (presumably the U valleys in figure 5.4(b), where the conduction band minimum is located in the bulk material). The amplitude of the phonons in the regions around the M and K points build up in about 0.6–1.0 ps. They suggest the dominant electron scattering is to in-plane phonons. The decay of the nonequilibrium phonons was observed to be on a much longer time scale than the above rise time, but no lifetime was inferred from the data, although the lifetime is likely close to the build-up time observed for the acoustic modes.

Transient changes in WS_2 were studied using broad band femtosecond pump–probe spectroscopy [250]. Here, the sample was placed on a fused silica substrate. The pump was provided by 250 fs pulses from an amplified fiber laser at 2.4 eV, which primarily excited the B-exciton, although a smaller number of electron–hole pairs will be excited at the A-exciton. The pump strength was varied to produce pairs in the range $0.9–13.0 \times 10^{12}$ cm^{-2}. The probe was a white light continuum from passing part of the pump through a YAG crystal. As in the III–Vs, the photo-excited carriers emit a cascade of phonons as they relax to the band minima. The authors observed both a many-body response from the electrons and holes and then transfer of this energy to the phonon system. An increase in temperature of the distribution appears to peak at about 10 ps, and then decays; this temperature depends upon the pump fluence as expected. They conclude that a large fraction of the pump energy winds up in the phonons, with transfer from optical to acoustic phonons in a few ps.

Later studies of WS_2 utilized time-resolved THz spectroscopy [251], with the same system discussed above for MoS_2 [87]. However, the response is different. In WS_2, the authors find a photoinduced conductivity that they attribute to the ultrafast formation of positive trions (a charged exciton, in this case two holes and one electron). There is a fast (~500 fs) and a slow (~2–4 ps) relaxation of the system following the probe. The fast decay is attributed to trion–phonon interactions, with the subsequent slow relaxation perhaps due to defects.

The evolution of the phonon populations and lattice dynamics of $MoTe_2$ were studied with UED and exciting photons of 800 and 400 nm [252]. It was found that the phonon populations at M, K, and Λ (near U) were heavily excited by the 800 nm (near the main gap) illumination and rose to their nonequilibrium values in ~2 ps. With 400 nm illumination, slightly slower amplitudes in these modes were obtained, but also excitation of phonons near Γ were observed, confirming both inter-valley and intra-valley relaxation of the electrons. The diffraction is shown in figure 5.43. It may be seen that there is strong excitation of the Γ point phonons, which are intra-valley scatterers, and other excitations (highlighted in the figure) at the zone edge and midway between Γ and K. The zone center interaction is thought to be to both LO–TO pairs (see figure 5.5). It is thought that the initial relaxation is via the intra-valley phonons until the excitation of the inter-valley modes is increased by the electron–phonon interactions. As the minimum of the conduction band in this multi-layer sample is at Λ (U), inter-valley scattering can be through the Λ phonons or others. The latter are thought to be the zone-edge LA modes at M or K.

Study of the ultrafast build-up of the excitons in single-layer MoS_2 was studied with ultrafast differential reflectivity [253]. It is found that, by pumping at 2.29 and 2.75 eV, obtained from an optical parametric amplifier excited with 1.55 eV photons, excited electron–hole pairs are generated that then decay into excitons by a cascade of optical phonons.

5.5.4 Scaling the optical phonon lifetime

In condensed matter theory (and experiment), one often finds that a variety of material parameters scale with some critical dimension, usually the inter-atomic spacing or the lattice constant (which are, of course, intimately connected). For example, coupling between different wave functions in tight-binding theory is believed to scale as d^{-2} in band structure calculations, where d is the inter-atomic spacing [254]. For the anharmonic interaction, the phonon lifetime depends upon the third-order elastic constant for the material, at least for the diamond and zinc-blende structures. Weinrich has pointed out this this should have the same order of magnitude as the typical elastic constant and scale with the bulk modulus [122]. Harrison [254], in turn, has suggested that the bulk modulus should scale as d^{-5}. As the lifetime varies as the square magnitude of the matrix element, this would give a scaling of the lifetime as d^{-10}. In figure 5.44, the most likely value of the lifetime (based upon the data presented in the preceding sections) at room temperature is plotted as a function of a lattice dimension. Because some of the materials have a doping dependence, the values for the lowest density (typically $\leqslant 10^{17}$ cm^{-3}) is used.

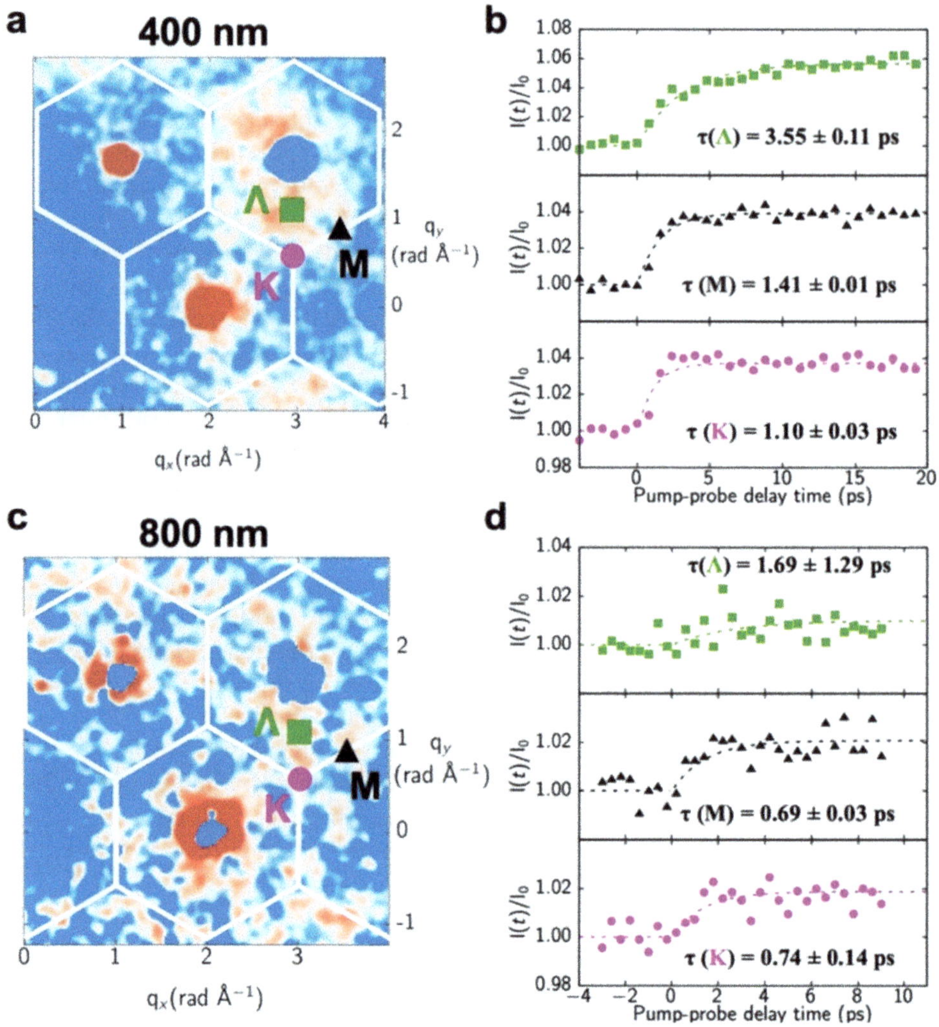

Figure 5.43. Diffuse scattering around the {110} and {100} planes highlighting three high-symmetry wave number points: K, M, and Λ (or U in earlier notation), during 400 nm excitation (a,b) and 800 nm excitation (c,d). The variation in time constant suggest anisotropic coupling to different nonequilibrium phonons. The 400 nm excitation leads to slower population build-up at the zone edges than the 800 nm excitation. Reprinted from Krishnamoorthy *et al* [252], copyright (2019) by the American Chemical Society.

For the III–V zinc-blende structures, the cube edge is used, which of course varies linearly with the inter-atomic spacing. For the wurtzite structures, the length a of the basal plane lattice constant is used. Even if we actually used the inter-atomic spacing for both lattices, there would be separation between the two groups, so the choice of length is to spread them further apart. The straight lines follow a d^{-10} behavior to illustrate the extent of the scaling. We note that for the wurtzite materials, the plot is for the A_1(LO) mode, while for the zinc-blende materials, it is the normal polar LO mode. The data for the zinc-blende material seems to fit this behavior quite well,

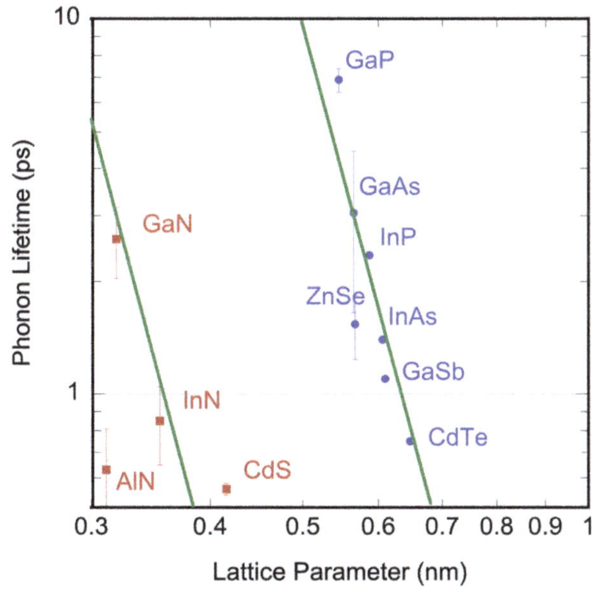

Figure 5.44. The experimentally measured nonequilibrium LO phonon lifetime, taken from the experiments mentioned in this chapter, for a number of zinc-blende and wurtzite semiconductors. The green lines are a d^{-10} behavior based upon dependence of the lifetime on the bulk modulus. Variation in zinc selenide is a range of measurements, while that for GaN arises from an unexpected dependence on the carrier density.

although it is not possible to say that the slope is exactly −10, but lies close to this value.

The data for the wurtzite materials seems to be relatively far off such a curve, particularly for both AlN and CdS. The lifetime in AlN was determined from the Raman linewidth, and it is not known whether this gave a valuc that was too small or not. What is known is that CdS has a strong piezoelectric effect, as was discussed in the introduction to this chapter. Probably of more importance is that the group III–N materials often are grown with heterostructure strain. In addition, like CdS, they generally have much higher piezoelectric coefficients, so that there is a reasonable chance that the second-order piezoelectric constant may contribute significantly to LO phonon decay. In addition, it is known that the strain can modify these constants, both the anharmonic and piezoelectric forms. This would not change the temperature dependence (figure 5.26) as the latter is governed by the Bose–Einstein factors of the secondary phonons and not by the coupling constants. At any rate, there needs to be more study of the nature of the phonon decays in these materials.

References

[1] Van Vleck J H 1941 *Phys. Rev.* **59** 724

[2] Orbach R 1961 *Proc. R. Soc. London* A **264** 458

[3] Scott P L and Jeffries C D 1962 *Phys. Rev.* **127** 32

[4] Stesmans A, Yue W and Witters J 1984 *Sol. State Commun.* **52** 525

[5] Hutson A R, McFee J H and While D L 1961 *Phys. Rev. Lett.* **7** 237
[6] Hutson A R and While D L 1962 *J. Appl. Phys.* **33** 40
[7] Hutson A R 1962 *Phys. Rev. Lett.* **9** 296
[8] Ishida A and Inuishi Y 1967 *IEEE Trans. Electron Dev.* **14** 600
[9] Tehon S W and Wanuga S 1964 *Proc. IEEE* **52** 1113
[10] Challis L J, Ghazi A A and Jeffries D J *et al* 1978 *Phys. Rev. Lett.* **40** 519
[11] Challis L J 1983 *Contemp. Phys.* **24** 229
[12] Kent A J, Newton M I and Rampton V W *et al* 1987 *Jpn. J. Appl. Phys.* **26** 1757
[13] Vella-Coleiro G P 1969 *Phys. Rev. Lett.* **23** 697
[14] Shah J, Leite R C C and Scott J F 1970 *Sol. State Commun.* **8** 1089
[15] Shah J 1978 *Sol.-State Electron.* **21** 43
[16] Dharma-Wardana M W C 1994 *Sol. State Commun.* **86** 83
[17] Zhou X Q, Rühle W W and Schullatz P *et al* 1994 *Semicond. Sci. Technol.* **9** 704
[18] Xu Z Y and Tang C L 1984 *Appl. Phys. Lett.* **44** 692
[19] Shah J, Pinczuk A, Gossard A C and Wiegmann W 1985 *Phys. Rev. Lett.* **54** 2045
[20] Ryan J F, Taylor R A, Turberfield A J and Worlock J M 1986 *Surf. Sci.* **170** 511
[21] Murdin B N, Heiss W and Langerak C J G M *et al* 1997 *Phys. Rev.* B **55** 5171
[22] Lugli P and Goodnick S M 1987 *Phys. Rev. Lett.* **59** 716
[23] Tsen K T, Joshi R P, Ferry D K and Morkoc H 1989 *Phys. Rev.* B **39** 1446
[24] Aksamija Z and Ravaioli U 2006 *J. Comp. Electron.* **5** 431
[25] Barker J M, Ferry D K and Goodnick S M *et al* 2004 *J. Vac. Sci. Technol.* B **22** 2045
[26] Sadi T and Kell R W 2007 *IEEE Electron Dev. Lett.* **28** 787
[27] Leach J H, Zhu C Y and Wu M *et al* 2009 *J. Appl. Phys.* **95** 223504
[28] Ashok A, Vasileska D, Hartin O L and Goodnick S M 2010 *IEEE Trans. Electron Dev.* **57** 562
[29] Auer C, Schürrer F and Ertler C 2007 *J. Comp. Electron.* **6** 325
[30] Weinrich N 1965 *Solids: Elemenary Theory for Advanced Students* (New York: Wiley)
[31] Holm-Kennedy J W and Champlin K S 1972 *J. Appl. Phys.* **43** 1878
[32] Canali C, Jacoboni C and Nava F *et al* 1975 *Phys. Rev.* B **12** 2265
[33] Ando T, Stern F and Fowler A B 1982 *Rev. Mod. Phys.* **54** 437
[34] Born M and Huang K 1954 *Dynamical Theory of Crystal Lattices* (London: Oxford University Press)
[35] Akis R, Ayubi-Moak J and Ferry D K *et al* 2008 *J. Phys.: Cond. Matter* **20** 384201
[36] Ferry D K 2019 *Semicon. Sci. Technol.* **34** 044001
[37] Ferry D K 2020 *Semiconductor Bonds and Bands* 2nd edn (Bristol: IOP Publishing)
[38] Fröhlich H 1954 *Adv. Phys.* **3** 325
[39] Lyddane R H, Sachs R G and Teller E 1941 *Phys. Rev.* **59** 673
[40] Fröhlich H 1937 *Proc. R. Soc.* A **160** 280
[41] Waugh J L T and Dolling G 1963 *Phys. Rev.* **132** 2410
[42] Phillips J C 1973 *Bonds and Bands in Semiconductors* (New York: Academic)
[43] Siegle H, Kaczmarczyk G and Philippidis L *et al* 1997 *Phys. Rev.* B **55** 7000
[44] Wirtz L and Rubio A 2004 *Sol. State Commun.* **131** 141
[45] Fischetti M V, Kim J and Narayanan S *et al* 2013 *J. Phys. Cond. Matter* **25** 473202
[46] Ferry D K 2016 *Semicond. Sci. Technol.* **31** 11LT01
[47] He G, Nathawat J and Kwan C-P *et al* 2017 *Sci. Rep.* **7** 11256
[48] Kaasbjerg K, Thygesen K S and Jacobsen K W 2012 *Phys. Rev.* B **85** 115317

[49] Lucovsky G, Brodsky M H and Burstein E 1968 *Localized Excitations in Solids* (New York: Plenum Press) p 592
[50] Barker A S Jr. and Sievers A J 1975 *Rev. Mod. Phys.* **47** S1
[51] Jusserand B and Sapriel J 1981 *Phys. Rev.* B **24** 7194
[52] Bonneville R 1981 *Phys. Rev.* B **24** 1987
[53] Tsu R, Kawamura H and Esaki L 1972 *Proc. 11th Intern. Conf. Phys. Semicond., Warsaw* (Warsaw: PWN-Polish-Scientific) p 1135
[54] Kash J A, Jha S S and Tsang J C 1987 *Phys. Rev. Lett.* **58** 1869
[55] Tsen K T, Ferry D K, Salvador A and Morkoc H 1998 *Phys. Rev. Lett.* **80** 4807
[56] Shifren L, Ferry D K and Tsen K T 1999 *Physica* B **272** 419
[57] Shifren L, Ferry D K and Tsen K T 2000 *Phys. Rev.* B **62** 15379
[58] Pearsall T P, Carles R and Portal J C 1983 *Appl. Phys. Lett.* **42** 436
[59] Kuan H and Su Y K 1995 *Semicond. Sci. Technol.* **10** 540
[60] Brodsky M H and Lucovsky G 1968 *Phys. Rev. Lett.* **30** 990
[61] Hudait M K, Clavel M B and Saluru S *et al* 2018 *AIP Adv.* **8** 115119
[62] Zhang B, Chen C and Han J *et al* 2018 *AIP Adv.* **8** 045021
[63] Drabble J R and Brammer A I 1967 *Proc. Phys. Soc. London* **91** 959
[64] Gubanov A I and Davydov S Y 1972 *Sov. Phys.—Sol. State* **14** 1020
[65] Spears D L 1970 *Phys. Rev.* B **2** 1931
[66] Akimov A V, Kaplyanskii A A and Kozub V I *et al* 1988 *J. Luminescence* **40–41** 711
[67] Kent A J, Hardy G A and Hawker P *et al* 1988 *Phys. Rev. Lett.* **61** 180
[68] Challis L J, Kent A J and Rampton V W 1990 *Semicond. Sci. Technol.* **5** 1179
[69] Baumberg J J, Williams D A and Kohler K 1997 *Phys. Rev. Lett.* **78** 3358
[70] Onishchenko E E, Bagaev V S and Galkina T I *et al* 2002 *Physica* E **12** 450
[71] Msall M E and Wolfe J P 2002 *Phys. Rev.* B **65** 195205
[72] Naber W J M, Fujisawa T and Liu H W *et al* 2006 *Phys. Rev. Lett.* **96** 136807
[73] Ferry D K 2020 *Transport in Semiconductor Mesoscopic Devices* 2nd edn (Bristol: IOP Publishing) Sec. 8.2
[74] Khrapai V S, Ludwig S and Kotthaus J P *et al* 2008 *J. Phys. Conden. Matt.* **20** 454295
[75] Prokudina M G, Khrapai V S and Ludwig S *et al* 2010 *Phys. Rev.* B **82** 201310
[76] Vanacore G M, Hu J and Liang W *et al* 2017 *Struc. Dyn.* **4** 044034
[77] Shank C V, Fork R L, Leheny R F and Shah J 1979 *Phys. Rev. Lett.* **42** 112
[78] Shank C V, Fork R L and Yen R *et al* 1979 *Sol. State Commun.* **47** 981
[79] Bordone P, Vasileska D and Ferry D K 1996 *Phys. Rev.* B **53** 3846
[80] Shah J 1989 *Sol.-State Electron.* **32** 1051
[81] Fork R L, Brito-Cruz C H, Becker P C and Shank C V 1987 *Opt. Lett.* **12** 483
[82] Becker P C, Fragnito H L and Brito-Cruz C H *et al* 1988 *Appl. Phys. Lett.* **53** 2089
[83] Gaumnitz T, Jain A and Pertot Y *et al* 2017 *Opt. Express* **25** 27506
[84] Rice A, Jin Y, Ma X-F and Zhang X-C 1994 *Appl. Phys. Lett.* **64** 1324
[85] Hartl I, Li X D and Chudoba C *et al* 2001 *Opt. Lett.* **26** 608
[86] Ponomarev D S, Gorodetsky A and Yachmenev A E *et al* 2019 *J. Appl. Phys.* **125** 151605
[87] Cunningham P D, McCreary K M and Hanbicki A T *et al* 2016 *J. Phys. Chem.* C **120** 5819
[88] Ulbricht R, Hendry E and Shan J *et al* 2011 *Rev. Mod. Phys.* **83** 543
[89] Deng H, Chen L and Qu W *et al* 2020 *Opt. Engr.* **59** 056101
[90] Papalazarou E, Faure J and Mauchain J *et al* 2012 *Phys. Rev. Lett.* **108** 256808
[91] Kanasaki J, Tanimura H and Tanimura K 2014 *Phys. Rev. Lett.* **113** 237401

[92] Tanimura H, Kanasaki J and Tanimura K *et al* 2016 *Phys. Rev.* B **93** 161203
[93] Sjakste J, Tanimura K and Barbarino G *et al* 2018 *J. Phys. Cond. Matt.* **30** 353001
[94] Volckaert K, Ostami H and Biswas D *et al* 2019 *Phys. Rev.* B **100** 241406
[95] Johnson Q, Mitchell A, Keeler R N and Evans L 1970 *Phys. Rev. Lett.* **25** 1099
[96] Johnson Q, Mitchell A and Evans L 1971 *Nature* **231** 310
[97] Johnson Q and Mitchell A 1972 *Phys. Rev. Lett.* **29** 1369
[98] Wark J S, Whitlock R L and Hauer A A *et al* 1989 *Phys. Rev.* B **40** 5705
[99] Larson B C, Tischler J Z and Mills D M 1986 *J. Mater. Res.* **1** 144
[100] Murnane M M, Kapteyn H C, Rosen M D and Falcone R W 1991 *Science* **251** 531
[101] Schoenlein R W, Leemans W P and Chin A H *et al* 1996 *Science* **274** 236
[102] Cao J, Hao Z and Park H *et al* 2003 *Appl. Phys. Lett.* **83** 1044
[103] King W E, Campbell G H, Frank A and Reed B 2005 *J. Appl. Phys.* **97** 111101
[104] Musumeci P, Moody J T and Scoby C M *et al* 2010 *Rev. Sci. Instrum.* **81** 013306
[105] Chatelain R P, Morrison V R, Godbout C and Siwick B J 2012 *Appl. Phys. Lett.* **101** 081901
[106] Inoue I, Deguchi Y and Ziaja B *et al* 2021 *Phys. Rev. Lett.* **126** 117403
[107] Manz S, Casandruc A and Zhang D *et al* 2015 *Faraday Disc.* **177** 467
[108] Li D, Wang Z L and Wang Z 2018 *J. Phys. Chem. Lett.* **9** 6795
[109] Morrison V R, Chatelain P T, Godbout C and Siwick B J 2013 *Opt. Express* **21** 21
[110] René de Cotret L P, Pöhls J-H and Stern M J *et al* 2019 *Phys. Rev.* B **100** 214115
[111] Welland I and Ferry D K 2019 *Semicond. Sci. Technol.* **34** 064003
[112] Peterson C L, Frei M R and Lyon S A 1989 *Sol-State Electron.* **32** 1919
[113] van Driel H M 1979 *Phys. Rev.* B **19** 5928
[114] Tsen K T, Abramsohn D A and Bray R 1982 *Phys. Rev.* B **26** 4770
[115] Shah J 1986 *IEEE J. Quantum Electron.* **22** 1728
[116] Deveaud B, Shah J and Damen T C *et al* 1989 *Sol.-State Electron.* **31** 435
[117] Faist J, Capasso F and Sirtori C *et al* 1994 *Sol.-State Electron.* **37** 1273
[118] Conibeer G, Zhang Y, Bremner S and Shrestha S 2017 *Proc. SPIE* **10099** 10990K-1
[119] Conibeer G, Zhang Y, Bremner S and Shrestha S 2017 *Jpn. J. Appl. Phys.* **56** 091201
[120] Xia H, Patterson R and Smyth S *et al* 2017 *Appl. Phys. Lett.* **110** 043102
[121] Ziman J 1963 *Electrons and Phonons* (Oxford: Oxford University Press)
[122] Weinrich N 1965 *Solid: Elementary Theory for Advanced Students* (New York: Wiley)
[123] Cowley R A 1965 *J. Phys. (Paris)* **26** 659
[124] Klemens P G 1966 *Phys. Rev.* **148** 845
[125] Ridley B K and Gupta R 1991 *Phys. Rev.* B **43** 4939
[126] Vallée F and Bogani F 1991 *Phys. Rev.* B **43** 12049
[127] Ferry D K 1974 *Phys. Rev.* **9** 4277
[128] Flytzanis C 1972 *Phys. Rev. Lett.* **29** 772
[129] Ambacher O, Smart J and Shealy J R *et al* 1999 *J. Appl. Phys.* **85** 3222
[130] Tsen K T, Ferry D K and Botchkarev A *et al* 1998 *Appl. Phys. Lett.* **72** 2132
[131] Tsen K T, Kiang J G and Ferry D K 2007 *Appl. Phys. Lett.* **90** 152107
[132] Kennedy C J, Matter J C and Smirl A L *et al* 1974 *Phys. Rev. Lett.* **32** 419
[133] Auston D H and Shank C V 1974 *Phys. Rev. Lett.* **32** 1120
[134] Smirl A L, Matter J C, Elci A and Scully M O 1976 *Opt. Commun.* **16** 118
[135] Trigo M, Fuchs M and Chen J *et al* 2013 *Nature Phys.* **9** 790
[136] Zhu D, Robert A and Henighan T *et al* 2015 *Phys. Rev.* B **92** 054303

[137] Zürch M, Chang H-T and Borja L J *et al* 2017 *Nature Commun.* **8** 15734
[138] Kishimura H, Yazaki A and Kawano H *et al* 2002 *J. Chem. Phys.* **117** 10239
[139] Harb M, Ernstorfer R and Dartigalongue T *et al* 2006 *J. Chem. Phys.* B **110** 25308
[140] Harb M, Peng W and Sciaini G *et al* 2009 *Phys. Rev.* B **79** 094301
[141] Kar A, Upadhya P C and Dayeh S A *et al* 2011 *IEEE J. Sel. Top. Quantum Electron.* **17** 889
[142] Ciano C, Persichetti L and Montanari M *et al* 2020 *Phys. Rev.* B **102** 205302
[143] Jacoboni C, Nava F, Canali C and Ottaviani G 1981 *Phys. Rev.* B **24** 1014
[144] Chang R K, Ralston J M and Keating D E 1969 *Light Scattering Spectra of Solids* ed G B Wright (New York: Springer) paper E-3
[145] Leheny R F, Shah J and Fork R L *et al* 1979 *Sol. State Commun.* **31** 809
[146] Ulbrich R 1971 *Phys. Rev. Lett.* **27** 1512
[147] Southgate P D, Hall D S and Dreeben A B 1971 *J. Appl. Phys.* **42** 2868
[148] Inoue M, Takenaka N, Shirafuji A and Inuishi Y 1978 *Sol.-State Electron.* **21** 29
[149] Inoue M, Takenaka N and Inuishi Y 1978 *Phys. Lett.* **69A** 228
[150] von der Linde D, Kuhl J and Klingenberg H 1980 *Phys. Rev. Lett.* **44** 1505
[151] Kash J, Tsang J C and Hvam J M 1985 *Phys. Rev. Lett.* **54** 2151
[152] Tsen K T and Morkoç H 1986 *Phys. Rev.* B **34** 4412
[153] Zhou X Q, van Driel H M, Rühle W W and Ploog K 1992 *Phys. Rev.* B **46** 16148
[154] Ryan J F, Taylor R A and Turberfeld A J *et al* 1984 *Phys. Rev. Lett.* **53** 1841
[155] Yang C H, Carlson-Swindle J M, Ryan S A and Worlock J M 1985 *Phys. Rev. Lett.* **55** 2359
[156] Pan J L, Hopfel R A and Shah J 1986 *J. Appl. Phys.* **59** 3925
[157] Sapega V F, Chamberlain M P and Ruf T *et al* 1995 *Phys. Rev.* B **52** 14144
[158] Langot P, Del Fatti N and Christofilos D *et al* 1996 *Phys. Rev.* B **54** 14487
[159] Pellemans H P M and Planken P C M 1998 *Phys. Rev.* B **57** R4222
[160] Vallée F 1994 *Phys. Rev.* B **49** 2460
[161] Höpfel R A, Shah J and Gossard A C 1986 *Phys. Rev. Lett.* **56** 765
[162] Knox W H, Hirliman C and Miller D A B *et al* 1986 *Phys. Rev. Lett.* **56** 1191
[163] Ruf T, Wald K and Yu P *et al* 1993 *Superlatt. Microstruc.* **13** 203
[164] Tsen K T and Morkoc H 1988 *Phys. Rev.* B **38** 5615
[165] Tatham M C, Ryan J F and Foxon C T 1989 *Sol.-State Electron.* **32** 1497
[166] Tatham M C, Ryan J F and Foxon C T 1989 *Phys. Rev. Lett.* **63** 1637
[167] Maciel A C, Freyland J M and Rota L *et al* 1998 *Appl. Phys. Lett.* **68** 1519
[168] Zhang X, Nie Z and Wang B *et al* 2020 *Phys. Chem. Chem. Phys.* **22** 25819
[169] Caville S A, Kent A J and Challis L J *et al* 1999 *Physica* B **272** 171
[170] Tsen K T and Morkoc H 1988 *Phys. Rev.* B **37** 7137
[171] Tsen K T 1992 *Semicond. Sci. Technol.* **7** B191
[172] Rühle W W, Leo K and Bauser E 1989 *Phys. Rev.* B **40** 1756
[173] Kuhl J and Bron W E 1984 *Sol. State Commun.* **49** 935
[174] Bron W E, Kuhl J and Rhee B K 1986 *Phys. Rev.* B **34** 6961
[175] Snow P A, Maly P, Westland D J and Ryan J F 1989 *Sol.-State Electron.* **32** 1485
[176] Snow P A, Westland D J and Ryan J F *et al* 1989 *Superlatt. Microstruc.* **5** 595
[177] Maly P, Maciel A C and Ryan J F *et al* 1994 *Semicond. Sci. Technol.* **9** 719
[178] Smith D C, O'Sullivan E D and Rota L *et al* 1997 *Phys. Stat. Solidi (b)* **204** 110
[179] Takeuchi H, Mizoguchi K and Nakayama M *et al* 2001 *J. Phys. Soc. Japan* **70** 2598
[180] Canonico M, Poweleit C and Menéndez J *et al* 2002 *Phys. Rev.* B **88** 215502

[181] Elsaesser T, Bäuerle R J and Kaiser W 1989 *Phys. Rev.* B **40** 2976
[182] Grann E D, Tsen K T and Ferry D K 1996 *Phys. Rev.* B **53** 9847
[183] Markelz A G, Asmar N G, Gwinn E G and Brar B 1998 *Appl. Phys. Lett.* **72** 2439
[184] Esmaielpour H, Whiteside V R and Piyathilaka H P *et al* 2018 *Sci. Rep.* **8** 12473
[185] Westland D J, Ryan J F and Scott M D *et al* 1988 *Sol.-State Electron.* **31** 431
[186] Ulrici B and Jahne E 1976 *Phys. Stat. Solidi (b)* **74** 601
[187] Sakamoto R, Kohno T and Kamiyoshi T *et al* 1992 *Semicond. Sci. Technol.* **7** B271
[188] Irmer G, Wenzel M and Monecke J 1996 *Phys. Stat. Solidi (b)* **195** 85
[189] Kozin I E, Ignatiev I V and Nair S V *et al* 2000 *J. Luminescence* **87–89** 441
[190] Clady R, Tayegjee M J Y and Aliberti P *et al* 2012 *Prog. Photovolt.* **20** 82
[191] http://ioffe.ru/SVA/NSM/Semicond/
[192] Borcherds P H, Alfrey G G, Saunderson D H and Woods A D B 2022 *J. Phys. C: Sol. State Phys.* **8** 2022
[193] Trigo M, Chen J and Viswanath V H *et al* 2010 *Phys. Rev.* B **82** 235205
[194] Hoffmann M C, Hebling J and Hwang H Y *et al* 2009 *Phys. Rev.* B **79** 161201(R)
[195] Lindenberg A M, Kang I and Johnson S L *et al* 2000 *Phys. Rev. Lett.* **84** 111
[196] Feuer T, Morak A and Uschmann I *et al* 2001 *Appl. Phys.* B **72** 15
[197] Larsson J, Allen A and Bucksbaum P H *et al* 2002 *Appl. Phys.* A **75** 467
[198] Grann E, Chen Y and Tsen K T *et al* 1996 *Phys. Rev.* B **80** 3840
[199] Akimov A V, Scherbakov A V and Zhmodikov A L *et al* 1997 *Phys. Rev.* B **56** 12100
[200] Anand S, Verma P, Jain K P and Abbi S C 1996 *Physica* B **226** 331
[201] Kutzer V, Siegle H and Thompson C *et al* 1997 *Mater. Sci. Engr.* **B43** 46
[202] Mehendale M, Sivananthan S and Schroeder W A 1997 *Appl. Phys. Lett.* **71** 1089
[203] Bergman L, Alexson D and Murphy P L *et al* 1999 *Phys. Rev.* B **59** 12977
[204] Kuball M, Hayes J M, Shi Y and Edgar J H 2000 *Appl. Phys. Lett.* **77** 1958
[205] Meneses E A, Januzzi N and Ramos J G P *et al* 1975 *Phys. Rev.* B **11** 2213
[206] Saito H and Göbel E O 1985 *Phys. Rev.* B **31** 2360
[207] Juršénas S, Žukaskus A and Baltramiejûnas R 1992 *J. Phys.: Cond. Matter* **4** 9987
[208] Klimov V, Bolivar P H and Kurz H 1995 *Phys. Rev.* B **52** 4728
[209] Shank C V, Auston D H, Ippen E P and Teschke O 1978 *Sol. State Commun.* **26** 570
[210] Junnarkar M R and Alfano R R 1986 *Phys. Rev.* B **34** 7045
[211] Masumoto Y and Sasaki F 1991 *J. Luminesc.* **48–49** 189
[212] Vengurlekar A S, Prabhu S S, Roy S K and Shah J 1994 *Phys. Rev.* B **50** 15461
[213] Vengurlekar A S, Prabhu S S, Roy S K and Shah J 1995 *Phys. Rev.* B **51** 14233
[214] Sheih S J, Tsen K T and Ferry D K *et al* 1995 *Appl. Phys. Lett.* **67** 1757
[215] Tsen K T, Joshi R P and Ferry D K 1997 *Phys. Stat. Solidi (b)* **204** 106
[216] Matulionis A, Liberis J and Matulionienè I *et al* 2003 *Phys. Rev.* B **68** 035338
[217] Ramonas M, Matulionis A and Liberis J *et al* 2005 *Phys. Rev.* B **71** 075324
[218] Tsen K T, Kiang J G, Ferry D K and Morkoç H 2006 *Appl. Phys. Lett.* **89** 112111
[219] Beechem T and Graham S 2007 *J. Appl. Phys.* **103** 093507
[220] Wang K, Simon J, Goel N and Jena D 2006 *Appl. Phys. Lett.* **88** 022103
[221] Xu G, Tripathy S K and Mu X *et al* 2009 *Laser Phys.* **19** 745
[222] Davydov V Y, Emtsev V V and Goncharuk A N *et al* 1999 *Appl. Phys. Lett.* **75** 3297
[223] Liang W, Tsen K T and Ferry D K *et al* 2004 *Appl. Phys. Lett.* **84** 3849
[224] Pomeroy J W, Kuball M and Lu H *et al* 2005 *Appl. Phys. Lett.* **86** 223501
[225] Tsen K T, Kiang J G and Ferry D K *et al* 2007 *J. Phys. Cond. Matter* **19** 236219

[226] Tsen K T and Ferry D K 2009 *J. Phys. Cond. Matter* **21** 174202

[227] Yang M D, Chen Y P and Shu G W *et al* 2008 *Appl. Phys.* A **90** 123

[228] Yang M D, Liu Y W and Shen J L *et al* 2009 *J. Appl. Phys.* **105** 013526

[229] Aliberti P, Feng Y and Shrestha S K *et al* 2012 *Appl. Phys. Lett.* **99** 223507

[230] Kampfrath T, Perfetti L and Schapper F *et al* 2005 *Phys. Rev. Lett.* **95** 187403

[231] Newson R W, Dean J, Schmidt B and van Driel H M 2009 *Opt. Express* **17** 2326

[232] Breusing M, Ropers C and Elsaesser T 2009 *Phys. Rev. Lett.* **102** 086809

[233] Yan H, Song D and Mak K F *et al* 2009 *Phys. Rev.* B **80** 121403

[234] Wang H, Strait J H and George P A *et al* 2010 *Appl. Phys. Lett.* **96** 081917

[235] Huang L, Hartland G V and Chu L-Q *et al* 2010 *Nano Lett.* **10** 1308

[236] Kang K, Abdoula D, Cahill D G and Shim M 2010 *Phys. Rev.* B **81** 165405

[237] Lui C H, Mak K F, Shan J and Heinz T F 2010 *Phys. Rev. Lett.* **105** 127404

[238] Wu S, Liu W-T and Liang X *et al* 2012 *Nano Lett.* **12** 5495

[239] Chatelain R P, Morrison V R, Klarenaar B L M and Siwick B J 2014 *Phys. Rev. Lett.* **113** 235502

[240] Harb M, Enquist H and Jurgilaitis A *et al* 2016 *Phys. Rev.* B **93** 104104

[241] Golla D, Brasington A, LeRoy B J and Sandhu A 2017 *Apl. Mat.* **5** 056101

[242] Sullivan S, Vallabhaneni A and Kholmanov I *et al* 2017 *Nano Lett.* **17** 2049

[243] Zihlmann S, Makk P and Castilla S *et al* 2019 *Phys. Rev.* B **99** 075419

[244] Li S-L, Tsukagoshi K, Orgiu E and Samori P 2016 *Chem. Soc. Rev.* **45** 118

[245] Nyby C, Sood A and Zalden P *et al* 2000 *Adv. Func. Mat.* **30** 2002282

[246] Xiao D, Liu G-B and Feng W *et al* 2012 *Phys. Rev. Lett.* **108** 196802

[247] Mannebach E M, Li R and Duerloo K-A *et al* 2015 *Nano Lett.* **15** 6889

[248] Zhu W, Low T and Lee Y-H *et al* 2013 *Nature Commun.* **5** 3087

[249] Waldecker L, Bertoni R and Hübener H *et al* 2017 *Phys. Rev. Lett.* **119** 036803

[250] Ruppert C, Chernikov A and Hill H M *et al* 2017 *Nano Lett.* **17** 644

[251] Gustafson J K, Cunningham P D and McCreary K M *et al* 2019 *J. Phys. Chem.* C **123** 30676

[252] Krishnamoorthy A, Lin M-F and Zhang X *et al* 2019 *Nano Lett.* **19** 4981

[253] Trovatello C, Katsch F and Borys N J *et al* 2020 *Nature Commun.* **11** 5277

[254] Harrison W 1980 *Electronic Structure and the Properties of Solids* (San Francisco, CA: Freeman)

IOP Publishing

Hot Carriers in Semiconductors

David K Ferry

Chapter 6

Seeking the distribution function

In the previous chapters, hot electrons (and holes) and the properties they exhibit, as seen in experiments, have been discussed. Now, it is time to begin to examine the underlying theoretical basis for this behavior. Naturally, there are many approaches and even more approximations that have appeared in the literature. The key factor in all such approaches is to reach a balance between the driving forces, provided by the electric field or the excess optical excitation, and the relaxation forces, provided primarily by the phonons. But, this does not ensure a stable steady-state, as seen by the Gunn effect in chapter 2. But, this balance is required for a steady-state. Does a steady-state always exist? It might be said to exist even if it is oscillatory as in the Gunn effect. But, it might not exist in a case in which not all possible scattering processes are included. A classic example is SiO_2, which surprisingly is a low mobility semiconductor [1]. The dominant scattering mechanism is a pair of polar optical phonons [2], but these do not provide sufficient scattering to force a steady-state at some very high electric fields. Polar runaway, as this is called, has been known for some time [3], but is usually not seen. Rather, in studies with Monte Carlo techniques (to be discussed in chapter 7), it was found that these phonons could only stabilize the velocity for fields below ~1.2–2 MV cm^{-1} [4] (breakdown is observed at ~7 MV cm^{-1}). Thus, another scattering mechanism is required if stability is to be explained above these fields. One suggestion was Umklapp scattering of near zone-edge acoustic modes [5], while another invoked scattering to subsidiary valleys of the conduction band [6]. Either process is capable of stabilizing the distribution equally well [7], and experiments have not shown which may be the crucial one. The point is that not all theoretical treatments are equal. They become useful only when they contain all the necessary physical processes in an adequate manner.

Essentially all treatments of hot carriers, and other transport conditions, in semiconductors are based upon a one-electron transport equation, which usually begins with the Boltzmann transport equation. As with most transport equations, this equation determines the distribution function under the balance described above. How do we arrive at a one-electron (or one-hole) transport equation when there are some 10^{15}–10^{20} carriers per cubic centimeter in the device? It turns out that we can achieve this if we ignore correlations between the electrons. There are formal ways of showing this, but they boil down to ignoring two-particle correlations. Boltzmann called this the *Stosszahl* Ansatz, or molecular chaos, approximation. Even in so doing, the distribution function is not the end product, as transport coefficients are achieved from integrals over this distribution. What are these integrals, and how are they determined? Some of them are easy, while others are difficult. And this is the case where the distribution only deviates from equilibrium by a small amount. The heart of this chapter is the various solution techniques that have been given for this transport equation and how solutions are obtained. Then, some of the not so normal scattering processes, such as plasmon scattering are discussed along with the equation of motion used to define the relaxation processes.

The Boltzmann equation yields a one-electron distribution function that describes how the carriers are distributed in position and momentum (or energy) as a function of time. This distribution function gives the probability for the occupation of a state of given energy and momentum and position. Just what is meant by this probability? There are various ways this quantity can be defined. For example, it is possible to say that the distribution function $f(\mathbf{v}, \mathbf{x}, t)$ for an ensemble of N particles is the probability of finding a particle in the box of volume $\Delta\mathbf{x}$, centered at $\mathbf{x}$, and $\Delta\mathbf{k}$, centered at $\mathbf{k}$, at time t. This involves energy as, classically, the momentum is a well-defined function of the energy. Here, $\mathbf{k}$ is the particle wave number (velocity is obtained through the definition of an effective mass) and $\mathbf{x}$ is the position. In this sense, the distribution function is described in a six-dimensional phase space (or 6+1 dimensions when time is included), and the quantities $\mathbf{x}$ and $\mathbf{v}$ do not refer to any single carrier but to the position in this phase space.

The previous view is to be contrasted with the idea that the N particles are described by a $6N$-dimensional configuration space, where we have $3N$ velocity variables and $3N$ position variables, all a function of time. It is important to differentiate between these two spaces. The former can be thought of as the one-electron *phase* space, which is a normal 6+1 'laboratory' space, while the latter is clearly a many electron *configuration* space that lives only in theory. It is the transition from this configuration space to the physical phase space that normally yields a hierarchy of distributions: one-electron, two-electron, etc. And, it is via the molecular chaos approximation that we decouple the one-electron distribution from the other distributions. The Boltzmann equation is already a projection from the configuration space to the laboratory space, usually achieved by e.g. the BBGKY hierarchy (after Bogoliubov, Born, Green, Kirkwood, and Yvon, described in [8]). Yet, there remain some issues, notably how this distribution should be normalized. One approach, common with probability distributions, is

$$\iint d^3\mathbf{x}d^3\mathbf{k}f(\mathbf{x}, \mathbf{k}, t) = 1. \tag{6.1}$$

As with all probability functions, the integral over the measure space must sum to unity. However, this is not the only definition that can be made.

An alternative definition is to define the distribution function as the 'average' (the concept of this average will be defined below) number of particles in a phase space box of size $\Delta \boldsymbol{x} \Delta \boldsymbol{v}$ located at the phase space point $(\boldsymbol{x}, \boldsymbol{v})$ at time t. Then, the distribution function satisfies

$$\int\int d^3\mathbf{x} d^3\mathbf{k} f(\mathbf{x}, \mathbf{k}, t) = N(t). \tag{6.2}$$

In this case, $N(t)$ is the *total number of particles* in the system at time t.

It might be supposed that the Fermi–Dirac distribution satisfies (6.1) and in fact defines the Fermi energy level as a function of time. However, this is not correct for two reasons. First, the normalization is wrong; recall that the Fermi–Dirac distribution has a maximum value of unity for energies well below the Fermi energy, but the integral over the Fermi–Dirac distribution is certainly not unity. A second objection is more serious. The Fermi–Dirac distribution is a point function, and its application to inhomogeneous systems must be handled quite carefully. The Fermi energy is connected to the electrochemical potential, which may vary (relative to one of the band edges) with position relative to a point where the band edge is taken to be the reference in energy. Then the Fermi energy is position dependent in this view, and this is especially the case when the density is position dependent (more properly, the Fermi energy is constant in the absence of current, and the band edges vary in position). Consequently, when dealing with the Fermi–Dirac distribution, the relevant approximations must be stated. For the most part of this chapter, the non-degenerate Maxwell–Boltzmann form will be used. The extension to degenerate systems is usually straightforward (and hopefully obvious).

The variation of the distribution function in position, momentum, and time, is governed by what is called an equation of motion. It is this equation that is called the Boltzmann equation and is of interest here. In equilibrium, no transport takes place since the distribution function is symmetric in **v**-space (more properly, **k**-space, since our energy bands are described in this space and the conversion to a velocity is usually nonlinear). Since there are an equal number of carriers with oppositely-directed momenta, the net current is zero. Hence, for transport, the distribution function must be modified by the applied fields (and made asymmetric in momentum space). Alternatively, when the semiconductor is optically excited from an external source, the density is usually inhomogeneous and this can lead to transport as well. It is these modifications that must now be determined. The forcing functions, such as the applied field, are themselves reversible quantities and the evolution of the distribution function in phase space is unchanged by these fields. It is only the presence of the scattering processes that can change this evolution, and the classical statement of this fact is (the right-hand side represents changes due to scattering processes)

$$\frac{df(\mathbf{x}, \mathbf{v}, t)}{dt} = \left.\frac{df(\mathbf{x}, \mathbf{v}, t)}{dt}\right|_{\text{collisions}}. \tag{6.3}$$

By expanding the left-hand side via the chain rule of differentiation, the Boltzmann transport equation becomes

$$\frac{\partial f}{\partial t} + \mathbf{v} \cdot \nabla f + \frac{d\mathbf{k}}{dt} \cdot \frac{\partial f}{\partial \mathbf{k}} = \left. \frac{df(\mathbf{x}, \mathbf{v}, t)}{dt} \right|_{\text{collisions}} . \tag{6.4}$$

The first term is the explicit time variation of the distribution function, while the second term accounts for transport induced by a spatial variation of the density (and distribution function). The third term describes the field-induced transport. These three terms on the left-hand side are collectively known as *streaming* terms. Here, the third term has been written with respect to the momentum wave vector rather than the velocity to account for the role of the former in the crystal momentum. Still, in keeping with the discussion above, the change of the distribution function with position must be sufficiently slow that the variation of the wave function itself is very small in a single unit cell (in essence, this is the condition for which an effective mass can be defined, and the carriers can be defined as semi-classical). The third term on the left-hand side, is just the Lorentz force in the presence of electric and magnetic fields, and this term must be sufficiently small that it does not introduce any mixing of wave functions from different bands. In optical processes, this assumption is usually broken, and no single effective mass exists. Finally, the time variation must be sufficiently slow that the distribution function evolves slowly on the scale of either the mean-free time between collisions or on the scale of any hydrodynamic relaxation times. This latter requirement rests upon the Stosszahl Ansatz of Boltzmann.

The term on the right-hand side of (6.4) describes the scattering processes. Any scattering process induces carriers to make a transition from some initial state **k** into a final state **k**′ with a probability $P(\mathbf{k}, \mathbf{k}')$. Then, the number of electrons scattered depends on the latter probability as well as on the probabilities of the state **k** being full [given by $f(\mathbf{k})$] and the state **k**′ being empty [given by $1 - f(\mathbf{k}')$]. If the scattering process can flip the spin then a factor of 2 is also included for the spin degeneracy of the final state. This can be seen as mixing the definitions for the distribution function given above, and point us toward the Fermi–Dirac distribution. But, it is (6.2) that is the preferred approach. And, for most of this chapter, it will assumed that the distribution function is small compared to unity. With these ideas, the scattering term can be written as

$$\left. \frac{df(x, v, t)}{dt} \right|_{\text{collisions}} = \sum_{k'} \{P(k', k) f(k')[1 - f(k)] - P(k, k') f(k)[1 - f(k')]\} . \tag{6.5}$$

The terms $P(\mathbf{k}, \mathbf{k}')$ also contain a summation over all possible scattering mechanisms by which electrons (or holes) can move from **k** to **k**′. It must also be considered that the first term on the right involves carriers coming *into* the state by emitting a phonon from a higher energy state or by absorbing a phonon from a lower energy state. The second term includes terms for electrons leaving (going *out*

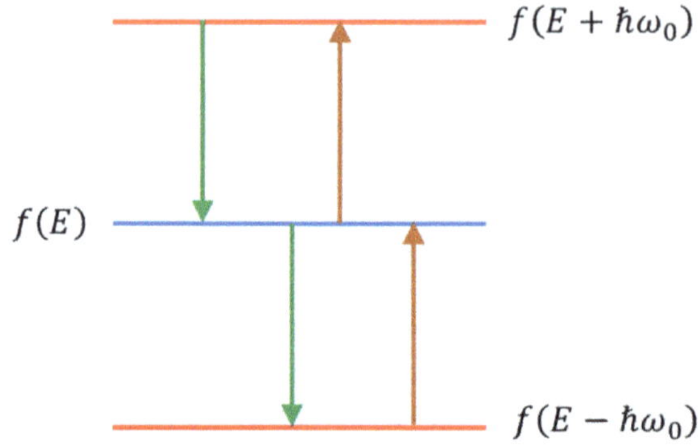

Figure 6.1. Scattering events that contribute to the time variation of the distribution function. Carriers can *leave* the energy level of interest by emission to a lower energy level or by absorption to a higher energy level. Similarly, they can *enter* the energy level of interest by emission from a higher energy level or by absorption from a lower energy level. The emission of phonons is shown by the green arrows and the absorption of phonons is shown by the brown arrows.

of) the state to a lower energy state by emitting a phonon or to a higher energy state by absorbing a phonon. These two terms are commonly referred to as *in*-scattering and *out*-scattering. These processes are shown in figure 6.1, where an energy level *gains* particles (*in*-scattering) from higher energy levels by phonon emission and from a lower energy level by phonon absorption. Similarly, the given energy state loses (*out*-scattering) particles by emission to a lower energy state and by absorption to a higher energy level.

When the distribution function is driven out of equilibrium by the *streaming* forces on the left-hand side of (6.4), the collision terms work to restore the system to equilibrium. Interactions between the carriers within the distribution work to randomize the energy and momentum of the distribution by redistributing these quantities within the distribution. This is known to lead to a Maxwellian distribution in the non-degenerate case through a process that can be shown to maximize the entropy of the distribution. However, it is the phonon interactions that cause the overall distribution function to come into equilibrium with the lattice. If the lattice itself is in equilibrium, it may be considered as the *bath* and the phonons serve to couple the electron distribution to this thermal bath. It is also possible for the phonon distribution to be driven out of equilibrium (as we have seen in the preceding chapters) and this makes for a very complicated set of equations to be solved.

6.1 The relaxation time approximation

If no external fields are present, the collisions tend to randomize the energy and the momentum of the carriers and return them to their equilibrium state. In linear response, it is often useful to assume that the rate of relaxation is proportional to the deviation from an equilibrium distribution function and that the distribution function decays to its normal equilibrium value in an exponential manner. For this approximation, a relaxation time τ may be introduced by means of the equation

$$\left.\frac{df(\mathbf{x}, \mathbf{v}, t)}{dt}\right|_{\text{collisions}} = -\frac{f - f_0}{\tau}. \tag{6.6}$$

This is an extremely simple approximation, but it will serve a useful purpose in illustrating how the Boltzmann equation is solved, and how averages over the distribution function are obtained. These basic averages will hold true, in their mathematical format, for all forms of the distribution function that are obtained by other techniques.

In (6.6), f_0 is the equilibrium distribution function, either a Fermi–Dirac distribution or the Maxwellian approximation to this distribution. Equation (6.6) is a fairly common approximation that is easily made if the scattering processes are either elastic, or are isotropic and inelastic. Then, in the homogeneous situation, (6.3) and (6.6) lead to

$$f(t) = f_0 + \left[f(0) - f_0\right]e^{-t/\tau}. \tag{6.7}$$

This is the equation for the natural decay of the distribution function due to scattering, so $f(0)$ is the nonequilibrium initial state at the beginning of the decay process. On the other hand, if an external force is applied, the distribution function will approach the nonequilibrium steady state with this same exponential behavior, due to the assumption of linear response.

While the relaxation time approximation (6.6) has been made, it is necessary to be able to determine τ from the scattering rates. As may be inferred from (6.5), this entails an integration over the distribution function. To see this, consider the case of the elastic scattering process in further detail. If the scattering process is elastic, the states $\mathbf{k}$ and $\mathbf{k}'$ lie on the same energy shell [that is, $E(\mathbf{k}) = E(\mathbf{k}')$], and it is feasible to assume that $P(\mathbf{k}, \mathbf{k}') = P(\mathbf{k}', \mathbf{k})$, so that the relaxation term in (6.5) becomes

$$\left.\frac{df(\mathbf{x}, \mathbf{v}, t)}{dt}\right|_{\text{collisions}} = \sum_{\mathbf{k}'} P(\mathbf{k}', \mathbf{k})[f(\mathbf{k}') - f(\mathbf{k})]. \tag{6.8}$$

This form holds whether we are dealing with the Fermi–Dirac distribution or the Maxwellian distribution.

Before proceeding with (6.8), it is necessary to actually partially 'solve' the Boltzmann equation. It will be assumed that the system is spatially homogeneous, and is in a temporally steady state. Then, the Boltzmann equation is simplified to

$$\frac{e}{\hbar}\mathbf{E} \cdot \frac{\partial f}{\partial \mathbf{k}} = -\frac{f - f_0}{\tau}, \tag{6.9}$$

and this is easily rewritten as

$$f(\mathbf{k}) = f_0 - \tau\mathbf{E} \cdot \mathbf{v}\frac{\partial f_0}{\partial E}, \tag{6.10}$$

where it has been assumed that the derivative of f may be approximated by the derivative of f_0. That is, it is assumed that the second term is small. This works fine for this approximation except if there is a magnetic field. Its effects would vanish in this approximation, and a more detailed approach must be followed [9]. (Note that there is an obvious conflict between the electric field and the energy notation here.) One should take care not to confuse the two, the field will always be a vector while the energy is a scalar. (Where this confusion may not be easily recognized, the symbol **F** will be used for the field.) This form is now introduced into (6.8), so that

$$\left.\frac{df(\mathbf{x},\mathbf{v},t)}{dt}\right|_{\text{collisions}} = \int_{S_k} dS_k P(\mathbf{k}',\mathbf{k})\mathbf{E}\cdot[\mathbf{v}(\mathbf{k}') - \mathbf{v}(\mathbf{k})]\frac{\partial f_0}{\partial E}. \tag{6.11}$$

In this last form, the summation over the three-dimensional momentum has been reduced to an integration over the energy surface as the scattering has been taken to be elastic (initial and final states lie on the same energy shell, as discussed above). In the circular argument that is being used, this result is now returned to (6.10) to give

$$\begin{aligned}\left.\frac{df(\mathbf{x},\mathbf{v},t)}{dt}\right|_{\text{collisions}} &= -(f - f_0)\int_{S_k} dS_k P(\mathbf{k}',\mathbf{k})\left[1 - \frac{\mathbf{v}(\mathbf{k}')\cdot\mathbf{v}(\mathbf{k})}{\mathbf{v}^2}\right]\\ &= -(f - f_0)\int_{S_k} dS_k P(\mathbf{k}',\mathbf{k})[1 - \cos\vartheta]\end{aligned}. \tag{6.12}$$

In this last form, the angle between the two velocities has been utilized, and an angular weighting is found. In truly elastic, isotropic scattering such as acoustic phonons in spherically symmetric bands, the cos(θ) term integrates to zero with the shell integration. However, for anisotropic scattering that arises from Coulombic scattering centers, such as impurities and polar optical phonons, this cos(θ) term becomes quite important. It will be shown in chapter 8 that this term is very hard to recover in quantum transport. Equation (6.12) now allows the *momentum* relaxation time for elastic scattering processes to be introduced as

$$\frac{1}{\tau(E)} = \int_0^{\pi} d\vartheta \int_0^{2\pi} d\varphi \sin\vartheta P(\vartheta,\varphi)[1 - \cos\vartheta]. \tag{6.13}$$

In this last equation, the substitution $P(\vartheta, \varphi) = k^2 P(\mathbf{k}, \mathbf{k}')$ has been made, and $P(\mathbf{k}, \mathbf{k}')$ is normally determined from the Fermi golden rule [8]. In this last form, the importance of the angle variation becomes obvious; Coulombic scattering does not randomize the momentum completely, so the weighting term in square brackets converts from the simple scattering time to the momentum relaxation time. In measurements of e.g. Shubnikov–de Haas, it is the simple scattering time that is recovered, rather than the momentum relaxation time, whereas mobility depends upon the momentum relaxation time.

Once, the relaxation time is determined, then distribution function (6.10) can be averaged to find the mobility as the current density is

$$\boldsymbol{J} = -e^2\int dE\rho(E)\mathbf{v}(E)\tau\mathbf{E}\cdot\mathbf{v}\frac{\partial f_0}{\partial E}. \tag{6.14}$$

The first term on the right-hand side of (6.10) does not contribute to the current, as there is no current in equilibrium, and any simple integral (as above) over this term vanishes.

It was pointed out in chapter 2 that the current density may not be parallel to the electric field in materials with multiple equivalent band valleys, such as Si and Ge. That will lead to a tensor form for (6.14), but as long as the material is in equilibrium, or the field is along a principle axis, the cubic symmetry will be retained and (6.14) will be scalar. Inter-valley transfer between equivalent valleys of course violates this assertion. As our approach here is tutorial at the moment, it will be assumed that (6.14) is a scalar, and the vector velocity will be taken in the direction of the electric field. Then, only the term in the square of the velocity is retained in integrals over Maxwellians, so that (6.14) may be rewritten as

$$\boldsymbol{J} = -e^2\mathbf{E}\int dE\rho(E)\tau v_{\mathbf{E}}^2\frac{\partial f_0}{\partial E}. \tag{6.15}$$

At this point, a series of relationships will be introduced. First, when the field is small,

$$v_{\mathbf{E}}^2 \sim \frac{v^2}{3} = \frac{2E}{3m*}, \tag{6.16}$$

where $m*$ is the effective mass, and the carrier density is

$$n = \int dE\rho(E)f_0(E), \tag{6.17}$$

as the second term in (6.10) is assumed small. When these relationships are used in (6.15), it is found that (6.15) becomes

$$\boldsymbol{J} = -\frac{2}{3}\frac{ne^2}{m_c}\frac{\int dE\rho(E)E\tau\frac{\partial f_0}{\partial E}}{\int dE\rho(E)f_0(E)}\mathbf{E}. \tag{6.18}$$

Using normal relationships for near equilibrium (low field) transport [9], the *average* momentum relaxation time can be found to be

$$\langle\tau_m\rangle = -\frac{2}{3}\frac{\int dE\rho(E)E\tau\frac{\partial f_0}{\partial E}}{\int dE\rho(E)f_0(E)}. \tag{6.19}$$

It is apparent that the scattering time τ is different from the momentum relaxation time $\langle\tau_m\rangle$. This difference can be seen in figure 6.2, which is for impurity scattering in graphene at low temperatures. The difference arises from the very anisotropic nature of ionized impurity scattering at low temperatures and the unique density of states in graphene [10].

As an example of how exponents get into the temperature dependence, assume that $\tau_m = AE^{-s}$, where A and s are constants that are different for the different scattering mechanisms. For parabolic bands, the density of states in three dimensions varies as $CE^{1/2}$, so that if the denominator is integrated by parts, one obtains

$$\langle\tau_m\rangle = \frac{\int dEE^{3/2}\tau_m(E)\frac{\partial f_0}{\partial E}}{\int dEE^{3/2}\frac{\partial f_0}{\partial E}}. \tag{6.20}$$

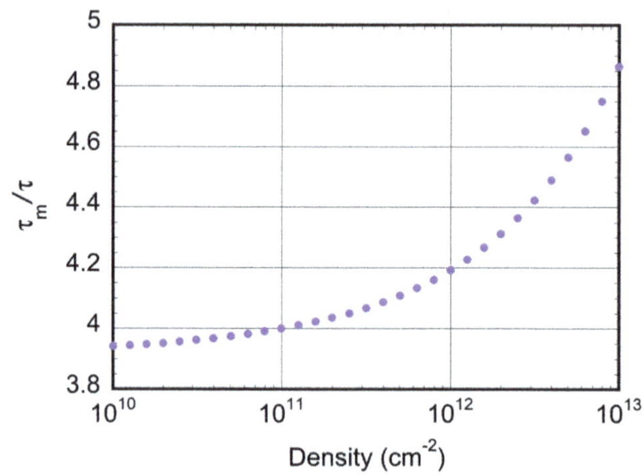

Figure 6.2. The ratio of the average momentum relaxation time $\tau_m = \langle \tau \rangle$ given in (6.19) for electrons in graphene at 4.2 K. The difference arises primarily from the anisotropic nature of scattering by ionized impurities in the oxide adjacent to the graphene. Adapted from Ferry *et al* [10]. The density dependence arises from the complicated two-dimensional screening of the impurities as well as from the Fermi energy shift.

Introducing the assumed form for the relaxation time gives

$$\langle \tau_m \rangle = A \frac{\int dE (E^{5/2-s}) \frac{\partial f_0}{\partial E}}{\int dE E^{3/2} \frac{\partial f_0}{\partial E}}. \tag{6.21}$$

For the Maxwellian,

$$\frac{\partial f_0(E)}{\partial E} \sim -\frac{1}{k_B T} \exp\left(-\frac{E}{k_B T}\right), \tag{6.22}$$

and (6.21) can be written as

$$\langle \tau_m \rangle = \frac{A}{(k_B T)^s} \frac{\Gamma[(5/2) - s]}{\Gamma[(3/2) + 1]}, \tag{6.23}$$

where the $\Gamma(\cdot)$ are the Gamma functions. So, a particular energy dependence of the scattering rate transfers to a particular temperature dependence. This becomes more complicated when the Bose–Einstein factors in the scattering rates are present, as that introduces two temperatures into the system, as the temperature here comes from the distribution function and thus is the hot carrier temperature.

In figure 6.3, the use of impurity scattering to determine the electron temperature is shown for a case of degenerate InSb at 1.8 K [11]. Here, the sample is heated optically via free-carrier absorption of a CO_2 laser at 0.117 eV. In this approach, discussed in chapters 2 and 3, as a method to provide a thermometer for the hot carriers, the Shubnikov–de Haas amplitude is determined as the temperature is varied, and then as the laser power is varied. By comparing the amplitude of the Shubnikov–de Haas oscillations for the two different methods of heating the system,

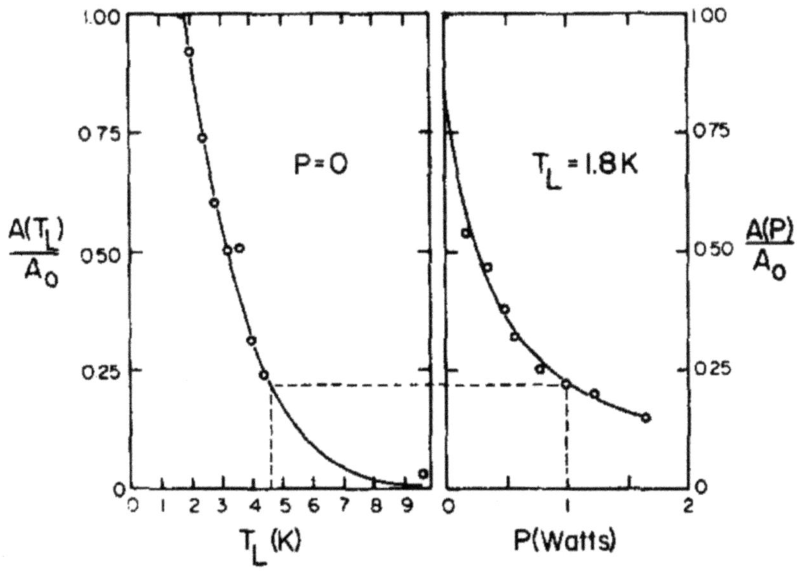

Figure 6.3. The Shubnikov–de Haas amplitude is plotted, relative to that with no laser excitation, as a function of temperature in the left panel, and as a function of power of the CO_2 laser in the right panel. The dashed line shows how the electron temperature is determined from these two panels. Reprinted from Moore *et al* [11], copyright (1978) with permission from Elsevier.

the electron temperature can be determined. The reference value was taken at 1.8 K in the absence of laser heating using the peak near 0.23 T. In the figure, a dotted line illustrates the process of determining the equivalent electron temperature.

6.2 Expanding in Legendre polynomials

It is clear from (6.14) that the electric field direction introduces a preferred axis in momentum space for the distribution function. When carrier streaming occurs, it is along the electric field. So, it is natural then to consider orienting the coordinate system such that the electric field is the polar axis in three dimensions. Once this is done, then the distribution function may be expanded in Legendre polynomials whose argument is the polar angle, as

$$f(E, \cos\vartheta) = \sum_{l=0}^{\infty} f_l(E) P_l(\cos\vartheta). \tag{6.24}$$

For isotropic and elastic collisions, it is reasonable even to terminate this series as $l=1$. This would lead to the relaxation-time approximation discussed in the previous section. If the scattering is not of this form, then further terms in the expansion are obviously needed. This expansion was used earlier in section 2.4 for a study of impact ionization, but it was assumed there that the asymmetric part of the distribution was a delta function in the polar angle.

6.2.1 Legendre differential equations

To begin, it will be assumed that the energy surfaces are spherically symmetric and the momentum dependence is parabolic. There is no reason that these approximations need to be made, except that it makes the approach simpler and less tedious (and easier to understand). The approach will use (6.9), in which the left-hand side becomes [12]

$$\begin{aligned}\frac{e}{\hbar}\mathbf{F}\cdot\frac{\partial f}{\partial \mathbf{k}} &= \frac{eF}{\hbar}\sum_{l=0}^{\infty}\left[\frac{lk^{l-1}}{(2l-1)}\frac{\partial}{\partial k}\left(k^{1-l}f_{l-1}\right)+\frac{l+1}{k^{l+2}(2l+3)}\frac{\partial}{\partial k}\left(k^{l+2}f_{l+1}\right)\right]P_l(\cos\vartheta) \\ &= \frac{eF}{\hbar}\left[\frac{2}{3k}f_1(E)+\frac{\hbar v}{3}\frac{\partial f_1}{\partial E}+\hbar v\frac{\partial f_0}{\partial E}\cos\vartheta\right]+\cdots\end{aligned} \quad (6.25)$$

On the right side, the term F is the electric field, which has been used to avoid confusion with the energy terms. Using the latter form, the low-order terms can be inserted into the Boltzmann equation to give

$$\frac{eF}{\hbar}\left[\frac{2}{3k}f_1(E)+\frac{\hbar v}{3}\frac{\partial f_1}{\partial E}+\hbar v\frac{\partial f_0}{\partial E}\cos\vartheta\right]=\left.\frac{\partial f_0}{\partial t}\right|_{sc}+\left.\frac{\partial f_1}{\partial t}\right|_{sc}\cos\vartheta. \quad (6.26)$$

It is trivial to show that the second term on the right is just the relaxation-time approximation for elastic scattering, and the pre-factor to the cosine term becomes just $-f_1/\tau_m$.

It is clear that (6.26) can be separated into two equations, one which contains the pre-factor of the cosine term and one which lacks any angular dependence. These two equations can be written as

$$f_1(E) = -e\tau_m vF\frac{\partial f_0}{\partial E}, \quad (6.27)$$

and

$$\left.\frac{\partial f_0}{\partial t}\right|_{sc} = \frac{2eF}{3\hbar k}\left[f_1(E)+E\frac{\partial f_1}{\partial E}\right]. \quad (6.28)$$

The first (6.27) is exactly the relaxation-time approximation (but this is expected from the form assumed for the first-order collision term above). However, the relaxation-time approximation does not arise from this equation, but from the neglect of the second equation (6.28) and the assumption that the zero-order distribution is known already. This is not a good assumption. Here, the zero-order distribution must be found by solving a differential equation of some complexity. This zero-order distribution is often referred to as the *energy distribution function.*

To proceed further, the collision operator for the energy distribution function must be evaluated for a specific scattering mechanism. The particular one is not important at this point, as the procedure can be followed for almost any process, and the differential equation must be solved via computer if too many processes are invoked. The basic form for the collision integral is

$$\left.\frac{\partial f_0}{\partial t}\right|_{sc} = \frac{2\pi}{\hbar}\sum_{\mathbf{q}}\{|M_e|^2[f_0(\mathbf{k}-\mathbf{q}) - f_0(\mathbf{k})] + |M_a|^2[f_0(\mathbf{k}+\mathbf{q}) - f_0(\mathbf{k})]\}, \quad (6.29)$$

where the terms M are the matrix elements, and the subscripts e and a stand for emission and absorption of a phonon. The energy conserving delta functions that follow from the Fermi golden rule have been left out for clarity, but they assert that

$$E(\mathbf{k} \pm \mathbf{q}) = E(\mathbf{k}) \pm \hbar\omega_0, \quad (6.30)$$

where the latter term is the phonon energy.

For the current example, the emission or absorption process will be taken to be inelastic acoustic phonon scattering which, in the equipartition limit, has the matrix elements [9]

$$|M|^2 = \frac{E_1^2 k_B T}{2\rho_m v_s^2}, \quad (6.31)$$

where E_1 is the acoustic deformation potential, ρ_m is the mass density, and v_s is the acoustic sound velocity ($\omega_0 = qv_s$), usually for the LA mode. The summation in (6.29) can be converted into an integral in three-dimensional momentum space, and the integral over the polar angle involves the delta functions [9], which leads to a factor of $m^*/\hbar^2 kq$ that arises from this integration as well as the limits on q of

$$0 < q \leqslant k \pm \frac{2m^* v_s}{\hbar}. \quad (6.32)$$

Since, the energy exchange is small for these acoustic phonons, the distribution function can be expanded as

$$f_0(\mathbf{k} \pm \mathbf{q}) \approx f_0(E) \pm \hbar q v_s \frac{\partial f_0(E)}{\partial E} + \frac{(\hbar q v_s)^2}{2}\frac{\partial^2 f_0(E)}{\partial E^2} + \cdots. \quad (6.33)$$

The scattering rate (6.29) then becomes

$$\begin{aligned}\left.\frac{\partial f_0}{\partial t}\right|_{sc} &= \frac{2\pi}{\hbar}\frac{|M|^2}{4\pi^2}\frac{m^*}{\hbar^2 k}\\ &\left\{\int_0^{k+2m^*v_s/\hbar} q\left[\hbar q v_s \frac{\partial f_0(E)}{\partial E} + \frac{(\hbar q v_s)^2}{2}\frac{\partial^2 f_0(E)}{\partial E^2}\right]dq\right. \quad . \quad (6.34)\\ &\left. + \int_0^{k-2m^*v_s/\hbar} q\left[-\hbar q v_s \frac{\partial f_0(E)}{\partial E} + \frac{(\hbar q v_s)^2}{2}\frac{\partial^2 f_0(E)}{\partial E^2}\right]dq\right\}\end{aligned}$$

When the integration is carried out, a variety of constants will be found, many of which give rise to the actual scattering rate for acoustic phonons (in the elastic approximation) to be [9]

$$\frac{1}{\tau_{ac}} = \frac{E_1^2 k_B T m^*}{\pi \rho_m v_s^2 \hbar^2} k. \tag{6.35}$$

Here, it may be seen that the quantity $k\tau_{ac}$ is a function only of a set of constants and unchanging material properties. Thus, the mean-free path, which is constant in this system, may be written as

$$\lambda = v\tau_{ac} = \frac{\hbar k}{m^*} \tau_{ac}. \tag{6.36}$$

It may be noted that the second and third terms are connected by the fundamental definition of the effective mass, which equates the quasi-particle momentum to the wave momentum [9]. This now leads to the desired result for the scattering rate to be

$$\left.\frac{\partial f_0}{\partial t}\right|_{sc} = \frac{4m^* v_s}{k_B T \lambda} \sqrt{\frac{2E}{m^*}} \left[f_0 + \left(\frac{E}{2} + k_B T \right) \frac{\partial f_0(E)}{\partial E} + \frac{E k_B T}{2} \frac{\partial^2 f_0(E)}{\partial E^2} \right]. \tag{6.37}$$

This may now be used in (6.28), along with (6.27) to give the resulting differential equation for the energy distribution function as

$$(E + S k_B T) \frac{\partial^2 f_0(E)}{\partial E^2} + \left(\frac{E}{k_B T} + 2 + \frac{S k_B T}{E} \right) \frac{\partial f_0(E)}{\partial E} + \frac{2}{k_B T} f_0(E) = 0. \tag{6.38}$$

Here, the reduced electric field has been introduced with the constants as

$$S = \frac{(eF\lambda)^2}{6 m^* v_s^2 k_B T} \equiv \left(\frac{F}{F_0} \right)^2. \tag{6.39}$$

To facilitate solving this differential equation, it is convenient to introduce a reduced energy through the change of variables $x = E/k_B T$, with which (6.38) may be rewritten as

$$(S + x) \frac{\partial^2 f_0}{\partial x^2} + \left(x + 2 + \frac{S}{x} \right) \frac{\partial f_0}{\partial x} + 2 f_0 = 0. \tag{6.40}$$

With a little intuition, this can be again rewritten as

$$\frac{1}{x} \frac{\partial}{\partial x} \left[x(S + x) \frac{\partial f_0}{\partial x} + x^2 f_0 \right] = 0, \tag{6.41}$$

which may immediately be integrated once to give

$$x(S + x) \frac{\partial f_0}{\partial x} + x^2 f_0 + A = 0. \tag{6.42}$$

The properties of the energy distribution are that both its value and the value of its derivative are well behaved at $x = 0$, so that A can be set to zero without loss of generality. The remaining terms may be rearranged and integrated again to yield

$$f_0(x) = B(S + x)^S e^{-x}. \tag{6.43}$$

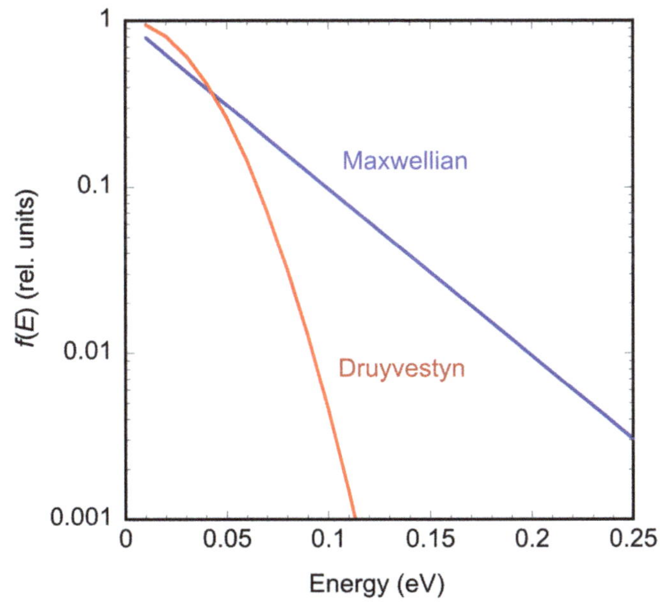

Figure 6.4. Comparison between the Druyvestyn distribution and a Maxwellian, both at an electron temperature of 500 K.

If the electric field is small compared to F_0, then S is small, and the distribution is simply the equilibrium Maxwellian distribution. However, if the field is large, the appearance of the S^S pre-factor leads to a distribution of the form

$$f_0(x) \sim \exp\left[-\left(\frac{EF}{2F_0 k_B T}\right)^2\right], \tag{6.44}$$

which has the characteristics of the Druyvestyn distribution [13] already encountered in section 2.4. In figure 6.4, the Druyvestyn distribution is compared to a Maxwellian, both at a carrier temperature of 500 K. The Druyvestyn is larger than the Maxwellian at low energies (both are normalized to 1 at the zero of energy), but falls off much more rapidly as the energy rises.

6.2.2 Extension to optical modes

The scattering by optical modes is very inelastic, and the simple approach above cannot be followed in detail. Here, the approach will be almost straightforward once the detailed physics is discussed [14, 15]. On the other hand, their scattering remains isotropic in the case of the non-polar optical modes, which also include the inter-valley scattering phonons. The general matrix elements for the optical modes are given by

$$|M|^2 = \frac{\hbar D^2}{2V\rho_m\omega_0} \tag{6.45}$$

plus the appropriate Bose–Einstein distribution function and an energy conserving delta function. D is the optical coupling energy (in units of eV cm^{-1}), and the over variables have previously been defined. When this is inserted into (6.29), with the Bose–Einstein and delta functions, the integration over the polar angle again gives us the factor $m^*/\hbar^2 kq$ and also a set of limits on the momentum integration. The fact that we are dealing with the spherically symmetric energy distribution function is significant, the limits on integration tell us that the four terms that appear in (6.29) have to be carefully evaluated. Particles can come into $f_0(E)$ by the emission of a phonon from a higher energy state or by the absorption of a phonon from a lower energy state. Similarly, they can leave to a lower energy state by the emission of a phonon or to a higher energy state by the absorption of a phonon. Limits on the momentum q are given by the range of energies for which the argument of the delta function actually has a zero. These values then yield [8]

$$k - \sqrt{k^2 - \frac{2m^*\omega_q}{\hbar}} < q < k + \sqrt{k^2 - \frac{2m^*\omega_q}{\hbar}}, \tag{6.46}$$

with the additional requirements

$$k^2 > \frac{2m^*\omega_q}{\hbar},\ E_k > \hbar\omega_q \tag{6.47}$$

for phonon emission, and

$$\sqrt{k^2 + \frac{2m^*\omega_q}{\hbar}} - k < q < \sqrt{k^2 + \frac{2m^*\omega_q}{\hbar}} + k \tag{6.48}$$

for phonon absorption. It is important to note that these terms are evaluated at *the energy level for which the absorption or emission initiate*, and not just at the state of interest. For example while the state at $E + \hbar\omega_0$ is emitting a phonon, and will have $N + 1$ as a factor, the final state at E is actually absorbing that phonon and will have the appropriate limits for that fact. Thus, when these scattering features are put into (6.29), the scattering term becomes

$$\left.\frac{\partial f_0}{\partial t}\right|_{sc} = \frac{m^* D^2}{4\pi\rho\hbar^2\omega_0 k}\left\{\int_{q_a^-}^{q_a^+}\left[(N+1)f_0(\mathbf{k}+\mathbf{q}) - Nf_0(\mathbf{k})\right]dq + \int_{q_e^-}^{q_e^+}\left[Nf_0(\mathbf{k}-\mathbf{q}) - (N+1)f_0(\mathbf{k})\right]dq\right\}. \tag{6.49}$$

The integrations are relatively straightforward [8], using the spherical symmetry of the distribution, and the results can be written as

$$\left.\frac{\partial f_0}{\partial t}\right|_{sc} = \frac{N(y)}{\Lambda}\sqrt{\frac{2E}{m^*}}\left\{\sqrt{1+\frac{y}{x}}\left[e^y f_0(E+\hbar\omega_0) - f_0(E)\right] + \sqrt{1-\frac{y}{x}}\left[f_0(E-\hbar\omega_0) - e^y f_0(E)\right]\right\}. \tag{6.50}$$

Here, the reduced energy has been introduced in the square roots in the curly brackets with the notations and relations

$$y = \frac{\hbar\omega_0}{k_B T}, \quad (N+1) = e^y N, x = \frac{E}{k_B T}, \tag{6.51}$$

and the quantity

$$\Lambda = \frac{2\pi\rho_m \hbar^3\omega_0}{m^{*2}D^2} \tag{6.52}$$

has been introduced as a rough estimate of the optical phonon mean-free path.

At this point a difficulty arises. Even if only the lowest-order term is retained from (6.28), the differential equation will arise from

$$\left.\frac{\partial f_0}{\partial t}\right|_{sc} = -\frac{2(eF)^2}{3\sqrt{2m^*E}}\frac{d}{dE}\left(v\tau_m E\frac{df_0}{dE}\right), \tag{6.53}$$

and the optical and acoustic phonon effects are already apparent if we were to ignore the optical phonons in τ_m. But, this should not be done. Rather, it will be necessary to add the scattering term from the acoustic modes as well. To proceed, it is desired to integrate over the scattering functions, to get rid of the derivative on the right-hand side of (6.53). The desired integral for the acoustic modes is

$$\Omega_{ac} = \int_0^E \sqrt{E\left.\frac{\partial f_0}{\partial t}\right|_{sc,\,ac}}\,dE = \frac{2^{3/2}\sqrt{m^*}v_s^2}{\lambda k_B T}E^2\left(f_0 + k_B T\frac{df_0}{dE}\right). \tag{6.54}$$

The same integration can be done for the optical modes, but first one wants to make the change of variables $\xi = E$ in the first integral and $\xi = E + \hbar\omega_0$ in the second integral. This allows the two integrals to be combined into one. Using (6.53), (6.54), and the modified (6.50) leads to the integro-differential equation for f_0 as

$$\frac{2m^*v_s^2}{\lambda k_B T}E^2\left(f_0 + k_B T\frac{df_0}{dE}\right) + \sqrt{\frac{2E}{m^*}}\frac{(eF)^2}{3\tau_m} = \frac{N}{\Lambda}\int_{E-\hbar\omega_0}^{E}\sqrt{\xi(\xi+\hbar\omega_0)}\left[e^y f_0(\xi+\hbar\omega_0) - f_0(\xi)\right]. \tag{6.55}$$

Here, it must be remembered that the relaxation time τ_m is also a function of energy, so that this is a complicated equation to evaluate.

In general, no closed form has been found for the solution to (6.55) that is valid over all ranges of energy. If the carrier energy is smaller than that of the optical phonon, the latter have little effect and the solution can be taken as (6.43). On the hand, if the energy is higher than that of the optical phonon, then the latter play a major role. In this latter case, one can make expansions in the small quantity $\hbar\omega_0/E$. To make this more feasible, the limits of the integral (6.55) will be shifted to $E \pm \hbar\omega_0/2$ in order to symmetrize it. Then the expansion in the small quantity leads to the estimation for Ω_{op} being

$$\Omega_{\text{op}} \sim E\hbar\omega_0\left[(e^y - 1)f_0 + \frac{\hbar\omega_0}{2}(e^y + 1)\frac{df_0}{dE}\right]. \tag{6.56}$$

Using this result, a great many terms can be brought together to define the electron temperature through

$$k_B T_e = \frac{(eF)^2 \Lambda}{3\hbar\omega_0 \tau_m}\sqrt{\frac{2E}{m^*}} + \frac{\hbar\omega_0}{2}\coth\left(\frac{y}{2}\right). \tag{6.57}$$

If one expands the hyperbolic cosine for small y, the last term becomes just $k_B T$, so that the leading term in (6.57) is just the temperature increase due to the electric field. Introducing this temperature, and the approximation (6.56), gives the resulting differential equation for the high energy limit to be

$$\left(1 + \frac{2m^* v_s^2}{k_B T}\frac{\Lambda}{\lambda}\frac{E}{\hbar\omega_0}\right)f_0 + \left(\frac{2m^* v_s^2}{\hbar\omega_0}\frac{\Lambda}{\lambda}E + k_B T_e\right)\frac{df_0}{dE} = 0. \tag{6.58}$$

If the further approximation is made to ignore the first term in the second set of parentheses, the solution becomes

$$f_0(E) \sim \exp\left[-\frac{E}{k_B T_e}\left(1 + \frac{2m^* v_s^2}{k_B T}\frac{\Lambda}{\lambda}\frac{E}{\hbar\omega_0}\right)\right]. \tag{6.59}$$

If the second term in parentheses is small, the distribution is just a Maxwellian at an elevated electron temperature. When this term is not small, there is an evident expansion of the distribution at high energy that does not have a simple thermal equivalent and even can exhibit a 'kink' in its energy dependence. A comparison with the Maxwellian is shown in figure 6.5. Here, parameters for GaAs at 300 K have been used, and it is assumed that the ratio of mean-free paths is 5, with a carrier temperature of 500 K. For these parameters, the result has similarities to the Maxwellian distribution but falls off somewhat faster [13]. Nevertheless, the reader should not forget the large number of approximations that have been made to get to this closed form solution. This will not be a unique occurrence, as will be seen in subsequent discussions. But, this is no reason to revert to the Maxwellian as the differences are there in the resulting E^2 term that is Druvestyn-like and will magnify the differences.

In figure 6.6, the average energy of carriers in SiO_2 at high electric field is depicted [16]. This was determined using the spherical harmonic expansion, taking into

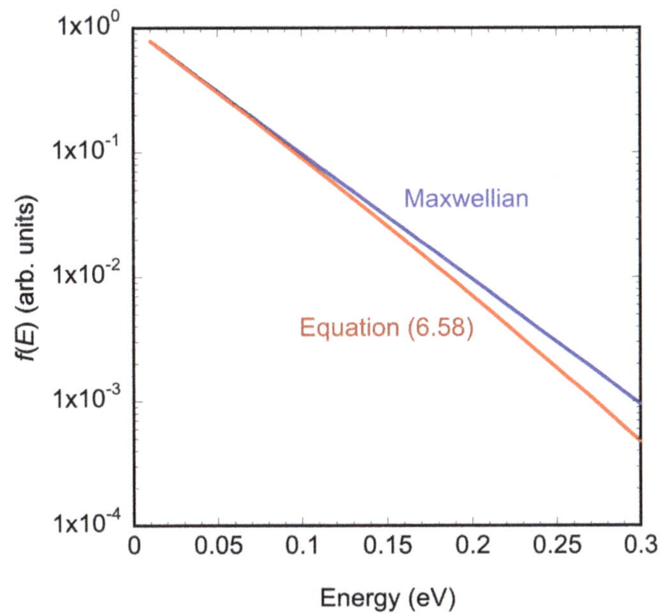

Figure 6.5. Comparison of the distribution function from (6.59) and the Maxwellian distribution, both at a carrier temperature of 500 K. Parameters suitable to GaAs at 300 K were used, along with the assumption that $\Lambda/\lambda = 5$.

account the full band structure of the material and the polar scattering determined from this full band structure. The calculation for a parabolic band is shown for comparison, and the data is taken from [17]. The data is not sufficient and doesn't clearly reflect the low field properties. Moreover, it is clear that there is a divergence around 4–5 MV cm^{-1} that will be further explained in chapter 5.

6.3 The drifted Maxwellian distribution

It seems to be clear now that one has to solve some difficult equations in order to find the distribution function in the presence of a high electric field. It is only slightly easier in the case of energetic photon excitation. While no electric field is usually present (except in a solar cell where the field is present), one still has to address the manner in which the detailed scattering processes affect the shape of the distribution. Indeed, the kink that appears in the energy distribution function in (6.59) has little to do with the electric field, as the latter appears only in the carrier temperature. Nevertheless, input from the photons will also affect the carrier temperature, but the kink arises from the difference in acoustic and optical scattering.

The treatment of hot carrier distributions can generally be broken into two distinct regions, which are differentiated by the concept of the electron temperature. Even the idea of defining an electron temperature is limited to those cases where the spherically symmetric part of the distribution evolves into a Maxwellian form such as $\exp(-E/k_BT_e)$. If the distribution differs from this form to any significant amount, then the idea of an electron temperature becomes rather vague. However, Fröhlich and Paranjape [18] demonstrated many years ago that this form would emerge if the

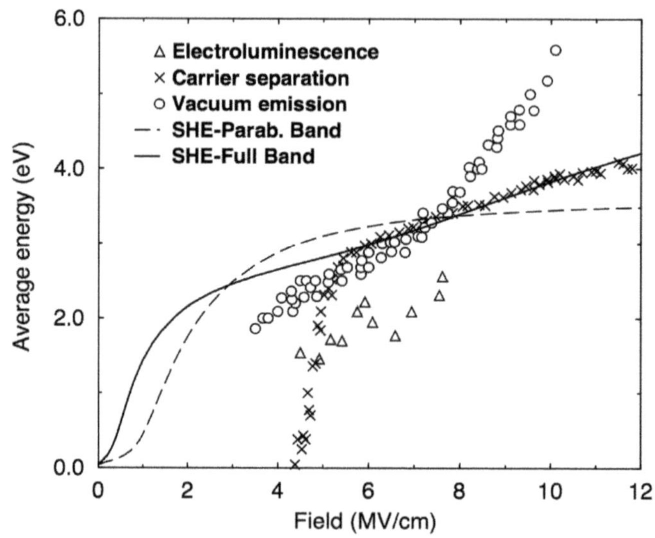

Figure 6.6. The average energy for electrons in SiO_2 as a function of the applied electric field, as determined with a spherical harmonic (Legendre polynominal) expansion and the full band structure of the material (solid curve). The dashed curve uses only parabolic bands. The experimental data is from Fischetti and DiMaria [17]. Reprinted from Gnani *et al* [16], copyright (2002) with permission from Elsevier.

carrier–carrier scattering was sufficiently strong to dominate other scattering processes (see, for example, figure 2.10). The carrier–carrier scattering provides a fast time scale that dominates the evolution of the distribution function and forces it into quasi-equilibrium *form* as described e.g. by Bogoliubov [19]. When this assumption is valid, the carrier distribution in momentum space will be in an internal steady-state with a carrier temperature that is a function of the electric field, or the excess energy of the photons in optical excitation, although it is often accompanied by other parameters that are related to various constants of the motion.

One of the early efforts to determine the electron density at which the carrier–carrier scattering becomes dominant was that of Hearn [20]. A more extensive treatment was given by Takanaka *et al* [21]. Following the procedure of the latter authors, the scattering rate for carrier–carrier scattering can be derived as [9]

$$\Gamma_{ee}(k) = \frac{ne^4m^*}{8\sqrt{\pi}\hbar^3 q_D^2 \varepsilon_s^2\, k} e^{-x^2} \int_0^\infty e^{-y^2} \frac{y^2}{y^2 + \lambda^2} \sinh(2xy) dy, \tag{6.60}$$

Where q_D is the Debye screening wave vector, and

$$y^2 = \frac{\hbar^2 \xi^2}{2m^* k_B T_e}, \; \xi = \mathbf{k}_2 - \mathbf{k}, \tag{6.61}$$

$$x^2 = \frac{\hbar^2 k^2}{2m^* k_B T_e}, \tag{6.62}$$

and

$$\lambda^2 = \frac{\hbar^2 q_D^2}{2m^* k_B T_e}. \tag{6.63}$$

Here, ξ is the scattered wave vector from the initial state $\mathbf{k}$ to the final state $\mathbf{k}_2$, and the distribution function is assumed to be a Maxwellian at the carrier temperature T_e. The subscripts have indicated electrons, but the same can be applied to holes. A plot of the scattering rates for electron–electron scattering for carriers in a Si MOSFET at an assumed temperature of 2500 K is given in figure 6.7 from calculations by Ferry *et al* [22]. The equivalent calculation for the energy gained or lost by a carrier in carrier–carrier scattering for a two-dimensional system has been given as well [23].

It is a general property of nonequilibrium systems that the distribution function can be described by a set of parameters prescribed by the integral invariants of the system [24]. The simplest form of this is what we refer to as a drifted Maxwellian, which accounts for conservation of density, momentum, and energy. For example, one such form for the drifted Maxwellian may be written as

$$f(\mathbf{r}, \mathbf{k}, t) = C \exp\left[-\frac{\hbar^2(\mathbf{k} - \mathbf{k}_d)^2}{2m^* k_B T_e}\right]. \tag{6.64}$$

There are obviously three parameters in this equation: the normalization constant C which is attributed to conservation of density, the drift momentum $\mathbf{k}_d$ which is attributed to conservation of momentum, and the electron temperature T_e which is attributed to conservation of energy. A set of equations must be generated in order to establish how these three quantities evolve with time, position, and driving forces such as electric fields or high energy photons. If we expand (6.64), the first two contributions (which tend to be the ones usually explored) are given by

$$f_0(E) = C \exp\left(-\frac{E}{k_B T_e}\right), \tag{6.65}$$

and

$$f_1(E) = v\left(\frac{m^* v_d}{k_B T_e}\right) f_0(E). \tag{6.66}$$

In reaching these forms, the fundamental definition of the effective mass $\hbar k = m^* v$ has been used to change the momenta to the velocities. Here, $\mathbf{v}_d$ is the drift velocity of the ensemble of carriers. It has also been assumed that the pre-factor in (6.66) is small.

6.3.1 The generic equation

The equations that are derived to lead to the balance in density, momentum, and energy are referred as the balance equations. They will be derived from the Boltzmann transport (6.4) assuming that the distribution itself is given by (6.64)

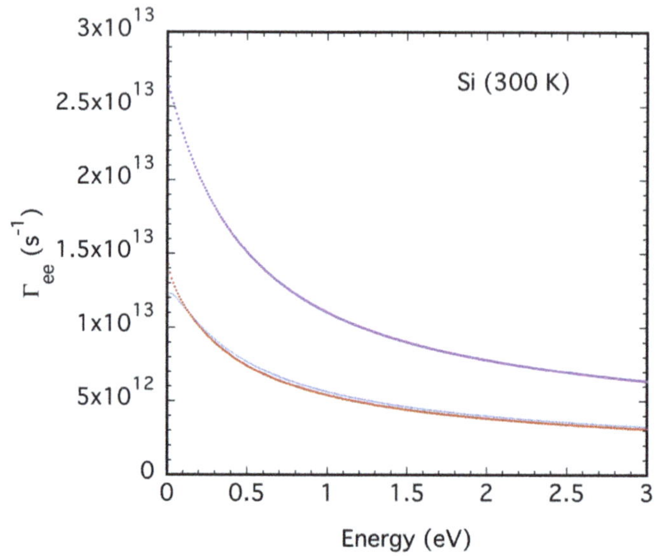

Figure 6.7. Electron–electron scattering rates for a Si MOSFET at 300 K. It is assumed that the carrier distribution is characterized by an electron temperature of 2500 K. Rates are shown for the loss of energy by the carrier (blue), gain of energy by the carrier (red) and total scattering rate (magenta). The data is adapted from Ferry *et al* [22].

(or by one of its extensions). Getting to the balance equations is relatively easy. The difficulty arises from evaluating the collision integrals that will derive from the right-hand side of (6.4). The equations will be derived in this section, and the carrier scattering integrals will be discussed for an example scattering process in the following sections.

The approach that will be followed is to consider a function of momentum $\varphi(\mathbf{p})$. When this function is integrated over momentum with the distribution function, its average is obtained, as

$$\langle\varphi(\mathbf{p})\rangle = \frac{1}{n}\int \varphi(\mathbf{p})f(\mathbf{p})d\mathbf{p}, \tag{6.67}$$

where n is the carrier density and provides the normalization through the equivalent of (6.17). Again, we proceed with the situation in which the system is subject to an applied electric field E. However, it will be considered that there are several valleys or sets of valleys in the band structure, so that the equations will be written for valley i. The parameters in various valleys will therefore carry a valley subscript. Then, the Boltzmann (6.4) becomes

$$\frac{\partial}{\partial t}(n_i\langle\varphi_i\rangle) + \frac{1}{m^*}\nabla\cdot\left(n_i\langle\mathbf{p}_i\varphi_i\rangle\right) - n_i e\mathbf{E}\cdot\left\langle\frac{\partial\varphi_i}{\partial\mathbf{p}_i}\right\rangle = \left\langle\varphi_i\frac{df(\mathbf{x},\mathbf{v},t)}{dt}\bigg|_{\text{coll.}}\right\rangle. \tag{6.68}$$

The last term on the left-hand side has been integrated by parts to move the momentum derivative from the distribution to the function that is being averaged. In general, all the terms on the left-hand side involve only the single valley *i*. But, the right-hand side will involve inter-valley scattering as well. Hence, the scattering term becomes more difficult. We can use (6.5), neglecting the filling factor terms, to write the right-hand side of (6.68) as

$$\left\langle \varphi_i \frac{df(\mathbf{x}, \mathbf{v}, t)}{dt}\bigg|_{\text{coll.}} \right\rangle \\ = \sum_{\mathbf{p}_i} \varphi(\mathbf{p}_i) \left\{ \sum_{\mathbf{p}_i'} \left[P(\mathbf{p}_i', \mathbf{p}_i) f_i(\mathbf{p}_i') - P(\mathbf{p}_i, \mathbf{p}_i') f_i(\mathbf{p}_i) \right] \cdot \right. \\ \left. + \sum_{\mathbf{p}_j'} \left[P(\mathbf{p}_j', \mathbf{p}_i) f_j(\mathbf{p}_j') - P(\mathbf{p}_i, \mathbf{p}_j') f_i(\mathbf{p}_i) \right] \right\} \tag{6.69}$$

In the last summation, the two terms represent scattering *in* to valley *i* from valley *j* and scattering *out* from valley *i* to valley *j*. This inter-valley scattering can, of course, carry particles from one valley to another, but it can also transfer momentum and energy. It may be noted that integrals are being performed over both momenta values, so that the result will not be dependent upon the order of integration of the momenta values in the scattering rates *P*. In fact, by interchanging the orders of integration, and redefining the two momenta as necessary, (6.69) can be rewritten as

$$\left\langle \varphi_i \frac{df(\mathbf{x}, \mathbf{v}, t)}{dt}\bigg|_{\text{coll.}} \right\rangle = \sum_{\mathbf{p}_j} f_j(\mathbf{p}_j') \varphi(\boldsymbol{\xi}_j) \sum_{\mathbf{p}_i} P(\mathbf{p}_i, \mathbf{p}_j') \\ - \sum_{\mathbf{p}_i} f_i(\mathbf{p}_i) \varphi(\mathbf{p}_i) \sum_{\mathbf{p}_j'} P(\mathbf{p}_i', \mathbf{p}_i) \quad . \tag{6.70}$$

Here, the change in momentum and energy in the inter-valley process requires that

$$E_j(\boldsymbol{\xi}_j) = E_i(\mathbf{p}_i) \pm \hbar\omega_0 \pm \Delta_{ij}, \tag{6.71}$$

where Δ is the energy separation between the minima of the two valleys. The last term of each of the two functions on the right-hand side of (6.70) is just the scattering rate

$$\Gamma_{j\to i}(\mathbf{p}_i) = \sum_{\mathbf{p}_j'} P(\mathbf{p}_i', \mathbf{p}_i). \tag{6.72}$$

Finally, the scattering rate can be rewritten as

$$\left\langle \varphi_i \frac{df(\mathbf{x}, \mathbf{v}, t)}{dt}\bigg|_{\text{coll.}} \right\rangle = n_j \left\langle \varphi(\boldsymbol{\xi}_j) \Gamma_{j\to i}(\mathbf{p}_i) \right\rangle_j - n_i \left\langle \varphi(\mathbf{p}_i) \Gamma_{i\to j}(\mathbf{p}_i) \right\rangle_i . \tag{6.73}$$

The subscript on the average tells us which valley distribution function is involved in the averaging process.

6.3.2 The balance equations

Each of the three equations arises from different forms for the function $\varphi(\mathbf{p}_i)$. As will be seen, these equations can be quite complicated in the presence of multiple scattering processes. To begin, consider the case for which $\varphi(\mathbf{p}_i) = C$; that is, it is just the normalization constant for the distribution function. Then, (6.68) becomes

$$\frac{\partial n_i}{\partial t} + \frac{1}{e}\nabla \cdot \boldsymbol{J} = n_j\langle \Gamma_{j\to i}(\mathbf{p}_i)\rangle_j - n_i\langle \Gamma_{i\to j}(\mathbf{p}_i)\rangle_i. \tag{6.74}$$

This is just the continuity equation (with $\boldsymbol{J} = ne\mathbf{v}_d$)with gain and loss terms. In a single valley situation, the right-hand side is zero, as carriers do not enter and leave the valley. But, in a multi-valley situation, the right-hand side is an evaluation of the inter-valley scattering and the redistribution of the carrier population. However, it has to noted that the second term on the left involves the drift velocity and therefore couples this equation to the momentum-balance equation. Such a result is known in statistical mechanics, where one-electron equations are coupled to two-electron equations in expansions such as the BBGKY formulation [8]. This is the simplest of the equations, merely confirming that the number of carriers must be conserved (in the absence of some carrier generation process). In optical excitation, the generation and recombination of electron–hole pairs would also appear on the right-hand side of (6.74).

The next equation in the hierarchy is the momentum-balance equation, for which $\varphi(\mathbf{p}_i) = \mathbf{p}_i$. In this case, (6.68) becomes

$$\begin{aligned}\frac{\partial(n_i m^* \mathbf{v}_d)}{\partial t} + \frac{1}{m^*}\nabla \cdot \left(n_i\langle \mathbf{p}_i\mathbf{p}_i\rangle\right) - n_i e\mathbf{E} = n_i\langle \mathbf{p}_i\Gamma_{i\to i}(\mathbf{p}_i)\rangle \\ + \sum_j n_i\left\langle\left[\mathbf{p}_j \pm \sqrt{\frac{2m^*}{\hbar^2}(\hbar\omega_0 \pm \Delta_{ij})}\right]\Gamma_{j\to i}(\mathbf{p}_j)\right\rangle - \sum_i n_i\langle \mathbf{p}_i\Gamma_{i\to j}(\mathbf{p}_i)\rangle\end{aligned}. \tag{6.75}$$

This equation is now coupled not only to the energy-balance equation through the second term on the left-hand side, but it is also coupled to the density balance through the first term on the left-hand side. If the time derivative of the density is inserted into this equation using (6.74), then this equation becomes quite complicated with multiple scattering terms. The second term on the left-hand side actually involves the tensor $\mathbf{p}_i\mathbf{p}_i$, which contains many terms coupling different directions in the crystal. Quite often, this term is approximated as a diagonal scalar term, so that it becomes $2m^* k_B T_e$. But, often this cannot be done, for example when a magnetic field is present (where it will still be diagonal, but not a scalar).

The last equation that will be treated here is the energy-balance equation, for which $\varphi(\mathbf{p}_i) = p_i^2/2m^*$. Using this in (6.68) leads to

$$\begin{aligned}\frac{\partial(n_i\langle E_i\rangle)}{\partial t} + \frac{1}{m^*}\nabla \cdot \left(n_i\langle \mathbf{p}_i E_i\rangle\right) - n_i e\mathbf{E} \cdot \mathbf{v}_d = n_i\langle E_i\Gamma_{i\to i}(\mathbf{p}_i)\rangle \\ + \sum_j n_j\langle [E_i(\mathbf{p}_i) \pm \hbar\omega_0 \pm \Delta_{ij}]\Gamma_{j\to i}(\mathbf{p}_j)\rangle - \sum_i n_i\langle E_i\Gamma_{i\to j}(\mathbf{p}_i)\rangle\end{aligned}. \tag{6.76}$$

Once more, the second term on the left-hand side couples to a higher-order equation as well as to both the density balance equation and the momentum-balance equation. The first term on the left-hand side couples also to the density balance equation. If there is a desire to stop with just these three equations, this second term on the left-hand side has to be approximated in a manner that breaks the coupling to the higher-order term. One such approximation is to take the average kinetic energy as $3k_B T_e/2$, and use this in the term so that this term becomes merely

$$\nabla \cdot \left(\frac{3n_i k_B T_e}{2} \mathbf{v}_d \right). \tag{6.77}$$

This still leaves three coupled equations that are going to be difficult to solve. Nevertheless, this approach has been used quite effectively with numerical techniques in which to solve the various integrals and equations [25, 26, 27].

6.3.3 Polar optical scattering

As an example of the use of the above equations, which are often referred to as the hydrodynamic equations, the saturation or non-saturation of polar optical phonon scattering will be investigated. It will be assumed that the system is homogeneous, time invariant, and only a single Γ valley (for a direct gap semiconductor such as GaAs) will be examined in order to simplify the equations and make them easier to follow. In this single valley, the equations for the steady-state response are merely the momentum-balance equation

$$-e\mathbf{F} = \frac{1}{n} \left\langle \mathbf{p}_i \Gamma_{i\to i}(\mathbf{p}_i) \right\rangle, \tag{6.78}$$

and the energy-balance equation

$$e\mathbf{F} \cdot \mathbf{v}_d = \frac{1}{n} \left\langle E_i \Gamma_{i\to i}(\mathbf{p}_i) \right\rangle. \tag{6.79}$$

Here, **F** is the electric field to differentiate from the energy. The polar optical phonon scattering process leads to a scattering rate that has already been given in (4.16–4.17) as

$$\begin{aligned} \left(\frac{dE}{dt} \right) \bigg|_{LO} = \hbar\omega_0 \left(\frac{1}{\tau_{ab}} - \frac{1}{\tau_{em}} \right) = \frac{m^* e^2}{2\pi\gamma\epsilon_0 \sqrt{2m^* E}} \left[N_q \sinh^{-1}\left(\frac{E}{\hbar\omega_0} \right)^{1/2} \right. \\ \left. - \left(N_q + 1 \right) \sinh^{-1}\left(\frac{E}{\hbar\omega_0} - 1 \right)^{1/2} \right] \end{aligned}, \tag{6.80}$$

where

$$\frac{1}{\gamma} = \omega_0^2 \left(\frac{1}{\epsilon_{r,\infty}} - \frac{1}{\epsilon_{r,0}} \right). \tag{6.81}$$

This form differs somewhat from that normally seen, as the inverse hyperbolic functions replace the equivalent logarithmic functions [8]. This form is easier for the computation of the integrals over the distribution function. The average of this energy relaxation rate is given by

$$\left\langle \frac{dE}{dt} \right\rangle = \frac{m^* e^2 N}{n 2\pi\gamma\epsilon_0} \int_0^\infty \frac{dE}{\sqrt{2m^* E}} \left\{ \sinh^{-1}\left(\sqrt{\frac{E}{\hbar\omega_0}}\right) \right. \\ \left. - e^y \sinh^{-1}\left(\sqrt{\frac{E}{\hbar\omega_0} - 1}\right) \right\} \rho(E) e^{-E/k_B T_e}, \tag{6.82}$$

where ρ is the density of states in three dimensions. Only the symmetric part of the distribution has been retained, as the momentum terms in the second term will vanish for the Maxwellian shape upon integration. To proceed, a change of variables in the second term in curly brackets will be made, using $E' = E - \hbar\omega_0$. Then the two terms can be combined as

$$\left\langle \frac{dE}{dt} \right\rangle = \frac{m^* e^2 N}{n 2\pi\gamma\epsilon_0} (e^{y - y_e} - 1) \int_0^\infty \frac{dE}{\sqrt{2m^* E}} \sinh^{-1}\left(\sqrt{\frac{E}{\hbar\omega_0}}\right) \rho(E) e^{-E/k_B T_e}, \tag{6.83}$$

where

$$y = \frac{\hbar\omega_0}{k_B T}, \quad y_e = \frac{\hbar\omega_0}{k_B T_e}. \tag{6.84}$$

The density of states is given by

$$\rho(E) = \frac{1}{2\pi^2} \left(\frac{2m^*}{\hbar^2} \right)^{3/2} E^{1/2}, \tag{6.85}$$

so that (6.85) can be written as

$$\left\langle \frac{dE}{dt} \right\rangle = \frac{m^{*2} e^2 N}{n 4\pi^3 \gamma \hbar^3 \epsilon_0} (e^{y - y_e} - 1) \int_0^\infty dE \sinh^{-1}\left(\sqrt{\frac{E}{\hbar\omega_0}}\right) e^{-E/k_B T_e}. \tag{6.86}$$

If the change of variables $x = E/\hbar\omega_0$, the integral becomes

$$\left\langle \frac{dE}{dt} \right\rangle = \frac{m^{*2} e^2 N \omega_0}{n 4\pi^3 \gamma \hbar^2 \epsilon_0} (e^{y - y_e} - 1) \int_0^\infty dx \sinh^{-1}(\sqrt{x}) e^{-x y_e}. \tag{6.87}$$

The integral is a standard form of the modified Bessel function of the second kind of zero order (K_0) [28]. Thus, the energy loss rate becomes

$$\left\langle \frac{dE}{dt} \right\rangle = \frac{m^{*2} e^2 N k_B T_e}{n 8\pi^3 \gamma \hbar^3 \epsilon_0} (e^{y - y_e} - 1) e^{-y_e/2} K_0\left(\frac{y_e}{2}\right). \tag{6.88}$$

To complete this, the normalization is given by replacing the density with

$$n = \int_0^\infty \rho_{3D}(E) e^{-E/k_B T_e} dE = \frac{1}{4} \left(\frac{2m^* k_B T_e}{\pi\hbar^2} \right)^{3/2}. \tag{6.89}$$

The momentum relaxation time must be considered for (6.78). This process is somewhat more difficult as the scattering rate has to be redeveloped with the momentum loss directly incorporated within the new formulation. The quantity of interest will have the general form

$$\frac{1}{n}\left\langle \mathbf{p}_i \Gamma_{i\to i}(\mathbf{p}_i)\right\rangle = \left\langle \frac{m^* \mathbf{v}}{\tau_{\mathrm{LO}}}\right\rangle, \tag{6.90}$$

but this form cannot be found in the easy method that arose for the energy relaxation process because the momentum exchanged is not constant. Hence, the beginning must be

$$\begin{aligned}\frac{dp_F}{dt} = \frac{2\pi}{\hbar}\sum_q \Big\{ & \hbar q_F \mid M \mid^2 N\delta\left(E_{k+q} - E_k - \hbar\omega_0\right) \\ & - \hbar q_F \mid M \mid^2 (N+1)\delta\left(E_{k-q} - E_k + \hbar\omega_0\right)\Big\}\end{aligned}, \tag{6.91}$$

where

$$\mid M \mid^2 = \frac{\hbar e^2}{2V\gamma\omega_0 q^2}, \tag{6.92}$$

and screening is being ignored for this approach. The quantities p_F and q_F correspond to the momentum exchange along the electric field. These are related to the normal angles via

$$\mathbf{q}\cdot\mathbf{F} = qF\left[\cos\vartheta\cos\vartheta_F + \sin\vartheta\sin\vartheta_F\cos\left(\phi - \phi_F\right)\right], \tag{6.93}$$

where ϑ and ϑ_F are the angles that **q** and **F** make with **k**, where the latter is taken to be the polar direction. The second term in (6.93) will vanish under integration over the azimuthal angle. Integration over the polar angle is more complicated as it involves the argument of the delta functions and yields the additional variables

$$\cos\vartheta = \mp\frac{q}{2k} \pm \frac{m^*\omega_0}{\hbar kq}. \tag{6.94}$$

Using these relations, (6.91) becomes

$$\begin{aligned}\frac{dp_F}{dt} = \frac{\hbar e^2 N}{4\pi\gamma\omega_0}\Bigg\{ & \int_{q_a^-}^{q_a^+}\left(\frac{m^*\omega_0}{\hbar kq} - \frac{q}{2k}\right)dq \\ & - e^y \int_{q_e^-}^{q_e^+}\left(\frac{m^*\omega_0}{\hbar kq} + \frac{q}{2k}\right)dq\Bigg\}\cos\vartheta_F\end{aligned}. \tag{6.95}$$

The second term in each integral can be integrated trivially while the first term is the normal integration for polar optical phonon scattering [9]. As a result, the integration can be done straight away, dropping the last cosine term as it is involved also with the term on the left, so that one obtains

$$\begin{aligned}\frac{dp_F}{dt} = -\frac{\hbar e^2 N}{4\pi\gamma\omega_0}\Bigg\{&k\sqrt{1-\frac{\hbar\omega_0}{E}} - \frac{2m^*\omega_0}{\hbar k}\sinh^{-1}\left(\sqrt{\frac{E}{\hbar\omega_0}}\right)\\ &+ e^y\left[k\sqrt{1+\frac{\hbar\omega_0}{E}} + \frac{2m^*\omega_0}{\hbar k}\sinh^{-1}\left(\sqrt{\frac{E}{\hbar\omega_0}-1}\right)\right]\Bigg\}.\end{aligned} \tag{6.96}$$

The integrations over the distribution function uses the first-order term in the drifted Maxwellian form (6.66) for the integration. The integration is relatively straightforward and yields the result [8]

$$\begin{aligned}\frac{1}{n}\left\langle\frac{dp_F}{dt}\right\rangle = m^* v_d \frac{e^2 N y_e^2}{6\pi\hbar}\sqrt{\frac{m^*\omega_0}{\pi\hbar}}\Bigg\{&(e^{y-y_e}+1)e^{y_e/2}K_1\left(\frac{y_e}{2}\right)\\ &+(e^{y-y_e}-1)e^{y_e/2}K_0\left(\frac{y_e}{2}\right)\Bigg\}.\end{aligned} \tag{6.97}$$

Here, the modified Bessel functions of the second kind of both order 0 and 1 have resulted from the integrations.

In figure 6.8, the use of the drifted Maxwellian equation to evaluate transport in the central valley of cubic SiC is illustrated [29]. Here, two inter-valley optical phonons, as well as the polar optical and acoustic intra-valley modes, were taken into account and the different curves are for different sets of values. Each of these different phonons each yield equations like (6.88) and (6.97), but all can be evaluated with a hand calculator and a spreadsheet, although, of course, a computer makes them easier once you have the evaluations of the Bessel functions in hand. In curve *a*, the values of 10^9 eV cm^{-1}, 10^9 eV cm^{-1}, and 11.9 eV are taken for $D_{ij,1}$, $D_{ij,2}$, and E_1, respectively, and an effective polar field of 250 kV cm^{-1} was used. The effective polar field is given as $F_0 = m^*e^2/\hbar\omega_0\gamma$. In curve *b*, the values of 10^9 eV cm^{-1}, 10^9 eV cm^{-1}, and 11.9 eV are taken for $D_{ij,1}$, $D_{ij,2}$, and E_1, respectively, and an effective polar field of 500 kV cm^{-1} was used. For curve *c*, the values of 10^9 eV cm^{-1}, 0.3×10^9 eV cm^{-1}, and 15.0 eV are taken for $D_{ij,1}$, $D_{ij,2}$, and E_1, respectively, and an effective polar field of 200 kV cm^{-1} was used. The mobility data is taken from Nelson *et al* [30]. The procedure used is to fit the low-field mobility with the various parameters and then use these to predict the high-field velocity. Both curves *a* and *b* give reasonable fits to the mobility in panel (a), so both are used in panel (b) for the high-field velocity.

In figure 6.9, domain motion is shown for an n^+-n-n^+ structure in GaAs at a high electric field [31]. The n-type active region is only 2.8 μm in the figure, and the heavily doped contact regions are not shown. The simulation is carried out using the hydrodynamic equations discussed in this section. It is clear that domains form in time, but they do not seem to propagate as expected. Rather, they remain pinned at the cathode contact, where they grow and die, and still manage to modulate the current. But, it is strange that they do not propagate. Normally, a pinned domain at

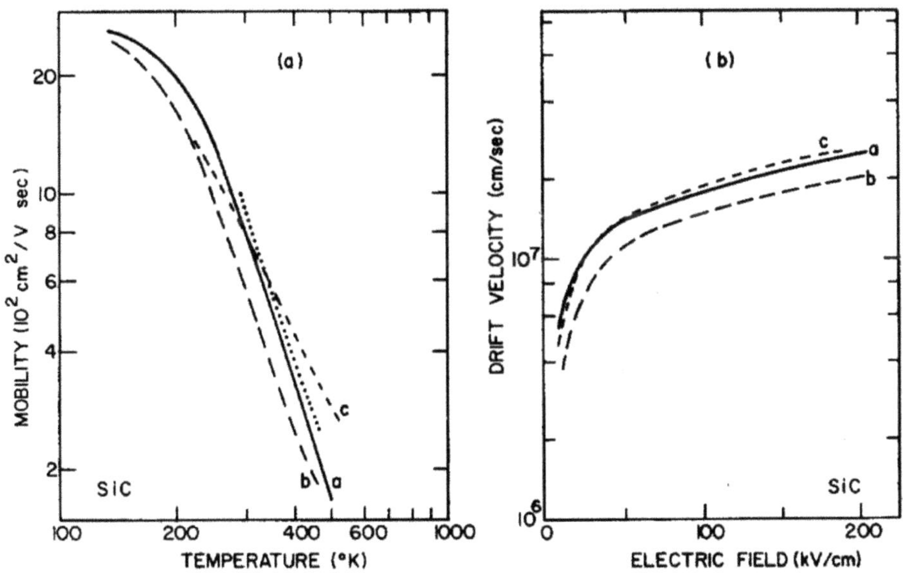

Figure 6.8. The hydrodynamic equations are used to fit to the mobility of SiC in panel (a) and then to predict the high-field velocity in panel (b). The parameters for each curve are given in the text. The dotted line in panel (a) is data from Nelson *et al* [30]. Reprinted with permission from Ferry [29], copyright (1975) by the American Physical Society.

the cathode does not give rise to oscillations, as there is not a clear mechanism for the decay of such a pinned domain.

To generate the answer to whether or not the velocity saturates for polar optical scattering, equations (6.78) and (6.79) may be rewritten as

$$eFv_d = \frac{1}{n}\left\langle \frac{dE}{dt} \right\rangle, \qquad eF = \frac{1}{n}\left\langle \frac{dp_F}{dt} \right\rangle. \tag{6.98}$$

Equations (6.88), (6.89), and the largest term (the last term in the curly brackets) of (6.97) are now inserted into these equations, and the asymptotic drift velocity may be isolated to show that

$$v_d^2 = -\frac{3k_B T_e}{m*}\tanh\left(\frac{\hbar\omega_0}{2k_B T_e}\right)\ln\left(\frac{\hbar\omega_0}{2k_B T_e}\right). \tag{6.99}$$

The negative sign assures that the electron temperature must be such that $k_B T_e > \hbar\omega_0/2$, so that the logarithm term is negative as well. Interestingly enough, (6.99) differs from (2.3) for non-polar optical scattering only by a factor of 4 and the logarithm term. Equation (2.3) shows saturation, but (6.99) does not saturate explicitly because of the logarithm term. As the electron temperature rises, the argument of the logarithmic term becomes increasingly smaller, and the value more

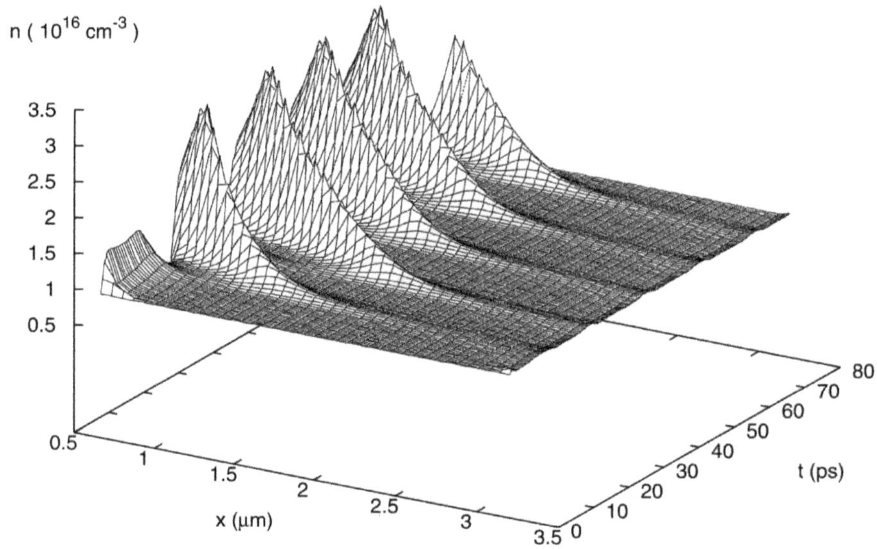

Figure 6.9. Domain formation in a GaAs n^{+}-n-n^{+} structure. These domains appear to be pinned at the cathode and do not propagate. Reprinted from Kao and Chen [31], copyright (2002) with permission from Elsevier.

negative. Thus, saturation does not occur (because of the logarithmic term). Rather, in the direct gap polar materials, inter-valley scattering and transfer usually occurs and the scattering in the satellite valleys limits the velocity, where the scattering is dominated by the non-polar scattering.

6.4 The energy diffusion equations

As should have become apparent by now, determination of the distribution function for hot carriers is quite difficult, and mathematically complicated. In the above, the Boltzmann transport equation has been used. But, there are other approaches in statistical mechanics. One approach, due primarily to Kurosawa [32], eases the problems somewhat, although it too can be mathematically complex. Although fully equivalent to solving the Boltzmann equation, it is formulated entirely in *energy* space. For mainly elastic scattering, Levinson [33] has shown that it is equivalent to a kinetic approach based upon the Fokker–Planck equation in energy space.

An electron in the conduction band (or a hole in the valence band) behaves quite similarly to a particle undergoing Brownian motion, if the energy change is small. It has a random velocity direction that is randomly changed by interactions, in this case scattering due to phonons or to other carriers. This fact gives rise to one of two equations that are used in the energy diffusion approach, and this is the diffusion equation in energy space itself. The second equation is a continuity equation which takes care of gain and loss terms when the energy exchange is not small, for example with optical phonons. These two equations may then be solved to yield the distribution function (naturally as a function of energy). This approach has been used to compute the generation rate for impact ionization in InSb and InAs [34], and in magneto-transport in strong magnetic fields [35]. It has also been used to study the generation of nonequilibrium phonons [36] and applied to degenerate carriers in the inversion layer of MOS semiconductor structures [37].

6.4.1 Equations in energy space

In energy space, the occupation of a particular state is given by the product of the probability for occupation of that state and the function that provides the number of states on that particular energy shell. Thus, the occupation of a given energy shell is given by

$$g(E) = \rho(E)f(E), \tag{6.100}$$

where the first term on the right is our normal density of states (which we will take to be that for a parabolic band) and the second term is the distribution function as a function of energy. A continuity equation can be written for this quantity, which also defines a number of terms, as

$$I(E) = -D(E)\frac{\partial g(E)}{\partial E} + V(E)\,g(E)[1 - f(E)]. \tag{6.101}$$

This equation is a continuity equation of sorts, and the term $I(E)$ represents the gain and loss of particles on a particular energy shell through large energy exchange collisions such as those of the optical phonons. The terms $D(E)$ and $V(E)$ represent the diffusion of energy through energy space, for example by elastic scattering events or small energy exchange scattering events, and the energy flow (energy velocity) in this energy space (this is not the normal carrier velocity). To more clearly illustrate the approach, the occupancy factor in (6.101) will be dropped, assuming $f(E) < 1$.

Each of the two quantities D and V can be expanded as

$$\begin{aligned} D(E) &= D_T(E) + D_F(E) \\ V(E) &= V_T(E) + V_F(E) \end{aligned}. \tag{6.102}$$

The first of each of these two expressions, with the subscript T, refers to the contributions that arise from interactions with various scattering centers that do not involve large energy exchanges. The latter two terms, with the subscript F, correspond to movement through energy space induced by an external force, such as the electric field. While the Fermi–Dirac form can be used easily, it introduces added complexity although it will be retained in a later section. In the absence of the flow term $I(E)$ and the forcing terms, (6.101) can be solved as [32]

$$g(E) \sim \exp\left(\int^{E} \frac{V_T(E')}{D_T(E')} dE'\right). \tag{6.103}$$

In normal classical mechanics and Brownian motion in equilibrium, this last term would require

$$\exp\left(\int^{E} \frac{V_T(E')}{D_T(E')} dE'\right) = \rho(E)\exp\left(-\frac{E}{k_B T}\right), \tag{6.104}$$

which requires

$$\frac{V_T(E)}{D_T(E)} = \frac{1}{\rho(E)}\frac{d\rho(E)}{dE} - \frac{1}{k_B T}. \tag{6.105}$$

Suppose that there is no flow, but also no thermally activated diffusion and energy drift. That is, the two terms with subscript T are set to zero as well as the flow I. This gives the equation

$$-D_F \frac{\partial g(E)}{\partial E} + V_F g(E) = 0. \tag{6.106}$$

In this limit, the population becomes

$$\exp\left[\int^{E} \left\{V(E')/D(E')\right\} dE'\right] = \rho(E), \tag{6.107}$$

and

$$\frac{V_F}{D_F} = \frac{1}{\rho(E)}\frac{d\rho(E)}{dE}. \tag{6.108}$$

This means that (6.105) and (6.108) point to the fact that the distribution function is of the form

$$g_0(E) = \rho(E)\exp\left[-\frac{1}{k_B T}\int^{E} \frac{dE'}{1 + (D_F/D_T)}\right] \tag{6.109}$$

at small fields.

The flow equation for gain and loss of particles is actually constant in equilibrium, and it's the derivative of this term that is important. This is described as

$$\frac{\partial I(E)}{\partial E} = \frac{dg(E)}{dt} = I_1 - I_2 \tag{6.110}$$

where

$$\begin{aligned} I_1(E) &= \Gamma_{em}(E + \hbar\omega_0)g(E + \hbar\omega_0) + \Gamma_{ab}(E - \hbar\omega_0)g(E - \hbar\omega_0), \\ I_2(E) &= [\Gamma_{em}(E) + \Gamma_{ab}(E)]g(E). \end{aligned} \tag{6.111}$$

The various Γ are the scattering rates for the emission and absorption of the optical phonons.

6.4.2 The coefficients

In general, it is relatively easy to show that D_F and V_F are related to the energy (or field) dependent mobility. To understand this, consider the following quantity [32]

$$\left(\frac{dE}{dt}\right)_F = \int_0^\infty E\left(\frac{\partial g}{\partial t}\right)_F dE = \int_0^\infty E\left\{-\frac{\partial}{\partial E}\left(-D_F\frac{\partial g}{\partial E} + V_F g\right)\right\}dE. \qquad (6.112)$$

This is just the rate of energy gain that the electric field gives to the system. By integrating this entire expression once by parts, and then the first term in curly brackets again by parts, this expression is transformed into

$$\left(\frac{dE}{dt}\right)_F = \int_0^\infty g\left(V_F + \frac{\partial D_F}{\partial E}\right)dE. \qquad (6.113)$$

So long as we are considering a single valley semiconductor, below the threshold for any ionization processes, the drift velocity arises from the mobility according to

$$v_d = F\int_0^\infty \mu(E)g(E)dE. \qquad (6.114)$$

It may be recalled from (6.98) that the power input is given by

$$\left(\frac{dE}{dt}\right)_F = ev_d F, \qquad (6.115)$$

so that we can use the last three equations to conclude that

$$eF^2\mu(E) = V_F + \frac{\partial D_F}{\partial E}. \qquad (6.116)$$

Using (6.108), this becomes

$$eF^2\mu(E) = \frac{1}{\rho}\frac{\partial(\rho D_F)}{\partial E}. \qquad (6.117)$$

This can be integrated to give us the result

$$D_F(E) = \frac{eF^2}{\rho(E)}\int_0^E \mu(E')\rho(E')dE' \qquad (6.118)$$

with the initial condition

$$D_F(0) = 0. \qquad (6.119)$$

Since the interest in these diffusive terms lies in the elastic scattering terms, we consider elastic acoustic phonon scattering. The scattering rate for the acoustic modes, in the equipartition limit is given in (6.34), and

$$\mu(E) = \frac{e\tau_{ac}}{m^*} = \frac{e\pi\rho_m v_s^2 \hbar^2}{E_1^2 k_B T m^{*2}} \sqrt{\frac{\hbar^2}{2m^* E}}, \tag{6.120}$$

and thus

$$D_F(E) = \frac{e\pi\rho_m v_s^2 \hbar^2}{E_1^2 k_B T m^{*2}} \sqrt{\frac{\hbar^2}{2m^*}} \frac{1}{2\pi^2} \left(\frac{2m^*}{\hbar^2}\right)^{3/2} E = \frac{e\rho_m v_s^2 E}{\pi E_1^2 k_B T m^*}, \tag{6.121}$$

$$V_F(E) = \frac{e\rho_m v_s^2}{2\pi E_1^2 k_B T m^*}. \tag{6.122}$$

In the last form, (6.108) has again been used. It should be noted that these results depend explicitly upon the energy dependence in the acoustic scattering rate, and other scattering rates with other energy dependences will give different values for these parameters.

The thermal diffusivity in energy space for the acoustic phonons is given by a modification of the Fermi golden rule, in which

$$D_T(E) = \frac{\pi}{\hbar} \sum_q (\hbar q v_s)^2 \frac{E_1^2 k_B T}{\rho_m V v_s^2} \delta(E - E') \to \frac{4m^* v_s^2}{3\tau_{ac}} E, \tag{6.123}$$

where the acoustic phonon scattering time is given in (6.34).

6.4.3 Non-trivial solutions

In the absence of the large energy exchanging optical phonons, the solutions are trivially given by (6.109). Hence, the distribution is given by a simple integration in energy space. In the non-trivial case, the optical phonons participate in the transport through the time dependence of the flow $I(E)$. Once this is added to the set of equations, it is no longer possible to simply integrate over energy to find a solution. Equation (6.111) is non-local in energy, as it involves results from higher energies in order to solve at a given energy.

Yamada and Kurosawa have suggested to add a continuity equation to the set given above, which is written as [32, 38]

$$\frac{\partial f}{\partial t} = -\frac{1}{\rho(E)} \frac{\partial I(E)}{\partial t} + \left[\frac{df(E)}{dt}\right]_{sc}. \tag{6.124}$$

Hence, if one now takes the first term on the right-hand side from the derivative of (6.101) with respect to energy, a pair of equations with time derivatives is obtained. This may be iterated forward in time from an equilibrium condition at $t = 0$ to find the temporally evolving distribution function [39]. The last term on the right-hand side is not taken from (6.111), but from the normal optical phonon scattering terms, such as (6.28). In figure 6.10, the electron temperature is shown for hot electron magneto-transport in InSb at a low temperature of 4.2 K [39]. The quantity Γ is the level broadening of the Landau levels, and is given as a temperature corresponding to the impurity scattering that dominates this quantity. The transport is studied including

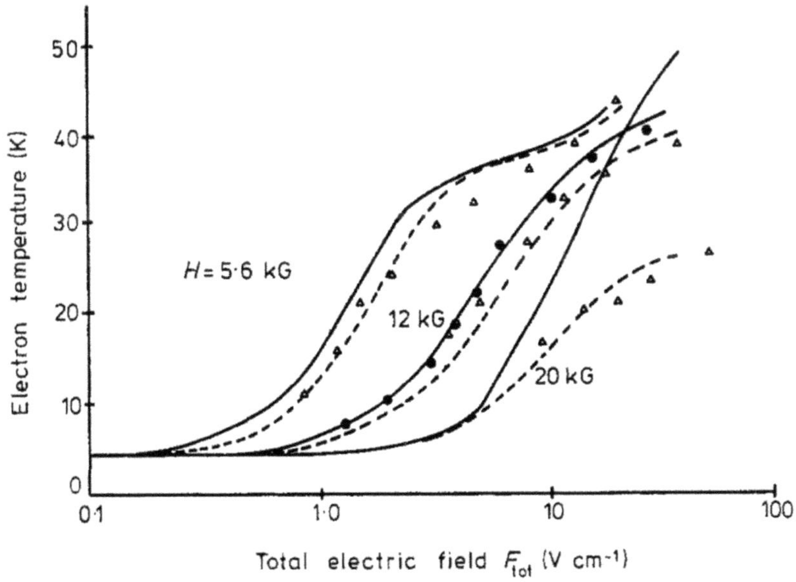

Figure 6.10. The average electron temperature (solid curve, circles) and relative temperature (dashed curve, based upon populations in the two lowest Landau levels, diamonds) as function of the total electric field (applied plus Hall field) for InSb. The experimental data is shown as circles and diamonds, for an electron density of 2.5×10^{13} cm^{-3} at 4.2 K, and various magnetic fields (10 kG = 1.0 T) and a level broadening of 10 K, corresponding to the number of impurities being 8×10^{14} cm^{-3}. Reprinted from Partl *et al* [39], copyright IOP Publishing. All rights reserved.

piezoelectric, acoustic and polar optical scattering contributions. The *relative* temperature corresponds to the ratio of populations in the first and zero Landau levels, which is determined experimentally. In figure 6.11, the transient behavior of the distribution function is plotted versus carrier energy (in terms of the cyclotron energy) for a total electric field that is switched on (or off) at $t = 0$, and has a peak value of 100 V cm^{-1}. Both the turn-on and turn-off transients are shown.

6.5 Low-dimensional systems

In the case of low-dimensional systems, evaluated at low temperatures, as discussed in chapter 3, the optical phonons are usually not excited and energy relaxation processes are dominated by low energy phonons whose interactions are almost elastic. In many cases, the solutions, such as in the last section can be found in closed form. In particular, it is fair to assume that the distribution function is a Fermi–Dirac at an elevated electron temperature [40]. The justification for this follows from the Kurosawa [32] energy diffusion equation of the last section. Price [40] rewrites these equations in the form

$$-A(E)\frac{df}{dE} + \int_0^E dE_1 \int_E^\infty dE_2\{W(1,2)f(1)[1-f(2)] - W(2,1)f(2)[1-f(1)]\} = 0, \tag{6.125}$$

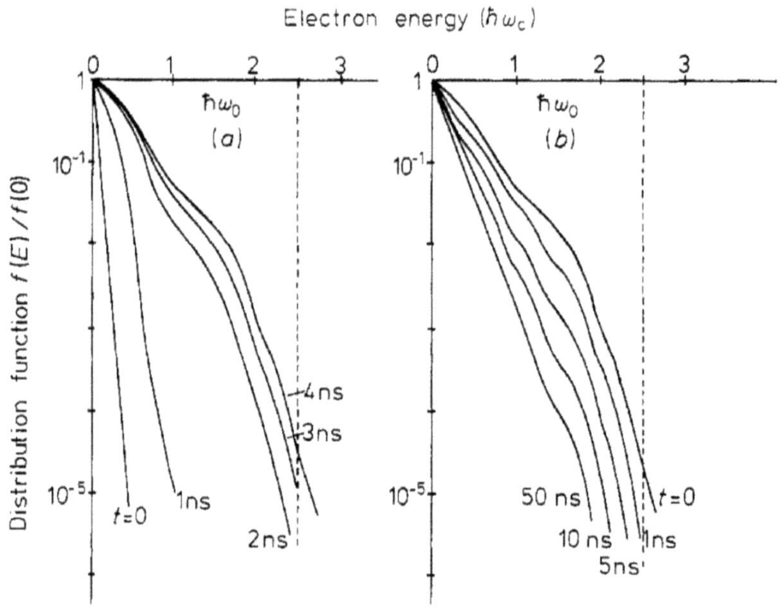

Figure 6.11. Transient behavior of the electron distribution function in InSb at 4.2 K for the field applied at $t = 0$ (a) and turned off at $t = 0$ (b). The magnetic field is 12 kG (1.2 T), and the energy is plotted relative to the cyclotron energy, which is chosen equation to the optical phonon energy, thus sitting on a magneto-phonon resonance. Here, the number of impurities is 8×10^{14} cm^{-3} and the carrier density is 1.5×10^{14} cm^{-3}. The peak total electric field is 100 V cm^{-1}. Reprinted from Partl *et al* [39], copyright IOP Publishing. All rights reserved.

where $f(E)$ is the distribution function, the numbers denote the two energies of the integrations, and W is the angle-averaged scattering process such that

$$\frac{1}{\tau(E_1)} = \int dE_2 W(1,2)[1 - f(E_2)]. \tag{6.126}$$

The coefficient $A(E)$ is given by $e^2F^2D(E)$, where F is the electric field, as above, and $D(E)$ is the diffusion coefficient $\lambda v/2$, where v is the velocity and λ is the mean-free path along the electric field. The drift mobility is then given by

$$\mu = e\frac{\int dED(E)f(1 - f)}{\int dEEf(1 - f)} = e\frac{\langle D\rangle}{\langle E\rangle}. \tag{6.127}$$

In this approach, the field is taken to be not large, so that we are dealing with the warm electron regime, primarily at low temperature where the optical modes are not involved. Nevertheless, it may be assumed that the thermal temperature is larger than the energy exchange in the various collision processes, so that it can also be said that the second term of (6.125) can be written as

$$-B(E)f(1 - f) + k_BT\frac{df}{dE}, \tag{6.128}$$

where

$$B(E) = \int_0^E dE_1 \int_E^\infty dE_2[W(2,1) - W(1,2)]. \tag{6.129}$$

With (6.128), the solution of (6.125) becomes

$$f(E) = \left[1 + \exp\left(\int_{E_F}^{E} \frac{dE'}{k_B T_e}\right)\right]^{-1}, \tag{6.130}$$

where

$$k_B T_e = k_B T + \frac{A(E)}{B(E)}. \tag{6.131}$$

If the electron temperature is such that the new thermal energy is small compared to the Fermi energy, then this will be a Fermi–Dirac function. That is, the electron temperature should not vary much over the Fermi–Dirac transition region.

The fact that the Fermi–Dirac distribution arises in this treatment is more or less a tautology, that arises once the $(1 - f)$ appears in (6.128). This can be seen simply from the derivative of the distribution with respect to energy. For a Maxwellian, the derivative is

$$\frac{df}{dE} = -\frac{1}{k_B T} f, \tag{6.132}$$

whereas for a Fermi–Dirac it becomes

$$\frac{df}{dE} = -\frac{1}{k_B T} f(1 - f). \tag{6.133}$$

What makes the nonequilibrium warm electron remain in the Fermi–Dirac form is the assumption that the electron (or hole) temperature remains almost constant across the energy range in which the distribution transitions from 1 to 0.

6.5.1 The energy dissipation equation

The equation will now be developed for the energy loss rate for the above distribution function of scattering rates that are nearly elastic—the energy exchange is small compared to the thermal energy. These scattering rates depend upon e.g. the phonon momentum, for which this momentum will be described for a momentum normal to the two-dimensional plane Q and a momentum in the two-dimensional plane q. Within the plane, the scattering momentum may be related to the particle momentum through [8]

$$q = 2k\cos(\vartheta/2), \tag{6.134}$$

where ϑ is the scattering angle between k and k', which both lie essentially on the energy shell within the above approximations—the energy width of this shell, given by the thermal energy, is larger than the energy exchange. The contribution to the scattering rate for a Q normal to the two-dimensional plane is proportional to the form factor

$$| I(Q)|^2 = \left| \int dz \mid \psi(z) \mid^2 e^{iQz} \right|^2 . \quad (6.135)$$

From this, the scattering rate, as a function of the angle is [40]

$$\Gamma_\pm(\vartheta) = \int dQ \mid I(Q)|^2 \frac{m^* V}{2\pi\hbar^3} \mid M \mid^2 S^2(Q)[1 - f(E \pm \hbar\omega)]\left(N + \frac{1}{2} \mp \frac{1}{2}\right). \quad (6.136)$$

Here, $| M |^2$ is the square of the scattering matrix element (minus the Bose–Einstein factor), and $S(Q)$ is the screening function. The upper sign is for phonon absorption and the lower sign for phonon emission. The energy loss rate is now given by

$$e\mu F^2 = \frac{1}{E_F} \int dQ \mid I(Q)|^2 \left\langle \frac{m^* V}{2\pi\hbar^3} \mid M \mid^2 S^2(Q)\hbar\omega\Phi(\hbar\omega) \right\rangle_\vartheta , \quad (6.137)$$

where

$$\begin{aligned}\Phi(\beta) = \int dE\{&(N + 1)f(E)[1 - f(E - \beta)] \\ &- Nf(E)[1 - f(E + \beta)]\}\end{aligned} , \quad (6.138)$$

and Φ is merely shorthand for the messy integral of (6.138), but has no obvious physical meaning. On substituting the Bose–Einstein function for N and the Fermi–Dirac function for f, it can be shown that [40]

$$\Phi(\beta) = N\left(\frac{\beta}{k_B T_e}\right) - N\left(\frac{\beta}{k_B T}\right). \quad (6.139)$$

Using this result in (6.137) gives

$$e\mu F^2 = F(T_e) - F(T), \quad (6.140)$$

where

$$F(T_i) = \frac{1}{E_F} \int dQ \mid I(Q)|^2 \left\langle \frac{m^* V}{2\pi\hbar^3} \mid M \mid^2 S^2(Q)\hbar\omega N\left(\frac{\hbar\omega}{k_B T_i}\right) \right\rangle_\vartheta . \quad (6.141)$$

where the angular average is to be evaluated at the Fermi energy.

6.5.2 Scattering exponents

The three-dimensional matrix element for acoustic modes may be found in many textbooks. The form to be used here is for parabolic energy bands for which the overlap integral is unity. Then, the three-dimensional matrix element of (6.136) is [9]

$$| M |^2 = \frac{\hbar E_1^2(q^2 + Q^2)}{2\rho_m V}. \tag{6.142}$$

With this matrix element, the angle average becomes

$$\langle\ldots\rangle_\vartheta = \frac{1}{\pi}\frac{m^* E_1^2}{4\pi\rho_m \hbar^3}\int_0^\pi (q^2 + Q^2)S^2(q)(\hbar\omega)^2 N\left(\frac{\hbar\omega}{k_B T_i}\right)\delta(E - E')d\vartheta. \tag{6.143}$$

At this point, we have to look at the dispersion relation for the phonons, which is

$$\omega = v_s\sqrt{q^2 + Q^2}. \tag{6.144}$$

The conservation of energy gives (6.134), by which q goes from 0 to $2k_F$ as ϑ goes from 0 to π. In the equipartition limit, we can expand the Bose–Einstein function as

$$N\left(\frac{\hbar\omega}{k_B T_i}\right) \sim \frac{k_B T_i}{\hbar\omega}, \tag{6.145}$$

where T_i stands for the electron or lattice temperature in (6.140), and (6.143) becomes

$$\langle\ldots\rangle_\vartheta = \frac{1}{\pi}\frac{m^* E_1^2 v_s k_B T_i}{4\pi\rho_m \hbar^2}\int_0^\pi (q^2 + Q^2)^{3/2} S^2(q)\delta(E - E')d\vartheta. \tag{6.146}$$

In the unscreened case, $S = 1$, so that using the properties of the delta function gives us

$$\langle\ldots\rangle_\vartheta = \frac{1}{\pi}\frac{m^* E_1^2 v_s k_B T_i}{4\pi\rho_m \hbar^2}(4k_F^2 + Q^2)^{3/2}. \tag{6.147}$$

This can now be used in (6.141) to give

$$F(T_i) = \frac{1}{E_F}\frac{1}{\pi}\frac{m^* E_1^2 v_s k_B T_i}{4\pi\rho_m \hbar^2}\int dQ(4k_F^2 + Q^2)^{3/2} \,| I(Q)|^2. \tag{6.148}$$

Generally, the wave function is normalized although, one typically has a characteristic distance [40]

$$\frac{1}{z} = \frac{1}{\pi}\int dQ \,| I(Q) |^2. \tag{6.149}$$

Now, to normalize the overall integral (6.148), reduced variables are introduced, following the form of (6.145), as

$$\xi = \frac{\hbar v_s k_F}{k_B T_i}, \quad \kappa = \frac{\hbar v_s Q}{k_B T_i}, \tag{6.150}$$

and

$$F(T_i) \sim \frac{1}{E_F}\frac{m^* E_1^2 b k_B T_i^5}{4\pi\rho_m \hbar^6 v_s^3} \int d\kappa (4\xi^2 + \kappa^2)^{3/2} \mid I(\kappa) \mid^2 . \tag{6.151}$$

The integral is merely a numerical factor that depends upon the exact nature of the wave functions in the z direction. But, it may be seen that the factor p from

$$e\mu F^2 \sim T_e^p - T^p, \tag{6.152}$$

discussed in chapter 3, has the value 5 for acoustic mode energy relaxation.

In piezoelectric scattering, the dependence on momentum is quite different. The ratio of the matrix elements for piezoelectric and deformation potentials, in addition to different coupling constants has [8]

$$\frac{\mid M \mid_{pz}^2}{\mid M \mid_{ac}^2} \sim \frac{1}{(q^2 + Q^2)} \tag{6.153}$$

in the unscreened case and

$$\frac{\mid M \mid_{pz}^2}{\mid M \mid_{ac}^2} \sim \frac{(q^2 + Q^2)}{\left(q^2 + Q^2 + q_D^2\right)^2} \sim \frac{1}{q_D^4}(q^2 + Q^2) \tag{6.154}$$

for the strongly-screened case. For strong screening, the Debye screening wave number is usually much larger than the scattering wave number, so that the value of p is reduced to 3 (the 3/2 power in (6.148) is reduced to 1/2) for the unscreened piezoelectric scattering and raised to 7 for the strongly-screened piezoelectric scattering (the 3/2 power in (6.148) is increased to 5/2). In figure 6.12, a comparison of acoustic phonon and piezoelectric scattering is shown for a GaAs/AlGaAs interface quasi-two-dimensional electron system [41]. Because the piezoelectric scattering is anisotropic (in the coupling constant), and the piezoelectric effect is described by a tensor, the authors have separated the longitudinal (lPZ) and the transverse (tPZ) scattering rates (relative to the electric field). The carrier density is 5.5×10^{11} cm^{-2}, and the phonons are taken to have a lattice temperature at the mK level, well below any of the electron temperatures observed. Clearly, above about 1.5 K, the acoustic modes begin to dominate the piezoelectric modes. In figure 6.13, the measured energy loss rate is plotted against $T_e^3 - T^3$, for quasi-two-dimensional electrons at the interface of a GaAs/AlGaAs heterostructure at 1.8 K [42]. Once again, it is clear that the data fit $p = 3$, as expected for unscreened piezoelectric scattering. But, it may be noticed from the right-hand scale that the electron temperature only goes up to about 6 K. Data is taken from two samples: one with $n = 6.75 \times 10^{11}$ cm^{-2} and mobility of 1.49×10^5 cm^2 Vs^{-1}, and the other with $n = 4.5 \times 10^{11}$ cm^{-2} and mobility of 4.4×10^3 cm^2 Vs^{-1}.

It should not be assumed that the energy loss to the piezoelectric interaction always dominates the system. It can change due to many different effects. In figure 6.14, the energy loss rate for electrons is plotted versus $T_e^3 - T^3$, for three

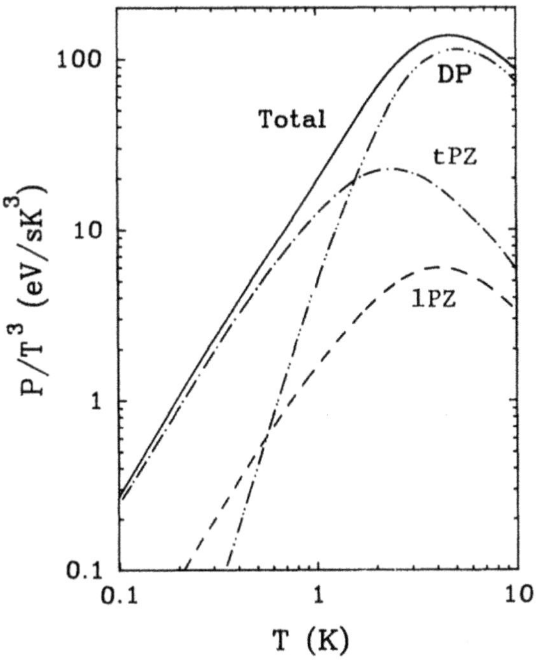

Figure 6.12. The calculated energy loss rate, divided by the cube of the electron temperature, as a function of the electron temperature. The various dashed lines show the contributions from acoustic deformation potential (DP), longitudinal (l) piezoelectric (lPZ), and transverse (t) piezoelectric (tPZ) scattering. Reprinted with permission from Ma *et al* [41], copyright (1991) by the American Physical Society.

different samples at various magnetic fields [43]. The samples again include electrons at the interface of a GaAs/AlGaAs heterostructure, with densities in the range 1.7–7.2 × 10^{11} cm^{-2} and mobilities in the range 1.03–0.75 × 10^6 cm^2 Vs^{-1} at 4 K, although the data in the figure comes from a base temperature of 0.5 K. It appears that the scattering is dominated by piezoelectric scattering at low electron temperatures, below 2–6 K, depending upon the magnetic field. At higher temperatures, the value of p increases dramatically, and apparently with different values. At the lowest magnetic fields, this may be due to screened piezoelectric scattering, while at higher magnetic fields, it looks more like acoustic deformation potential scattering. In any case, the authors believe that the enhanced scattering arises from inter-Landau level scattering.

6.5.3 Graphene

While graphene is a two-dimensional material, its unique band structure causes it to have different properties with constant electron (or hole) velocity $v_F \sim 10^8$ cm s^{-1}, and a different density of states

$$\rho_{gr}(E) = \frac{2E}{\pi(\hbar v_F)^2} \tag{6.155}$$

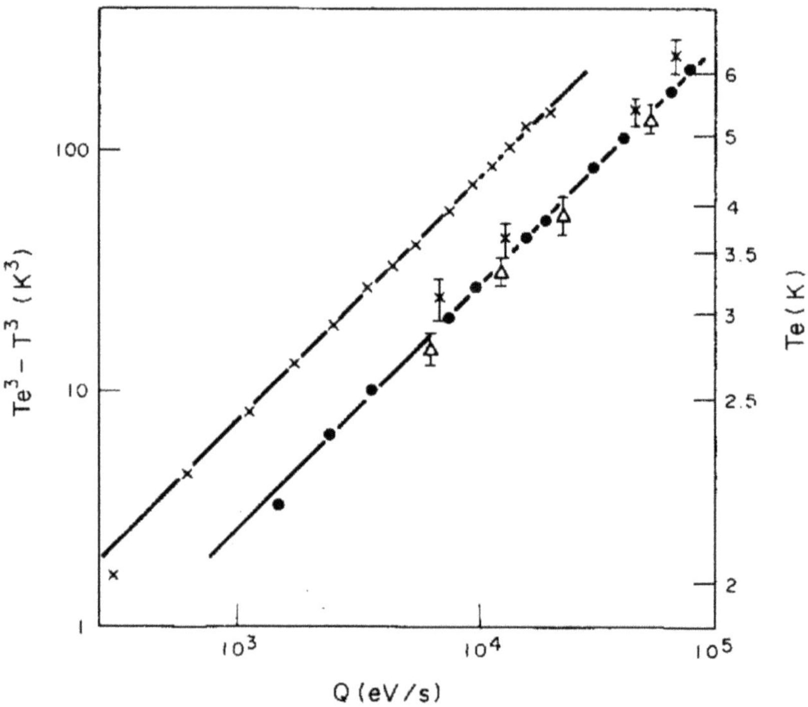

Figure 6.13. The quantity $T_e^3 - T^3$ plotted against the power loss per electron for two different samples (given in the text). Data was obtained from Shubnikov–de Haas (crosses), magnetoconductivity at low magnetic field (triangles), and temperature dependent conductance (dots). Reprinted from Kreschuk *et al* [42], copyright (1988) with permission from Elsevier.

that is different from other two-dimensional materials. Moreover, there is no momentum normal to the plane of the graphene, so some of the integrals above do not appear. Nevertheless, the relation (6.140) remains valid with the function now given by [44]

$$F(T_i) = \frac{E_1^2 E_F}{n_s \rho_m \pi^2 \hbar^4 v_s^2 v_F^3} \int_0^\infty dq (\hbar\omega)^3 \sqrt{1 - \left(\frac{\gamma}{E_F}\right)^2} N\left(\frac{\hbar\omega}{k_B T_i}\right). \tag{6.156}$$

Here, $\gamma = \hbar v_F q/2$ and must be smaller than the Fermi energy, as is obvious from the square root term. The other symbols have their normal meaning [9]. In the limit of small q, the Bose–Einstein function can be approximated as in (6.145) and, after normalizing as done (6.150), the integral is a simple ζ-function, and (6.156) becomes [44]

$$F(T_i) = \frac{6E_1^2 E_F (k_B T_i)^4}{n_s \rho_m \pi^2 \hbar^5 v_s^3 v_F^3} \zeta(4). \tag{6.157}$$

Now, it is apparent that $p = 4$ for acoustic deformation potential scattering in graphene. However, as was seen in figure 3.29, this does not fit with the observed value for energy loss rate in high electric field behavior in graphene. This will be discussed further in the section below on plasmon scattering.

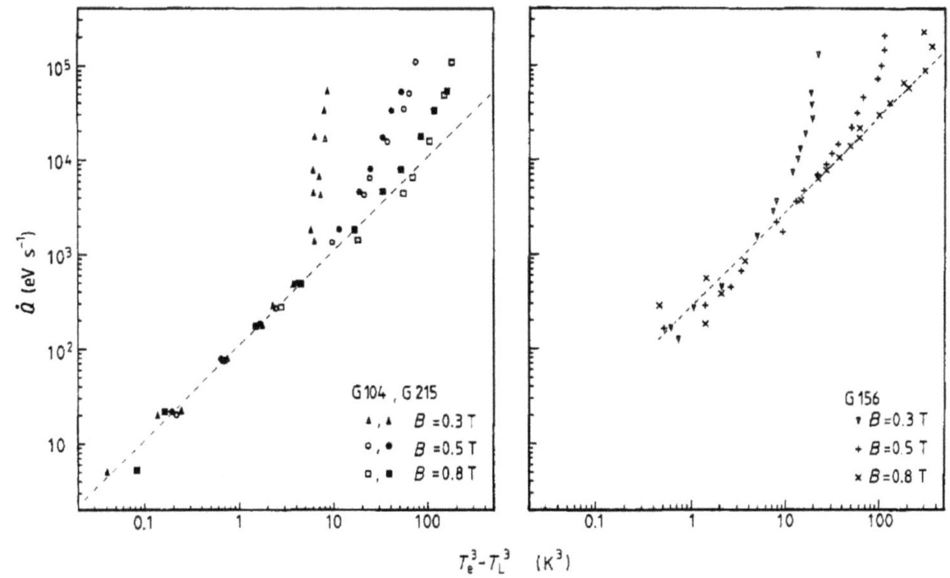

Figure 6.14. Dependence of the power loss per electron on $T_e^3 - T^3$, for three different samples at different magnetic fields. The broken lines are guides to the eye for the given value of p. At low electron temperatures, the curves follow the piezoelectric expectation, but break at higher temperatures. The break appears earlier for lower magnetic field. Reprinted from Leadley *et al* [43], copyright IOP Publishing. All rights reserved.

Nevertheless, some experiments have found $p = 4$ in their evaluation of the power loss. Using mid-infrared lasers in ultrafast optical studies of graphene, it was found that the cooling was dominated by loss to the acoustic deformation potential modes of the graphene lattice, with the proper dependence on electron temperature [45]. Then, the study of graphene with small electric fields applied, and measurements on the GHz noise emitted from the sample [46]. At sufficiently high bias voltage V, they found that the noise temperature varied as $T_e \sim \sqrt{V}$, which they conclude is appropriate to the two-dimensional phonons in graphene. In figure 6.15, the noise temperature is plotted as a function of the electric field for a variety of densities of both electrons and holes. Also shown is a $V^{2/5}$, which would be appropriate for a T_e^5 behavior, so the value of $p = 4$ is the best fit. While the experiment uses an electric field, the noise signal is quite likely dominated by the carriers in the main body of the distribution—the energy distribution function, where the energy loss is expected to be dominated by the acoustic phonons. This will be discussed further in the following discussion of plasmon scattering.

6.6 Plasmon interactions

The interaction among electrons or holes, or even each other, is a complicated and long-range process. Certainly, the interaction is a typical Coulomb interaction as has been known for a great many years. Indeed, the optical dielectric constant differs from that of free space due to the contributions of the valence electrons and their plasmon description (as will be shown in appendix A). By plasmon is meant the

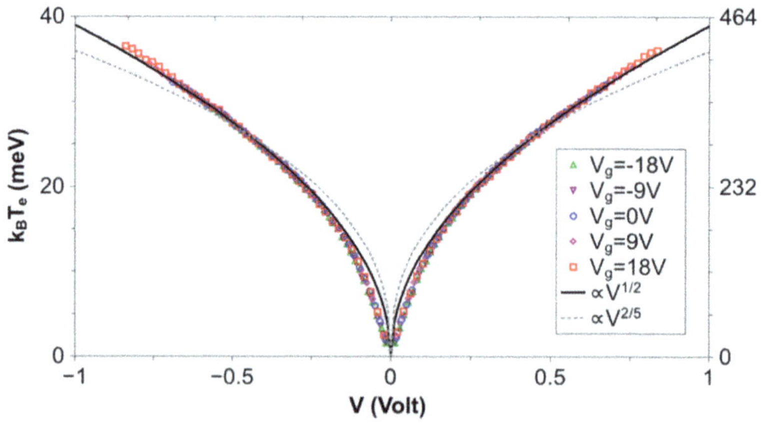

Figure 6.15. Effective carrier temperature in graphene, determined from the microwave noise measurements, as a function of the bias voltage, for a variety of gate voltages. The two lines (solid and dashed) correspond to different dependences on the gate voltages, and to T_e^4 and T_e^5, respectively. Only the former fits the data. Reprinted with permission from Betz *et al* [46], copyright (2012) by the American Physical Society.

long-range part of the Coulomb interaction. Typically, the Coulomb interaction depends upon q, in momentum space, as $1/q^2$. In scattering integrals such as (6.156), this leads to an unfortunate divergence as $q \rightarrow 0$, a well-known limitation. To counter this, it is common to assume that the Coulomb potential itself is screened by the other electrons, either given a Debye screening length [47] or a Fermi–Thomas screening length [48, 49]. When the electron–electron interaction is considered through a screened potential, then the scattering rate can be computed as indicated in figure 6.7. But, this is not the complete carrier–carrier interaction!

Generally, in order to treat the complete carrier–carrier interaction, the entire dielectric function must be considered [9]. Within the linear response formalism, this dielectric function is the Lindhard dielectric function [50] (the Lindhard dielectric function and its various approximations relevant to this section will be discussed in appendix A). From this, the various contributions to the dielectric function can be obtained, including those of the valence electrons [9]. Certainly, this linear response function is good for most of the applications of interest here, but certainly it does not describe systems that are nonlinear in the polarization, such as those of interest in nonlinear optics. Nor does it provide much insight into biological liquids, such as water whose polarizability has been the topic of decades of work. Nevertheless, the Lindhard function describes well the interaction of carriers with each other, although it will be seen that the full momentum-dependent dynamic screening function is often required.

The full interaction is described as well by Bohm and Pines [51], in which a cut-off wave number q_c is introduced. This cut-off corresponds reasonably well to the inverse of the Debye length

$$\lambda_D = \sqrt{\frac{\epsilon k_B T_e}{ne^2}}, \tag{6.158}$$

where n is the carrier density and ϵ is the dielectric function. Corresponding to (6.158), there is a Debye wave vector $q_{\mathrm{D}} = 1/\lambda_{\mathrm{D}}$. The introduction of the cut-off wave number splits the interaction into two parts, which may roughly be defined as a short-range ($q > q_{\mathrm{D}}; r < \lambda_{\mathrm{D}}$), and a long-range part ($q < q_{\mathrm{D}}; r > \lambda_{\mathrm{D}}$). Obviously, the short-range interaction is described by the screened Coulomb interaction, which is often called the *single-particle* interaction. The importance of the long-range part of the interaction is that a given carrier interacts with a great many other carriers at the same time, and this is usually described as the carrier interacting with the *collective modes* of the carrier gas—the plasmons, with the plasmon frequency given by (in three dimensions)

$$\omega_p^2 = \frac{ne^2}{m^* \epsilon}. \tag{6.159}$$

The connection between these two parts is governed by the dielectric function and its singularities [9]. The separation into the two parts allows each part to be treated separately with the Fermi golden rule as two different scattering processes. The strength of the interaction is governed by the imaginary part of the reciprocal of the dielectric function as

$$\mathrm{Im}\left\{\frac{1}{\epsilon(q, \omega)}\right\}. \tag{6.160}$$

One might think that the strength of the plasmon interaction would be given by the so-called plasmon-pole approximation

$$\epsilon = \epsilon_s\left(1 - \frac{\omega_p^2}{\omega^2}\right), \tag{6.161}$$

where ϵ_s is the high-frequency permittivity of the semiconductor. The important point is that scattering broadens the plasmon pole, and it is this broadened function that becomes important, as well as the momentum dependence of the dielectric function. Hence the momentum dependence must be kept in order to properly assess the scattering (beyond the plasmon-pole approximation).

When the full dielectric function is retained through the Lindhard form, the important terms for three-dimensional plasmon scattering can be identified as [9]

$$\epsilon(q) = \epsilon(0)\left[1 - \frac{\omega_p^2}{(\mathbf{q} \cdot \mathbf{v})^2}\right], \tag{6.162}$$

where $\epsilon(0)$ is the low-frequency permittivity and includes the phonon contribution to the polarization in compound semiconductors, and v is the carrier velocity. The inverse dielectric function can now be written as

$$\frac{1}{\pi}\mathrm{Im}\left\{\frac{1}{\epsilon(q)}\right\} = \frac{1}{2\epsilon(0)}\left[\delta(\mathbf{q} \cdot \mathbf{v} + \omega_p) + \delta(\mathbf{q} \cdot \mathbf{v} - \omega_p)\right], \tag{6.163}$$

which includes terms for both the emission and absorption of a plasmon by the carrier. This may now be used in the Fermi golden rule, which leads to an integral

$$\Gamma_{pl} = \frac{1}{2\pi\hbar}\int q^2 dq\left(\frac{\hbar\omega_p}{q^2}\right)\int \sin\vartheta d\vartheta \frac{1}{\pi}\mathrm{Im}\left\{\frac{1}{\epsilon(q)}\right\}. \tag{6.164}$$

The integration over the angles involves the delta functions and leads to two terms, including the Bose–Einstein occupation functions for the plasmons as bosons, and [52]

$$\begin{aligned}\Gamma_{pl} = \frac{ne^2\omega_p}{4\pi\hbar^2 k}&\left\{N\ln\left(\frac{q_D/k}{\sqrt{1+\hbar\omega_p/E}-1}\right)\right. \\ &\left.+(N+1)\ln\left(\frac{q_D/k}{1-\sqrt{1-\hbar\omega_p/E}}\right)\right\}\end{aligned}. \tag{6.165}$$

The form is completely reminiscent of that for optical phonon scattering and the limits on the integrations arise in the same way except for the cut-off functions. In high electric fields, it is found that when the plasmon temperature and the free-carrier temperature are computed separately, they nevertheless converge to the same value [53], as one would expect.

6.6.1 Low-dimensional material

Carrier–carrier scattering through single-particle scattering conserves energy and momentum between the two interacting particles. Energy relaxation processes must occur through the plasmon scattering, as discussed above. This is true regardless of the dimensionality, but new concepts and problems arise in low-dimensional systems. The three-dimensional system, discussed above, has a well-defined plasmon mode with a frequency that depends mainly on the electron density, and only weakly on the momentum. And, this frequency remains well defined even when $q = 0$. In low-dimensional systems, this is no longer the case. In two-dimensional systems, the density becomes a sheet density, and formulas such as (6.159) are no longer dimensionally correct. In one dimension, it becomes even more acute.

In order to bring (6.159) into the proper dimension, the loss of a factor of cm^{-1} due to the sheet density appearing in the formula means that the plasmon definition will have to involve the momentum q. This falls naturally out of the Lindhard dielectric function [9], and the *two-dimensional* plasmon frequency is given by [54]

$$\omega_p^2 = \frac{n_s e^2}{2m^*\epsilon(0)}q \equiv \chi_2 q. \tag{6.166}$$

Here, χ_2 is the two-dimensional susceptibility, and n_s is the two-dimensional sheet density of carriers. As the momentum vector goes to zero, the plasma mode disappears. Even for a non-zero momentum vector, the plasma frequency is a dispersive quantity and the plasmon-pole approach becomes a nebulous concept, and certainly not one for use in scattering theory. Consequently, a more complete dielectric function is required, somewhat more involved than that of the last section. As a result, both frequency (energy) and momentum variables will have to be retained in the computation, not least because the so-called plasmon frequency entering into the emission and absorption calculation will be momentum dependent.

Another important fact about the following equations is that the lower limit for the momentum integration cannot be set to zero, as this integral will then diverge. A lower limit is taken from a natural cutoff of the interaction, that arises from the inverse of the mean-free path, the longest allowed real space quantity in the integral. This follows the assumption that other scattering would break up any process that would lead to plasmon scattering with a smaller wave number. Thus, the minimum momentum wave number is $q_{\min} \sim 1/v\tau = 1/\sqrt{2D\tau}$, in the assumption that the carriers are not too hot.

The process of energy loss by an energetic electron in a quasi-two-dimensional electron gas is similar to that of the previous paragraph, except that the plasmon energy will likely be small, so that we do not need to differentiate between emission and absorption. Then, we can write the Bose–Einstein factors as

$$N_{pl} + \left(1 + N_{pl}\right) = \coth\left(\frac{\hbar\omega}{2k_{\mathrm{B}}T_e}\right). \tag{6.167}$$

No subscript is used on this frequency, as the actual plasmon frequency is dispersive and depends upon the actual scattering wave momentum as given in (6.166). We can now write the scattering rate in analogy with (6.163) as

$$\begin{aligned}\Gamma_{pl}(k) &= \frac{2\pi}{\hbar}\sum_q \int_{-\infty}^{\infty} \frac{d\omega}{2\pi}\coth\left(\frac{\hbar\omega}{2k_{\mathrm{B}}T_e}\right)\frac{e^2}{2q} \\ &\times \mathrm{Im}\left(\frac{1}{\epsilon(\mathbf{q},\omega)}\right)\delta\left(\omega - \frac{E(k+q) - E(k)}{\hbar}\right)\end{aligned}. \tag{6.168}$$

Here, it should be noted that the actual Coulomb potential is given by [54]

$$\frac{e^2}{2q\epsilon(\mathbf{q},\omega)}, \tag{6.169}$$

so that the scattering arises from the imaginary part of this interaction. The dielectric function comes from the momentum-dependent screening, and may be approximated as [54]

$$\epsilon(\mathbf{q},\omega) = \epsilon(0)\left[1 - \frac{\chi_2 q\tau^2}{(1 + i\omega\tau)^2 + Dq^2\tau}\right]. \tag{6.170}$$

In the low frequency limit, the imaginary part of the inverse dielectric function, along with the bare Coulomb potential, may be written as

$$V(q)\mathrm{Im}\left(\frac{1}{\epsilon(\mathbf{q},\,\omega)}\right) \sim -\frac{e^2}{2\epsilon(0)q}\frac{2\omega}{Dq^2+\chi_2 q\tau}. \tag{6.171}$$

Two possible limits can be found from (6.168), depending upon the size of the plasmon energy relative to the thermal energy. First, it will assumed that the argument of the hyperbolic cotangent is small (high temperature), so that it may be expanded as

$$\coth\left(\frac{x}{2}\right) \sim 2/x, \tag{6.172}$$

and the integral over the energy can be written as

$$\begin{aligned}\int_{-\infty}^{\infty}\frac{d\omega}{2\pi}\frac{e^2}{2\epsilon(0)\hbar q}\frac{4k_BT_e}{Dq^2+\chi_2 q\tau}\delta\left(\omega-\frac{E(k+q)-E(k)}{\hbar}\right)\\ =\frac{e^2k_BT_e}{\pi\hbar\epsilon(0)q(Dq^2+\chi_2 q\tau)}\end{aligned}. \tag{6.173}$$

With this result, the scattering rate can be evaluated, using the lower cutoff momentum value discussed above, to be

$$\begin{aligned}\Gamma_{pl}(k) = \frac{2\pi}{\hbar}\sum_q\frac{e^2k_BT_e}{\pi\hbar\epsilon(0)q(Dq^2+\chi_2 q\tau)} = \frac{e^2k_BT_e}{\pi\hbar^2\epsilon(0)}\int_{q_{\min}}^{\infty}\frac{dq}{Dq^2+\chi_2 q\tau}\\ \sim\frac{e^2k_BT_e}{\pi\hbar^2\epsilon(0)\chi_2\tau}\ln\left(1+\chi_2\tau\sqrt{\frac{2\tau}{D}}\right)\end{aligned}. \tag{6.174}$$

It should be noticed that this scattering rate has a characteristic role-off with temperature as $\tau_{pl} \sim 1/T$, which is quite typical, and seen experimentally, for the dephasing rate of the electrons in a quasi-two-dimensional semiconductor at low temperature, as discussed in chapter 3.

At very low temperatures, the large argument limit of the hyperbolic cotangent may be taken, and in this case

$$\coth\left(\frac{x}{2}\right) \sim 1. \tag{6.175}$$

Then, the result will be independent of the electron temperature, and that is seen very clearly at the lowest temperatures. With this approximation, (6.168) becomes, using (6.171),

$$\Gamma_{pl}(k) = \frac{2\pi}{\hbar}\sum_q\int_{-\infty}^{\infty}\frac{d\omega}{2\pi}\frac{e^2}{2\epsilon(0)q}\frac{2\omega}{Dq^2+\chi_2 q\tau}\delta\left(\omega-\frac{E(k+q)-E(k)}{\hbar}\right). \tag{6.176}$$

To simplify the process, the angular integration part of the summation over momentum will be done first, and this leads to

$$\int d\vartheta\delta(\omega - qv\cos\vartheta) = \frac{1}{\sqrt{q^2v^2 - \omega^2}}, \tag{6.177}$$

which then gives

$$\Gamma_{pl}(k) = \frac{2\pi}{\hbar}\int\frac{qdq}{4\pi^2}\int_{-\infty}^{\infty}\frac{d\omega}{2\pi}\frac{e^2}{\epsilon(0)q}\frac{2\omega}{Dq^2 + \chi_2 q\tau}\frac{1}{\sqrt{q^2v^2 - \omega^2}} \quad . \tag{6.178}$$

The frequency integration will diverge at the upper limit, although a natural cut-off at $1/\tau$ can be imposed without loss of generality. Then, the result after this integration is

$$\Gamma_{pl}(k) = \frac{e^2}{4\pi^2\epsilon(0)\hbar v\tau}\int\frac{dq}{q^2(Dq + \chi_2\tau)}\left[qv - \sqrt{q^2v^2 - (1/\tau)^2}\right]. \tag{6.179}$$

Expanding the square root term gives the approximate result

$$\Gamma_{pl}(k) = \frac{e^2}{8\pi^2\epsilon(0)\hbar v\tau\chi_2}\left[\sqrt{\frac{D}{\tau}} + \frac{2D}{\chi_2\tau^2}\ln\left(\frac{\sqrt{D\tau}}{\chi_2}\right)\right]. \tag{6.180}$$

In this result, there is no temperature, as discussed above, so this low temperature limit is independent of temperature as found in chapter 3.

In figure 6.16, the measured phase-breaking time in a pair of open quantum dots (non-tunneling transport) found in a GaAs/AlGaAs heterostructure at low temperature is plotted [55]. To get the effective electron temperature, the resistance of the structure was measured as a function of lattice temperature and then as a function of the applied electric field, measured at the lowest temperature. This gives an effective electron temperature at each electric field, and thus provides the thermometer for the system. The electron density in the array was $3.6 \pm 0.1 \times 10^{11}$ cm^{-2}. In the figure, it is clearly seen that the phase-breaking time falls with 1/T at higher temperatures and is constant at low temperature as expected from the above discussions. While the open quantum dot may be considered as almost a zero-dimensional system, the fact that they are open precludes this assumption. In fact the closed dots are closed systems and there will be no loss, nor any applied transport. In the open dots, the carriers within the dots are actually losing their energy via interaction with carriers outside the dots in the two-dimensional material in which the dots are embedded.

In the case of a nanowire, which is a quasi-one-dimensional system, a different form of the dielectric function will appear. While one thinks of these as one-dimensional, they are usually created by a narrow quasi-two-dimensional system by the use of, for example, lateral gates to create a narrow constriction for transport in the two-dimensional system. In such a case, the one-dimensional density is created from the two-dimensional density by $n_l = n_s W$, where W is the width of the constriction. The Fermi level is usually then set by that of the 2D regions away from

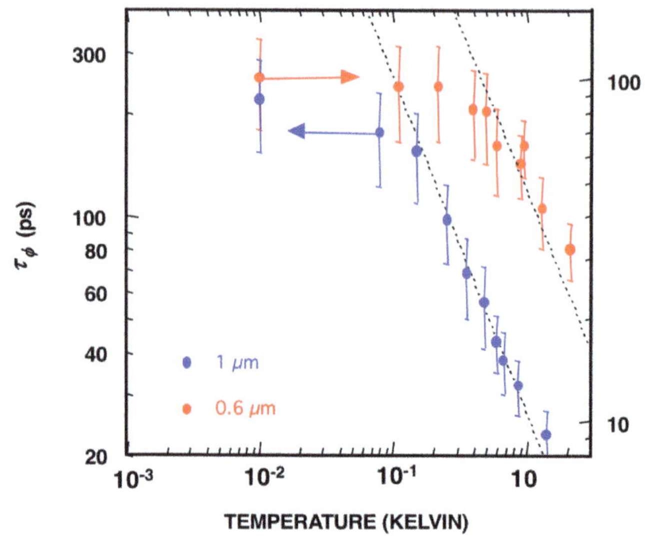

Figure 6.16. The measured phase-breaking time in two different open quantum dots of different dimension as a function of carrier temperature. The dotted lines are guides to the eye for a 1/T variation in the phase-breaking time. Reprinted with permission from Ferry *et al* [55], copyright IOP Publishing. All rights reserved.

the constriction. Within a quasi-one-dimensional nanowire, the Coulomb interaction is further modified from the normal form to be [54]

$$V(q) = \frac{e^2}{4\pi\epsilon(0)\ln(1 + q_0^2/q^2)}. \tag{6.181}$$

The multi-sub-band possibilities will be ignored and consider only the transport in the lowest sub-band. The logarithmic factor in (6.181) will eventually cancel with an equivalent factor, and so it will drop out in the following development. In one dimension, the plasmon dispersion curve becomes

$$\omega_p^2 = \frac{n_1 e^2}{4\pi m^* \epsilon(0)} q^2 \equiv \chi_1 q^2. \tag{6.182}$$

Then, the momentum-dependent dielectric function becomes

$$\epsilon(\mathbf{q}, \omega) = \epsilon(0)\left[1 - \frac{\chi_1 q^2 \tau^2}{(1 + i\omega\tau)^2 + Dq^2\tau}\right]. \tag{6.183}$$

With this dielectric function, the potential is now

$$V(q)\mathrm{Im}\left(\frac{1}{\epsilon(\mathbf{q}, \omega)}\right) \sim -\frac{e^2}{2\pi\epsilon(0)q^2}\frac{\omega}{D + \chi_1\tau}. \tag{6.184}$$

Here, χ_1 is the one-dimensional susceptibility. This may now be used in (6.168). In evaluating (6.168) for the one-dimensional case, it will be more convenient to carry out the integration over q prior to the integration over the frequency. The first integration still involves the delta function and some cutoffs for the limits of integration on the subsequent frequency integration will be required. In fact, frequencies that lie between $1/\tau_{pl}$ and $1/\tau$ (going from the lower frequency to the higher one) are of interest. The momentum integration is now

$$\sum_q \frac{e^2\omega}{2\pi\epsilon(0)q^2(D+\chi_1\tau)}\delta\left(\omega - \frac{E(k+q)-E(k)}{\hbar}\right) = \frac{m_*e^2}{2\pi\hbar\epsilon(0)(D+\chi_1\tau)}\sqrt{\frac{\hbar}{2m_*\omega}}. \tag{6.185}$$

The high temperature regime will be pursued first, using (6.172), and

$$\Gamma_{pl}(k) = \frac{e^2k_BT}{\hbar^{3/2}\sqrt{2m_*}\,\epsilon(0)(D+\chi_1\tau)}\sqrt{\tau_{pl}}. \tag{6.186}$$

In the result, the two cutoff frequencies have been used, although only the lower one was used to avoid the divergence. The reciprocal of the left-hand side is the same term in the square root. Solving for the electron–plasmon scattering rate gives

$$\Gamma_{pl} = \frac{1}{\tau_{pl}} = \left[\frac{e^2k_BT}{\hbar^{3/2}\sqrt{2m_*}\,\epsilon(0)(D+\chi_1\tau)}\right]^{2/3}. \tag{6.187}$$

This shows that the plasmon induced dephasing time varies as $T^{-2/3}$, which is different from the rate in other dimensionalities. This resulting temperature dependence has been found in other approaches as well [56, 57].

Turning to the low temperature limit, and using (6.175) changes the last frequency integral to become

$$\Gamma_{pl} = \frac{e^2}{2\pi\sqrt{2\hbar m_c\tau}\,\varepsilon(0)(D+\chi_1\tau)}. \tag{6.188}$$

just as in the quasi-two-dimensional case, the electron–plasmon scattering time becomes independent of the temperature at very low temperatures, as found in experiment [58, 59]. This functional form has also been found by [57].

6.6.2 Graphene

In graphene, as has been pointed out above, the nature of the scattering processes are somewhat more complicated due to the unusual band structure of this material. Generally, the optical phonons in graphene have a very high energy, and are not particularly important for transport except at room temperature and above [60]. At lower temperatures, some have thought that the major cooling process of carriers in the high energy elongation of the distribution function was via the acoustic phonons. Due to momentum conservation constraints, however, this process is quasi-elastic, and so is very inefficient at mediating energy transfer. And, the exponent for acoustic mode scattering doesn't match the one seen in experiment (see sections 3.5.4 and 6.5.3). One process that can explain the energy relaxation is through the emission of plasmons. In comparing results with experiments, it is convenient to express the power input per carrier, which gives also the power loss in steady state, in the form (6.140). It has already been shown above that acoustic phonons give $p = 4$, while the experiments give $p = 3$.

Basically, graphene is a two-dimensional semiconductor, but the density of states is different from normal two-dimensional semiconductors due to its Dirac-like bands. However, the starting place is still (6.171), except the term in D will be dropped, so that the leading expression is

$$V(q)\mathrm{Im}\left(\frac{1}{\varepsilon(\mathbf{q},\,\omega)}\right) \sim -\frac{e^2}{2\varepsilon(0)q}\frac{2\omega}{\chi_2 q\tau}. \tag{6.189}$$

Because the interest is in the energy loss, the development will depend upon the *difference* between the emission and absorption terms, while $\hbar\omega$ is the energy that is exchanged in each interaction with the plasmons. Then, the power loss per electron can be written as

$$P = -2\pi\sum_q\int\frac{\omega d\omega}{2\pi}\frac{e^2\omega}{\varepsilon(0)\chi_2 q^2\tau}\delta_\omega\Phi(\hbar\omega), \tag{6.190}$$

where Φ is given by (6.138), but in the case of graphene yields

$$\Phi(x) = \frac{xE_F}{\pi(\hbar v_F)^2}[N(x,\,T_e) - N(x,\,T_L)], \tag{6.191}$$

As discussed in (6.139). The remaining integrals can now be formulated. The delta function in (6.190) provides the energy conservation for the emission process in terms of the magnitudes of the various vectors, using the linear dispersion for graphene, as

$$\hbar v_F(k - |\,\mathbf{k} - \mathbf{q}\,|) = \hbar\omega = \hbar\sqrt{\chi_2 q}, \tag{6.192}$$

where the plasma frequency from (6.167) has been used. The proper dynamic mass at the Fermi energy has been used in the last term (the susceptibility). Expanding the terms on the left-hand side yields the angle variation as

$$\cos\vartheta = -\left[\frac{q}{2k} - \frac{\chi_2}{2kv_F^2} + \frac{1}{v_F}\sqrt{\frac{\chi_2}{q}}\right]. \tag{6.193}$$

The terms inside the square brackets are supposed to yield a positive quantity. The equivalent expression for the absorption term is

$$\cos\vartheta = \left[-\frac{q}{2k} + \frac{\chi_2}{2kv_F^2} + \frac{1}{v_F}\sqrt{\frac{\chi_2}{q}}\right]. \tag{6.194}$$

It is clear that there are solutions for the above equations with the proper scattering angle for emission and absorption of plasmons. In the emission case, scattering lies in a very small range of angles, of the order of 6–10 degrees from back-scattering, although direct back-scattering is forbidden by chiral symmetry. For absorption, a very large range of angles are allowed, but the momentum exchange is quite high. There is almost no density dependence for the scattering when the density is in the range 10^{11}–10^{12} cm^{-2}, but it gets stronger near the Dirac point, although the plasmons may be heavily damped in this region by the single-particle scattering. From the above results, the integrals in (6.190) may be put into the form

$$F_i(T_i) = \frac{e^2 E_F (k_B T_i)^3}{2\pi \hbar^4 v_F^2 \chi_2 \tau} \int_0^{u_{\max}} \frac{u^2 du}{e^u - 1}, \tag{6.195}$$

where

$$\begin{aligned} u &= \frac{\hbar q v_F}{k_B T_i} \\ u_{\max} &= \frac{2\hbar v_F k_F}{k_B T_i} \end{aligned}. \tag{6.196}$$

In general, the upper limit of the integral is density and temperature dependent, but the integral approaches a value of about 2.4 as the upper limit gets large. Hence, the average energy loss becomes

$$n_s e \mu F^2 = F_e(T_e) - F_L(T_L) = \frac{2.4 k_B^3 (T_e^3 - T_L^3)}{2\pi \hbar^2 v_F^2 \tau}. \tag{6.197}$$

With the unique band structure of graphene, the explicit density dependence has dropped out of this final expression for the power loss, but will return when this expression is divided by the density to get the power loss per electron. The second important point is that the loss rate is characterized by the exponent $p = 3$, consistent with experiments. The only parameter in (6.197) is the scattering time τ, and this may be determined by measuring the mobility as a function of temperature and density. However, the mobility is related to the momentum relaxation time and not just the scattering time. In general for graphene at low temperature, it has been found that the reduction of the scattering time to get the momentum relaxation time varies from about a factor of 3.95 at a density of 10^{10} cm^{-2} up to a factor of 4.85 at a density of 10^{13} cm^{-2} [61].

In figure 6.17, the computed power loss per electron are compared to the experimental values of Somphonsone *et al* [62]. Here, the measured mobility at each electron density and temperature is used to compute the relaxation time τ. This is then used to evaluate (6.197) at each data point from the experiments. Then, the computed power loss is plotted with the actual data points, and the exponent is shown to compare very favorably. Using this approach to determine the actual scattering time means that there are no adjustable parameters in the fits to the data shown in figure 6.17. Other figures showing the fit to the data are given in section 3.5.4.

Experiments were cited above that gave power loss with $p = 4$. This is possible in optical experiments, where the measurements give information on the spherically

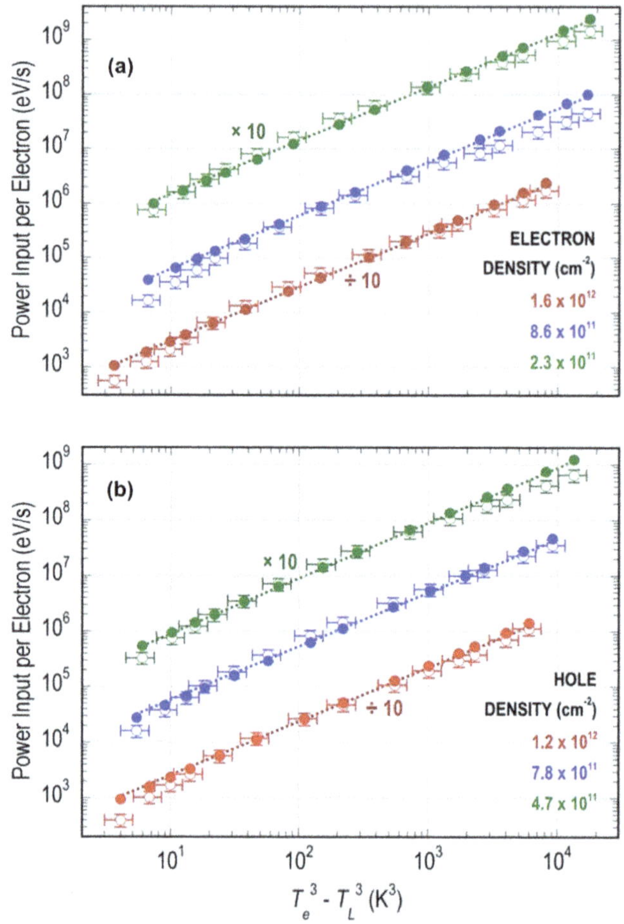

Figure 6.17. The energy loss per carrier for (a) electrons and (b) holes in graphene. In each case, the open symbols are experimental data from [62], and the solid symbols are the theory from (6.197). Each color refers to a different density of carriers, and error bars are shown for the experimental data only. Dotted lines through the theory points are guides to the eye that indicate a linear variation for with $p = 3$. Reprinted from Ferry *et al* [61], copyright (2016) with permission from Springer Nature.

symmetric part of the distribution. The plasmon discussed here affects mainly the asymmetric part of the distribution produced by the high electric field. This process drops the carriers back into the symmetric distribution, but does not convey the energy to the lattice. The latter has to be done by a phonon process, which leads to the $p = 4$ in optical experiments.

Appendix A. The Lindhard function

In this chapter, the dielectric function has been invoked several times in order to evaluate scattering processes that are dependent upon momentum and energy transfer. This dielectric function relates the various charged species, that move in response to external (or internal) potentials, and provides an idea about the screening of those potentials. In compound semiconductors, such as the III–V materials, these charged species include the atoms themselves as they have an effective charge arising from the polarization within the material. This contributes to the total polarization within the semiconductor, and it is this total polarization that differentiates the semiconductor from free space. As a result, the total dielectric function can often be written as

$$\epsilon(q, \omega) = \epsilon_0 + \delta\epsilon_L + \delta\epsilon_e, \tag{A.1}$$

where ϵ_0 is the permittivity of free space. The second and third terms of (A.1) are contributions from the lattice and from electrons (or later holes as well) to the total permittivity, respectively. In the last term, the electrons of interest are both the bonding electrons (the valence electrons) and the free carriers, and these two different sets contribute importantly to the dielectric function. Importantly, it is the bonding electrons which contribute to yield the high frequency, or *optical*, dielectric 'constant' ε_∞ of the semiconductor.

In the absence of free carriers, the permittivity must transition between its high-frequency value and the static dielectric 'constant.' In the non-polar material, there is no distinction between these two, and the dielectric constant is ε_s. But, with compound semiconductors that are of polar nature, the lattice polarization contributes to the total permittivity through [9]

$$\delta\varepsilon_L(\omega) = [\varepsilon(0) - \varepsilon_\infty]\frac{\omega_{\mathrm{TO}}^2}{\omega_{\mathrm{TO}}^2 - \omega^2}, \tag{A.2}$$

where the Lyddane–Sachs–Teller [63] condition gives

$$\frac{\omega_{\mathrm{LO}}^2}{\omega_{\mathrm{TO}}^2} = \frac{\varepsilon(0)}{\varepsilon_\infty}. \tag{A.3}$$

Generating the dielectric function is a linear, self-consistent approach using perturbation theory [64]. It begins with the assumption of a perturbing potential

$$\delta U(\mathbf{q}, \omega) = U_0 e^{i(\mathbf{q}\cdot\mathbf{r} - \omega t) + \alpha t}, \tag{A.4}$$

where α is a damping constant to account for the dissipative part of the carrier–carrier interaction, and allows one to adiabatically turn on the perturbation at $t \to -\infty$. The initial wave function state at momentum $\boldsymbol{k}$ is coupled to one at $k \pm q$ by the external potential. The sinusoidal excitation by the external potential thus picks out only a few selected couplings to the initial wave function, and the perturbed wave function is

$$\psi_k = | k\rangle + b_{k+q} | k + q\rangle + b_{k-q} | k - q\rangle. \tag{A.5}$$

The second and third terms represent the deviation from the initial state that is introduced by the perturbing potential. First-order time-dependent perturbation theory then gives the coefficients as

$$b_{k+q} = \frac{\langle k+q \mid e\delta U \mid k \rangle}{E(k+q) - E(k) + \hbar\omega - i\hbar\alpha} = \frac{eU_0 e^{-i(\omega t)+\alpha t}}{E(k+q) - E(k) + \hbar\omega - i\hbar\alpha}. \quad (A.6)$$

Only a single state couples to the initial wave function. The change in electron density that is produced by this perturbation is

$$\delta\rho = -e\sum_k f(k)(|\psi_k|^2 - 1), \quad (A.7)$$

where the squared magnitude of the wave function is the probability that the state actually occurs in the summation. The distribution function gives the occupancy of that state and the distribution of states over the momentum and/or energy. Weighting the wave function by the distribution function actually gives the density at a given value of momentum. Linear response means keeping only the lowest order in the summation (A.7), and the density perturbation becomes

$$\delta\rho = -e\sum_k f(k)\left[\frac{1}{E(k+q) - E(k) + \hbar\omega - i\hbar\alpha} + \frac{1}{E(k-q) - E(k) + \hbar\omega + i\hbar\alpha}\right]eU_0 \quad (A.8)$$
$$+ \text{c.c.}$$

This perturbation gives a self-consistent potential which is found from a solution of Poisson's equation and then is added to any applied potential V that is external to the system, as

$$U_0 = V + \Phi \equiv \frac{\epsilon_0}{\epsilon(q, \omega)} V. \quad (A.9)$$

This process gives the Lindhard dielectric function as

$$\epsilon(\mathbf{q}, \omega) = \epsilon_0 + \frac{e^2}{q^2}\sum_k f(k)\left[\frac{1}{E(k+q) - E(k) + \hbar\omega - i\hbar\alpha} + \frac{1}{E(k-q) - E(k) + \hbar\omega + i\hbar\alpha}\right] \quad (A.10)$$

The sum over the carrier distribution in (A.10) is over *all* the electrons, free carriers as well as valence electrons that form the covalent bonds. At this point, the sum over the valence electrons will be separated from the rest by splitting the summation term. The denominators in the above equations indicate that we are dealing with optical excitations between at least two energy levels. With the valence electrons, one of those levels lies in the large energy spread of the valence band, while the other energy is very high in the conduction band and approaching free space levels. These excitations are at much higher energy than the band gap itself. In some sense, the excitation is between the *average* energy of the valence band and the *average* energy of the conduction band. These excitations typically involve a momentum shift $\boldsymbol{k}$ that is the order of a reciprocal lattice vector. Consequently, the excitation gap between the two energies is much larger than the optical frequency with which we are probing the dielectric function. First, the two terms in the square bracket are combined as

$$\varepsilon(\mathbf{q}, \omega) = \varepsilon_0 + \frac{e^2}{q^2}\sum_k f(k)\left[\frac{2E(k) - E(k-q) - E(k+q)}{[E(k+q) - E(k) + \hbar\omega - i\hbar\alpha][E(k) - E(k-q) - \hbar\omega - i\hbar\alpha]}\right]. \tag{A.11}$$

The terms in the denominator may be rewritten using

$$E(k \pm q \pm G) - E(k) \sim E_{\text{gap}} \gg \hbar\omega, \tag{A.12}$$

where the gap is not the band gap, but the effective optical gap discussed above. The numerator is expanded as

$$2E(k) - E(k-q) - E(k+q) \approx -q^2\frac{\partial^2 E}{\partial k^2} + \cdots \approx -\frac{\hbar^2 q^2}{m_{\text{opt}}}, \tag{A.13}$$

where a Taylor series expansion gives the leading term, and the effective mass is that appropriate to the joint optical transition. With these approximations, (A.13) can be rewritten as

$$\epsilon(\mathbf{q}, \omega) = \epsilon_0 + \frac{e^2}{q^2}\frac{\hbar^2 q^2}{m_{\text{opt}}}\sum_k f(k)\frac{1}{E_{\text{gap}}^2} = \varepsilon_0\left[1 + \left(\frac{\hbar\omega_P}{E_{\text{gap}}}\right)^2\right] \equiv \varepsilon_\infty. \tag{A.14}$$

Here, the valence plasma frequency (note the capital P here to distinguish from the free-carrier plasma frequency) is given as

$$\omega_P^2 = \frac{Ne^2}{m_{\text{opt}}\varepsilon_0}. \tag{A.15}$$

and the last term on the right of (A.14) is the high-frequency dielectric permitivitty. The expression in square brackets is the Penn dielectric function [8]. The quantity N is the total density of valence electrons. Typically, the valence plasma frequency corresponds to an energy of the order of 10–20 eV. This is typically in the far ultraviolet range of optics. When the lattice contribution (A.2) is added, the dielectric function for free carriers, at low frequency (below the infrared response that leads to the lattice polarization) becomes

$$\epsilon(\mathbf{q}, \omega) = \epsilon(0) + \frac{e^2}{q^2}\sum_k f(k)\left[\frac{1}{E(k+q) - E(k) + \hbar\omega - i\hbar\alpha} + \frac{1}{E(k-q) - E(k) + \hbar\omega + i\hbar\alpha}\right], \tag{A.16}$$

where the first term on the right is now the low-frequency permittivity. There are many approximations that can be made to (A.16) to give static screening, the plasmon-pole approximation and so on [9]. The approximation to be pursued here is called the dynamic screening approximation.

To proceed to the dynamic screening approximation, the two terms in the square brackets of (A.16) are combined by a change in the momentum variable [9] to give

$$\epsilon(\mathbf{q},\omega)=\epsilon(0)+\frac{e^2}{q^2}\sum_k\left[\frac{f(k)-f(k+q)}{E(k+q)-E(k)+\hbar\omega-i\hbar\alpha}\right]. \tag{A.17}$$

The numerator and denominators are then expanded as

$$\begin{aligned} E(\mathbf{k}+\mathbf{q})-E(\mathbf{k}) &\sim \mathbf{q}\cdot\frac{\partial E}{\partial \mathbf{k}}+\cdots=\hbar\mathbf{q}\cdot\mathbf{v}+\cdots \\ f(\mathbf{k})-f(\mathbf{k}+\mathbf{q}) &\sim -\mathbf{q}\cdot\frac{\partial f}{\partial \mathbf{k}}+\cdots=-\hbar\mathbf{q}\cdot\mathbf{v}\frac{\partial f}{\partial E}+\cdots \end{aligned}. \tag{A.18}$$

Then, (A.17) becomes

$$\epsilon(\mathbf{q},\omega)=\epsilon(0)+\frac{e^2}{q^2k_BT}\sum\nolimits_k f(k)\frac{\mathbf{q}\cdot\mathbf{v}}{\mathbf{q}\cdot\mathbf{v}+\omega-i/\tau}. \tag{A.19}$$

The summation can be converted to an integral with the momentum (velocity) taken as the polar axis, so that the dot products introduce the polar angle. Integration over this angle is relatively complicated, so to ease this the last term is expanded as

$$\frac{\mathbf{q}\cdot\mathbf{v}}{\mathbf{q}\cdot\mathbf{v}+\omega-i/\tau}\sim 1-\left[1-\frac{\mathbf{q}\cdot\mathbf{v}}{\omega-\frac{i}{\tau}}+\frac{(\mathbf{q}\cdot\mathbf{v})^2}{\left(\omega-\frac{i}{\tau}\right)^2}+\cdots\right]. \tag{A.20}$$

In carrying out the integration, the second term in the square brackets will always vanish. In three dimensions, the summation goes to an integral in which the differential has a term sin $\vartheta d\vartheta$, while the dot product adds a cos ϑ. This integrates to zero, and similar effects occur in both one and two dimensions. On the other hand, the third term will involve the square of the cosine, which does not vanish in the integration, but adds a factor of $1/d$, where d is the dimensionality of the system. Then, the expansion in (A.20) can be reversed after integration, with the integration over the distribution function merely giving the density, with the result

$$\epsilon(\mathbf{q},\omega)=\epsilon(0)\left[1-\frac{\omega_p^2\tau^2}{(1+i\omega\tau)^2+Dq^2\tau}\right], \tag{A.21}$$

and the diffusion coefficient is $D=v^2\tau/d$, although it must be identified with an average velocity in bringing this quantity outside the integration. Note that if $\omega\to 0$, the Debye screening result is recovered, while if the limit $q\to 0$, $\omega\gg 1/\tau$, D/τ can be replaced with k_BT/m^* and the free-carrier plasma result is recovered. Hence, (A.21) satisfies the requisite limits. While this result has been obtained for a non-degenerate semiconductor, the same result will be recovered for the degenerate case [9]. Various forms of this result have been used in the discussions of scattering above.

References

[1] Hughes R C 1975 *Phys. Rev. Lett.* **35** 449

[2] Ferry D K J 1979 *J. Appl. Phys.* **50** 1422

[3] Stratton R 1958 *Proc. R. Soc. Lond.* A **246** 406
[4] Fitting H J and Frieman J U 1982 *Phys. Status Solidi* A **63** 349
[5] Fischetti M V 1984 *Phys. Rev. Lett.* **53** 1775
[6] Porod W and Ferry D K 1985 *Phys. Rev. Lett.* **53** 1775
[7] Ferry D K 1988 *The Physics and Technology of Amorphous SiO_2* ed R A B Devine (Geneva, Springer) p 365
[8] Ferry D K 1991 *Semiconductors* (New York: Macmillan) sec. 5.11.1
[9] Ferry D K 2020 *Semiconductors: Bonds and Bands* 2nd edn (Bristol: IOP Publishing)
[10] Ferry D K, Somphonsone R, Ramamoorthy H and Bird J P 2015 *J. Comp. Electron.* **15** 144
[11] Moore B T, Seiler D G and Kahlert H 1978 *Sol.-State Electron.* **21** 247
[12] Paige E G S 1964 *The Electrical Conductivity of Germanium* (New York: Wiley) vol 8 of *Progress in Semiconductors* ed A F Gibson and R E Burgess
[13] Druyvesteyn M J 1930 *Z. Phys.* **64** 790
[14] Yamashita J and Watanabe M 1954 *Prog. Theor. Phys.* **12** 443
[15] Reik H G and Risken H 1961 *Phys. Rev.* **124** 777
[16] Gnani E, Reggiani S and Rudan M 2002 *Physica* B **314** 193
[17] Fischetti M V and DiMaria D J 1988 *Phys. Rev. Lett.* **36** 629
[18] Fröhlich H and Paranjape V V 1956 *Proc. Phys. Soc. London* B **69** 21
[19] Bogoliubov N N 1962 *Studies in Statistical Mechanics* ed J de Boer and G E Uhlenbeck (Amsterdam: North Holland)
[20] Hearn C J 1965 *Proc. Phys. Soc.* **86** 881
[21] Takanaka N, Inoue M and Inuishi Y 1979 *J. Phys. Soc. Jpn.* **47** 861
[22] Ferry D K, Goodnick S M and Hess K 1999 *Physica* B **272** 538
[23] Ferry D K 1977 *Phys. Lett.* A **60** 243
[24] Zubarev D N 1974 *Nonequilibrium Statistical Mechanics* (New York: Consultants Bureau)
[25] Ferry D K 1975 *Phys. Rev.* B **12** 2361
[26] Colalongo L, Valdinoci N, Gnudi A and Rudan M 1998 *VLSI Des.* **6** 283
[27] Taschini S, Rudan M and Brunetti R 1999 *Phys. Rev.* B **60** 13582
[28] Abramowitz M and Stegun I A 1964 *Handbook of Mathematical with Formulas, Graphs and Mathematical Tables* (Washington, DC: Nat. Bur. Standards)
[29] Ferry D K 1975 *Phys. Rev.* B **12** 2361
[30] Nelson W E, Holder F A and Rosenbloom A 1966 *J. Appl. Phys.* **37** 333
[31] Kao C-H and Chen L-W 2002 *Sol.-State Electron.* **46** 915
[32] Kurosawa T 1965 *J. Phys. Soc. Jpn.* **20** 937
[33] Levinson 1965 *Sov. Phys. Sol. State* **6** 1665
[34] Dumke W P 1968 *Phys. Rev.* **167** 783
[35] Yamada E and Kurosawa T 1968 *Proc. 9th Int. Conf. Phys. Semicond.* (Moscow: Nauka)
[36] Ferry D K 1973 *Phys. Rev.* B **8** 1544
[37] Ferry D K 1978 *Surf. Sci.* **73** 136
[38] Yamada E and Kurosawa T 1973 *J. Phys. Soc. Jpn.* **34** 603
[39] Partl H, Müller W, Kohl F and Gornik E 1978 *J. Phys. C Sol. State Phys.* **11** 1091
[40] Price P 1982 *J. Appl. Phys.* **53** 6863
[41] Ma Y, Fletcher R and Zaremba E *et al* 1991 *Phys. Rev.* B **43** 9033
[42] Kreschuk A M, Martisov M Y and Polyanskaya T A *et al* 1988 *Sol. State Commun.* **65** 1189
[43] Leadley D R, Nicholas R J, Harris J J and Foxon C T 1989 *Semicond. Sci. Technol.* **4** 879
[44] Kubakaddi S S 2009 *Phys. Rev.* B **79** 075417

[45] Sun D, Divin C and Berger C *et al* 2011 *Phys. Status Solidi* C **8** 1194

[46] Betz A C, Vialla F and Brunel D *et al* 2012 *Phys. Rev. Lett.* **109** 056805

[47] Debye P and Hückel E 1923 *Phys. Zeit.* **24** 185

[48] Fermi E 1928 *Zeit. Phys.* **48** 73

[49] Thomas L H 1927 *Math. Proc. Camb. Phil. Soc.* **23** 542

[50] Lindhard J 1954 *Dan. Mat.-Fiz. Medd.* **28** 1

[51] Bohm D and Pines D 1953 *Phys. Rev.* **92** 609

[52] Lugli P and Ferry D K 1985 *Appl. Phys. Lett.* **46** 594

[53] Lugli P and Ferry D K 1985 *IEEE Electron Dev. Lett.* **6** 25

[54] Ferry D K, Goodnick S M and Bird J P 2009 *Transport in Nanostructures* 2nd edn (Cambridge: Cambridge University Press)

[55] Ferry D K, Akis R and Bird J P 2005 *J. Phys. Cond. Matt.* **17** 21017

[56] Altshuler B L, Aronov A G and Khmelnitskii D E 1982 *J. Phys. C Sol. State Phys.* **15** 7367

[57] Golobev D S and Zaikin A D 1998 *Phys. Rev. Lett.* **37** 2578

[58] Ikoma T, Odagiri T and Hirakawa K 1992 *Quantum Effects Physics, Electronics, and Applications* ed K Ismail, T Ikoma and H I Smith (Bristol: IOP Conf. Series) 127 p 346

[59] Mohanty P, Jariwala E M Q and Webb R A 1997 *Phys. Rev. Lett.* **78** 3366

[60] Fischetti M V, Kim J and Narayanan S *et al* 2013 *J. Phys. Cond. Matter* **25** 473202

[61] Ferry D K, Somphonsane R, Ramamoorthy H and Bird J P 2016 *J. Comp. Electron.* **15** 144

[62] Somphonsane R, Ramamoorthy H and Bohra G *et al* 2013 *Nano Lett.* **13** 4305

[63] Lyddane R H, Sachs R G and Teller E 1941 *Phys. Rev.* **59** 673

[64] Ziman J 1964 *Principles of the Theory of Solids* (Cambridge: Cambridge University Press)

IOP Publishing

Hot Carriers in Semiconductors

David K Ferry

Chapter 7

The ensemble Monte Carlo method

As was intrinsically obvious in the last chapter, the use of the Boltzmann transport equation and other analytical approaches, are quite difficult to evaluate carefully in the real situation of a semiconductor with non-parabolic energy bands and complicated scattering processes. Even when a first-order Legendre expansion is suitable, it still remains a problem to evaluate the scattering integrals in a real-world situation. For this reason, an alternative approach is to use the computer to completely solve the transport problem with a stochastic methodology, and several methods have arisen to do this, two of which are an integral iteration technique [1, 2] and the Monte Carlo method [3]. Today, the most widely used approach is the latter, although the Monte Carlo method began basically as a single-particle approach. The ensemble Monte Carlo (EMC) technique, which uses an ensemble of particles propagating in parallel, has been used now for more than five decades as a numerical method to simulate far-from-equilibrium transport in semiconductor materials and devices. It has been the subject of many reviews [4–7]. The methodology of the EMC approach is to spend time and effort on characterizing the correct physics, perhaps even to the use of full-zone pseudopotential band structures [8], and the scattering properties, and let the computational burden be handled by a modern parallel processing computer. Hence, no attempt is made to arrive at an *ab initio* analytical form, although the output of the EMC computation can yield the form of the distribution function quite well. These approaches have become extremely sophisticated, and a study of many factors such as degeneracy [9], discrete impurities [10] in real space, and many-body effects [11] can be incorporated.

The EMC is built around the general Monte Carlo technique. Simply put, an ensemble of carriers is allowed to evolve under the accelerating forces of the electric field, and any inter-particle forces or magnetic field. These evolutions are interrupted by scattering events, and a random walk is generated to simulate the stochastic motion of particles subject to the collision processes. The collisions provide both the momentum relaxation process and the energy relaxation process, as well as

doi:10.1088/978-0-7503-3947-6ch7

inter-valley scattering. Even generation and recombination can be accommodated, even on the ultrafast time scale [12]. Random walks, and stochastic techniques, may be used to evaluate complicated multiple-dimensional integrals [13, 14]. In the Monte Carlo transport approach, a path integral is evaluated by the multi-step process outlined above. The length of each free flight, and the selection of the appropriate scattering process, are selected by weighted probabilities, with the weights adjusted according to the physics of the transport process. As a result, very complicated physics can be introduced without any additional complexity of the formulation (although more extensive computer time may arise). At appropriate times through the simulation, averages are computed to determine quantities of interest, such as the drift velocity, average energy, and so forth. By simulating an ensemble of carriers, rather than the single carrier normally used in a Monte Carlo procedure, the non-stationary time-dependent evolution of the carrier distribution, and the appropriate ensemble averages, can be determined quite easily without resorting to any need for time averages. As there are other types of Monte Carlo processes [13, 14], it is important to note that the approach here is a kinetic Monte Carlo for the evolving dynamics of the ensemble of particles.

In the following sections, the path integral equation for the distribution function is derived first, primarily due to the fact that the kinetic Monte Carlo being discussed is a solution to the resulting Fredholm integral equation. Following that, the various statistical approaches to determining stochastic evaluations of the two principle parts of the integral equation will be outlined, and then many applications will be described.

7.1 The path integral

To begin, the Boltzmann equation will be converted into a path integral as a method to illustrate the steps in the EMC process. In this approach, the streaming terms on the left-hand side will be written as partial derivatives of a general derivative of the time motion along a 'path' in a six-dimensional phase space; this is used to develop a closed-form integral equation for the distribution function, which is a form of a Fredholm integral. This integral has itself been used to develop an iterative technique [1], but provides a basis for the connection between the Monte Carlo procedure and the Boltzmann equation. To begin, the Boltzmann equation is written following the form in chapter 6 as

$$\left(\frac{\partial}{\partial t} + \mathbf{v} \cdot \nabla + \frac{e}{\hbar}\mathbf{E} \cdot \frac{\partial}{\partial \mathbf{k}}\right) f(\mathbf{k}, r, t) = -\Gamma_0 f(\mathbf{k}, r, t) + \int d^3\mathbf{k}' P(\mathbf{k}', \mathbf{k}) f(\mathbf{k}', r, t), \tag{7.1}$$

where

$$\Gamma_0 = \int d^3\mathbf{k} P(\mathbf{k}, \mathbf{k}') \tag{7.2}$$

is the total *out-scattering rate*. That is, (7.2) provides the entire rate of decrease of population from the state described by $f(\mathbf{k}, \mathbf{r}, t)$ due to scattering of particles out of this state. The last term on the right in (7.1) provides the complementary *in-scattering* of particles from other points in momentum (and energy) space. These two terms are defined by the corresponding processes in (6.5) and figure 6.1.

At this point it is convenient to transform to a variable that describes the motion of the distribution function along a trajectory in phase space. It is usually difficult to think of the motion of the distribution function, but perhaps easier to think of the motion of a typical particle that characterizes the distribution function. That is, the distribution function itself will be represented by the ensemble of particles that are utilized for the Monte Carlo simulation. This ensemble may be considered as existing in phase space, and in the absence of scattering, the phase-space volume is conserved. Each particle thus has unique path in phase, and our coordinates will be transformed to be along this path defined by the position and momentum of the given particle. In the distribution function, each path is considered to represent a fraction of the distribution. For this, the motion is described in a six-dimensional phase space, which is sufficient for the one-particle distribution function being considered here [1]. The coordinate along this trajectory is taken to be s, and the trajectory is rigorously defined by the semi-classical trajectory, which can be found by any of the techniques of classical mechanics (i.e., it corresponds to that path which is an extremum of the action). It is as easy to remember, however, that it follows Newton's laws, where the forces arise from all possible potentials—including any self-consistent ones that arise in device simulations. Each normal coordinate can be parameterized as a function of this variable as

$$\mathbf{k} \rightarrow \mathbf{x}^*(s), \mathbf{k} \rightarrow \mathbf{k}^*(s), t \rightarrow s, \tag{7.3}$$

and the partial derivatives are constrained by the relationships

$$\frac{d\mathbf{x}^*}{ds} = \mathbf{v}, \quad \mathbf{x}^*(t) = \mathbf{r}, \quad \frac{d\mathbf{k}^*}{ds} = \frac{e}{\hbar}\mathbf{E}, \quad \mathbf{k}^*(t) = \mathbf{k}. \tag{7.4}$$

With these changes, the Boltzmann equation becomes simply

$$\frac{\partial f}{\partial s} + \Gamma_0 f = \int d^3\mathbf{k}^{*\prime} P(\mathbf{k}^{*\prime}, \mathbf{k}^*) f(\mathbf{k}^{*\prime}, \mathbf{x}^*, s). \tag{7.5}$$

This is now a relatively simple equation to solve. It should be recalled at this point that $P(\mathbf{k}^{*\prime}, \mathbf{k}^*)$ is the probability per unit time that a collision scatters a carrier from state $\mathbf{k}^{*\prime}$ to $\mathbf{k}^*$, and these variables will be retarded due to the phase-space variations described above. The form (7.5) immediately suggests the use of an integrating factor $\exp(\Gamma_0 s)$, so that this equation becomes

$$\frac{d}{ds}[f(\mathbf{k}^*)e^{\Gamma_0 s}] = \int d^3\mathbf{k}^{*\prime} P(\mathbf{k}^{*\prime}, \mathbf{k}^*) f(\mathbf{k}^{*\prime}, \mathbf{x}^*, s) e^{\Gamma_0 s}, \tag{7.6}$$

and, if we restore the time variables appropriate to the laboratory space, we arrive at

$$f(\mathbf{k}, t) = f(\mathbf{k}, 0)e^{-\Gamma_0 t} + \int_0^t dt' \int d^3\mathbf{k}' P\left(\mathbf{k}' - \frac{e}{\hbar}\mathbf{E}, \mathbf{k}\right) f\left(\mathbf{k}' - \frac{e}{\hbar}\mathbf{E}, \mathbf{r}, t'\right) e^{-\Gamma_0(t-t')}. \tag{7.7}$$

This last form is often referred to as the Chambers–Rees path integral [15], and is the form from which an iterative solution or the Monte Carlo solution can be developed.

The integral (7.7) has two major components. The first is the scattering process by which the carriers described by $f(\mathbf{k}')$ are scattered (by the processes within P). The second is the ballistic drift under the influence of the field, with a probability of the drift time given by $\exp[-\Gamma_0(t - t')]$. These are the two parts of the Monte Carlo algorithm, and it is from such an integral that we recognize that the Monte Carlo method is merely evaluating the integral stochastically. The problem is that there is no retardation in the scattering process, so that the scattering rate and energy are supposed to respond instantaneously to changes in the momentum along the path $k' = (e/\hbar)Et'$. The particles represented by the distribution function within the integral instantaneously respond during the previous drift. In essence, this is the Markovian assumption, and is true only in the long-time limit (Boltzmann's Stosszahl Ansatz). The EMC process can be used in the short-time, transient regime without modification, but then it is a solution of a non-Markovian version of the Boltzmann equation, which is called the Prigogine–Resibois equation [16].

7.2 The Monte Carlo process

It should be noted that the path integral (7.7) begins with the equilibrium distribution at $t = 0$. It is not necessary to do this, but it ensures that the transient evolution of the distribution connects with reality. The initialization will be different for transport under an electric alone and that under optical excitation. But, both have quite similar behavior. The development of the initial distribution follows similar paths for both forms of excitation. To achieve this initialization, the optical process will be introduced here. For example, the photons coming into the semiconductor will create electrons and holes. For sufficiently high photon energy, the electrons can be excited from both the light- and heavy-hole bands, as well as from the spin split-off band. Only the heavy-hole excitation will be discussed, as the other two follow from it in a straightforward manner. As discussed in chapter 4, for $\hbar\omega > E_G$, the energy in excess of the gap energy will split as kinetic energy between the valence and conduction band, with the factors for each determined by the effective masses of the two bands. That is, the photon energy splits according to

$$\hbar\omega = E_G + \frac{\hbar^2 k^2}{2}\left(\frac{1}{m_e} + \frac{1}{m_{hh}}\right), \tag{7.8}$$

and the fraction that goes into e.g. the conduction band is

$$\frac{m_{hh}}{m_{hh} + m_e}. \tag{7.9}$$

Thus, the electrons appear in the conduction band with the excess energy

$$E_0 = \frac{m_{hh}}{m_{hh} + m_e}(\hbar\omega - E_G). \tag{7.10}$$

This is the *average* energy of the carrier distribution that is created. There will be a distribution function for these electrons that spreads around this average energy. The width of this distribution will be determined by the line width of the photon source and the temperature at which the semiconductor is held. Generally, this produces an electron distribution that has a characteristic temperature T_0 which may or may not be larger than the lattice temperature depending upon the above characteristics. But, one can assume that the distribution is something like a Maxwellian around this distribution.

With ultrafast excitation, the temporal shape of the photon pulse must also be considered. Often the laser pulse width (in time) is determined as a Gaussian from auto-correlation experiments, but such a distribution is not easily handled in a simulation due to the very slow decay of the pulse at $t - t_0 \to \pm\infty$, where t_0 is the time at the center of the laser pulse. So, either the Gaussian has finite cut-offs, or something like a $\cos^2[\eta(t - t_0)]$, where η is a parameter adjusted to give the pulse the right temporal width. If N is the total number of particles to be used in the Monte Carlo process, then these particles are generated at random times $t_0 - \pi/2\eta \leqslant t \leqslant t_0 + \pi/2\eta$. However, these times cannot be uniformly distributed but the times are weighted by the temporal line shape. To accomplish this, select a random number r $(0 \leqslant r \leqslant 1)$, which then yields the time

$$t = t_0 \pm \frac{t_0}{\pi}\cos^{-1}(\sqrt{r}). \tag{7.11}$$

One way to easily implement this is to initialize the N electrons into the distribution around E_0. It is easy to see that when $r = 1$, $t = t_0$, and when $r = 0$, $t = t_0 \pm t_0$. The inverse cosine ensures that more numbers are centered around zero, than elsewhere. The choice of the plus or minus sign is made with a second random number. Then, in addition to assigning each electron with a position, momentum, and energy, it is also assigned a time t_i (for the ith electron) at which it is *activated* into the ensemble. Thus, at $t = 0$, there are no active electrons, and at $t = 2t_0$ all N electrons are activated. When the distribution is a thermal distribution at the bottom of the conduction band, then one merely sets $E_0 = 0$, and chooses all $E_i > 0$. And, the duration of the starting 'pulse' is also zero.

The second helpful consideration at the beginning of the procedure is to create a table of all scattering values. In the second part of the Monte Carlo procedure of (7.7) is the scattering process. To compute all scattering probabilities at each time step is quite computer intensive. To this, a scattering table can remove this computation requirement and is helpful in also picking which scattering process is

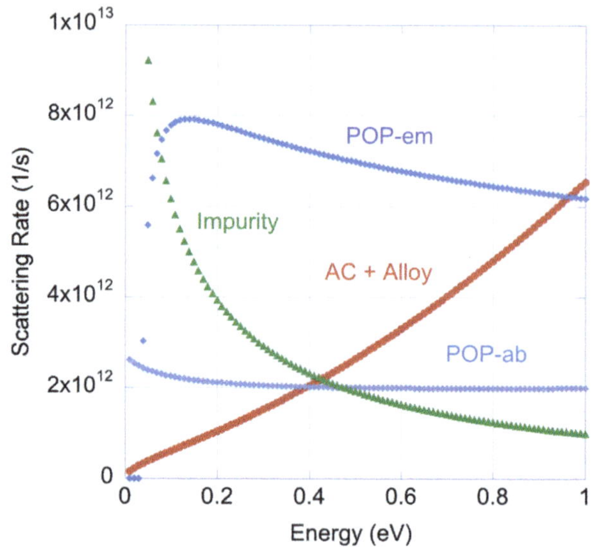

Figure 7.1. Intra-valley scattering rates for $In_{0.65}Al_{0.35}As$ as a function of energy for electrons in the Γ valley at 300 K.

to be implemented. To accomplish this requires gridding the energy. At a lattice temperature, usually a 1 meV energy step is enough, but at low temperatures, a finer grid will likely be necessary. The average thermal energy of an electron at room temperature is $3k_BT2 \sim 39$ meV. This is much larger than the 1 meV energy step, so the noise associated with this step will be small. In figure 7.1, the scattering rates for the polar optical phonon emission and absorption scattering, impurity scattering, and combined acoustic and alloy scattering are plotted for $In_{0.65}Al_{0.35}As$, whose band structure and parameters are given in Welland and Ferry [17]. Only the intra-valley scattering for the Γ valley are shown. The acoustic and alloy scattering are combined as they are both elastic and have the same energy dependence [18]. These few scattering rates are sufficient to fully discuss the EMC approach. However, inter-valley scattering to the L valleys sets in at about 0.53 eV.

In the initialization routine, several matrices are going to be required in order to keep tabs on the particles as they evolve. Obviously, matrices $\overline{\overline{KX}}$, $\overline{\overline{KY}}$, $\overline{\overline{KZ}}$ will be required to store the momentum values, $\overline{\overline{E}}$ will be required to store the energy values. Each of these matrices will be of order N, where the index gives the particle number. If we are interested in anything other than a homogeneous semiconductor, then matrices $\overline{\overline{X}}$, $\overline{\overline{Y}}$, $\overline{\overline{Z}}$ give the positions of each of the particles. In optical excitation, the density in the z-direction (normal to the surface) will not be homogeneous due to the $\exp(-\alpha z)$, where α is the absorption coefficient, decay of the photon intensity. All of the above matrices are real valued. Another important matrix is the $\overline{\overline{val}}$, which carries integer values denoting the valley index; e.g., 1, 2, 3 may stand for the Γ, L, and X valleys, respectively. It will be found to be useful, especially for parallelizing any code, to carry another few matrices that are denoted by $\overline{\overline{EN}}$, $\overline{\overline{delE}}$, $\overline{\overline{flag}}$, where the first two carry real numbers and the last one carries integer values. These three represent the energy after scattering, the change in energy during scattering, and a flag used to denote the type of scattering. They will be explained more fully in the following discussions. In the initialization stage, $\overline{\overline{val}}$ and $\overline{\overline{flag}}$ are set uniformly to 1. The matrices $\overline{\overline{EN}}$, $\overline{\overline{delE}}$ are initialized to 0.

7.2.1 Free flight generation

The path integral (7.7) contains two separate processes. The first part of the integral addresses the dynamics of the particle motion that is assumed to consist of free flights interrupted by instantaneous scattering events. The latter change the momentum and energy of the particle according to the physics of the particular acceleration and drift processes. Of course, precisely how long a carrier will drift before scattering is a random process, as it continuously interacts with the lattice as indicated by the term

$$e^{-\Gamma_0(t-t')}. \tag{7.12}$$

Here, t is the final time of the drift and t' is the time at which the drift started, as it is the time at which the distribution function is known. In fact, as may be seen from figure 7.1, each and every scattering process is a function of energy. As a result, this term is actually of the form

$$\exp\left(-\int_0^{t'} \Gamma_0[E(t'')]dt''\right), \tag{7.13}$$

as the scattering rate Γ_0 is a function of the energy E, and the energy E is changing with time due to the accelerating field in (7.4). This is a complicated integral to evaluate. Within the stochastic approximations, the actual transport is simulated by introducing a probability density $P(t)$, where $P(t)dt$ is the joint probability that a carrier will both arrive at time t without scattering (after its last scattering event at $t = 0$), and then will actually suffer a scattering event at this time (i.e., within a time interval dt centered at t). This is the probability (7.13). The probability of actually scattering within this small time interval at time t may be written as $\Gamma[E(t)]dt$, where $\Gamma[E(t)]$ is the total scattering rate of a carrier of wave vector $\mathbf{k}(t)$ and energy E(t). This scattering rate represents the sum of the contributions of each scattering process that can occur for a carrier of this wave vector (and energy). The explicit time dependence indicated is a result of the evolution of the wave vector (and energy) under any accelerating electric (and magnetic) fields. In terms of this total scattering rate, the probability that a carrier has not suffered a collision after time t is given by (7.13). This function has not been normalized as a good probability would be, as the magnitude of (7.13) will vary from zero to unity, a proper range for a random number.

To generate the random flight times for each particle, a random number generator, which gives us a uniform distribution of real numbers between zero and one (including these two values) may be used. Hence, the normalization factor is extraneous, and the random flight time is sampled from $P(t)$ according to the random number r as

$$r = 1 - \exp\left[-\int_0^t \Gamma_0(E(t'))dt'\right]. \tag{7.14}$$

The exponential term varies from 0 to 1, and since $1 - r$ is statistically the same as r, this latter expression may be simplified as

$$\ln(r) = -\int_0^t \Gamma_0(E(t'))dt'. \tag{7.15}$$

This remains a difficult integral, but the sum of the scattering processes from figure 7.1 are shown in figure 7.2. While the curve is relatively flat, it contains a number of rapid variations, especially at low energy, which complicates the integral.

Equation (7.15) is the fundamental equation used to generate the random free flight for each carrier in the ensemble. If there is no accelerating field, the time dependence of the wave vector vanishes, and the integral is trivially evaluated. In the general case, however, this simplification is not possible, and it is expedient to resort to a mathematical *trick*. Here, a fictitious scattering process that has no effect on the carrier will be introduced. This process is called *self-scattering*, and the energy and momentum of the carrier are unchanged under this process [19]. An energy dependence will be assigned to this process in just such a manner that the total scattering rate is a constant, as

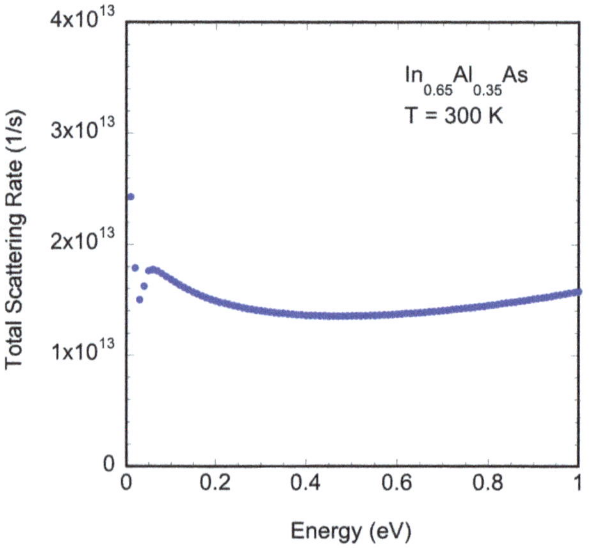

Figure 7.2. The sum of the intra-valley scattering rates of figure 7.1 are plotted versus the electron energy. The peak at low energy is obviously due to Coulomb scattering which can be seen to increase rapidly at low energy already in figure 7.1.

$$\Gamma_T - \Gamma_{\text{self}}(E) = \sum_i \Gamma_{0,i}(E), \tag{7.16}$$

where the sum runs over all the *real* scattering processes as a function of the energy E. The energy dependence of the self-scattering term is determined so that the quantity Γ_T is independent of energy and therefore independent of the acceleration time. Because the self-scattering process has no effect on the energy, momentum, or position of the particle, it will have no effect upon the transport properties being evaluated by the Monte Carlo process. But, the introduction of this imaginary scattering process simplifies the temporal integration in (7.15). Consider figure 7.3. Here, the sum of the scattering rates is $<2.5 \times 10^{13}\ \text{s}^{-1}$. Hence, it is convenient to set $\Gamma_T = 2.5 \times 10^{13}\ \text{s}^{-1}$. Then, the energy dependent self-scattering is the value between this number and the black curve. With this, the time of flight may be determined to be

$$t = -\frac{1}{\Gamma_T}\ln(r). \tag{7.17}$$

In an ensemble Monte Carlo process, each individual particle in the ensemble carries its own flight time, and this time is selected at the start of each flight. This creates another problem in that the ensemble averages have to be computed at particular time steps, and the manner in which the particles are synchronized will be discussed below.

Following the free flights, the momenta, energies, and positions are updated according to the general classical mechanics of Hamiltonian systems, as

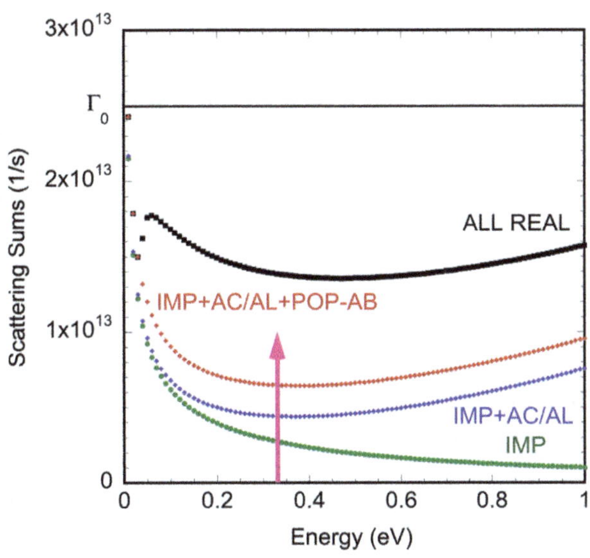

Figure 7.3. Selection of the scattering process is achieved by the following process. A random number gives the magenta arrow whose length is $r_2\Gamma_T$. Since it lies above the red curve, but below the black curve, polar emission is selected as the scattering process.

$$\mathbf{k}_i(t) = \mathbf{k}_i(0) + \frac{e\boldsymbol{F}}{h}t_i$$
$$E_i(t) = \frac{\hbar^2 k_i^2(t)}{2m^*} \quad . \tag{7.18}$$
$$\mathbf{r}_i(t) = \mathbf{r}_i(0) + \frac{\hbar\mathbf{k}_i(t)}{m^*}t_i$$

In this, parabolic bands have been assumed, but the hyperbolic non-parabolic bands could just as well be (and usually are) used. This is a detail that is easily incorporated into (7.18). Indeed, the basic definition of the effective mass is used in the last line to give the carrier velocity. Equation (7.18) is the basic acceleration equation that will be used many times in the EMC simulations.

7.2.2 Scattering

The second part of (7.7) is the scattering of the particles. This is the major part of (7.7), and in fact occupies the majority of the simulation code and the time in the EMC process. It is represented by the factors

$$\int d^3\mathbf{k}' P\left(\mathbf{k}' - \frac{e}{\hbar}\mathbf{E}, \mathbf{k}\right) f\left(\mathbf{k}' - \frac{e}{\hbar}\mathbf{E}, \mathbf{r}, t'\right). \tag{7.19}$$

In fact, this represents all of the in-scattering terms for the left-hand side of (7.7), but for the distribution function in (7.19), it is actually out-scattering. So, only the normal scattering processes need be described, and these are given in figure 7.1. To capitalize on parallel processing, all of the particles need to be done in parallel, and that requires considering each step in the process at the same time. There is a step between the acceleration of the previous section and the scattering of this section, and that involves the existence of a time grid for taking averages. The intervening step involves determining whether or not a scattering event will occur within the current cell of the time grid. This will be discussed below, but the important point is that if the scattering event is to occur, the flag $\overline{\overline{(flag)}}_i$ will have been set to 2, to indicate that a scattering event has to be carried out for particle i. Only particles with the flag set to 2 are considered in the scattering process.

In the scattering event, the important process is selecting the scattering process for the particular particle. This is, of course, done with another random number (an important point is that most modern compilers include the ability to select a large matrix of random numbers at the same time). The selection of the scattering event will be demonstrated with the intra-valley scattering events shown in figure 7.1. It is easily expanded to cover any number of scattering events, including inter-valley scattering. Nevertheless, it will be demonstrated with the simple case. To begin, the scattering rates are replotted as partial sums, as indicated in figure 7.3. For the particular particle in question, it is assumed that at the end of the acceleration stage, this particle has an energy of 0.27 eV. Another random number, say r_2 is chosen. This number is multiplied by Γ_T, which is the same number given in (7.17) as the maximum total scattering rate. This result is indicated by the magenta arrow in figure 7.3. Notice that it extends above the line corresponding to the sum of impurity and acoustic/alloy scattering and polar optical absorption, so these processes are *not* the source of the scattering. On the other hand, it lies below the sum of the previous three events plus polar phonon emission (the black line includes all scattering processes shown in figures 7.1 and 7.2). Thus, polar phonon emission is chosen to be the scattering process. If the arrow had been below the red line, then polar optical absorption would have been selected. If the arrow had extended above the thin black line, then self-scattering would have been selected.

Since the selected scattering process is polar phonon emission, the energy change matrix element is set to $(\overline{\overline{delE}})_i = -\hbar\omega_0$. In addition, some recognition of whether or not the scattering is isotropic or anisotropic is required, and the flag will be used for this, with the flag element set to $(\overline{\overline{flag}})_i = 3$, which for this example indicates that the scattering is anisotropic. Only piezoelectric, impurity, and the polar optical phonons have this property. If the process had been isotropic, as in acoustic or inter-valley scattering, the flag element would have been set to 4. If this flag had remained at 2, it would have indicated that self-scattering occurred for that particle, and the energy and momentum would remain unchanged. The need for the flag lies in the fact that determining the final state energy and momentum involves quite different processes for isotropic and anisotropic scattering. If the scattering is also elastic, the energy change element remains at 0.

For the isotropic scattering, all states on the final energy shell are available. The energy $E_i(t)$ is incremented by the energy change to give $\overline{\overline{EN_i}}$ and this is used to choose the new magnitude of the momentum as

$$k_i = \sqrt{\frac{2m^* \overline{\overline{EN_i}}}{\hbar^2}}, \tag{7.20}$$

and the projection onto the coordinates is done with two additional random variables for the spherical geometry as

$$\begin{aligned} \vartheta &= r_3\pi \\ \varphi &= 2r_4\pi \end{aligned}. \tag{7.21}$$

These are now used to compute the momentum in the three directions.

With anisotropic scattering, determining the final angle ϑ is more difficult. The azimuthal angle remains as given in the second line of (7.21). Moreover, it depends upon the energy of the carrier as to what the scattering angle will become. Consider impurity scattering which is anisotropic. The Coulomb scattering potential in three dimensions varies with the quantity

$$\frac{1}{q^2 + 4k^2 \sin^2(\vartheta/2)}. \tag{7.22}$$

But, the scattering rate depends upon the square of the matrix element, thus upon the square of the Coulomb scattering potential as

$$\left[\frac{1}{q^2 + 4k^2 \sin^2(\vartheta/2)}\right]^2. \tag{7.23}$$

Hence, a scattering angle ϑ_{sc} can be selected using the integral

$$r_3 = \frac{\int_0^{\vartheta_{sc}} [q^2 + 4k^2 \sin^2(\vartheta/2)]^{-2} \sin\vartheta d\vartheta}{\int_0^{\pi} [q^2 + 4k^2 \sin^2(\vartheta/2)]^{-2} \sin\vartheta d\vartheta}. \tag{7.24}$$

The integrals are messy, but not complicated, and it is possible to then find

$$\cos\vartheta_{sc} = \left[1 - \frac{2r_3 x}{1 + x - r_3}\right], \tag{7.25}$$

where

$$x = \frac{q_D^2}{4k^2}, \tag{7.26}$$

and q_D is the reciprocal of the Debye length. The scattering angle is plotted as a function of the random number in figure 7.4, with the value of x as a parameter. It may be seen that almost 75% of the scattering is into the forward direction.

A similar procedure is followed for the anisotropic polar optical phonons, except in this case, there is a significant energy change due to the phonon energy. Now, several matrices come into play, as

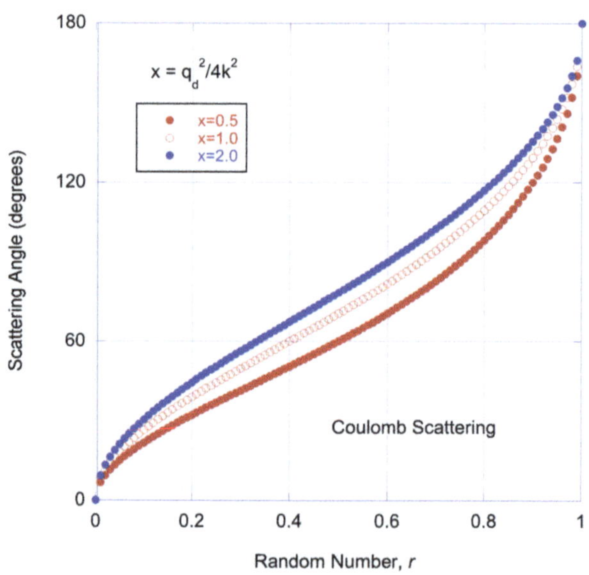

Figure 7.4. The scattering angle as a function of the random number, which is uniformly distributed between 0 and 1. The parameter x is given by (7.26).

$$
\begin{aligned}
E_i &= (\overline{\overline{E}})_i \\
(\overline{\overline{delE}})_i &= \pm\hbar\omega_0 \\
(\overline{\overline{EN}})_i &= (\overline{\overline{E}})_i + (\overline{\overline{delE}})_i
\end{aligned} \quad . \tag{7.27}
$$

In the example above, only the positive sign in the second line is used, as absorption of the phonon was chosen. The magnitude of the momentum values of both the initial and the final state are needed as

$$
\begin{aligned}
k^2 &= \frac{2m^*}{\hbar^2}E_i \\
k'^2 &= \frac{2m^*}{\hbar^2}(\overline{\overline{EN}})_i
\end{aligned} \quad . \tag{7.28}
$$

The scattering rate is proportional to the square of the Coulombic term, so that in analogy with (7.23), this becomes

$$
(k^2 + k'^2 - 2kk'\cos\vartheta)^{-2}. \tag{7.29}
$$

And, as before, the scattering angle ϑ_{sc} can be selected by using

$$
r_3 = \frac{\int_0^{\vartheta_{sc}} [k^2 + k'^2 - 2kk'\cos\vartheta]^{-2} \sin\vartheta d\vartheta}{\int_0^{\pi} [k^2 + k'^2 - 2kk'\cos\vartheta]^{-2} \sin\vartheta d\vartheta}. \tag{7.30}
$$

To proceed, the parameter

$$
x = \frac{2kk'}{k^2 + k'^2} \tag{7.31}
$$

is introduced, which leads to

$$
\cos\vartheta_{sc} = \frac{1 + x - 2r_3}{1 + x(1 - 2r_3)}. \tag{7.32}
$$

Again, it may be seen that this has the correct limits for the range of the random number. The scattering angle as a function of the random number is shown in figure 7.5, with the parameter x. Once, more it may be seen that forward scattering is predominant in the process.

An important point in these discussions of the anisotropic scattering is that the scattering angle has been determined relative to the vector **k**, and not to the electric field which is oriented in the z-direction. This means that, after selecting the final angles and momentum values, these latter must be rotated into the plane of the laboratory coordinates where the electric field is the polar axis. Failure to do this leads to quite erroneous results for the transport simulation. This was not required for the isotropic scattering because of the fact that any point on the final energy

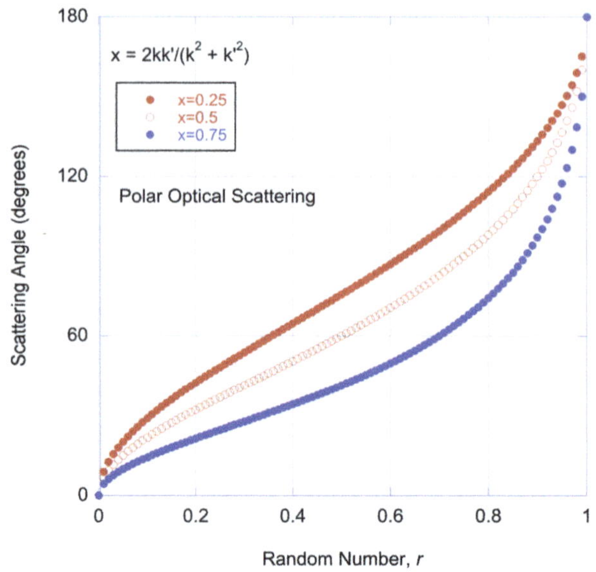

Figure 7.5. Scattering polar angle as a function of the random number selected. This has a larger variation with the parameter x, than the impurity scattering. Nevertheless, the dominant scattering has a forward direction. The strength of the scattering is set by the energy dependence shown in figure 7.1, but this angular variation is used to select the polar angle of scattering.

surface was equally likely, so that the angles could be described as actually being in the laboratory coordinates.

If inter-valley scattering is present, then there are additional processes that have to be considered. For example, if scattering from the Γ valley to the L valley is being considered, then further changes include

$$\begin{aligned} (\overline{\overline{delE}})_i &= \pm\hbar\omega_{iv} - \Delta_{\Gamma L} \\ (\overline{\overline{val}})_i &= 2 \end{aligned}. \tag{7.33}$$

The last term of the first line is the energy separation between the two valleys. In the reverse transfer, this energy would be added instead of subtracted. Then, the above considerations of the properties of the final state involve the mass of the valley in which the final state lies.

Following determination of the state of the particles after scattering, the EMC procedure returns to the acceleration process of the previous section, and new final times are chosen for all the electrons. However, this process becomes slightly more complicated when the laboratory time grid is present.

7.2.3 Time synchronization

As mentioned above, there are usually two time scales in the EMC procedure. The first is the local time that each particle carries. Then there is the global, or laboratory, time scale, that is a key problem in treating an ensemble of particles.

Ensemble averages are computed for such quantities as the drift velocity, position, and average energy, etc. A particular average, for example the velocity, is computed as

$$\mathbf{v}_d(t) = \frac{1}{N}\sum_i \mathbf{v}_i(t). \tag{7.34}$$

Notice that this is an ensemble average, and not a time average. In the far-from-equilibrium situation, and particularly when transients are involved, there is no ergodic theorem [20]. So, only ensemble averages have statistically valid properties. To achieve such an average, all the particles need to be aligned at the same time t, which here runs from the beginning of the simulation. Thus, the system is overlaid with a global time scale, with which each local particle time scale can be synchronized. In practice, this is achieved by introducing the global time variable T, which is discretized into steps as $n\Delta T$. Then at integer multiples of this time step, all particles are stopped in their free flights, and ensemble averages are computed. As described by (7.7), each particle has a flight which is composed of accelerations and scattering processes. Usually, the arrival at a synchronization time T is during one of the free flights. Thus, the free flight is stopped at this time point, and the parameters computed, and then included in the averages over the particle. In this sense, the imposition of the second time scale synchronizes the distribution and gives the global, or laboratory, time scale of interest in experiments. Hence, this time scale can be called the *laboratory* time scale. This process is illustrated in figure 7.6. Once this is done, the particle is sent on its way to continue its free flight until it reaches its

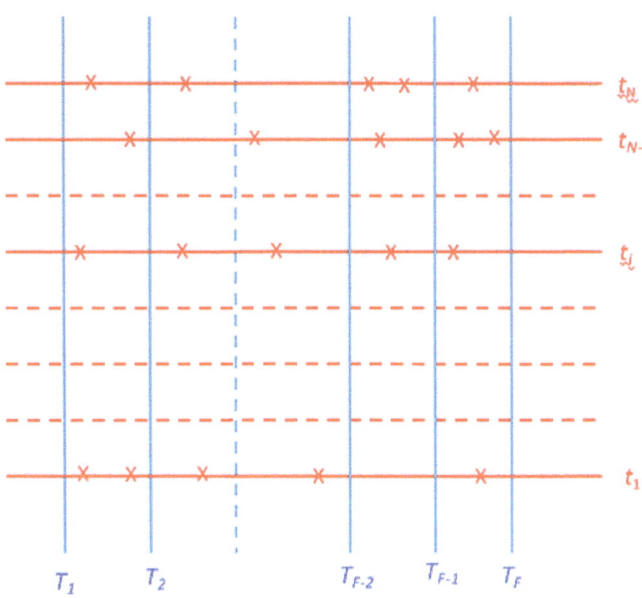

Figure 7.6. Alignment of particle time scales t with the global time scale T. The simulation and particle movement is stopped at each time step T_i and ensemble averages updated. Reprinted from Ferry [18], copyright IOP Publishing. All rights reserved.

scattering time. This process is efficient, although at the expense of more book-keeping in the algorithm.

In figure 7.6, the horizontal lines represent the individual particle time scales, while the vertical lines are the global time steps at which the motion is stopped and the ensemble averages computed. The crosses on the particle time lines are scattering events. While these global time steps are shown as being longer than the individual scattering times, this does not have to be the case. Any ΔT may be used, and is usually chosen for convenience. It can be larger than the mean drift time of a particle, or shorter. The procedure still works fine.

If one is incorporating nonlinear effects, such as the electron–electron interaction in real-space molecular dynamics, updating Poisson's equation, or nonequilibrium phonons, or degeneracy induced filling of the final states after scattering, then these processes are updated on the pauses of the T time scale as well. In device simulations, the global time steps are the points where the particle density is projected onto the simulation grid (for Poisson's equation and the resulting potential). In this latter case, the total electric field, from the potential V found from Poisson's equation and the inter-particle fields found from the molecular dynamics are updated on these global time scales. In many, if not most cases, the time steps required for Poisson's equation and the inter-particle forces are the shortest time steps in the entire simulation [21]. With molecular dynamics and Poisson's equation working together, there is a danger of double counting the forces. This is solved by a potential splitting procedure, which is discussed in section 7.3.

7.2.4 Rejection technique and state filling

While it was possible to actually integrate the angular probability functions for the anisotropic scattering, it may not always be the case for some unusual scattering processes, and one has therefore to resort to other statistical methods. Two examples of this are accounting for nonequilibrium phonons and degenerate statistics and the occupation of the final state. One method of treating these is the rejection technique. Suppose the probability density function for the process is quite nonlinear and not easily integrated to get the total probability. Then one can use a pair of random numbers ($r_{3,1}$, $r_{3,2}$) to evaluate the angle. Consider figure 7.7, in which we plot a complicated probability density function. Here, the abscissa could, for example be the polar angle θ, which has been normalized by π. In this manner, the maximum of the abscissa x is unity, so that the range of the function's argument is from zero to one. The maximum value of the ordinate (y-axis) is also normalized to unity, but one can always renormalize this to the span of the random numbers if they are not 0 to 1. The first random number is taken to correspond to the abscissa of the plot (the x-axis). This determines the argument of the function that is to be evaluated. For example, let us assume that this is $r_{3,1} = 0.75$ (this is indicated by the vertical dashed line in the figure). We now use the second random number to determine whether $f(r_{3,1}) > r_{3,2}$. If this relationship holds, then the value $r_{3,1}$ is accepted for the argument of the function, and the scattering process proceeds with this value. If, on the other hand, the relationship is not valid, then two new random numbers are chosen and

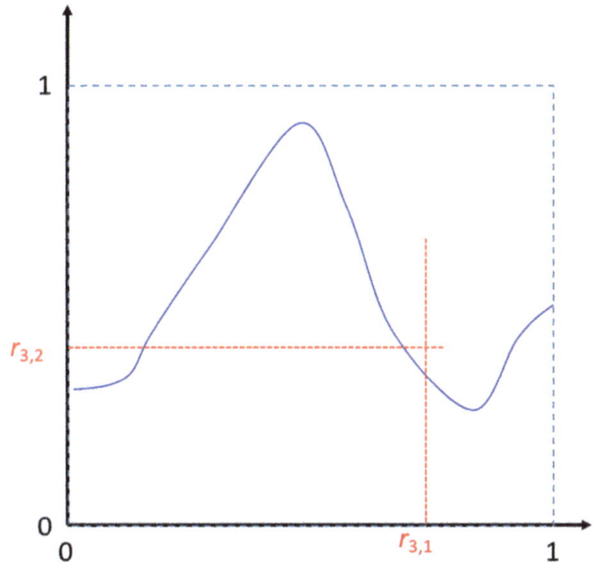

Figure 7.7. The rejection technique uses two random numbers. The first selects a horizontal value at which to evaluate the nonlinear function. The second determines whether or not to use the selected value, depending upon whether or not this value is larger or smaller than the value of the function at the first number. In the case shown, the pair will be rejected. For acceptance, the value of the second random number must lie below the blue curve. Reprinted from Ferry [18], copyright IOP Publishing. All rights reserved.

the process repeated. This repetition is continued until a pair of random numbers is found that satisfy the above inequality. Certainly, values of $r_{3,1}$ for which the function is large are more heavily weighted in this rejection process. We consider this in more detail for two important processes: (1) state filling due to the degeneracy of the electron gas, and (2) nonequilibrium phonons.

Consider the case of state filling, and consider an example of the EMC process to illustrate its efficacy [9]. Consideration of state filling appears in the selection of the final state after scattering, discussed above in section 7.2.2. Degeneracy and Fermi–Dirac statistics are introduced through the concept of a secondary self-scattering process [9, 22] based upon the rejection technique. It is termed secondary self-scattering, because if the condition $f(r_{3,1}) > r_{3,2}$ is not satisfied, the rejection proceeds exactly as a self-scattering process, which was introduced earlier. Each of the scattering processes must include a factor of $[1 - f(E)]$, where $f(E)$ is the time-dependent distribution function, and represents the probability that the final state after scattering is empty. Rather than recompute the scattering rates as the distribution function evolves in order to incorporate the degeneracy, all scattering rates are computed as if the final states were always empty. A grid *in momentum space* is maintained and the number of particles in each state is tracked (each cell of this grid has its population divided by the total number of states in the cell, which depends on the cell size, to provide the value of the distribution function in that cell). The scattering processes themselves are evaluated, but the acceptance of the process depends on the rejection technique. That is, an additional random number is used to accept the process if

$$r < 1 - f(\mathbf{p}_{\text{final}}, t). \tag{7.35}$$

Thus, as the state fills, most scattering events into that state are rejected and then are treated as an additional self-scattering process.

The most delicate point of the degeneracy method involves the normalization of the distribution function $f(\boldsymbol{p})$. The extension of the secondary self-scattering method to the ensemble Monte Carlo algorithm involves the fact that there are N electrons in the simulation ensemble, which represent an electron density of n. The effective volume V of 'real space' being simulated is N/n. The density of allowed wave vectors of a single spin in **k**-space is just $V/(2\pi)^3$. In setting up the grid in the three-dimensional wave vector space, the elementary cell volume is given by $\Omega_k = \Delta k_x \Delta k_y \Delta k_z$. Every cell can accommodate at most N_c electrons, with $N_c = 2\Omega_k V/(2\pi)^3$, where the factor of 2 accounts for the electron spin. For example, if the density is taken to be 10^{17} cm^{-3}, $N = 10^4$, and $\Delta k_x \Delta k_y \Delta k_z = (2 \times 10^5 \text{ cm}^{-1})^3$ ($k_F = 2.4 \times 10^6$ cm^{-1} at 77 K), then $V = 10^{-13}$ cm^3 and $N_c = 6.45$. N_c constitutes the maximum occupancy of a cell in the momentum space grid. (Obviously, a more careful choice of parameters would have N_c come out to be an integer, for convenience.) A distribution function is defined over the grid in momentum space by counting the number of electrons in each cell. The distribution function is normalized to unity by dividing the number in each cell by N_c for use in the rejection technique. It should be noted that N_c should be sufficiently large that round-off to an integer (if the numbers do not work out properly, as in the case above) does not create a significant statistical error.

The second situation mentioned above in using rejection techniques lies in the study of nonequilibrium phonons, such as discussed in chapter 4. The nonequilibrium phonon distributions can be incorporated into the EMC process, through the use of a rejection technique [23]. In the derivations presented earlier in this chapter, the assumption was made that the phonons are in equilibrium and characterized by N_q. Now, for example, a maximum value for this distribution $N_{q,\max}$ can be utilized in evaluating the scattering table. This latter quantity is an estimate of the maximum phonon population that will encountered in the calculation. As with degeneracy, this distribution is projected onto a momentum space grid, so that an individual cell in this discretized space has volume $\Delta q_x \Delta q_y \Delta q_z$. This small volume has available a number of states given by $V/(2\pi)^3$, where V is determined by the effective simulation volume N/n, as previously. The difference between state filling for carrier degeneracy and phonon state filling is that there is no limit to the number of phonons that can exist within a given state, as they obey the Bose–Einstein distribution. Then, within the simulation, the number of phonons emitted, or absorbed, with wave vector $\boldsymbol{q}$ is carefully monitored. At the synchronization times of the global time scale, the phonon population in each cell of momentum space is updated from the emission/absorption statistics that have been gathered during that time step. One must also include phonon decay, which is through a three-phonon process (chapter 5) to other modes of the lattice vibrations, so that the update algorithm is simply

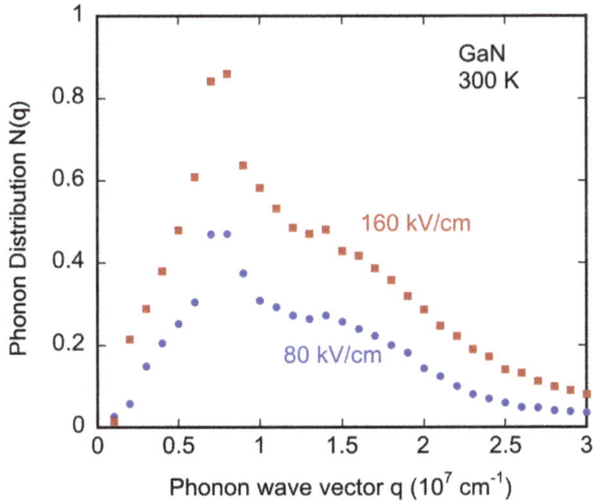

Figure 7.8. The nonequilibrium phonon distribution at two high values of the electric field as determined by Monte Carlo simulations for a GaN/AlGaN two-dimensional layer at the heterojunction interface.

$$N(\boldsymbol{q}, t + \Delta t) = N(\boldsymbol{q}, t) + G_{\mathrm{net},\Delta t}(\boldsymbol{q}) - \left[\frac{N(\boldsymbol{q}, t) - N_0}{\tau_{\mathrm{phonon}}}\right]\Delta t, \tag{7.36}$$

where N_0 is the equilibrium phonon distribution, $G_{\mathrm{net},\Delta t}(\boldsymbol{q})$ is the net (emission minus absorption) generation of phonons in the particular cell *during the global time step*, and τ_{phonon} is the phonon lifetime. During the simulation, each phonon scattering process is evaluated as if the maximum assumed phonon population were present. Then, a rejection technique is used, by which the phonon scattering process is rejected (and assumed to be a secondary self-scattering process) if

$$r_{\mathrm{test}} > \frac{N(\boldsymbol{q}, t)}{N_{\mathrm{q,max}}}. \tag{7.37}$$

This approach fails if $N(\boldsymbol{q}, t) > N_{\mathrm{q,max}}$, this has to be checked, and the simulation may have to be restarted with a larger value for $N_{\mathrm{q,max}}$.

In figure 7.8, the phonon distribution function is plotted at two values of the electric field for a GaN/AlGaN heterojunction device [24] at room temperature. Electrons in these structures show an experimental peak velocity near 3.2×10^7 cm s^{-1} at a field of 150 kV cm^{-1}, which is supported by detailed ensemble Monte Carlo simulations [24, 25]. The results in figure 7.8 indicate that the electrons interact only with a set of relatively low wave number phonons, but these can be driven quite far out of equilibrium.

7.3 Building a code

In the preceding sections, a discussion of the parts of a program were outlined. Here, it is desired to block together a diagram of how such a code is constructed. No actual code will be described, as that would require adopting a particular program, such as Fortran. It is true that nearly all scientific computing is done with this language, but not all. Compilers for Fortran, as well as the Fortran itself, have been optimized over the more than 70 years since it appeared. But, the language is not 70 years old, as it has varied and changed as new innovations have arrived. A 70 year old code will still run on most machines, but not optimally. For example, one advantage to Fortran is that memory does not have to be allocated and deallocated, as this is done automatically in the compiler. Furthermore, most supercomputers are evaluated with routines that are mostly Fortran in nature. If one wants to know what the currently fastest computer is, these are detailed at *top500.org*, with new listings each June and November. As of June, 2021, the fastest computer is the Fugaku machine at RIKEN in Japan, boasting a peak speed of ~537 PetaFLOPS (or 537×10^{15} floating-point operations per second). Number two is IBM's Summit machine at Oak Ridge National Laboratory at a measly ~201 PetaFLOPS.

In the following, several blocks of processing will be outlined to show how such a program is developed. The actual coding will be left to the reader. A broad blue line will define the flow direction of the code. The code described here uses a virtual gather-scatter process, which used to be used to speed the machine when the first pipelined architecture arrived. Today, no actual gather-scatter is performed, but is implied by the flags that are used. Parallelism is foremost in the mind of the programmer these days, and this means that jumps, branches (such as IF statements) need to be carefully monitored to make sure that they do not hinder the parallelization carried out by the compiler. The author's code follows these guidelines and is almost completely parallelized under the 'work station' software and compiler that is used (a commercial variety). The work station environment allows debugging, error detection, and compiling and running by a simple touch of a button. With built-in text editing, it greatly simplifies programming and computing.

7.3.1 The blocks

The first major block is shown in figure 7.9. The process of initialization means defining all fundamental constants and all materials constants, whether or not these are read from a file. In addition, all variables are assigned the size (number of bytes) and dimensions (for matrices). The last part of this is to generate the initial distribution function, whether it is a thermal equilibrium Maxwellian or an excited distribution arising from an ultrafast photon source. The valley index for each particle is set to 1 in this process, as the carriers are in the Γ valley for this example. The second part of this block is the generation of the scattering tables, one for each scattering process in a particular valley. As the masses vary from one valley to the next, this means the scattering rates for e.g. acoustic scattering are different in the different valleys, so three such tables are required for each scattering process. If the energy has a step of 1 meV and extends to 2 eV, this requires only 16 kB of main memory for each scattering process. If there are 50 various scattering processes, it still requires less than 1 MB for the entire table. This was a problem when computers had only 64 kB, but is no problem these days with fast GB memories.

The second step is to assign a scattering time for each particle in the simulation, according to (7.17). The following equations that describe the processes in the

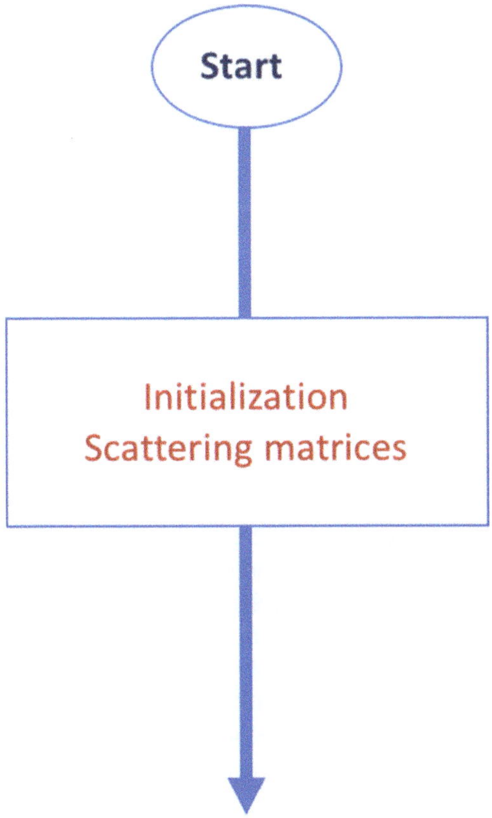

Figure 7.9. Initialization and pre-processing data generation and storage. The two steps are discussed in the text.

various boxes of the code will be written for hyperbolic bands, that are used in most semiconductors, and arise from $k \cdot p$ theory and often include the spin–orbit interaction. Focus will be on electrons in the conduction band, for which

$$\frac{\hbar^2 k^2}{2m_c} = E\left(1 + \frac{E}{E_G}\right) \tag{7.38}$$

or

$$E = \frac{E_G}{2}\left[\sqrt{1 + \frac{2\hbar^2 k^2}{m_c E_G}} - 1\right], \tag{7.39}$$

where E_G is the direct energy gap and m_c is the curvature mass at the bottom of the conduction band. The curvature mass here is used, as the basic definition of the effective mass

$$\frac{1}{m^*} = \frac{v}{\hbar k} = \frac{1}{\hbar^2 k}\frac{\partial E}{\partial k} \tag{7.40}$$

has an indeterminant value at the bottom of the band where both k and the derivative vanish. This has to be evaluated by L'Hopital's rule, which leads to

$$\frac{1}{m^*} = \frac{1}{\hbar^2}\frac{\partial^2 E}{\partial k^2} \tag{7.41}$$

at the band edge. This latter equation is often found in books, but it is forgotten that it only applies at the band edge. In general, the effective mass increases with energy as

$$m^* = m_c\left(1 + \frac{2E}{E_G}\right) \tag{7.42}$$

in hyperbolic bands.

In setting the scattering time, which will be taken as $t_s(i)$, where i denotes the particle of interest, it will be assumed that in this block time will run from 0 to ΔT, as described in section 7.2.3. Now, a decision is required as to whether or not the particle will undergo a collision in this time step of the global time, as shown in figure 7.10. If the decision is yes, then the momentum is updated with acceleration determined by $t_s(i)$. On the other hand, if $t_s(i) \geqslant \Delta T$, the acceleration is to last longer than the time step. However, the particle is now accelerated only up to the end of this time block. It has to stop at the end of the block so that ensemble averages can be taken, and forces updated if Poisson or molecular dynamics are included. This process will bring all particles to a common time at the end of the time step. Note that, if the scattering process is to occur, this is signaled by the setting of the flag variable to 2 for that particle.

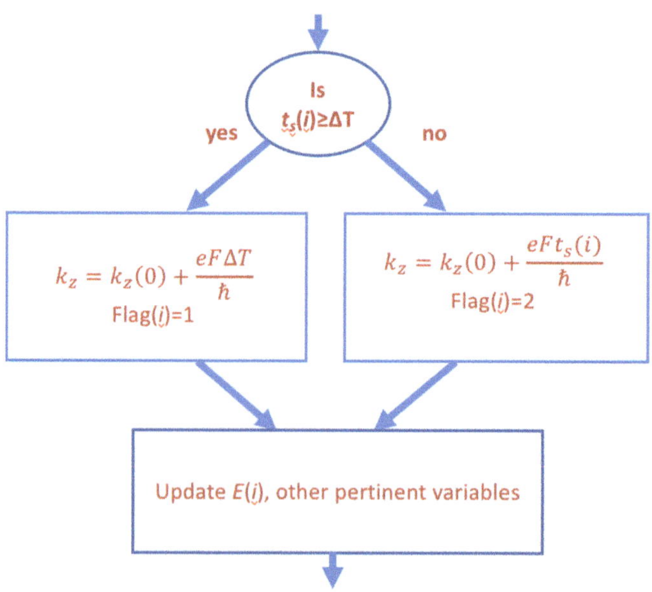

Figure 7.10. A decision has to be made as to whether the particle scatters within this time step of the global time.

The next step in the process is to choose the scattering process for each particle. These scattering processes are different in the three sets of valleys, so a check on the valley index has to be made. Only valley 1 will be describe here, as the process is the same for each set of valleys, and the example block is repeated three times, once for each valley. The outline for the process is shown in figure 7.11 (the entry point W will be described later). So, the first step is to check the valley index and select only those particles in the particular valley. Other electrons are shunted to the end of the block. The value of the flag also has to be checked, as those particles with a flag value of 1 do not undergo any scattering and are shunted to the end of the block. In a gather-scattering approach, those particles that have a valley index of 1 and a flag value of 2 would be gathered into a smaller matrix to optimize pipelining. With modern parallelism, this is not required. Once these particles are determined, then the scatterer selection follows figure 7.3, and the discussion at that figure. It should be concluded from figure 7.11 that, if the flag value remains at 2, the selected scattering process is self-scattering, as any real scatterer will change the flag value to a value depending upon whether the process is isotropic or anisotropic. The self-scattered particles will remain with their current momentum and energy, but they have to be distinguished from those whose flag is 1, as they have not yet been accelerated to the end of the time step. This will be accomplished in a later step. So, this step is the first part of the scattering and the final energy and momentum after the scattering is still to be determined.

The final scattering angles are selected based upon whether or not the scattering is isotropic. This is outlined in figure 7.12. Following this process, the final state energy is updated depending upon the *delE* value and the various momenta determined from the selected angles. This must be followed by the rotation in space, as the scattering angle is with respect to the initial momentum and not the electric field. This rotation is in textbook [26]. Nevertheless, there are some subtleties in this process, so it will be illustrated here. The momentum vector after the scattering

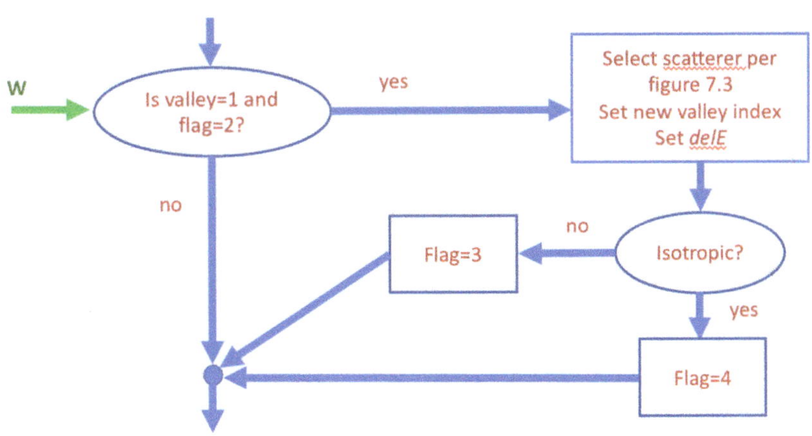

Figure 7.11. The first part of the scattering process is to select a scatterer from the tables. The physics of the selected process now dictates the energy gained or lost and whether or not the process is isotropic or anisotropic.

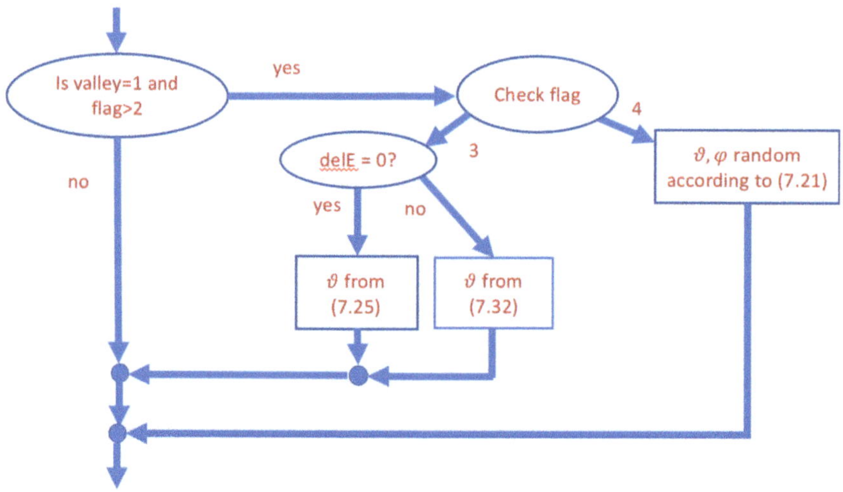

Figure 7.12. Selection of the final angles for the scattered particle.

process of figure 7.12, lies in a rotated coordinate system in which the new angles ϑ', φ' (the primes are used to denote the angles determined in this latter figure) are with respect to a coordinate system in which the initial momentum is the polar axis. In fact, the initial momentum is defined by angles ϑ, φ with respect to the laboratory coordinates in which the electric field is oriented along the z-axis. In the rotated coordinates, k' is defined as

$$K' = \mathbf{k}' = \begin{bmatrix} k'_x \\ k'_y \\ k'_z \end{bmatrix} = \begin{bmatrix} k'\sin\vartheta'\cos\varphi' \\ k'\sin\vartheta'\sin\varphi' \\ k'\cos\vartheta' \end{bmatrix}. \tag{7.43}$$

The rotation into the laboratory frame has two parts. First, the momentum vector is rotated around the z-axis to bring it into the (y, z)-plane. This rotation angle is $(\pi/2) - \varphi$. Note that these rotations work on the unprimed angles. As

$$\begin{aligned} \cos\left(\frac{\pi}{2} - \varphi\right) &= \sin\varphi \\ \sin\left(\frac{\pi}{2} - \varphi\right) &= \cos\varphi \end{aligned}, \tag{7.44}$$

this first rotation has the matrix

$$R_1 = \begin{bmatrix} \sin\varphi & \cos\varphi & 0 \\ -\cos\varphi & \sin\varphi & 0 \\ 0 & 0 & 1 \end{bmatrix}. \tag{7.45}$$

The second rotation involves a rotation around the x-axis by the angle $-\vartheta$. This produces the rotation matrix

$$R_2 = \begin{bmatrix} 1 & 0 & 0 \\ 0 & \cos\vartheta & -\sin\vartheta \\ 0 & \sin\vartheta & \cos\vartheta \end{bmatrix}. \tag{7.46}$$

Now, the new values for the components of the momentum are found from

$$K_{\text{new}} = R_2 R_1 K'. \tag{7.47}$$

The next step is a reacceleration of the particles that were selected for scattering. The process is a variant of figure 7.10, and becomes a little more delicate requiring some careful book-keeping. This is because each particle already has a value for $t_s(i)$, yet a new one is going to be determined. As an intermediate step, a value is determined for

$$\tau(i) = \Delta T - t_s(i). \tag{7.48}$$

A new $t_s(i)$ is determined from (7.17) and added to the old value, as

$$t_s(i) = t_{s,\text{old}}(i) + t_{s,\text{new}}(i). \tag{7.49}$$

With these values, we can repeat figure 7.10 as in figure 7.13. But, carefully note the difference in the acceleration times, as these are critical. In addition, those electrons that still have not reached the end of the time block must be returned to the scattering process, and the exit W is the entrance W that was mentioned in figure 7.11. Thus, the scattering blocks will be repeated until all the electrons reach the end of the time block (ΔT). Once this has occurred, the global time and the values of $t_s(i)$, are updated according to

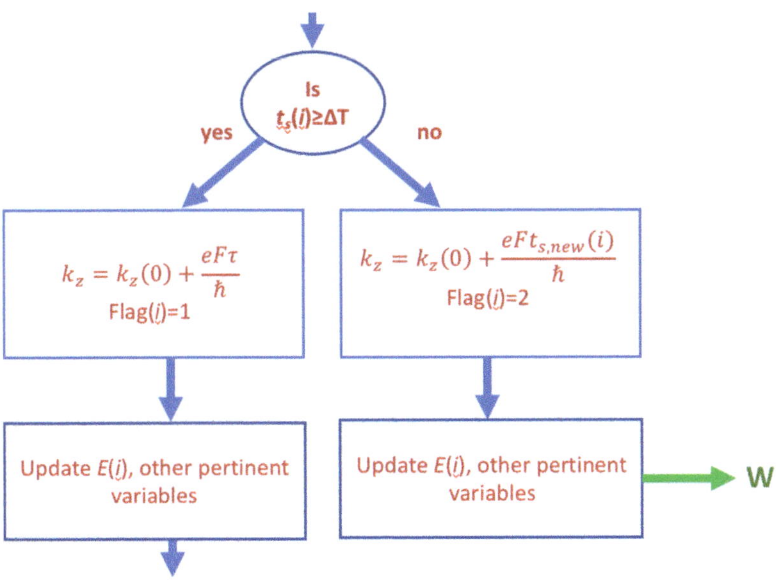

Figure 7.13. The particles are re-accelerated after the scattering process. If they haven't reached the end of the time block, they are returned to undergo another scattering process.

$$
\begin{aligned}
T &= T + \Delta T \\
t_s(i) &= t_s(i) - \Delta T^{\cdot}
\end{aligned}
\tag{7.50}
$$

The last line puts all of the scattering times back into the new time block. At the end of the code, any number of data sets can be collected and plotted to show what has occurred during the program.

7.3.2 Examples

To illustrate the data that can be determined from the EMC process, an example will be used to show the behavior of transport in $In_{0.53}Ga_{0.47}As$ at room temperature. This material has a direct band gap of approximately 0.77 eV, and is a multi-valleyed semiconductor in that the Γ, L, and X valleys are all important. The material is used in a variety of electronics applications, including both lasers and hot carrier solar cells. Here, however, it will be assumed that the interest is in the transport properties of bulk semiconductors. The velocity as a function of the applied electric field is shown in figure 7.14. The most obvious feature of the curve is the large peak at ~2.5 kV cm^{-1}, and then the drop. This is the well-known onset of inter-valley transfer, and the drop is due to the significant reduction of carriers remaining in the Γ valley. The mobility is significantly lower in the satellite valleys, and this is the reason the drift velocity drops rapidly. At fields above ~15–20 kV cm^{-1}, the velocity in the satellite valleys is saturated at a value slightly below 10^7 cm s^{-1}, and this gives the valley velocity seen in the curve at these fields. The energy and the populations in the various valleys are shown in figures 7.15 and 7.16. The energy in this case is the total average energy per carrier referenced to the bottom of the conduction band at Γ. For carriers in the satellite valleys, this energy

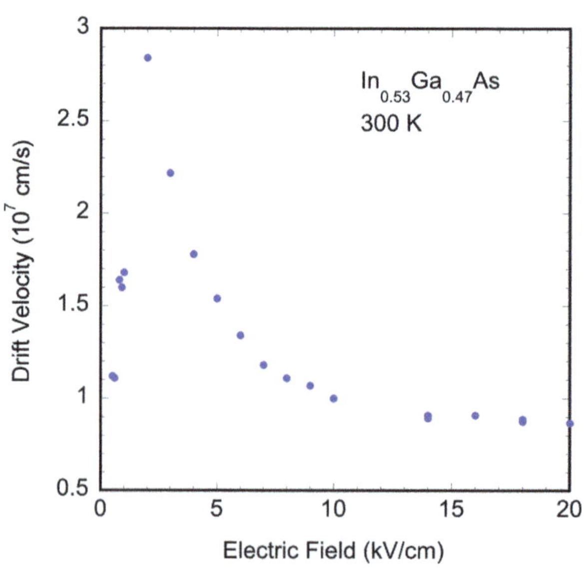

Figure 7.14. The velocity as a function of the electric field applied to it. The shape is discussed in the text.

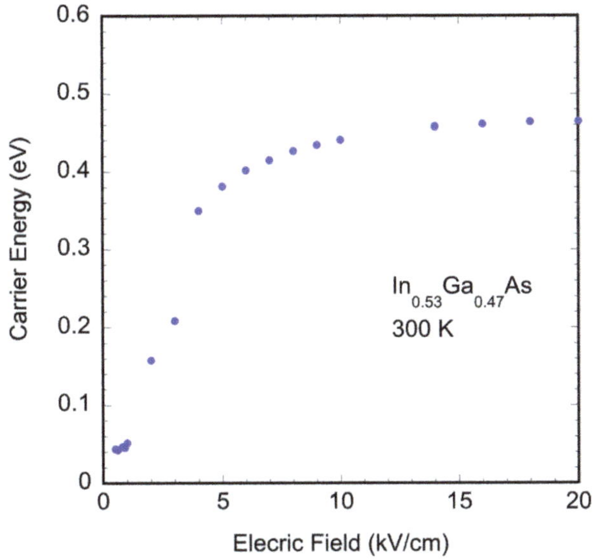

Figure 7.15. The variation of the total average energy as a function of the electric field.

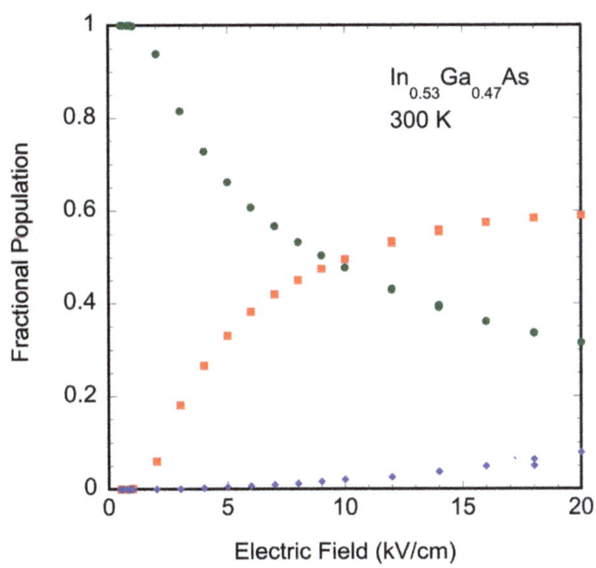

Figure 7.16. The fraction of carriers in the three sets of conduction band valleys.

includes their potential energy due to the higher band minima for these valleys. Also obvious is the fact that the carriers are largely distributed between the Γ and the L valleys, with occupation of the X valleys occurring only at the higher values of the field.

In figure 7.17, the transient velocity is shown as a function of time for several values of the electric field. At 10 kV cm^{-1}, it is quite clear that the velocity has a substantial overshoot effect, as discussed in chapter 2. Were the carriers to remain in

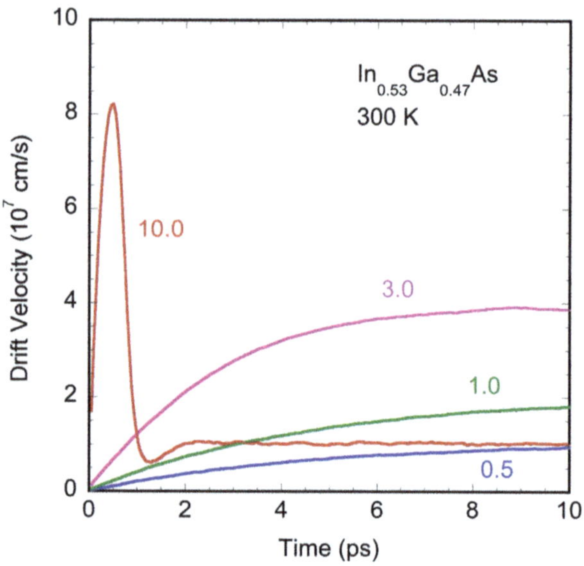

Figure 7.17. The transient velocity as a function of the time after the field is applied. The parameter is the electric field in kV cm^{-1}.

the Γ valley, they would unlikely have the peak velocity as their steady-state, as the energy relaxation process is slower than the momentum process, even with polar phonon scattering dominating. Even the 3 kV cm^{-1} curve is showing some overshoot, as the largest value shown in range of the plot is some 50% higher than the value in figure 7.14. However, this field is just over the peak of the curve, and it may take 15–20 ps before the velocity settles to its steady-value shown in figure 7.14. Obviously, the response is much quicker at 10 kV cm^{-1}. The transient response of the valley population is shown in figure 7.18 for a field of 10 kV cm^{-1}. One can see that the transfer is already well underway at 1 ps, and this is past the peak velocity at this field in figure 7.17. It may also be noted that there is a delay of ~0.5 ps before the transfer begins. This is simple physics, as it is assumed that the field is applied at $t = 0$. If we assume $k(0) = 0$ in figure 7.10, then the momentum rises ballistically according to

$$k(t) = \frac{eFt}{\hbar}, \tag{7.51}$$

and using this in (7.38) for hyperbolic bands, the time required to reach 0.5 eV of this order of magnitude. So, the delay is required to accelerate the carriers to sufficient energy to scatter to the satellite valleys. This delay is strictly seen in the case for electric field generation of hot carriers. Were the hot carriers introduced through an intense optical pulse, there is would be no need for an accelerative delay in the transfer, as was obvious in figure 4.1, where inter-valley transfer was already observed only 20 fs after the lsaer pulse.

Another data collection process gives the distribution function. For this, a grid is created in energy space and then the particles are projected onto this grid to obtain

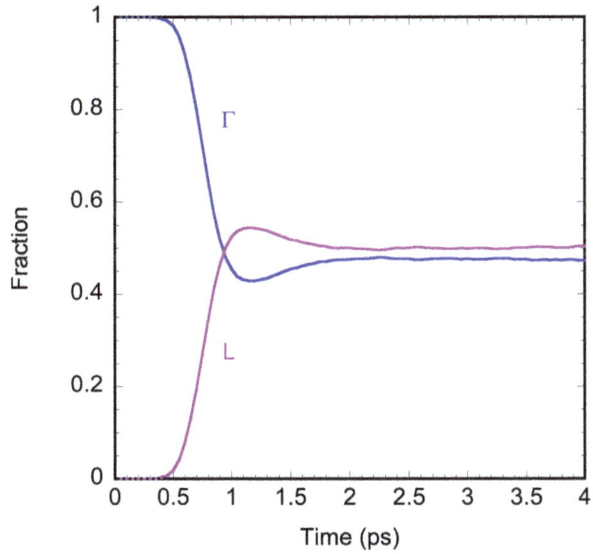

Figure 7.18. The temporal response of the valley population in the Γ and L valleys for an applied field of 10 kV cm^{-1}. The delay time before transfer occurs is attributed to the need to accelerate the carriers to sufficient energy to begin to scatter to the L valleys.

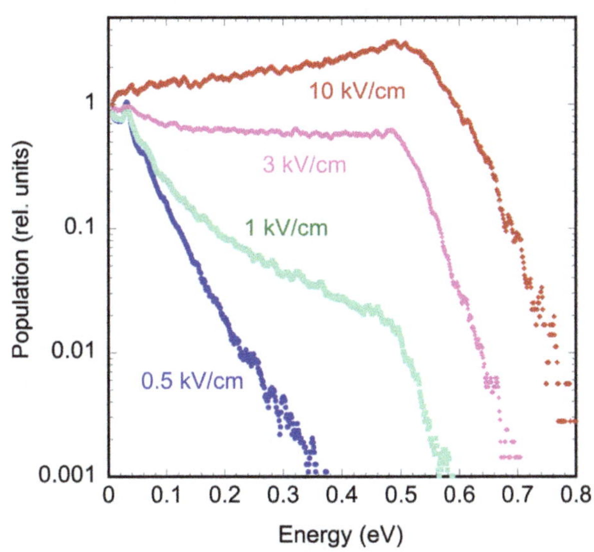

Figure 7.19. The normalized population as a function of energy for several values of the electric field.

the population of the carriers. This population for the Γ valley is shown in figure 7.19 for several values of the electric field. Note that this is not the distribution function but is the population discussed in section 6.4, which is the product of the distribution function and the density of states. The latter is complicated in the hyperbolic bands, due to the energy dependence of the effective mass, but it can be found in a

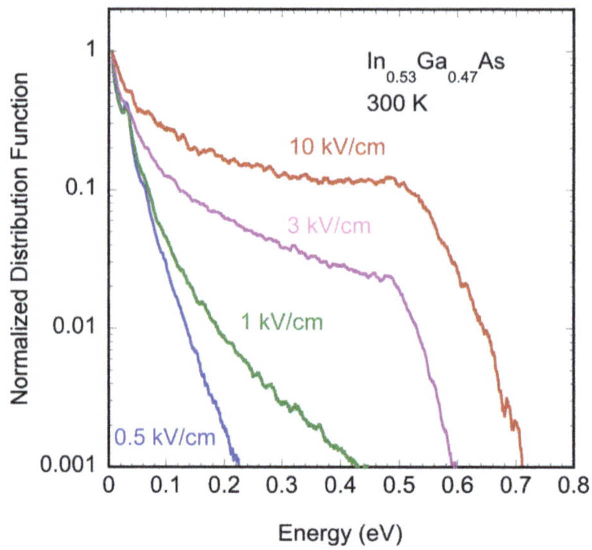

Figure 7.20. The normalized carrier distribution function as a function of energy for the same field values as in figure 7.19. The difference in this curve from the previous is the density of states, as discussed in the text.

straightforward manner. For the hyperbolic bands of (7.38), the standard parabolic band density of states is enhanced by a factor of

$$\sqrt{1 + \frac{E}{E_G}}\left(1 + \frac{2E}{E_G}\right). \tag{7.52}$$

The population data can be divided by the density of states and renormalized to give the actual distributions. These distribution functions are shown in figure 7.20 for the same values of electric field as used in figure 7.19. In both cases, there is a small 'bump' at low energy (32 meV), which corresponds to the polar optical phonon energy. While there are likely some artifacts at low temperature, due to both the low population and low density of states at low energy, this small peak is likely real and is well-known to originate from this phonon emission onset. The more obvious kink at 0.5 eV is due to the onset of inter-valley scattering and the reduction of carriers in the Γ valley when this process begins. While one might be able to assign a temperature for the lower electric fields, this becomes less possible for fields above the peak of the velocity curve. However, even for only 1 kV cm^{-1}, the distribution function does not decay as a simple exponential. And the deviation from a simple exponential becomes more pronounced as the field is raised.

7.4 Molecular dynamics and Poisson

In the above, the carrier–carrier interaction, and the carrier–impurity interaction have been treated as a normal perturbative scattering process in momentum space, but doing so in real space has been mentioned. That is, it is possible to treat the carrier–carrier interaction in real space and time. In this latter approach, no approximations or perturbative expansions are required, although other adaptations may seem useful. Of course, problems arise which have a somewhat different nature and cause these adaptations. Such problems can be discussed as they come along. The real-space approach is an adaptation of molecular dynamics, in which the forces between the particles are updated after each time step. That is, the force on each carrier, due to the other carriers and the charge impurities is computed anew at each time step in order to maintain the temporal evolution of these forces. These forces are directly computed from the Coulomb interaction in real space [11, 27]

$$e\boldsymbol{F} = -\sum_{j\neq i}\frac{e^2}{4\pi\varepsilon_s(\mathbf{r}_i - \mathbf{r}_j)^2} + \sum_{j}\frac{eq_j}{4\pi\varepsilon_s(\mathbf{r}_i - r_j)^2}. \tag{7.53}$$

The first term on the right is the carrier–carrier interaction. Note that this is the bare Coulomb interaction, as the screening will naturally occur due to the Coulomb interactions as the number of electrons increases. The second term on the right is the background charge due to the ionized impurities. This background charge may often be taken as a uniform background charge if the system is sufficiently large, but often the discreteness of this charge should not be neglected [10]. Once computed at the time step, this force is applied to the carriers during the next time step.

As in other molecular dynamics calculations [28, 29], the electrons are confined to a box in real space. If N is the number of particles in the EMC and n is the electron density for the simulation, then the box has size $\Omega = N/n$. In fact, two boxes are used. The first corresponds to the laboratory coordinates for the simulation. The second box is centered upon the carrier at position i in (7.53), and the force is computed using this box. Even for thousands of carriers, there is a need to use this second box to avoid introducing a bias in the force due to the fact that computation in the first box is not likely to average to zero, particularly if the particle of interest is not located at the center of the first box. Treating a large number of particles via molecular dynamics is traditionally a difficult problem [28] and a computationally *hard* problem, meaning that it is time consuming. Nevertheless, it can be done for a reasonable number of particles with modern parallel computers [30]. In dense systems, with degenerate levels of particles, there is also a need to correct for the inter-particle exchange interaction [31]. This degree of difficulty will not be discussed here, as it is usually not required, but the interested reader may refer to this latter reference where it is discussed in detail.

The problem becomes simplified if the system is inhomogeneous and requires a solution to Poisson's equation. This is because, the latter already includes the Hartree interaction arising from the carrier–carrier interactions, e.g., the force arises from a uniform, or nearly so, background spatially inhomogeneous charge density. It is the short-range particle correlations that arise in (7.53) that are not included with the Poisson equation. This is because, the mesh potential is not accurate on length scales smaller than the grid spacing. So, one brute force method to overcome this double counting is to separate the total Coulomb force and the mesh force to yield a short-range molecular dynamics force. The separation of the inter-particle

Coulomb force from the Poisson forces has been discussed for quite some time. Such a split was used by Kohn for electronic structure calculations, where inter-electronic forces have to be added to the potential solutions from Poisson's equation for the atomic potentials in the density-functional approach [32]. The split was also used earlier in order to treat molecular dynamics problems in plasma physics where the short-range interaction also appears between ions and electrons [33]. This general splitting of the Coulomb force has been shown to be useful in general many-body problems [34]. The general approach is similar to dividing a delta function, as

$$\delta(r) \rightarrow \delta(r) + f(r) - f(r). \tag{7.54}$$

In a similar manner, the Coulomb potential can be split into short-range and long-range parts as

$$\frac{1}{r} \rightarrow [1 - \mathrm{erf}(r)]\frac{1}{r} + \frac{1}{r}\mathrm{erf}(r). \tag{7.55}$$

The first term on the right-hand side is a short-range function that vanishes as the magnitude of r increases. On the other hand, the last term on the right-hand side is a long-range function that vanishes at short distances. This is, of course, the entire principle of the potential splitting in which the long-range function is found from Poisson's equation. Using a splitting of the form of (7.55) makes the Poisson's equation more difficult as the actual charge density has the more cumbersome multiplicative function. Nevertheless, the time saving from computing the molecular dynamics forces over only a short distance easily makes up the additional work for the Poisson equation.

When we actually compute the mesh force and the proper Coulomb force, using the short-range difference actually accomplishes the split more effectively. The new short-range force has several advantages. First, it is a relatively universal force curve. Secondly, because it is relatively universal, it can be tabulated for use in the actual molecular dynamics program. Third, the sum no longer runs over all electrons, but only those that lie within a short range of the particle of interest, because the short-range force dies off so quickly. There is also the advantage that, at high densities where the grid size usually has to be small, a larger grid size can be used with the corrections coming from the short-force. This makes both the Poisson solution and the molecular dynamics more efficient in computer resources [35]. In the following, the homogeneous case will be dealt with first. The Poisson solutions and the split force will follow thereafter.

7.4.1 A homogeneous semiconductor

First, when charged impurities are included, there are two terms in (7.53) and (7.55). The first term represents the force on the particle from other free particles, and the second is the force on the particle from the charged impurities. In a homogeneous semiconductor, we can ignore the second term if we treat the impurity scattering by the normal momentum space interaction. Or, we can also include the impurities in the molecular dynamics simulation [10]. As mentioned, using a finite number of

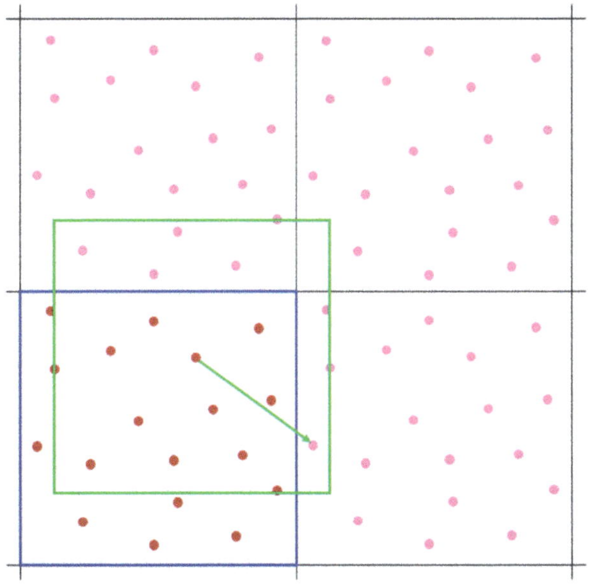

Figure 7.21. Various boxes for determining the inter-particle forces in molecular dynamics. The proper simulation box is outlined in dark blue with red particles. The centered box for computing the forces on a particle is the green box that yields a proper centered potential. Replicas of the simulation box have the magenta particles. The green arrow shows the inter-particle vector for computing the force on the particle within the blue box.

particles introduces a three-dimensional box, whose size is determined by the number of particles and the doping level considered in the simulation. The volume of this box is given as $\Omega = N/n$. The appearance of this induced box can affect the simulation by generating artefacts, and this must be avoided as much as possible. The artefacts arise because the box introduces an artificial periodicity into the simulation, and the effects of this periodicity have to be removed.

The Coulomb force is considered only within the shortest interconnecting length between any two particles, and this means that we need two boxes in the simulation. This is illustrated in figure 7.21. First is the box introduced by the size of the volume given in the previous paragraph (this box is the dark blue box in figure 7.21). The second box has to be centered on the particle at which the force is being computed, so that artificial non-centrosymmetric forces are not generated in the force calculation (this is the green box in figure 7.21). Thus, when the inter-particle force is computed at each time step, each of the other simulated particles interact with the particle of interest only through the shortest possible vector, which may be shifted by a lattice vector of the artificial periodic structure. Replicas of the original box and the particles in this box also are shown in figure 7.21 as well. Since the simulated volume may not correspond to each electron interacting through the shortest distance with the carrier of interest, replicas of the electrons that lie within the second neighbor boxes must be used in computing the forces through the well-known form for the electric field at the particle of interest, determined from (7.55), as

$$F_i(\mathbf{r}_i) = F_{\text{ext}}(\mathbf{r}_i) - \frac{e}{4\pi\varepsilon_s}\sum_{j\neq i}\frac{1}{r_{ij}^2}\mathbf{a}_{ij}, \tag{7.56}$$

where r_{ij} and $\mathbf{a}_{ij}$ are the length and unit vector direction of the vector between the two particles. The unit vector points *from* the particle of interest *toward* the interacting particle.

If a sum were to be computed over all the particles *and* their replicas, it would introduce superlattice effects, due to the size L of the box. This may be seen from

$$\begin{aligned}\sum_{\text{cells}}\frac{1}{r} = \frac{1}{r} + \sum_{L\neq 0}\frac{1}{|\mathbf{r}+L|} \sim \frac{1}{r}\\ +\sum_{L\neq 0}\frac{1}{L} - r\sum_{L\neq 0}\frac{1}{L^2} + r^2\sum_{L\neq 0}\frac{1}{L^3} + \cdots\end{aligned} \tag{7.57}$$

The denominator has been expanded, since in general $L > 3r$ when the centered cell is used for the direct-field calculation. The summations over the lattice vectors are sometimes called Ewald sums (although these were developed in Fourier space) [36], since they were developed for X-ray scattering in periodic structures. The method of evaluating these sums has been studied for quite some time in connection with molecular dynamics in other fields [37–39], and have well-developed formulas. This approach has been shown to be very effective in determining the inter-carrier scattering in femtosecond laser excitation of semiconductors [30]. Notice that the third- and higher-order terms add radially increasing terms to the summation, which complicates the artifacts themselves.

When (7.57) is multiplied by L, the expansion is in terms of the normalized distance L/r. Then, the resulting sums are independent of r, appearing only by the normalized distance r/L. The sum terms then become sums over the defining vectors for the box, and these sums have been worked out for many configurations. With this expansion then, the force needs to be calculated only over the actual number of particles that are in the simulation box N, provided that the particle-centered box is used in each case. To these sums is added the correction terms, which tend to be universal. If the number of actual particles used is large, than the correction terms are small. One rule of thumb is to make the edge of the box larger than the computed screening length, and use accordingly the number of particles to make this happen.

Once the electron distribution has achieved its steady-state, some information can be found from the radial distribution function. This function is the probability of finding carriers at a distance r from any individual carrier; that is, it is the number of carriers lying in a thin shell of radius r centered upon the carrier of interest. It is usually averaged over all of the carriers. It may be defined as

$$\rho_2(r) = \frac{1}{4\pi r^2}\int_0^{2\pi}\int_0^{\pi}\int_r^{r+\Delta r}\rho(\mathbf{r}-\mathbf{r}_i)r^2 dr\,\sin\vartheta d\vartheta d\varphi. \tag{7.58}$$

The normalizing factor in front is the area of the shell itself. If the distribution is homogeneous, the radial distribution function should be uniform at a value of unity.

With screening, however, other particles are pushed away. But, the function can be computed easily within the EMC.

In figure 7.22, the radial distribution function is computed for Si electrons at 77 K and with a density of 10^{16} cm^{-3}. Curves are shown for three different electric fields. While the number of electrons used here is small [27], strong effects can already be seen in the function. At the lowest field, 10 V cm^{-1}, the electron system shows no heating effects and it is very nearly in thermal equilibrium. The Debye length λ_D is indicated in the figure, and it is clear that the central particle exhibits Debye screening effects. At 5 kV cm^{-1}, the field has already reduced the screening, which apparently cannot keep up with the motion of the particle, which suggests the screening is a relatively slow effect. At the highest field, 50 kV cm^{-1} which is well into the saturation region of the velocity-field curve, a strong peak is observed in the radial distribution function, followed by a dip. This latter curve has a lot of similarity to the overshoot velocity in Si, but is likely the result of descreening the inter-particle interactions in the high electric field. The inset to the figure shows the first Legendre polynomial expansion coefficient on the radial distribution function showing that it is quite asymmetric in space. At a higher carrier concentration, the carrier–carrier interaction is much stronger and the asymmetry of the radial distribution function is strongly reduced. In figure 7.23, the function is shown for the same fields and temperature, but at a carrier density of 10^{18} cm^{-3}. Even the

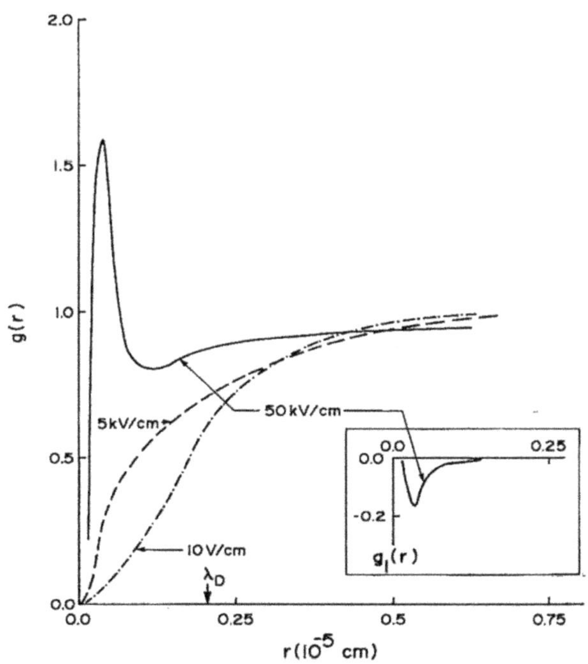

Figure 7.22. The radial distribution function for electrons in Si at 77 K, with a density of 10^{16} cm^{-3}. The parameter is the electric field applied to the system. The inset shows the first Legendre polynomial coefficient for a spherical harmonic expansion of the 50 kV cm^{-1} curve, illustrating the asymmetry of the function. Reprinted with permission from Lugli and Ferry [27], copyright (1986) by the American Physical Society.

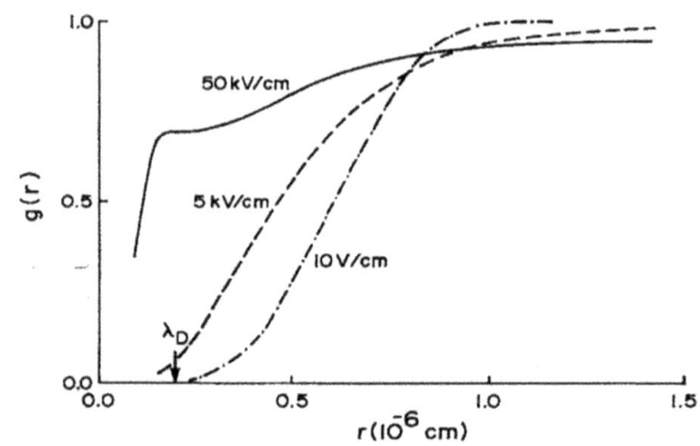

Figure 7.23. The radial distribution function for electrons in Si at 77 K, with a carrier density of 10^{18} cm^{-3}. The parameter is the electric field applied to the system. Reprinted with permission from Lugli and Ferry [27], copyright (1986) by the American Physical Society.

screening at low electric field is much stronger, extending well beyond the indicated Debye length. Moreover, the reduction in screening created by the electric field is much reduced, even at 50 kV cm^{-1}.

7.4.2 Solutions with Poisson's equation

One of the most significant uses of transport and simulation is in modeling the real semiconductor devices that are being produced by industry, whether they are the ubiquitous MOSFET, or other devices such as microwave HEMTs or solar cells for energy conversion, and EMC has been heavily involved [40, 41]. In these simulations, the internal self-consistent potential, that arises as charges (free carriers) move (or not when the impurities are included), is updated regularly, using solutions of Poisson's equation. Generally, this means that a grid is superimposed on the device, and the potential is determined by Poisson's equation on these grid points (the solution is often referred to as the mesh potential, as discussed in the previous section). Here, the grid is *not* part of a periodic structure, but must replicate the actual device structure. We will discuss this only in terms of the grid used to develop finite-difference equations, although commercial software more often uses finite elements to solve Poisson's equation [42]; the differences in these approaches is not important to the points we want to discuss in this section.

The potential that comes from solving Poisson's equation in the confined dimensions of the actual device leads to additional problems. First, one must incorporate the complicated set of boundary conditions that are imposed upon the solution. These boundary conditions may be actual potentials at some boundaries and forced derivatives of the potential at other boundaries. The potentials (and their derivatives) are usually different at different boundaries. In some cases, boundaries may be specified by boundary and/or surface charges, and these charges must be incorporated within the general Poisson equation scheme. Second, the internal charges, as well as any boundary/surface charge, seldom sits exactly upon the grid point at which Poisson's equation will be saved. Hence, one must use a method of projecting the charge onto the grid point in an equitable method that does not introduce artificial forces. There are many such approaches, and we will not discuss those here [43].

As was discussed above, the existence of the mesh introduces a cutoff at short distance for the actual potential variations. That is, the grid spacing provides an upper limit to the momentum wave vectors that can be simulated, just as it does in band structure. The periodicity of the grid itself introduces an artificial band structure that is superposed upon the real one. The importance of this limitation is that the mesh potential is no longer the complete carrier–carrier interaction potential that would be found by detailed molecular dynamics. This provides an easy way to implement the force splitting discussed in (7.55). The problem is now reduced to determining the best cutoff distance for which to separate the forces. One way to find this distance is to place a fixed particle in a position that is not one of the grid points. Then use a test particle that is moved around in the grid, and past the fixed particle, solving Poisson's equation for each position of the test particle. In figure 7.24, the results of such a test are plotted in terms of the magnitude of the electric field acting upon the test particle [44]. The true Coulomb field has a singularity at the position of the fixed particle because the distance in (7.55) goes to zero. However, the mesh field deviates from the true Coulomb field near the fixed particle, primarily because of the cutoff in allowed momentum states. The difference in the two potentials is a very short-range potential, and it is this potential that should be used in the molecular dynamics calculation.

In the source and drain regions of a device, one has to be careful about the implementation of the molecular dynamics treatment of the short-range force. In these regions, the charged impurities are of the opposite sign to the particles

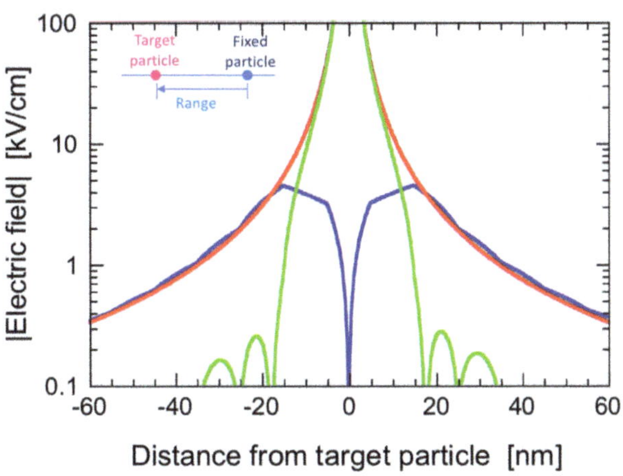

Figure 7.24. Forces acting upon a 'target particle' from a 'fixed particle'. The red curve is the exact Coulomb force, while the blue curve is the mesh force from a Poisson solution to the potential with a grid size of 10 nm. The difference is the green curve which is the short-range potential to be used in the molecular dynamics calculation in situations in which Poisson's equation is also solved. Adapted from Gross *et al* [44], copyright (1999) by the IEEE.

being considered. This means that the force between the impurities and the particles is positive and can lead to an unphysical attractive force. This will inevitably lead to unphysical trapping of the particles by the charged impurities, because the impurity doesn't respond to the thermal excitation of the particle. It has to determined just how to cut off this attractive potential to properly simulate thermal excitation and avoid trapping. To handle this problem, one needs to modify the short-range force in the proximity of the charged impurity, usually when the range is below 2 nm, which is approximately the Bohr radius of the charged impurity.

Effectively, one needs to truncate the short-range force in the region close to the impurity, and there are various ways to do this. Three methods to accomplish this truncation have been discussed in the literature [45]. These include: (a) a sharp cutoff of the potential at the Bohr radius, (b) a linear decrease of the force, and (c) setting the force to a constant for the small values of the range. All of these approaches have been evaluated, and (a) and (b) seem to be the most effective [45]. Selecting the most reasonable approach is based upon the average energy of the particles at very low values of the applied electric field. A free particle traveling through the semiconductor should have an average energy of $3k_BT/2 \sim 38.8$ meV. Merely using the mesh force from the Poisson solver, gives an average energy well above the physical value, which is often a problem with EMC techniques that do not include the carrier–carrier forces. The three approaches to the truncated inter-particle forces all come reasonably close to the desired average energy per particle, but only the linear force (b) gives a truly small noise about the proper energy. On the other hand, if corrected for this attractive force is allowed to stand without correction, the average energy drops down to only about 2 meV, which clearly indicates a trapped particle.

However, it turns out that a more physical approach can be utilized that actually gives somewhat better results. In modern nano-scale semiconductor devices, physical dimensions in some cases are of the order of 10–20 nm or less. Quantum effects have begun to appear both in experiment and in theoretical simulations [46]. But, one must begin to ask just how large the electron (or hole) is quantum mechanically [47]. Certainly, in classical physics, the electron is incredibly small. But, this is not the case in quantum mechanical systems, particularly in semiconductor devices. In the small devices mentioned, the key question is how to fit the electrons (or holes) into these small structures. The size of the electron has been argued for some time, and becomes important in determining an effective potential [48]. This discussion is incredibly important not only for fitting the electrons into a small device, but in determining how to cut off the impurity's attractive potential. The idea is to use an effective potential.

Effective potentials have appeared in the literature for many decades, particularly in regard to the question of how quantum effects modify equilibrium thermodynamics. As early as 1926, Madelung developed a set of hydrodynamic equations from the Schrödinger equation and showed that an additional quantum potential was present in the formulation [49] (later made famous by Bohm [50]). Shortly after this, Kennard demonstrated that quantum particles would exactly follow classical potential forces plus those from any additional quantum potentials [51]. Wigner followed by developing the Wigner function and showing that the first-order correction to the classical thermodynamic potential was a term that depended upon the spatial second derivative of this classical potential [52]. Somewhat later, Feynman and Hibbs showed that the normal partition function would have a quantum correction that involved the second derivative of the classical potential, but could be much more complicated in nature [53]. Finally, it was shown that the effective potential could be found from an effective size of the particle [47].

The connection between the size of the particle and a correction to the thermodynamic partition function arises easily when it is recognized that the effective quantum size of the particle depends upon the statistical properties [48]. If the particle is part of a dense high density Fermi gas, then the effective size of the particle is approximately

$$\delta r = 2\Delta r \sim \frac{\lambda_F}{\pi}, \tag{7.59}$$

where λ_F is the Fermi wavelength and $\Delta r \sim 1/k_F$ can be considered to replicate the physical extent of the wave function. On the other hand, if the carrier density is non-degenerate in the semiconductor, then the corresponding size of the particle can be thought to be

$$\delta r = 2\Delta r \sim \sqrt{\frac{3}{8}}\lambda_D \sim 0.61\lambda_D, \tag{7.60}$$

where λ_D is the Debye length used previously. This becomes important when one writes the total energy of a potential term in a system Hamiltonian. This takes the form

$$H_V = \int d\mathbf{r} V(\mathbf{r}) n(\mathbf{r}). \tag{7.61}$$

It may be assumed that the size, or extent of $|\psi(r)|^2$, of each particle can be described as a Gaussian with a standard deviation (one half the particle size) α. Then, the density can be written as

$$n(\mathbf{r}) = \sum_i n_i(\mathbf{r}) = \sum_i \int d\mathbf{r}' \exp\left(-\frac{|\mathbf{r} - \mathbf{r}'|^2}{\alpha^2}\right)\delta(\mathbf{r}' - \mathbf{r}_i), \tag{7.62}$$

where r_i is the classical position of each particle. With this expression, (7.61) can be rewritten as

$$\begin{aligned} H_V &= \int d\mathbf{r} V(\mathbf{r}) n(\mathbf{r}) \\ &= \sum_i \int d\mathbf{r} \delta(\mathbf{r} - \mathbf{r}_i) \int d\mathbf{r}' V(\mathbf{r}') \exp\left(-\frac{|\mathbf{r} - \mathbf{r}'|^2}{\alpha^2}\right) \end{aligned}. \tag{7.63}$$

Here, the two integration variables have been interchanged, as both are essentially dummy variables in the double integral. The first integral and the summation treats the particles as true point charges and the Gaussian smoothing has been transferred

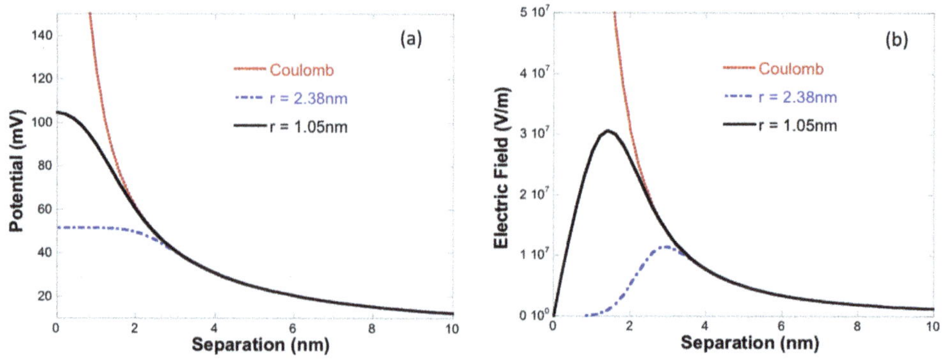

Figure 7.25. (a) The modified effective potential for the donor and (b) the electric field arising from that donor. Two different cutoff distances are used, and the smaller distance compares very well with the linear cutoff discussed in [45], although it is certainly smoothed by using the *effective potential*. This is described in the text and in [56, 57].

to the potential to give an *effective potential*. In the case of the MOSFET, the effective potential is a very smoothed version of the sharp high interface barrier between the oxide and the channel and the almost linear field in the Si itself due to the depletion charge. The result of the smoothing moves the classical charge away from the interface and raises the bottom of the conduction band just as the quantization in z-momentum would do. Both of these are observable effects in the characteristics of the MOSFET [54, 55].

As mentioned, there is a problem when treating ionized impurities that have the opposite sign of the particle, due to the strong attractive force and trapping that can occur. The potential truncation discussed above gives very good results, but it is ad hoc in nature. Using the effective potential to smooth the impurity potential gives a better way of handling the short-range force [56]. This approach has two steps in the process. First, the charge of the impurity atom is considered to be uniformly distributed within a sphere whose radius is the Bohr radius (approximately 2 nm as discussed above). When the charge is uniformly distributed, the field inside the sphere increases almost linearly with distance. Then, the total potential of this charged sphered is smoothed with the effective potential according to (7.63). The results of this approach are compared with the classical Coulomb potential in figure 7.25. Here, the smoothing is given by $\alpha_y = 0.52$ nm and $\alpha_x = 1.14$ nm. One can discern only a slight difference in the two quantum curves, but both peak at 50.7 nm, close to the ionization energy of the donor. In Si, the phosphorous donor ionization energy is 45 meV and that of arsenic is 54 meV. This approach is found to give excellent results for simulation of MOSFETs using Monte Carlo techniques with the molecular dynamics simulations [56, 57].

7.5 Real-space transfer

By real-space transfer, it is not meant to convey the normal motion of carriers in response to applied electric (and magnetic) fields, but to the transfer of carriers for example from a quantum well to the adjoining layers over a band offset barrier [58] (see the discussion in section 2.7). Such real-space transfer from GaAs quantum wells into AlGaAs barriers was treated first by EMC by Glisson *et al* [59]. In these simulations, one must also make sure the carriers carry a position variable along with the other dynamic variables, just as must be done in device simulations. Clear negative differential conductance was observed in these studies, as the carriers moved from the light mass GaAs to the heavy mass AlGaAs.

In simulations of GaAs/AlGaAs HEMTs, it was found that real-space transfer, as well as inter-valley scattering, occurred, and both of these effects limited the available velocity overshoot in the channel [60]. However, it was found that if the aluminum concentration was sufficiently high that the interfacial barrier was larger than the separation between the Γ and L valleys in GaAs, the real-space transfer was minimized. In subsequent work, it was demonstrated that gating of the channel would also minimize such real-space transfer [61].

Hess and colleagues also studied vertical transport through AlGaAs/GaAs/AlGaAs structures [62]. A condition for transfer from the GaAs to the AlGaAs (and the reverse transfer) required not only an energy above the barrier, but also conservation of energy and transverse momentum. They also studied real-space transfer HEMTs, in which carriers could transfer from a normal InGaAs channel over an AlGaAs barrier into a GaAs layer, to gain significant negative differential conductance [63].

7.6 Full band Monte Carlo

In the previous sections, the electric field was determined from the applied voltage and any self-consistent potential. However, the work was done in the effective mass approximation, usually with parabolic bands, but occasionally with the non-parabolic hyperbolic bands that arise in $k \cdot p$ theory. But, one can go further when it is desired to characterize completely electrons (or holes) that reach very high kinetic energies. In this case, the full band structure throughout the Brillouin zone can be used. The first to do this was apparently Shichijo and Hess [8]. The band structure for GaAs was determined with an empirical pseudopotential method [64]. This was a local pseudopotential without including the spin–orbit interaction [18]. The idea was to study impact ionization in GaAs, and the authors wanted a better description of the bands for carriers at the high energy required for this. The rationale is illustrated better in figure 7.26, where it can be seen that the actual band structure is quite far from an analytical band such as has been used previously. To implement the full band structure, it must be gridded within the Brillouin zone, but due to the symmetry of the zone only a 1/48 segment of the zone need be used. Nevertheless, scattering rates between all values of energy and momentum have to be computed and stored. While the accelerating force remains that of (7.18), the velocity is no longer a simple function of momentum. The band in which the particle is located must be kept as a parameter, and the velocity at any momentum determined from

$$\mathbf{v} = \frac{1}{\hbar}\frac{\partial E}{\partial \mathbf{k}}. \tag{7.64}$$

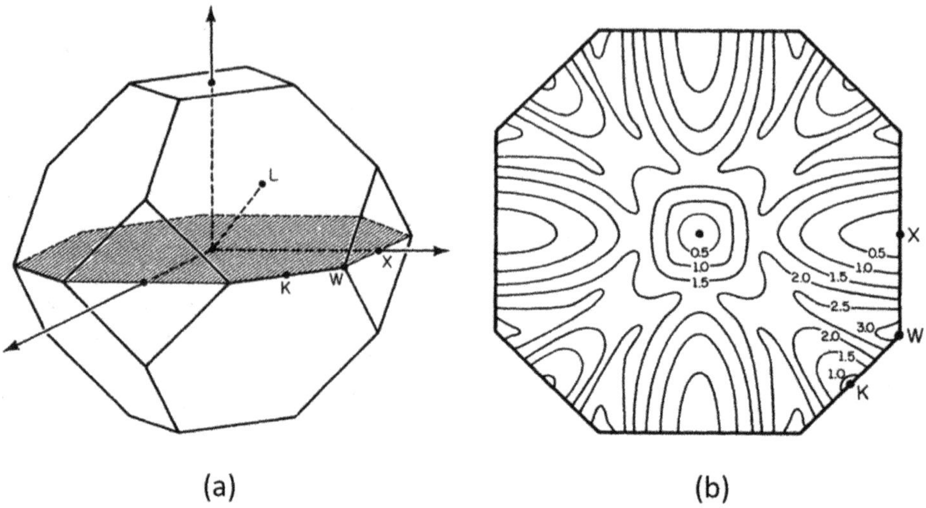

Figure 7.26. (a) The Brillouin zone for the zinc-blende lattice. (b) The conduction band energy isolines in the gray plane of (a). The lines are labeled with the energy measured from the conduction band minimum at Γ. Reprinted from Shichijo and Hess [8], copyright (1981) by the American Physical Society.

As a result, the gridding has to be fine enough to allow determination of the velocity with minimum noise in that determination. The major problem is the storage of the scattering tables, and the random selection of final momenta in the 48 different segments of the band structure. However, other problems arise where the bands split (or cross in silicon for example), such as at X in the valence band and near Γ in the second conduction band. The usual approach is to assert that the particle will follow the band with approximately the same slope in momentum space.

Perhaps the most published follow-on to the previous work is that of Fischetti and Laux, with the development of the DAMOCLES code at IBM [65, 66]. This effort began with the same local pseudopotential of Hess [8] without the spin–orbit correction, but was from the start aimed at simulation of semiconductor devices, which means coupling to the Poisson equation. Here, they showed that storing the entire Brillouin zone at 41 000 points would provide a speed advantage by avoiding the symmetry operations required by using just a 1/48 wedge. In addition, the advantage of using self-scattering is lost as the equations of motion cannot be simply integrated and must be numerically integrated. Nevertheless, using this code allowed the pair to improve the knowledge of band structure parameters, masses, and phonon coupling constants in the principal valleys of the conduction band. Further work has allowed them to extend this development to a wider set of semiconductors and semiconductor devices.

Another advance in full band calculations was taken in the development of what is known as cellular Monte Carlo, where the scattering rates are computed between cells in the Brillouin zone as a function of the initial momentum, rather than tabulating with energy cells [67]. This code has been adapted with some quantum corrections and is capable of device simulations, as it also incorporates the Poisson

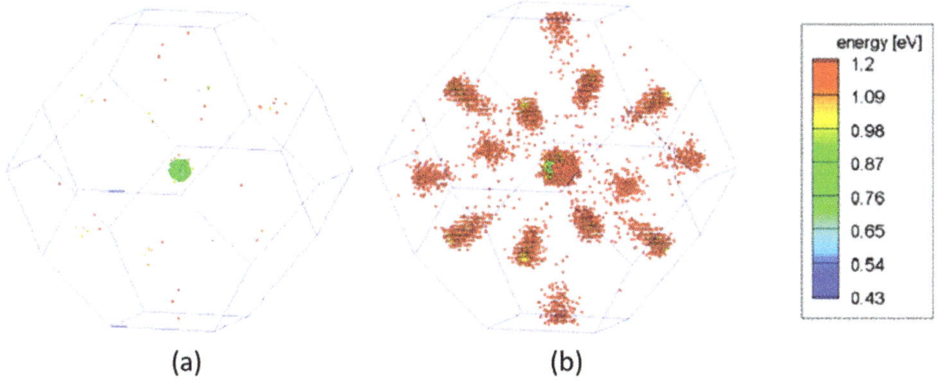

Figure 7.27. Depiction of the carrier position in the Brillouin zone. (a) With low bias, the carriers are in the Γ valley, although a few have moved out of this valley. (b) With high bias, the carriers occupy all the principal valleys and other positions as well. Reprinted from Akis *et al* [70], copyright IOP Publishing. All rights reserved.

equation [68, 69]. This method has been used extensively to study InGaAs strained channel HEMTs [41, 70]. Another rationale for the use of full band in the EMC may be seen in figure 7.27, where the position of the carriers is given in the Brillouin zone. In panel (a), the carriers reside in the Γ valley of the InGaAs. However, when significant bias is applied to the HEMT, the carriers scatter throughout the Brillouin zone, as shown in panel (b). Most of the carriers are in the principal valleys, the Γ, L, and X valleys. But, many are not in these valleys, and it is these that need the full band treatment in order to account for their effect in the transport.

7.7 Monte Carlo in device simulation

Probably the first EMC simulation of a semiconductor transistor was that of a HEMT [71]. This was from the same group that had verified the real-space transfer in the AlGaAs/GaAs system [59], so it was a small step to simulate the entire transistor. The authors compared both a normal single delta-doped structure with one that had delta-doping on both sides of the GaAs channel. By varying the doping level in the various layers, and the setback distance (of the doping layer from the channel), they were able to vary the threshold voltage of the device by a considerable amount. With these simulations, the authors were already suggesting that cutoff frequencies could be in the neighborhood of 50 GHz for a 0.5 μm gate length. Later simulations of InGaAs strained layer HEMTs moved this cutoff frequency to more than 140 GHz [72] for a 0.12 μm gate length device, which compared well with experiments giving a 114 GHz cutoff [73].

The first EMC simulations for Si MOSFETs were actually done with the full band structure program at IBM [65, 66]. These authors focused upon the hot carrier effects in these devices in subsequent work [40, 74]. Later work examined EMC simulations for Si MOSFETs, with a focus upon a strained Si on SiGe MOSFET [75]. Focus upon *p*-channel MOSFETs seems to have begun about the same time [76].

Study of a great many devices with EMC simulations has grown fairly rapidly since these early years. Such simulations have given a great deal of insight into the physics and the circuit operation of various devices, and enhanced the ability to progress to ever smaller dimensions and higher frequencies of operation. In the following discussion, the focus will be on a single device as an illustration of the type of information that can be obtained on the hot carrier transport in real devices. This example will be a microwave HEMT with a strained $In_{0.75}Ga_{0.25}As$ active channel bounded by two InAlAs layers that are lattice-matched to the InP substrate [41]. The upper InAlAs layer has delta doping at 3.5×10^{12} cm^{-2}. Here, the gate length began at 70 nm, but results have been found for smaller gate lengths, beginning at 35 nm and smaller. In figure 7.28, the current and transconductance of the 70 nm device is plotted as a function of the gate voltage. As is typical with HEMTs, the transconductance peaks at a particular gate voltage, which is 0, in this case as the device is a depletion mode device, and then decreases, as the gate begins to lose control over the channel. Nevertheless, the current still increases, although it begins to show saturation. The peak transconductance, almost 1300 mS mm^{-1} (the value is normalized to a width of 1 mm), is outstanding for a device of this configuration. Similarly, the peak current of 700 mA mm^{-1} (similarly normalized) is also very good. Studies on a 20 nm gate length device produced almost 1500 mA mm^{-1} when the delta-doping was placed only 2 nm from the active channel. The 35 nm gate length device was appropriately between these two values. The current in general is limited to some extent by the contact resistance in the real device, and in the simulation by the injection properties of the contacts. Specific control of the carrier injection in the contact regions can adjust the effective contact resistance and has an effect on the device performance.

In figure 7.29, the average energy and the average velocity along the channel are plotted versus distance for the 70 nm HEMT, for a drain voltage of 1 V and a gate voltage of 0.4 V. As the highest electric fields occur near the drain end of the gate, this is where the velocity peaks. The drop in velocity, however, occurs as the carriers

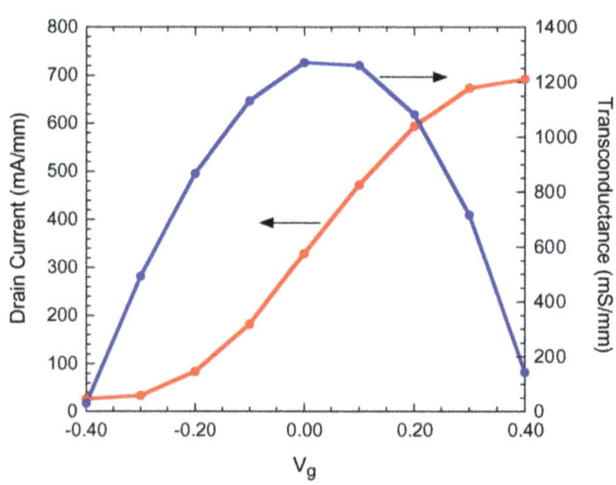

Figure 7.28. Transfer and current characteristics of the 70 nm HEMT, for a gate–drain spacing of 640 nm. Copyright (2007) by IEEE, reprinted with permission from Ayubi-Moak *et al* [41].

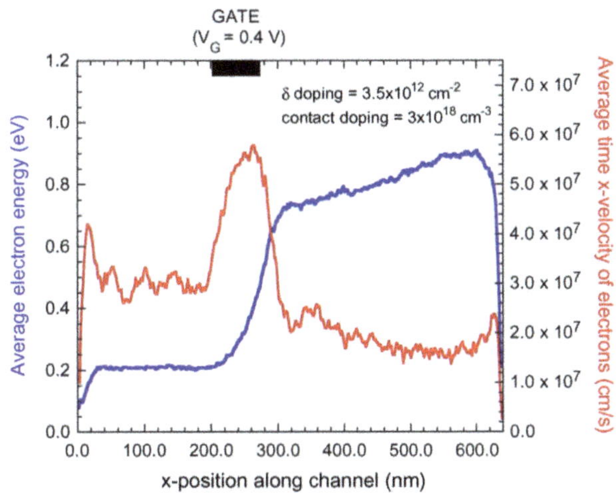

Figure 7.29. Variation of the average velocity and energy along the channel for the 70 nm HEMT. The high energy that remains all the way to the drain arises from the long storage time for carriers in the satellite valleys. Copyright (2007) by IEEE, reprinted with permission from Ayubi-Moak *et al* [41].

transfer to the L and X valleys of the conduction band (see figure 7.27). On the other hand, the carrier energy (which includes the potential energy of those carriers in the satellite valleys) remains very high due to the long return time of these carriers from the satellite valleys [77]. In fact, it is found that the carriers do not return to the Γ valley until they are well into the drain contact region. The fact that they are in the satellite valleys for most of the high field region, means that the breakdown voltage will be well above that expected for the narrow energy gap of the InGaAs. Note that the peak velocity under the gate is almost 6×10^7 cm s^{-1} indicates that significant transient overshoot velocity is involved in this region. This leads to a very short transit time through the gate region and hence a very high cutoff frequency, which will be discussed further below.

In order to examine how the transport under the gate affects the high-frequency performance, a slightly modified structure will be considered [78]. The same layer structure and doping are used, but the total device is now only 300 nm long, and the gate is placed 100 nm from the source contact. Gate lengths in range 10–50 nm are considered. To gain insight into the high-frequency performance, one first needs to establish an *effective* gate length. While the physical gate length is known, it represents a capacitive structure. The fact that it is separated from the channel by a distance that can be comparable to the physical gate length means that stray fields can be important. Indeed, this is true especially on the source end of the gate where the stray field begins to accelerate the carriers prior to their arrival at the physical gate. In figure 7.30, the average velocity and the carrier density are plotted as a function of distance for the region near the gates. These curves are shown for two different physical gate lengths of 10 and 50 nm. It may be noted that there is a small peak in the carrier density just before the physical gate, and this corresponds to a small dip in velocity and to the point where the velocity begins to rise in response to the longitudinal field. This point is designated as the start of the *effective* gate.

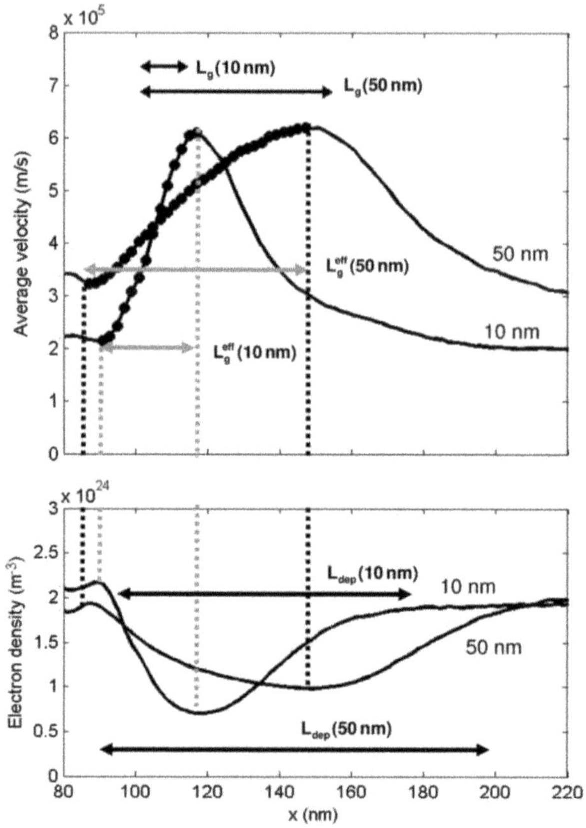

Figure 7.30. The average velocity (7.65) (top) and electron density (bottom) along the channel for two gate length devices. The actual gates are indicated by the black arrows at the top, while the determination of the effective gates is defined by the gray arrows. Copyright (2008) by IEEE, reprinted with permission from Akis *et al* [78].

Similarly, there is a point at which the velocity peaks, which may or may not be at the same point where the density is a minimum. This velocity peak is taken to be the end of the effective gate. The two effective gate lengths are shown by the gray arrows in the upper panel of the figure. The physical gate lengths and positions are indicated by the black arrows at the top of the upper panel. One can see that the gray arrows are always longer than the black arrows. One can see from the inset to figure 7.31 that the 50 nm physical gate has an effective length of 60 nm, while the 10 nm physical gate has an effective length of 17 nm. What actually determines the cutoff frequency is the transit time of the carriers through this effective gate length, and not that of the physical gate length. It must first be observed that the velocity is not constant across the depth (y) of the device, and a vertical average must first be taken as

$$v_{av}(x) = \int_0^d n(y)v(x, y)dy \Big/ \int_0^d n(y)dy, \tag{7.65}$$

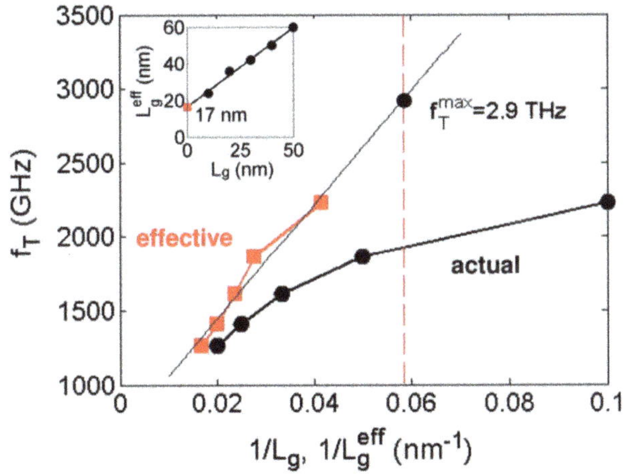

Figure 7.31. The cut-off frequency as a function of the inverse of the physical gate length (black) and the effective gate length (red). The inset shows the relationship between the two. Extrapolating to zero physical gate length gives the maximum cut-off of 2.9 GHz for this device structure. Copyright (2008) by IEEE, reprinted with permission from Akis *et al* [78].

where d is the thickness of the quantum well. Then, the effective transit time through the gate is given as

$$\tau_T = \frac{1}{2\pi f_T} = \int_0^{L_{\text{eff}}} \frac{dx}{v_{av}(x)}, \tag{7.66}$$

where the integral is over the portion of the x-axis defined by the effective gate parameters. In figure 7.31, the cutoff frequency f_T is plotted as a function of the gate length. Two curves indicate the physical gate lengths and the corresponding effective gate lengths. It may be seen that a linear behavior is found only by using the effective gate lengths. The ratio between them is indicated in the inset to the figure. By extracting the line in the inset to a zero physical gate length indicates that there is a minimum effective gate length for this device structure of 17 nm. Using this value in the main panel leads to the maximum cutoff frequency of 2.9 THz for this device structure. It is interesting to note that a 25 nm InP based HEMT with a similar design to that discussed here provided a 9 dB gain at 1 THz, with an f_{max} of 1.5 THz [79]. Using these parameters in figure 7.31 would give a theoretical estimate of f_T of ~1.8 THz and a comparable f_{max}.

References

[1] Budd H F 1967 *Phys. Rev.* **158** 798
[2] Thornber K K and Feynman R F 1970 *Phys. Rev.* B **1** 4099
[3] Kurosawa T 1966 *J. Phys. Soc. Jpn.* Suppl. **21** 424
[4] Jacoboni C and Reggiani L 1983 *Rev. Mod. Phys.* **65** 645
[5] Jacoboni C and Lugli P 1989 *The Monte Carlo Method for Semiconductor Device Simulation* (Vienna: Springer)

[6] Hess K 1991 *Monte Carlo Device Simulation: Full Band and Beyond* (Boston, MA: Kluwer Academic)
[7] Kosina H, Nedjalkov M and Selberherr S 2000 *IEEE Trans. Electron Dev.* **47** 1898
[8] Shichijo H and Hess K 1981 *Phys. Rev.* B **23** 4197
[9] Lugli P and Ferry D K 1985 *IEEE Trans. Electron Dev.* **32** 2431
[10] Joshi R P and Ferry D K 1991 *Phys. Rev.* B **43** 9734
[11] Jacoboni C 1976 *Proc. 13th Int. Conf. Phys. Semiconductors* (Rome: Tipo. Marves) p 1195
[12] Osman M A and Ferry D K 1987 *J. Appl. Phys.* **61** 5330
[13] Binder K (ed) 1979 *Monte Carlo Methods in Statistical Physics* (Berlin: Springer)
[14] Kalos M H and Whitlock P A 1986 *Monte Carlo Methods* (New York: Wiley)
[15] Rees H D 1972 *J. Phys.* C **5** 64
[16] Kreuzer H J 1981 *Nonequilibrium Thermodynamics and Its Statistical Foundations* (Oxford: Oxford University Press)
[17] Welland I and Ferry D K 2019 *Semicond. Sci. Technol.* **34** 064003
[18] Ferry D K 2020 *Semiconductors Bonds and Bands* 2nd edn (Bristol: IOP Publishing)
[19] Rees H D 1969 *J. Phys. Chem. Sol.* **30** 643
[20] Price P 1979 *Semiconductors and Semimetals* ed R K Willardson and A C Beer 14 (New York: Academic) p 249
[21] Vasileska D, Goodnick S M and Klimeck G 2010 *Computational Electronics: Semiclassical and Quantum Device Modeling and Simulation* (Boca Raton, FL: CRC Press)
[22] Bosi S and Jacoboni C 1976 *J. Phys.* C **9** 315
[23] Lugli P, Jacoboni C, Reggiani L and Kocevar P 1987 *Appl. Phys. Lett.* **50** 1251
[24] Barker J M, Ferry D K and Goodnick S M *et al* 2004 *J. Vac. Sci. Technol.* B **22** 2045
[25] Yu T-H and Brennan K F 2002 *J. Appl. Phys.* **91** 3730
[26] Goldstein H 1959 *Classical Mechanics* (New York: Addison-Wesley)
[27] Lugli P and Ferry D K 1986 *Phys. Rev. Lett.* **56** 1295
[28] Wood W W and Parker F R 1957 *J. Chem. Phys.* **27** 720
[29] Rahman A 1964 *Phys. Rev.* **136** A405
[30] Ferry D K, Kriman A M, Kann M J and Joshi R 1991 *Comp. Phys. Commun.* **67** 119
[31] Kriman A M, Kann M J, Ferry D K and Joshi R 1990 *Phys. Rev. Lett.* **65** 1619
[32] Kohn W, Meir Y and Makarov D E 1998 *Phys. Rev. Lett.* **80** 4153
[33] Kelbg G 1964 *Ann. Phys.* **468** 385
[34] Morawetz K 2002 *Phys. Rev.* E **66** 022103
[35] Gross W J, Vasileska D and Ferry D K 2000 *VLSI Des.* **10** 437
[36] Ewald P P 1921 *Ann. Phys.* **64** 253
[37] Brush S C, Salikin H L and Teller F 1964 *J. Chem. Phys.* **45** 2101
[38] Potter D 1973 *Computational Physics* (London: Wiley)
[39] Adams D J and Dubey G S 1987 *J. Comp. Phys.* **72** 156
[40] Fischetti M V and Laux S E 1991 *IEEE Trans. Electron Dev.* **38** 650
[41] Ayubi-Moak J S, Ferry D K and Goodnick S M *et al* 2007 *IEEE Trans. Electron Dev.* **54** 2327
[42] Weinbub J, Rupp K and Selberherr S 2014 *J. Comp. Appl. Math.* **270** 484
[43] Hockney R W and Eastwood J W 1981 *Computer Simulation Using Particles* (Maidenhead: McGraw-Hill)
[44] Gross W J, Vasileska D and Ferry D K 1999 *IEEE Electron Dev. Lett.* **20** 463
[45] Gross W J, Vasileska D and Ferry D K 2000 *IEEE Trans. Electron Dev.* **47** 2000

[46] Ferry D K and Grubin H L 1996 *Solid State Physics* vol 49 (New York: Academic) p 283
[47] Ferry D K 2000 *Superlatt. Microstruc.* **28** 419
[48] Ferry D K and Nedjalkov M 2018 *The Wigner Function in Science and Technology* (Bristol: IOP Publishing) ch 4
[49] Madelung E 1926 *Z. Phys.* **31** 876
[50] Bohm D 1952 *Phys. Rev.* **85** 166
[51] Kennard E H 1928 *Phys. Rev.* **31** 876
[52] Wigner E 1932 *Phys. Rev.* **40** 749
[53] Feynman R P and Hibbs A R 1965 *Quantum Mechanics and Path Integrals* (New York: McGraw-Hill) Sec. 10.3
[54] Vasileska D and Ferry D K 1999 *Nanotechnology* **10** 192
[55] Vasileska D, Akis R and Knezevic I *et al* 2002 *Microelectron. Eng.* **63** 233
[56] Ramey S M and Ferry D K 2003 *IEEE Trans. Nanotechnol.* **2** 193
[57] Ramey S M and Ferry D K 2004 *Semicond. Sci. Technol.* **19** S238
[58] Keever M, Shichijo H and Hess K *et al* 1981 *Appl. Phys. Lett.* **38** 36
[59] Glisson T H, Hauser J H and Littlejohn M A *et al* 1980 *J. Appl. Phys.* **51** 5445
[60] Wang T and Hess K 1985 *J. Appl. Phys.* **57** 5336
[61] Kizilyalli I C, Hess K and Iafrate G J 1987 *J. Appl. Phys.* **61** 2395
[62] Kim Y and Hess K 1988 *Sol.-State Electron.* **31** 877
[63] Patil M B, Ravaioli U, Hess K and Hueschen M 1992 *Semicond. Sci. Technol.* **7** B533
[64] Cohen M L and Bergstresser T K 1966 *Phys. Rev.* **141** 789
[65] Laux S E and Fischetti M V 1988 *IEEE Electron Dev. Lett.* **9** 467
[66] Fischetti M V and Laux S E 1988 *Phys. Rev.* B **14** 9721
[67] Rein A, Zandler G and Saraniti M *et al* 1994 *Proc. Int. Electron Dev. Mtg. (New York)* (Piscataway, NJ: IEEE) p 351
[68] Wigner S J and Goodnick S M 2000 *Superlatt. Microstruc.* **27** 417
[69] Yamakawa S, Goodnick S M, Aboud S and Saraniti M 2004 *J. Comp. Electron.* **3** 299
[70] Akis R, Ayubi-Moak J S and Ferry D K *et al* 2008 *J. Phys. Cond. Matt.* **20** 384201
[71] Tian H, Kim K W and Littlejohn M A 1991 *J. Appl. Phys.* **70** 4593
[72] Babiker S, Asenov A and Cameron N *et al* 1998 *VLSI Des.* **8** 313
[73] Cameron N, Taylor M R S and Mclelland H *et al* 1995 *IEEE Microwave Th. Tech. Sym. Digest* (New York: IEEE Press) p 435
[74] Fischetti M V and Laux S E 1989 *Sol.-State Electron.* **32** 1723
[75] Formicone G F, Vasileska D and Ferry D K 1997 *Sol.-State Electron.* **41** 879
[76] Formicone G F, Vasileska D and Ferry D K 1998 *VLSI Des.* **6** 167
[77] Grann E D, Tsen K T and Ferry D K 1996 *Phys. Rev.* B **53** 9847
[78] Akis R, Ayubi-Moak J S and Faralli N *et al* 2008 *IEEE Electron Dev. Lett.* **29** 306
[79] Mei X, Yoshida W and Lange M *et al* 2015 *IEEE Electron Dev. Lett.* **36** 327

IOP Publishing

Hot Carriers in Semiconductors

David K Ferry

Chapter 8

Quantum transport

The atoms of a semiconductor are held together by quantum mechanical forces. This quantum mechanics leads to the band structure for the electrons and the quantized vibrational spectra for the atoms, such as the phonon spectra of chapter 5. So far, the transport discussed in this book has been treated semi-classically, either with the Boltzmann equation in chapter 6 or the Monte Carlo method of chapter 7. Today, the critical lengths in, for example, the field effect transistor are in the nanometer range. It is clear that more detailed modeling of the quantum contributions in modern semiconductor devices is required. These contributions appear in many forms: (1) changes in the statistical thermodynamics for the carriers in the devices as well as in their connection and interaction with the external world, (2) new critical length scales, (3) an enlarged role for ballistic transport and quantum interference, and (4) new sources of fluctuations which will affect device performance. Indeed, many of these effects have already appeared in previous chapters in their connection with hot carriers.

One of the principles of classical physics is the strong connection between energy and momentum. For a free particle, it is clear that the relationship $E = p^2/2m$ is fundamental. Even when the quantum mechanics of the energy bands in semiconductors are discussed, the tenet that the energy is well defined at each and every value of momentum, in this case the crystal momentum, seems sacrosanct. But, the normal calculation of energy bands is done for the so-called empty lattice, in which the allowed energy levels are solved for a single electron [1]. This is the reason that density-functional theory calculations for the energy bands of semiconductors is notorious for not getting the energy gap (between the conduction and valence bands) correct. To go beyond this requires consideration of a variety of many-body corrections, principally a many-body self-energy correction from the Coulombic interactions among the valence electrons. This lowers the entire valence band through the cooperative interactions and provides a correction to the energy gap. The phrase 'self-energy' is a topic of interest for the transport carriers as well,

doi:10.1088/978-0-7503-3947-6ch8

especially as it is connected with going beyond the simple relationship $E = p^2/2m$. In three-dimensional momentum space, this latter relationship defines a spherical shell of the given energy, and all states with this energy lie on this shell. This connection between energy and momentum is introduced by the delta function

$$\delta(E - p^2/2m^*). \tag{8.1}$$

One of the most important properties in quantum transport, however, is the breaking of this constraining relationship. For most of quantum transport, energy and momentum are separate variables, although they do have a connection, just one not so strict as (8.1). In fact, the general rule in detailed quantum transport is that life is lived 'off the energy shell,' the shell in momentum space that is defined by (8.1). On this energy shell, the energy is strictly defined by the classical relation with momentum given in the delta function. Off the shell, this strict relationship becomes more nebulous due to quantum uncertainty.

Life off this shell is not so unusual a prospect. Indeed, it is a usual situation in electronic circuit theory. Consider a really simple R, L, C circuit from undergraduate circuit theory. It doesn't matter whether this is a parallel or a series circuit in principle, although the exact relationships will differ. Consider the series case. This circuit has a resonant frequency in the absence of the resistance, which is given by

$$\omega_0 = \frac{1}{\sqrt{LC}}. \tag{8.2}$$

However, when the resistance is present, the oscillation of the circuit is damped exponentially in time with the time constant $2LR^{-1}$, and the frequency of the oscillation is shifted to

$$\omega = \omega_0\sqrt{1 - \left(\frac{R}{2L\omega_0}\right)^2}. \tag{8.3}$$

The resonant frequency is reduced by the resistance. And, this means that the energy of the photons (from the electromagnetic field in the circuit) associated with the resonance is reduced. Essentially, the resistance represents part of the self-energy of the interacting carriers in the energy bands, although this analogy cannot be carried too far.

Two things happened when the resistance was added to the resonant circuit. First, damping appeared, and, second, the frequency was shifted by the non-infinite quality factor Q ($= \omega_0 L/R$) of the circuit. In the quantum world, the resistor represents interactions which take the system from a one-electron picture to what is called the many-body picture. These interactions can be among the many electrons that are present, or among the electrons and the lattice vibrations, or among the electrons and the impurities in the system, and so on. But, the interactions lead to the same two effects. However, they are defined as a single term, the *self-energy*. In the world of Schrödinger wave mechanics, the time variation for an energy eigenstate is the simple exponential $\exp(i\omega t) = \exp(iEt/\hbar)$. If the energy goes beyond (8.1) with a self-energy, it can be expressed as

$$E = \frac{p^2}{2m^*} - \Sigma(p), \tag{8.4}$$

where Σ is the self-energy. The real part of the self-energy corresponds to a shift downward of the energy (frequency shift). The imaginary part of the self-energy provides a damping of the state in time, which is the analog of the resistive damping of the resonant circuit. So, in situations when the self-energy is non-zero, states which do not lie on the energy shell can be important in the transport. The impact of these states is cleverly described as *off-shell effects*, which tends to sound very intellectual (and, perhaps, confusing) for such a straightforward effect.

The importance of the off-shell effects are illustrated in figure 8.1. In the semi-classical case of panel (a) where there is no self-energy, the relationship between momentum and energy is a well-defined curve, a parabola in this case according to (8.1). Of course, with full bands or with hyperbolic bands, the curve will be different, as the mass increases with momentum. On the other hand, in the quantum case of panel (b), the curve is rather diffuse, as there is no one-to-one connection between momentum and energy. This diffuseness of the curve is described by a *spectral function* $A(E, k)$ that relates the momentum to the energy. The difference now arises as the density of states is now determined by

$$\rho(E) = \frac{1}{\pi\hbar}\int \frac{d^3k}{(2\pi)^3} A(E, k) \tag{8.5}$$

in three dimensions. Normally, of course, there is a delta function relating the momentum to the energy in semi-classical physics. This spectral function describes the importance of the off-shell effects, particularly in scattering. Now, it has to be pointed out that not all formulations of quantum transport make clear the role of the spectral density. Indeed, in many formulations, it is sometimes obscure as to whether or not the fuzziness actually occurs, or is being considered. Nevertheless, a key distinction between quantum and classical transport is just the existence of the

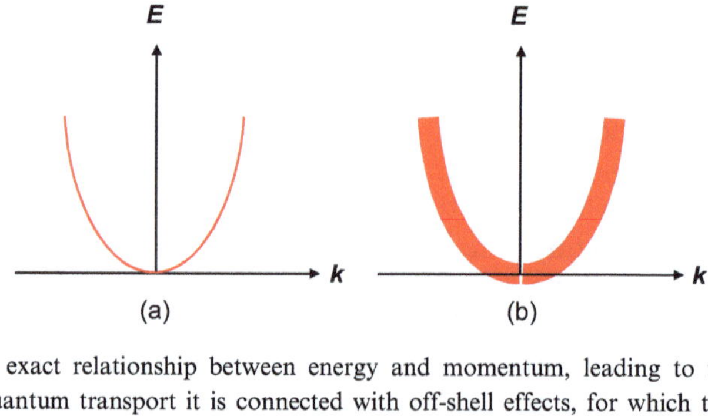

Figure 8.1. (a) The exact relationship between energy and momentum, leading to the delta function, is illustrated. (b) In quantum transport it is connected with off-shell effects, for which there is no one-to-one connection between momentum and energy, as is illustrated by the spread of the curve.

spread of states off the energy shell. In some sense, this arises from the uncertainty in momentum, but this connection has to be made carefully, as there generally is no uncertainty in the energy in non-relativistic quantum mechanics.

There are many approaches to dealing with quantum transport of carriers in semiconductors, but the number of approaches narrows considerably when the discussion turns to hot carriers. Yet, most of these start with the Schrödinger equation. The form of this equation is both reversible and dissipation free, while its use in actual devices is normally modified to incorporate the presence of dissipation, which of course is necessary with hot carriers, and approaches that begin and end with this equation can be used. Such approaches will be treated in section 8.2. Other approaches that find usage are Green's functions and the Wigner function. Both of these are variants of the density matrix, itself a representation that depends upon the Schrödinger equation, and which will be treated in section 8.3. Green's functions will be dealt with in section 8.4 and Wigner functions in section 8.5. This chapter is not an exhaustive treatment of quantum transport, but only an introduction to that important for hot carriers. A more exhaustive treatment may be found in [2].

It is important to remember that many hot carrier situations arise in real semiconductor devices, as is evident in the preceding chapters. This means that there is a need to consider very inhomogeneous situations, both in carrier densities and in local potentials. Approaches which ignore this are not destined to be very successful. It can be determined from a casual search of the literature that a great many applications of quantum transport deal only with ballistic transport, thus voiding the complications of scattering events in the semiconductor and in the device. This provides a number of misleading results that avoid some inconvenient connections between theory and reality. With this in mind, it is useful to review some simple basics about transport in ballistic structures and in devices. This is done in the next section 8.1.

8.1 Modes and the Landauer formula

Half a century ago, Rolf Landauer presented an approach to transport, that was dramatically different from the microscopic kinetic theory based on the Boltzmann equation that was discussed in chapter 6 [3, 4]. He suggested that one could compute the conductance by simply computing the transmission of a mode from an input reservoir to a similar mode in an output reservoir. In any semiconductor structure, and even more so in the small MOSFETs and HEMTs considered today, the transverse dimension is limited, and the Schrödinger equation leads to a set of transverse modes that arise from the size quantization. How many of these are occupied, for example, in the source depends upon the carrier population and the Fermi energy, whether or not the source is degenerately doped. Hence, the statistics of these two quantities determine how many of the modes are probably occupied. The transmission of this probability from one end of the device to the other was then very similar to the computation of a tunneling probability, except that there was no requirement that the process be one of tunneling. It could equally be scattering that moved carriers from one transverse mode to another. The only real constraint was that of lateral confinement so that the two reservoirs could be discussed in terms of their transverse modes, which were in a quasi-equilibrium with any applied potentials. That is, the electrons in the reservoirs were to be described by their intrinsic Fermi–Dirac distributions with any applied potentials appearing only as a shift of the relative energies (which would shift one Fermi level relative to the other), and possibly temperature differences between the two ends. While Landauer originally considered that the transport was ballistic, this is not required. Rather, the requirement is that a definitive mode can be assigned to each electron when it is in either of the two reservoirs, which means that if scattering is present, it must be described specifically as a transfer of the electron from one internal mode to another. This goes well beyond its initial interpretations and actually provides the ability to compute transport directly with solutions of the Schrödinger equation, which will be dealt with in section 8.2.

If it is assumed that ballistric transport occurs in the semiconductor structure, then a simple result can be obtained. In order to compute the current through the structure, the charge that occupies each mode must be determined for the energy range between E_{F} and $E_{\mathrm{F}} + eV$, where E_{F} is the Fermi level at the source and V is the applied potential (at the drain). If the applied bias is small, then the excess charge can be estimated, for a one-dimensional transverse mode, as

$$\delta Q \approx \frac{e^2 V}{2} \rho_{1d}(E_{\mathrm{F}}) = e^2 V \sqrt{\frac{m^*}{2\pi^2 \hbar^2 E_{\mathrm{F}}}}. \tag{8.6}$$

It should be noted that only one-half of the density is used, since only those electrons that are traveling from source to drain (those with positive momentum) are of interest. Also, because of the definition of the density of states, (8.6) is the charge per unit length. In order to compute the current, (8.6) need only be multiplied by the carrier velocity at the Fermi energy, which is

$$v_g = v_{\mathrm{F}} = \frac{1}{\hbar} \frac{dE}{dk}\bigg|_{E=E_{\mathrm{F}}} = \frac{\hbar k_{\mathrm{F}}}{m^*} = \sqrt{\frac{2E_{\mathrm{F}}}{m^*}}, \tag{8.7}$$

and this leads to the current

$$I_c = \delta Q v_g = e^2 V \sqrt{\frac{m^*}{2\pi^2 \hbar^2 E_{\mathrm{F}}}} \sqrt{\frac{2E_{\mathrm{F}}}{m^*}} = \frac{2e^2}{h} V. \tag{8.8}$$

A very interesting effect occurred in this development of the current, and that is the energy has dropped out of the equation. Hence, the current is completely independent of the energy, and this is true for each one-dimensional channel that flows through the constriction. This will become somewhat more complicated when multiple modes are considered. The total current is found from the number of modes that are occupied, which is

$$I_{\mathrm{total}} = N I_c = N \frac{2e^2}{h} V. \tag{8.9}$$

From this last expression, the conductance is obtained through the constriction as

$$G = \frac{I_{\text{total}}}{V} = N\frac{2e^2}{h}. \tag{8.10}$$

This last expression is usually called the Landauer formula. The conductance per mode is quantized in units of $2e^2/h$ (~77.28 μS) with a resulting magnitude that depends only upon the number of occupied channels. As is well-known in devices, the transport in saturation depends only upon the injection of carriers from the source to the channel. Hence, only the number of modes in the source region is important for saturation.

In the above, it was not considered that an electron in one mode in one reservoir might actually wind up in a different mode in the other reservoir. This is not reflected in the formula (8.10). Nor, does (8.10) offer the possibility of partial transmission of a mode. The transitions from one mode to another will become even more important when scattering within the channel region is taken into account. Consequently, it is more intuitive, if (8.10) is adapted to the real situation in which partial transmission of a mode is possible. To do this, (8.9) is rewritten as

$$I_{\text{total}} = \frac{2e^2}{h}VT(n), \tag{8.11}$$

where $T(n)$ is the transmission of the n channels in one reservoir, which is

$$T(n) = \sum_{m=1}^{\infty} T_{mn}. \tag{8.12}$$

Here, T_{mn} is the probability that electrons entering in channel n are scattered to channel m prior to exciting the device.

It might have occurred to some readers that, when purely ballistic transport is present, the conductance should have been infinite rather than the finite value given by the Landauer relation (8.11). In fact, the transition between the reservoir and the channel has a non-zero resistance which is a contact resistance between the one-dimensional channel and the reservoirs which give access to it. Normally, the reservoirs are very wide regions which are approximately in equilibrium, but even if not, the contact resistance arises from a mode mismatch between these latter regions and the channel. This occurs even when the transition is very smooth and adiabatic. This contact resistance goes directly to the question of just where the voltage drop across the device actually occurs. The voltage exists between the two reservoirs, but how does it vary between them. In a normal resistive conductor, this voltage should be dropped uniformly along the length of the constriction, just like a normal resistor, but this still requires unusual behavior in the contacts.

Since there is no scattering within the constriction, the voltage drop across it must correspond to a constant electric field, which can be quite low. In reality, the fact that the voltage is not dropped smoothly across the device means that charge must accumulate at the contacts, which leads to the variations in the voltage. As a result, the field in the center of the constriction is quite small, and the voltage drops are

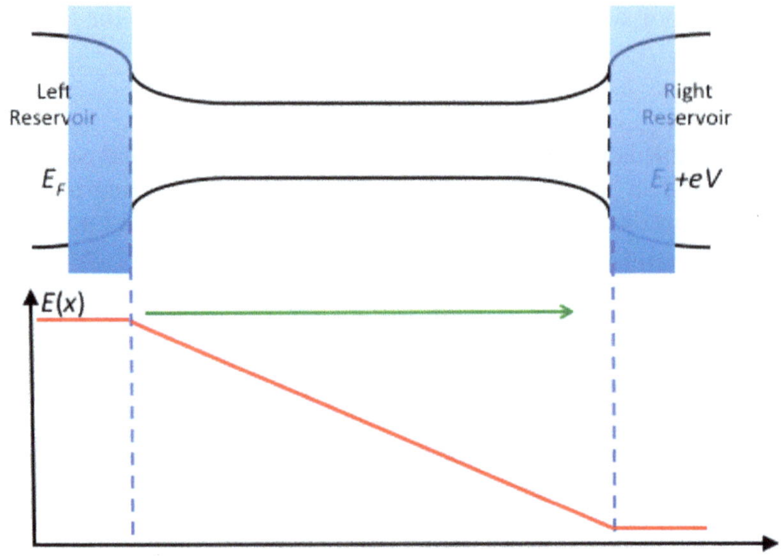

Figure 8.2. The energy $E(x)$ is plotted through a channel structure. An electron entering from the cathode at the left, and traveling ballistically, will gain energy as it moves through the channel, exiting with an additional energy defined by the applied voltage, as indicated by the green arrow.

forced to occur in the transition regions. Suppose we take the case of a linear voltage drop across the constriction so that there is a constant electric field in this region. Then the potential drop will look like that in the bottom panel of figure 8.2. But notice that, at the two points where the dashed line has been extended from the transition region, the electric field is discontinuous. Such a discontinuity in the electric field requires that a sheet of excess charge be present at this discontinuity. In fact, the charge at the left must be positive charge, while the charge at the right must be negative charge. So, there must be a depletion of electrons at the right and an accumulation of electrons at the left. If electrons are the major charge carrier, then the left-hand reservoir is the cathode, through which the electrons enter the structure. The accumulation of charge at the transition point contributes to the contact resistance.

There is still a problem with the potential as drawn in figure 8.2 The linear potential drop leads to a sizable electric field, which will accelerate the electrons. We can see this in the figure, where the electron energy is plotted. If the ballistic electron is injected from the left contact, it will gain energy from the electric field. That is, the ballistic electron travels at constant energy (green arrow in the figure), and the potential energy at the cathode is converted to kinetic energy as it travels. By the time it reaches the anode at the right-hand side, it will have gained kinetic energy given by eV. A more important aspect is introduced by Kirchoff's current law. As the electrons are accelerated by the electric field, their increase in velocity requires that fewer electrons are present as we move along the channel. That is, Kirchoff's current law requires the current to be uniform throughout the structure. Since the cross-sectional area does not change, this requires the product of density and velocity to be a constant value. As the velocity rises, the density must decrease. In a semiconductor, the density is usually set by the doping, but in a device this carrier depletion is a common occurrence. As the number of carriers drops, this creates an additional space charge in the channel, which in turn requires that the potential be nonlinear; e.g., *the potential drop cannot be a linear one.* The true form must be obtained from Poisson's equation. The result of this argument is that the linear potential drop shown in figure 8.2 cannot be correct if the carriers are moving via ballistic transport. The linear potential drop only can occur if there is sufficient scattering to ensure that the carriers move with a near-equilibrium energy. Our conclusion then can only be that the electric field must essentially be quite near zero in a channel of constant cross-section, if the carriers are to move via ballistic transport.

As a result, the potential drop must divide between the cathode and the anode transition regions, as shown in figure 8.3. While the two voltage drops are drawn as nearly the same, there is no real requirement that this be the case. In fact, it is usually assumed in the transport world that most of the discontinuity is related to the cathode. But now, ballistic transport can occur through the constriction without the carriers gaining excess energy from the applied bias. At the same time, we note that the potential drops in the transition regions now require that dipole charge densities exist at each transition. The potential drops can only occur through the existence of such dipole charge densities.

The end result of all of this discussion is that, in a device, for the potential to vary smoothly throughout the device, the density must be inhomogeneous, and this requires a solution to Poisson's equation. Thus, coupling Poisson's equation to the Schrödinger equation will allow the use of Landauer's equation in computing the

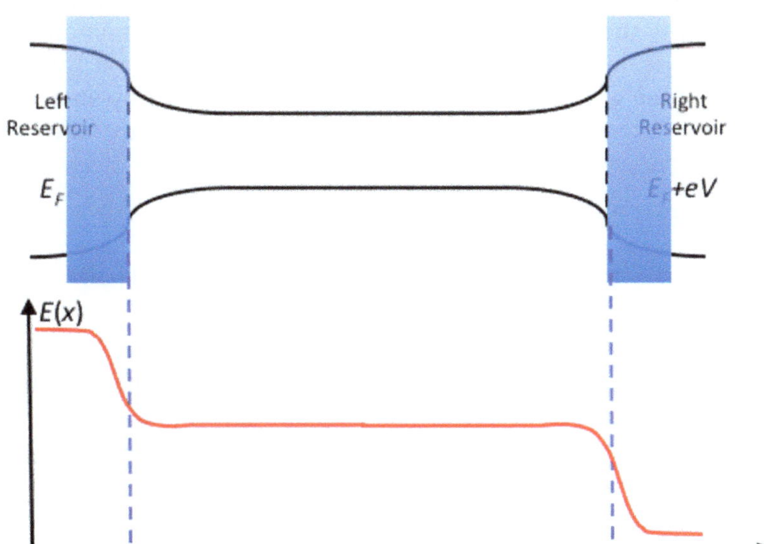

Figure 8.3. Potential drop required for ballistic transport through the a channel with constant cross-section. This form is required by Kirchoff's current law.

conductance, even under the far-from-equilibrium conditions of hot carriers in a device. The remark was made that, even with ballistic transport, the conductance would not be infinite. This can be illustrated by rewriting (8.11) as

$$\frac{1}{G} = \frac{h}{2e^2}\frac{1}{NT} = \frac{h}{2e^2 N} + \frac{h}{2e^2}\frac{1-T}{NT} \equiv \frac{1}{G_C} + \frac{1}{G_D}. \tag{8.13}$$

The resistance has been divided into a pair of resistances, one being the contact resistance $1/G_C$ which can never vanish, and a device resistance $1/G_D$ which can go away in the absence of scattering. The contact resistance is connected with the manner in which the actual voltage drop of figure 8.3 is proportioned across the charge regions in the transitions between the reservoirs and the channel.

While the Landauer formula was developed with a treatment of transport in which all the energy relaxation occurred in the reservoirs (true ballistic transport), this was not really required as mentioned above. The Landauer formula provides a rigorous framework for the analysis of nanostructures so long as current is conserved and the reservoirs can be analyzed in terms of modes. This allows one to determine transmission coefficients, as described above, for transport between these modes from, e.g. the Schrödinger equation, or from a classical approach such as the Duke tunneling formula in three dimensions [5]. As long as the scattering can be described as taking a particle from one channel to another, so that particle conservation accompanies current conservation, then any level of scattering can be incorporated within the Landauer formula. This does not mean that energy cannot be transferred to the crystal lattice by the scattering process. We need have only particle and current conservation for the validity of the Landauer formula. Further equations are required if we desire to examine energy flow within the nanostructure.

As an example, this formula can be connected to the common equation for the MOSFET. This will shed light on just how the behavior of the common MOSFET is not that different from the channel discussed here. The common formula for the MOSFET within the gradual channel approximation is given by an expression for the drain current in terms of the various potentials applied to the device, as

$$\begin{aligned} I_{\mathrm{D}} &= \frac{W\mu C_{\mathrm{ox}}}{L}\left(V_{\mathrm{G}} - V_{\mathrm{T}} - \frac{V_{\mathrm{D}}}{2}\right)V_{\mathrm{D}} \\ &= \frac{W\mu C_{\mathrm{ox}}}{2L}[(V_{\mathrm{G}} - V_{\mathrm{T}})^2 - (V_{\mathrm{G}} - V_{\mathrm{T}} - V_{\mathrm{D}})^2] \end{aligned}. \tag{8.14}$$

Here, W and L are the gate width and length (along the channel), C_{ox} is the gate capacitance per unit area, μ is the electron mobility, V_{T} is the threshold voltage where charge begins to accumulate in the inversion layer, and V_{G} and V_{D} are the gate and drain bias voltages (the source is usually taken to be grounded). In writing the equation in this way, contact with the Landauer formula becomes obvious. The first term in the square brackets represents particles entering the device from the source (left) electrode, while the second term represents particles entering the device from the drain (right) electrode. Thus, the source and drain are our reservoirs, and the particles flow exactly as in the Landauer formula.

In thermal equilibrium, with no bias applied to the gate and drain electrodes, there is no current flow through the device, since the two terms in the last line of (8.14) cancel each other. This cancellation remains the case even with gate voltage applied, but with the drain voltage set to zero. The gate voltage changes the properties of the channel, but does not upset the detailed balance of particle flow through the structure. However, when a small drain voltage is applied, the second term in the square brackets is reduced relative to the first term, and a small current begins to flow. As the drain bias is raised, the current increases. Once the second term is reduced to zero (it is not allowed to go negative, as this violates the conditions under which it was derived), the current saturates through the device at the normal value

$$I_{\mathrm{D,sat}} = \frac{W\mu C_{\mathrm{ox}}}{2L}(V_{\mathrm{G}} - V_{\mathrm{T}})^2. \tag{8.15}$$

When the drain bias is applied, the drain region is lowered in energy relative to the source region, just as shown in figures 8.2 and 8.3. Saturation occurs when this energy difference is sufficiently large that electrons in the drain can no longer surmount the barrier to reach the source. No matter whether the transport is ballistic or not (as when there is lots of scattering), the energy regions in the source and drain will be much the same. What will be different between these two cases is how the energy drops between the source and drain in response to solving Poisson's equation.

Now take the second line of equation (8.14) and write it in terms of an 'equilibrium' current that depends upon the gate voltage, as

$$I_{\mathrm{eq}}^{+} = I_{\mathrm{D,sat}} = \frac{W\mu C_{\mathrm{ox}}}{2L}(V_{\mathrm{G}} - V_{\mathrm{T}})^2. \tag{8.16}$$

Then, the second line of (8.14) becomes

$$I_{\mathrm{D}} = I_{\mathrm{eq}}^{+}\left[1 - \left(1 - \frac{V_{\mathrm{D}}}{V_{\mathrm{G}} - V_{\mathrm{T}}}\right)^2\right]. \tag{8.17}$$

For small values of the drain voltage, this result can be written in terms of a resistance (likely seen as a contact resistance) as

$$R_{\mathrm{C}} = \frac{V_{\mathrm{D}}}{I_{\mathrm{D}}} \sim \frac{V_{\mathrm{G}} - V_{\mathrm{T}}}{2I_{\mathrm{eq}}^{+}}. \tag{8.18}$$

Even for a very short device with perfect transmission, one cannot get around this minimum resistance. This resistance characterizes the transition region between the pure source and the channel (just as in the Landauer approach) and appears regardless of whether the transport is described by the mobility or is ballistic. In some sense, this resistance describes a barrier between the source and the channel [6]. The current flux of equation (8.16) is due to carriers from the source with sufficient kinetic energy to overcome this barrier between the source and the channel. It is important to note in equation (8.16) the flux is actually metered by the gate potential, so that this is a gate-controlled barrier. Once the carriers supplant this barrier, they are going to flow downhill (down the potential drop) to the drain. They will flow to the drain regardless of the nature of the transport. The only effect that ballistic transport can play is to change the shape of the potential landscape between the source and the drain through Poisson's equation. In this sense, scattering actually affects this potential landscape and serves to isolate the drain from the source. In a ballistic device, this isolation does not occur and the source–channel barrier can be affected by the drain voltage, an effect known as *drain-induced barrier lowering*.

8.2 Transport with the Schrödinger equation

There are many approaches to solve the Schrödinger equation in a finite-sized system, such as an electron device. When the system is large, it is often computationally more useful to use a recursive approach, in which the solution is generated by propagating a known contact solution through the structure with a slice-by-slice recursion. That is, the properties of a single slice are determined by those of the preceding slices starting from one reservoir or the other. The best known approach is the mode matching in which the wave function and its derivative are matched at each interfacial boundary between slices within the system. Unfortunately, this approach is known to be unstable for more than a few slices. An important variation of this approach is to switch the interface matching into the scattering matrix [7, 8], an adaptation borrowed from earlier microwave theory. This has been used for quite some time and works well for devices [9], even including the scattering processes, through the introduction of a self-energy whose imaginary part carries the scattering processes [10]. Stability in the recursion is based upon the scattering matrix and has long been a staple of microwave systems. It entered quantum mechanics through the Lippmann–Schwinger equation [11]. The strength of this approach lies in the fact that modal solutions to the Lippmann–Schwinger equation maintain their orthogonality through the scattering process [11]. In mesoscopic structures and nanostructures, the connection to microwave theory becomes stronger as the transport becomes dominated by the modes introduced by the lateral confinement. Scattering states, based upon the Lippmann–Schwinger equation, provide a method of building up an orthogonal ensemble, even with weighting of the states by e.g., a Fermi–Dirac distribution [12]. This is particularly useful for the initial conditions of the wave function in a particular semiconductor device [13]. In the recursive approach, it reaches new levels of viability when combined with the modal structure in small devices. The end of the recursion approach determines the transmission from one end to the other and evaluates the conductance via the Landauer formula.

8.2.1 The scattering matrix

The process begins with the discretization scheme for the Schrödinger equation. This will be illustrated for a uniform grid, in which the array has a slice width in the transverse direction of $M + 1$ points and a length of $N + 1$ slices. While a uniform system will be treated here, this is not necessary and a structure with different widths can be handled as easily [9]. A useful fact is that if we just limit the size, the solution automatically assumes that the wave function will satisfy

$$\psi(0, j) = \psi(M + 1, j). \tag{8.19}$$

As a result, the first and last points of the array are not of concern, and the slice wave function can be written as

$$\psi_s = \begin{bmatrix} \psi(M, s) \\ \vdots \\ \psi(2, s) \\ \psi(1, s) \end{bmatrix}. \tag{8.20}$$

Here, in $\psi(i, j)$, i defines the y-direction and j defines the x-direction, which is the current flow direction. A single subscript wave function is the *slice* wave function (8.20), which is a vector of site wave functions, in which the *site* wave function has two variables. Then, the Schrödinger equation can be rewritten as a matrix equation relating the slice vectors

$$H_{0,j}\psi_j - t\psi_{j+1} - t\psi_{j-1} = EI\psi_j, \tag{8.21}$$

where t is the diagonal matrix whose diagonal elements are the hopping energy

$$E_t = \frac{\hbar^2}{2m^* a^2}. \tag{8.22}$$

In (8.21), the Hamiltonian for the isolated slice is tri-diagonal and written as

$$H_{0,j} = \begin{bmatrix} (V_{Mj} - 4E_t) & E_t & 0 & & 0 & \\ E_t & (V_{M-1j} - 4E_t) & \cdot & & & \\ \cdot & \cdot & \cdot & & & \\ & & & \cdot & \cdot & 0 \\ & 0 & & \cdot & (V_{2j} - 4E_t) & E_t \\ & & & \cdot & E_t & (V_{1j} - 4E_t) \end{bmatrix} \tag{8.23}$$

where the upper right block and lower left block are identically zero, and I in (8.21) is an identity matrix. The individual elements of the slice wave function are the corresponding site wave functions as given in (8.20). Generally, the transverse wave functions for the width confinement are found from the corresponding slice Hamiltonian, e.g., (8.23) with $j = -1$. This process produces the site wave functions and the slice wave function, which is doubly degenerate with the ability to propagate both to the right and to the left. These wave functions may also be found through solving the equation

$$\det[T_0 - \lambda I] = 0, \tag{8.24}$$

where T_0 and λ are matrices whose rank is twice that of the Hamiltonian (8.23) and is given as

$$T_0 = \begin{bmatrix} U_+ & U_- \\ \lambda_+ U_+ & \lambda_- U_- \end{bmatrix}, \lambda = \begin{bmatrix} \lambda_+ \\ \lambda_- \end{bmatrix}. \tag{8.25}$$

$U_\pm$ are matrices with a rank that is equal to that of the Hamiltonian (8.23) and whose columns give the slice wave functions in the site representation. These are the modes of the transverse confinement and the positive and negative signs correspond to positive (right-directed) propagation and negative (left-directed) propagation. Similarly, the $\lambda_\pm$ are the corresponding set of eigenvalues. These $U_\pm$ matrices also give the site-to-mode transformations of the various wave functions. This is a key point. The solution of the Schrödinger equation normally gives the wave function at each point (the site) with an amplitude corresponding to the strength of the wave function at that site. In the Landauer formula, it is the modes that are important, so a transformation on the wave function has to be made so that we deal with the amplitude of each mode, rather than at each site. The difficulty lies in the fact that the transmission (conductance) is calculated for the *modes*, but the density used in the Poisson equation is calculated at each *site*.

There is a corresponding Fermi energy (the simulation is at zero temperature at this point), which enters the picture as modes whose eigenvalues are below the Fermi energy will be propagating, while those whose eigenvalues lie above the Fermi energy are evanescent. By starting at the −1 slice, it is propagated to slice 0. Once slice 0 is determined in the mode representation at the left 'contact,' the recursion proceeds from this. To carry out the recursion, everything has to be thought of in the scattering matrix form. The site to mode transformation creates slice wave functions from (8.20), in which each element corresponds to a mode amplitude in the system. These modes can be either propagating or evanescent, and it is important, especially in interfaces, to incorporate the evanescent modes. These can be evanescent in one part of space, but propagating in another part of space, such as in a tunneling problem. To continue, two new matrices, which are the 'scattering matrices' C_1^s and C_2^s at the general site $j = s$, are defined. These are the forward and backward amplitude corrections in the recursion. The recursion relation for these are now expressed through the defining relationship [14, 15]

$$\begin{bmatrix} C_1^{s+1} & C_2^{s+1} \\ 0 & 1 \end{bmatrix} = \begin{bmatrix} 0 & I \\ -I & t^{-1}(H_{0,s} - EI) \end{bmatrix} \begin{bmatrix} C_1^s & C_2^s \\ 0 & 1 \end{bmatrix} \begin{bmatrix} I & 0 \\ P_{1s} & P_{2s} \end{bmatrix}, \tag{8.26}$$

where the first matrix on the right-hand side is denoted as T_s, and H_{0s} is the slice Hamiltonian (8.23). For the zero temperature formulation, C_1^s is set to a unit matrix, although for a finite temperature simulation, the elements will be given by the Fermi–Dirac distribution for the energy of the corresponding eigenvalue *in the source region of the device*. Similar values are given to C_2^s *at the right-hand end of the system*. The equation (8.26) is required to be satisfied, and this allows determination of the propagation matrices P_{1s} and P_{2s} through the recursions

$$
\begin{aligned}
C_1^{s+1} &= P_{1s} = -P_{2s}(-I)C_1^s \\
C_2^{s+1} &= P_{2s} = (-IC_2^s + t^{-1}(H_{0,s} - EI))^{-1}.
\end{aligned} \tag{8.27}
$$

The unit matrix (the $-I$ term) is actually the $(T_S)_{2,1}$ element of the T_s matrix, which will be different from unity if there is a magnetic field applied, as it arises from $t^{-1}t$ and will be different when time-reversal symmetry is broken by the magnetic field. This will not be discussed here, but may be found in [14]. The second line of (8.27) may be recognized as the Dyson equation, which often appears in perturbative (or recursive) approaches, but is most common in the Green's function method. The Dyson equation will appear in more detail in later sections.

At the end of the sample, the conductance is determined from the modes, but the mode-to-slice transformation is used to get the site amplitudes for the density (the mode-to-slice transformation is the inverse of the site-to-mode transformation). At the end of the structure, the wave function at the output is determined in terms of the mode number and its values at the various sites. Thus, at site i and mode m the contribution to the wave function is $\psi_j(i, m)$. This can now be propagated in reverse to find the equivalent site wave functions at each slice via

$$
\psi_j(i, m) = P_1^j + P_2^j \psi_{j+1}(i, m). \tag{8.28}
$$

This recursion is a back-propagation that proceeds from right to left. The velocities for each mode are found from the imaginary part of the eigenvalue for that mode. At the same time, the density can be determined at each site by the square magnitude of the wave function at that site (initially, the normalization is to unity, but this can be understood to be in terms of the density in the initial contact). A fuller description of the approach can be found in Akis *et al* [15].

8.2.2 Self-energy

By its very form, it is quite simple to modify the recursion method to include scattering by an on-site self-energy that is added to the diagonal elements of (8.23) and represents the interactions with scattering centers. Generally, the self-energy has both real and imaginary parts and it is the latter that are of interest for dissipative scattering. In semiconductors, the scattering is generally quite weak, and is traditionally treated by first-order, time-dependent perturbation theory, which yields the common Fermi golden rule for scattering rates. With such weak scattering, the real part of the self-energy can generally be ignored for the phonon interactions, and the real part that arises from the carrier–carrier interactions is incorporated into the solutions of Poisson's equation. In the many-body formulations of the self-energy, the latter is usually expressed as a two site function [16]

$$
\Sigma(r_1, r_2). \tag{8.29}
$$

The two positions are usually three-dimensional vectors. Since transverse modes are used here in the recursion, this expression is rewritten as

$$
\Sigma(i, s; i', s'; x_1, x_2). \tag{8.30}
$$

As usual, the i and s (and their primed versions) are the transverse site and mode, respectively, and correspondingly the x's represent positions along the nanowire. We can then introduce a center-of-mass transformation [17], which is commonly used in both Green's functions and Wigner functions as well,

$$X = \frac{x_1 + x_2}{2}, \ \xi = x_1 - x_2, \tag{8.31}$$

and then Fourier transform on the difference variable to give

$$\Sigma(i, s; i', s'; X, k_x) = \frac{1}{2\pi} \int d\xi e^{ik_x\xi} \Sigma(i, s; i', s'; X, \xi). \tag{8.32}$$

The center-of-mass position X remains in the problem as the mode structure may change as one moves along the channel. At this point, the left-hand side of (8.32) is computed by the normal scattering rates, such as is done in quantum wells and quantum wires [18, 19]. As mentioned above, since scattering in semiconductors is relatively weak, it is sufficient to compute these using Fermi's golden rule, which is an evaluation of the bare self-energy in (8.32), rather than incorporating more elaborate many-body effects. Since the recursion is in the site representation, we have to reverse the Fourier transform in (8.32) to get the x-axis variation, and do a mode-to-site unitary transformation to get the self-energy in the form necessary for the recursion. That is, scattering in the semi-classical world is usually local in position, not in mode. The Fermi golden rule expression for each scattering process of interest is generated in a real space self-energy. The imaginary part of the self-energy is related to the scattering rate via

$$\mathrm{Im}\{\Sigma(i, s; i', s'; X, k_x)\} \equiv \hbar \left(\frac{1}{\tau}\right)_{i,s}^{i',s'}. \tag{8.33}$$

It is the latter scattering rate, which is a function of the x-directed momentum (which is related, in turn, to the energy of the carrier) in a cross-section of the device. Each cross-section can be thought of locally as a quantum wire. This scattering rate must be converted to the site representation with a unitary transformation. At site i,j, the correction due to scattering that gets added to the local potential $V_{i,j}$ is given by

$$\Gamma(i, s) = \mathrm{Im}\{\Sigma\} = U_s^\dagger \left(\frac{\hbar}{\tau}\right)_{i,j}^{i',j'} U_s. \tag{8.34}$$

Here, U_s is the mode-to-site transformation matrix discussed above. The actual scattering rates are computed in the normal manner [10]. The discussion above is for a two-dimensional self-energy, but the modifications for three dimensions is straightforward.

8.2.3 The MOSFET

This methodology, based upon the scattering matrix recursion technique, has been applied to a number of ballistic and diffusive devices [10]. The device to be considered here is a Si nanowire MOSFET structure with a cross-sectional area of 6 × 12 nm (depth by width). The x and z directions correspond to the length and the height of the device, respectively, while the y-direction is the width of the structure. The source and the drain contact regions are 10 nm in length and 18 nm in width, but have the same height as the active channel. In an actual device, the length of the source and the drain of a MOSFET would be much longer, but this length captures the important energy relaxation length. We implement open boundary conditions at the ends of the structure and on the sidewalls. The gate is applied only to the top surface of the nanowire, and has a length of 11 nm, which overlaps onto the source and drain. The actual channel length of the device used in these simulations is 9 nm. The entire structure is on a 10 nm buried oxide (BOX) layer. The gate oxide is taken to be 2 nm thick, and the channel region is doped to 5×10^{18} cm^{-3}. First, consider the ballistic transport. For example, in figure 8.4 is shown the local potential for a 12 nm wide, 10 nm gate length MOSFET at room temperature. The dips in the potential are the location of donors in the source and drain regions. There are few dopants in the channel [9]. The transport through the device is affected by the coherent interaction of the electrons with these donors and acceptors, and this coherence is exhibited by fluctuations in the current through the device, even though there is no scattering other than by these donor and acceptor atoms [20].

An important point relates to the crystal orientation of the device. As is normal, we assume that the device is fabricated on a (100) surface of the Si crystal, and we then orient the channel so that the current will flow along the [100] direction. This direction is chosen so that the principle axes of the conduction band valleys line up with the coordinate axes. By this, we mean that the [010] direction lines up along the y-direction and the [001] direction lines up with the z direction, and the six

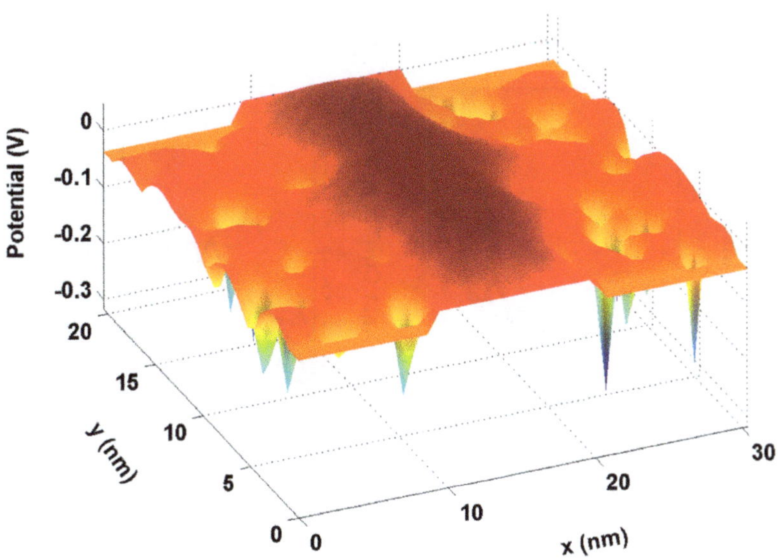

Figure 8.4. Potential profile for the wide-channel SOI MOSFET taken at a depth midway through the 7 nm thickness of the device. Copyright (2004) by the IEEE, reprinted from Gilbert and Ferry [20].

equivalent ellipsoids are oriented along the Cartesian coordinate axes. This is important so that the resulting quantization will split these ellipsoids into three pairs. Moreover, the choice of axes is most useful as the resulting Hamiltonian matrix will be diagonal. In contrast, if we had chosen the [110] direction to lie along the channel, the six ellipsoids would have split into a two-fold pair (those normal to the [100] plane) and a four-fold pair, but the Hamiltonian would not be diagonal since the current axis makes an angle with each ellipsoid of the four-fold pair. Using our orientation complicates the wave function, but allows for simplicity in terms of the amount of memory needed to store the Hamiltonian and to construct the various scattering matrices (as well as the amount of computational time that is required). The drain current for three gate biases is shown in figure 8.5. The fluctuations in current in both panels arise from the universal conductance fluctuations common from such random doping [14, 18]. However, the large current peaks in panel (b) correspond to longitudinal resonances due to discontinuities in the momentum at the ends of the channel. These may be considered to be similar to source–drain resonant tunneling.

In these approaches, it was possible to directly determine the diffusive to ballistic crossover length in the presence of a full set of scattering [10]. When the full set of scatterers is inserted into the device model discussed above, the low field mobility agrees well with various measurements for channel dopings from 10^{16} to 10^{20} cm^{-3}. Then, the diffusive to ballistic crossover can be determined by varying the channel length. For short channels, where the transport is governed by the fully ballistic Landauer formula (8.11), the resistance is independent of length as

$$R = \frac{1}{G} = \frac{\pi\hbar}{e^2 N}. \tag{8.35}$$

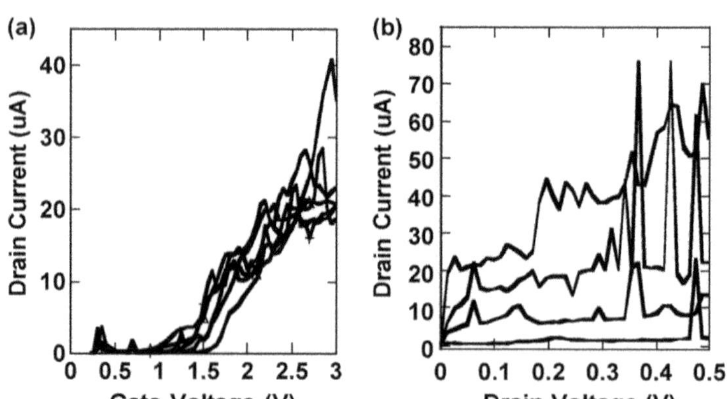

Figure 8.5. (a) curves for six different dopant distributions for the device of figure 8.4 for a drain voltage of 10 mV. Here, we see the results of the quantum interference arising from the donors. The electron density interacting with these causes the observed spikes. (b) Drain current curves MOSFET device. From bottom to top, the gate voltages are 1.5, 2, 2.5, and 3 V. The peaks are evidence of the formation of resonant levels in the channel. Copyright (2004) by the IEEE, reprinted from Gilbert and Ferry [20].

On the other hand, when the transport is diffusive, the resistance is given by the well-known formula

$$R = \frac{L}{\sigma A} = \frac{1}{ne\mu}\frac{L}{A}. \tag{8.36}$$

In figure 8.6, this is demonstrated for a device, such as that discussed above, but with two different widths of the nanowire channel [21]. In panel (a), we show the resistance at 100 K and 300 K for a nanowire cross-section of 4.3 × 4.3 nm^2. The small cross-section produces more quantization of the phonon modes, making them less effective. Here, at 300 K, it appears the crossover is at about 5 nm. In panel (b), we show the length effect for a channel cross-section of 6.5 × 6.5 nm^2. At a lattice temperature of 100 K, the resistance is still nearly independent of the length of the nanowire, at least up to 8 nm. But, at 300 K, the phonons are sufficiently excited that the resistance is almost linear in the length of the nanowire, indicating diffusive transport. So, for this nanowire, the ballistic to diffusive transport at 300 K occurs somewhere below 2 nm. The mobility in an MOS structure is of the order of 300 cm^2 Vs^{-1} or less. For this mobility, and transport along the (100) direction, this would give a momentum relaxation time of 32 fs, or an apparent mean free path of about 7 nm. However, the scattering rate is much faster than the momentum relaxation time, perhaps by as much as a factor of 4, and this lowers the real mean free path to under 2 nm, which is what is observed.

The above were for the simple channel structure and don't show the effects of the gate and drain potentials. For the above device with a channel width of 18 nm and a length of 10 nm, the effect is even more pronounced [10]. Here, the channel is undoped to reduce the impurity scattering effects and to study primarily the phonon effects. The channel has gates on the top and both sides, such as in a tri-gate device. The gate oxide thickness (SiO_2) on this device is 1 nm. As previously, this simulation is carried out at 300 K. The gate voltage of the device is held constant at $V_g = 0.5$ V.

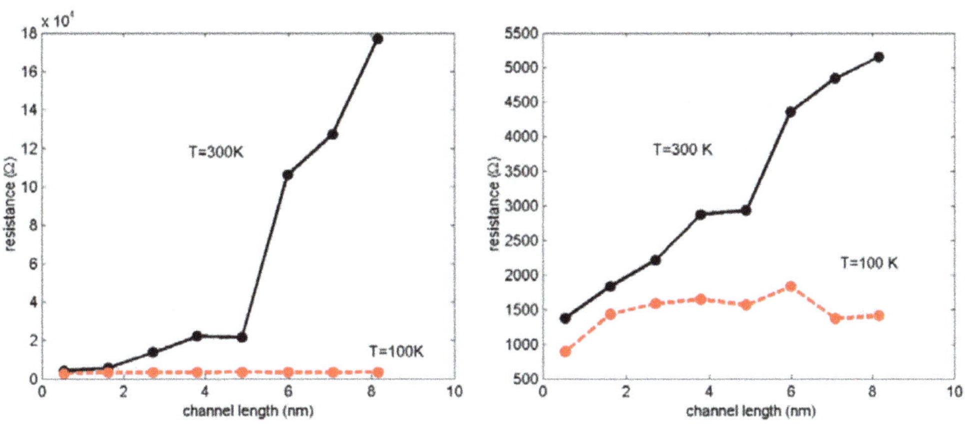

Figure 8.6. (a) Resistance versus the nanowire length for a device with a width of 4.3 nm at the indicated temperatures. (b) Resistance versus the nanowire length for a device with a width of 6.5 nm at the indicated temperatures. Reprinted from Akis *et al* [21], under the creative commons open access license.

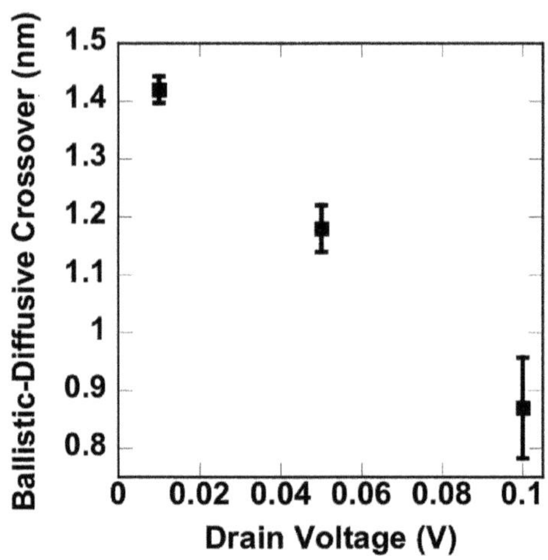

Figure 8.7. Ballistic to diffusive crossover length in a Si nanowire MOSFET, whose details are discussed in the text. Reprinted from Gilbert *et al* [10], copyright (2005) with permission of AIP Publishing.

The reduction of the ballistic to diffusive crossover is summarized in figure 8.7, where we plot the crossover length for three different drain voltages. There is a clear linear trend, within the error associated with the measurements, in the reduction of the ballistic to diffusive crossover by the enhanced phonon scattering in the drain of the silicon nanowire transistor. The more energy the carriers gain as they traverse the length of the nanowire transistor, the more scattering they see. Thus there is a linear drop in the ballistic to diffusive crossover length with increases in drain voltage. This is due to the fact that the carriers are accelerated, and heated, to the point that the phonon interaction is stronger in the drain than in the source, and therefore, accounts for the majority of the increase in channel resistance. The resistance of the quantum wire is dramatically increased as the drain voltage is increased, which may be attributed to considerable carrier heating (100 mV across 10 nm corresponds to 100 kV cm^{-1}, which is well into the hot electron region in silicon).

8.3 The density matrix

In going beyond the Schrödinger equation, there is a choice of a few methodologies. To be sure, each of these forms arises from the wave function that is introduced in this chapter, and each of these will be treated in more detail here and in later sections. Here, the most straightforward approach, which creates a matrix of modes, is treated. Generally, the Schrödinger equation is solved by assuming an expansion of the wave function in a suitable basis set so that each basis function is an energy eigen function according to

$$H\varphi_n = E_n\varphi_n. \tag{8.37}$$

Here, E_n is the energy level corresponding to the particular basis function φ_n. For example, in a quantum well, the basis functions are the set of wave functions corresponding to the bound states of the quantum well. In a quantum wire, these are the bound states arising from the quantization resulting from the wire width (and height). In both of these cases, the basis functions are the *mode* wave functions. They have values on the sites of the grids used in the computation. Then, the total wave function can be written as

$$\psi(r,\, t) = \sum_{n} c_n \varphi_n(r) e^{-iE_n t/\hbar}. \tag{8.38}$$

In general, the usefulness of the present form (8.38) lies in the existence of an entire set of basis functions, which are characteristic of the problem at hand. That is, generally, the set of basis functions is an infinite set, including occupied as well as empty levels, bound states and even unbound states in a finite depth quantum well. In general, a matrix can be constructed of the form

$$K(r,\, r';\, t,\, t') = \sum_{m,n} c_m^* c_n \varphi_m^\dagger(r) \varphi_n(r') e^{i(E_m t - E_n t')/\hbar}. \tag{8.39}$$

In many situations, the separation into two times is not necessary, as the processes which lead to these separate times are incredibly fast. For the general situation, one may then reduce (8.39) to a single time variable, and we then have the density matrix

$$\rho(r,\, r',\, t) = \sum_{m,n} c_{mn} \varphi_m^\dagger(r) \varphi_n(r') e^{i(E_m - E_n)t/\hbar}. \tag{8.40}$$

In this form, the two coefficients have been combined into a single one with two subscripts.

There are many different definitions of the density matrix, which differ only in the fine detail. One can also introduce the imaginary time $t \to -i\hbar\beta$, where $\beta = 1/k_B T$ is the inverse temperature (thermal energy). Then, the density matrix can be written in the thermal equilibrium form

$$\rho(r,\, r',\, T) = \sum_{n} e^{-\beta E} \varphi_n^\dagger(r) \varphi_n(r'). \tag{8.41}$$

8.3.1 The Liouville and Bloch equations

A different method of writing the density matrix is to express it in terms of the two wave functions as

$$\rho(r,\, r',\, t) = \psi^\dagger(r,\, t) \psi(r',\, t). \tag{8.42}$$

If this form is substituted into the Schrödinger equation, and the terms rearranged, one obtains the form for the time derivative of the density matrix

$$i\hbar \frac{\partial \rho}{\partial t} = H\rho - \rho H = [H,\, \rho], \tag{8.43}$$

a form known as the Liouville equation. In expanded form, this equation can be written as

$$i\hbar \frac{\partial \rho}{\partial t} = \left[-\frac{\hbar^2}{2m^*} \left(\frac{\partial^2}{\partial r^2} - \frac{\partial^2}{\partial r'^2} \right) + V(r) - V(r') \right] \rho. \tag{8.44}$$

While no dissipation is indicated here, it is quite easy to incorporate a dissipative term [22–24]. If the transformation leading to (8.41) is introduced, then the left-hand derivative becomes a derivative with respect to β, and that version of the equation is often called the Bloch equation.

There are some physical constraints on the density matrix that arise from the normalization of the wave function. For example, if (8.39) is integrated over the diagonal elements as

$$\begin{aligned}\int dr\rho(r, r, t) &= \int dr\sum_{m,n} c_m^* c_n \varphi_m^\dagger(r)\varphi_n(r)e^{\frac{i(E_m t-E_n t)}{\hbar}} \\ &= \sum_{m,n} c_m^* c_n \int dr\varphi_m^\dagger(r)\varphi_n(r)e^{i(E_m t-E_n t)/\hbar} = \sum_n |c_n|^2 = 1\end{aligned}. \tag{8.45}$$

8.3.2 The quantum kinetic equation

Equation (8.43) gives the equation of motion for the density matrix for an arbitrary potential imposed upon the system, either internally or externally. But, the equations do not contain any dissipation. On the other hand, the Boltzmann equation is a kinetic equation which describes the dynamic evolution of a distribution function under the influence of external fields and collisional forces. In this light, the resulting transport is a balance between the driving forces of the fields and the relaxation forces of the collisions. An analogous kinetic equation may be derived for the density matrix and describes the transport in equivalent fields and relaxation forces. Consider a semiconductor system composed of electrons and phonons, and the interaction between these two sub-systems, although interactions among the electrons could be added. The system is described by the Hamiltonian

$$H = H_0 + H_L + H_F + H_{eL}, \tag{8.46}$$

where the terms on the right-hand side describe the electrons, the lattice, the external fields, and the electron–lattice interaction, respectively. The external field will be taken in the scalar potential gauge $H_F = -eFr$, where F is the electric field, and the electron interaction could contain all of the appropriate scattering processes, typically evaluated in the semi-classical Fermi golden rule approach.

The total density matrix ρ is defined over the entire system. If there were no interactions between the electrons and the lattice, then the density matrix could be written as a tensor product of the density matrix for the electrons and the density matrix for the lattice. In this situation, a trace Tr_L over the lattice coordinates would yield just the density matrix for the electrons. Similarly, a trace Tr_e over the electron coordinates would yield just the density matrix for the lattice. When the interactions are present, then the simple divide into two distinct density matrices will not be possible. Nevertheless, the relevant density matrix for the electrons is achieved by the partial trace over the lattice variables as

$$\rho_e = Tr_L\{\rho\}. \tag{8.47}$$

The approach with the Hamiltonian (8.46) is followed by introducing this into the Liouville equation (8.43), and then tracing over the lattice variables to produce the partial density matrix for the electrons, as

$$i\hbar\frac{\partial\rho_e}{\partial t} = \left[H_0 + H_F, \rho_e\right] + Tr_L\{[H_{eL}, \rho]\}. \tag{8.48}$$

Clearly, the first term on the right-hand side is the electronic motion within the effective mass approximation under the influence of the applied field. That is, this first term represents simple ballistic transport. The second term represents the electron–phonon interaction and the effect that this has upon the resulting electronic density matrix. In the common understanding, this trace over the lattice variables corresponds to the summation over the phonon momentum vector q that was used in previous chapters. In the following, the short-hand $H_e = H_0 + H_F$ will be used. But, to proceed, another short-hand approach will be introduced and that is the use of commutator generating super-operators, with which (8.48) becomes

$$i\hbar\frac{\partial\rho_e}{\partial t} = \hat{H}_e\rho_e + Tr_L\left\{\hat{H}_{eL}, \rho\right\}. \tag{8.49}$$

In essence, these super-operators work in Liouvillian space, so that the super-operator has a dimension twice that of the Hamiltonian itself. If the density matrix is a matrix, then the super-operator is a fourth-order tensor so that the product implied in (8.49) yields matrices. The properties of these super-operators are straightforward and are reviewed in [2, 25]. Projection super-operators will also be used, which are just the super-operator form of a projection operator.

First, the density matrix equation (8.49) is Laplace transformed as

$$\left(s + \frac{i}{\hbar}\hat{H}_e\right)\tilde{\rho}_e = -\frac{i}{\hbar}Tr_L\{\hat{H}_{eL}\tilde{\rho}\} + \rho_e(0), \tag{8.50}$$

where the tilde over the density matrix refers to the Laplace transformed form, and the last term on the right is the initial condition on the density matrix at $t = 0$. Now, the projection super-operator [26–28], and its complement, are introduced as

$$\tilde{\rho}_1 = \hat{P}\tilde{\rho}_e, \; \hat{P}^2 = \hat{P}, \; \hat{Q} = \hat{1} - \hat{P}. \tag{8.51}$$

The last two expressions define the idempotent property of the projection super-operator and that there exists a projection operator onto the complement space. Now, what is meant with this projection operator and the one-electron density matrix? In general, there is a very large number of electrons in the semiconductor, say N. Then, the total phase space, or configuration space, has $6N$ coordinates (plus time). What is desired is a typical single-electron density matrix which has six coordinates plus time. This is usually achieved via the BBGKY hierarchy with one complication in that the equation for the one-electron density matrix results with an additional term that involves a two-electron function, that describes the correlations that remain. To proceed, one must make some approximations to this two-electron function. With the projection operators defined above, this term is not lost, but is buried within the scattering function. This particular projection super-operator commutes with the trace operation in (8.50), so that the scattering super-operator may be introduced through the definition

$$\hat{\Sigma}\tilde{\rho}_1 = Tr_L\{\hat{P}\hat{H}_{eL}\tilde{\rho}\}. \tag{8.52}$$

This form will be examined further below. First, however, using this definition, (8.50) becomes

$$\tilde{\rho}_1 = i\hbar\hat{P}\frac{1}{i\hbar s - \hat{H}_e - \hat{\Sigma}}\left[\hat{P}\rho_e(0) + \hat{Q}\rho_e(0)\right]. \tag{8.53}$$

The initial condition has been split into the projection and its complement for convenience in the following development.

In the following, a long expression will be obtained for the resolvent (the inverse term in the center of (8.53)) mainly to find the proper form for the collision operator, but also to give us the proper equation for the density matrix. The goal is to eventually produce a path integral that will allow the adoption of ensemble Monte Carlo (EMC) procedures. To proceed, this resolvent operator can be rewritten as

$$\hat{R}(s) = \frac{1}{i\hbar s - \hat{H}_e - \hat{\Sigma}} \equiv \frac{1}{i\hbar s - \hat{H}'} = \frac{1}{i\hbar s - (\hat{P} + \hat{Q})\hat{H}'(\hat{P} + \hat{Q})}, \tag{8.54}$$

for which it can be further expanded into two forms:

$$\begin{aligned}\hat{R}(s) &= \frac{1}{i\hbar s - \hat{P}\hat{H}'\hat{P}}\{1 - [(\hat{P} + \hat{Q})\hat{H}'Q + \hat{Q}\hat{H}'\hat{P}]\hat{R}(s)\} \\ &= \frac{1}{i\hbar s - \hat{Q}\hat{H}'\hat{Q}}\{1 - [(\hat{P} + \hat{Q})\hat{H}'\hat{P} + \hat{P}\hat{H}'\hat{Q}]\hat{R}(s)\}\end{aligned}. \tag{8.55}$$

The first form, on the top line, is pre-multiplied by the projection operator and the second form, on the bottom line, is pre-multiplied by the complement operator, so that the resolvent can be rewritten in the rather complex form

$$\begin{aligned}\hat{R}(s) = &\left(\hat{P} + \hat{Q}\frac{1}{i\hbar s - \hat{Q}\hat{H}'\hat{Q}}\hat{Q}\hat{H}'\hat{P}\right)\frac{1}{i\hbar s - \hat{P}\hat{H}'\hat{P} - \hat{C}}\left(\hat{P} + \hat{P}\hat{H}'\hat{Q}\frac{1}{i\hbar s - \hat{Q}\hat{H}'\hat{Q}}\hat{Q}\right) \\ &+ \hat{Q}\frac{1}{i\hbar s - \hat{Q}\hat{H}'\hat{Q}}\hat{Q}\end{aligned}, \tag{8.56}$$

where

$$\hat{C} = \hat{P}\hat{H}'\hat{Q}\frac{1}{i\hbar s - \hat{Q}\hat{H}'\hat{Q}}\hat{Q}\hat{H}'\hat{P} \tag{8.57}$$

is the *collision* operator. This expression involves the terms that connect a 'diagonal' element (in the projection first set of operators) to an off-diagonal element (the complement terms in the last set of operators) and reconnects them by the conjugate operation. This is just the type of term which leads to the Fermi golden rule, in that the first and last combinations in (8.57) basically are the matrix element and its conjugate, while the central term leads ultimately to the energy-conserving delta function in the semi-classical situation. Moreover, the Hamiltonian in the central term involves also the electronic Hamiltonian, so that the electron energy is incorporated as required to fit the Fermi golden rule. Recognizing all of this, (8.52) can now be rewritten in terms of the collision operator as

$$\hat{\Sigma}\tilde{\rho}_1 = Tr_L\left\{\hat{P}(\hat{H}_{eL} + \hat{H}_e)\hat{Q}\frac{1}{i\hbar s - \hat{Q}(\hat{H}_{eL} + \hat{H}_e)\hat{Q}}\hat{Q}\tilde{\rho}\right\}. \tag{8.58}$$

This complicated structure gives much information. First, the electron–electron interactions should be included in the Hamiltonian although the single-electron density matrix is being extracted. This is because the electron–electron interaction provides the screening of the appropriate interactions. Secondly, the electron–phonon interactions can lead to real energy shifts, which are the self-energy corrections to the single-particle energies. This is in addition to the normal dissipative scattering processes.

Using all of the pertinent information above, the equation of motion for the density matrix, in the Laplace domain, can be written as

$$\begin{aligned} i\hbar\tilde{\rho}_1(s) = i\hbar\rho_1(0) + \hat{P}\hat{H}_e\hat{P}\tilde{\rho}_1 + \hat{\Sigma}\tilde{\rho}_1 \\ + \hat{P}(\hat{H}_{eL} + \hat{H}_e)\hat{Q}\frac{1}{i\hbar s - \hat{Q}(\hat{H}_{eL} + \hat{H}_e)\hat{Q}}\hat{Q}\tilde{\rho}_e(0)^{\cdot} \end{aligned} \tag{8.59}$$

Note that the third term involves the entire many-electron initial density matrix, while the first term on the right involves only the single-electron density matrix. The last term on the right has been thought to produce a number of effects, including the random force used for the Langevin equation at the lowest order of the electron–phonon interaction [29] and a screening of the driving field in higher order [30]. The temporal equation can now be obtained by retransforming this last equation, which results in, for a force due to the electric field,

$$\frac{\partial\rho_1}{\partial t} + \frac{e}{\hbar}\boldsymbol{F}\cdot\frac{\partial\rho_1}{\partial\boldsymbol{k}} = -\frac{1}{\hbar^2}\int_0^t\hat{\Sigma}(t-t')\rho_1(t')dt', \tag{8.60}$$

disregarding the initial state decay. This form of the quantum kinetic equation was first derived by Barker [31], but its form is quite analogous to the Prigogine–Resibois equation [32], for which the EMC approach is quite well suited. Except for the convolution form of the collision term, this equation is essentially the same as the Boltzmann transport equation, which means that some quantum effects, such as quantization of the states, have been left out or are buried in the form of the one-electron density matrix. Nevertheless, temporal correlation effects are naturally included through the convolution integral for the scattering. Spatial correlations are included within the quantum nature of the self-energy term in the integral. An extensive investigation into the full spectrum of the density matrix in the Liouvillian space for nonequilibrium carriers also has been undertaken [33]. This has been applied to the sub-femtosecond initial state of transport [34], and for connecting to the correct Green's functions [35]. Because of the direct analogy of (8.60) with the Boltzmann equation, it can be used in the drifted Maxwellian approximation to generate a set of balance equations exactly as in section 6.3.2. This set has been generated and used to study transient transport in high electric fields [36]. A more effective approach, however, has been formed from adapting (8.60) into what has become called the Barker–Ferry equation.

8.3.3 The Barker–Ferry equation

In chapter 7, the EMC approach to solutions of the Boltzmann transport equation was discussed. This approach is based upon converting the equation into a path integral, which is then susceptible to the use of random number techniques with weightings based upon the nature of the physical processes involved [25, 37, 38]. The difference between the Boltzmann transport equation and the quantum equivalent (8.60) lies in the convolution integral for the scattering function on the right-hand side of the latter equation. However, this should not be a hindrance to the development of a path integral formulation that would allow for an EMC treatment of quantum transport. First, the retarded collision term in (8.60) is rewritten in a more common form as

$$\begin{aligned}\frac{\partial\rho_1(\boldsymbol{k},t)}{\partial t}+\frac{e}{\hbar}\boldsymbol{F}\cdot\frac{\partial\rho_1(\boldsymbol{k},t)}{\partial\boldsymbol{k}}=\int_0^t dt'\sum_{\boldsymbol{k}'}\{&S(\boldsymbol{k},\boldsymbol{k}';t,t')\rho_1(\boldsymbol{k}',t')\\&-S(\boldsymbol{k}',\boldsymbol{k};t,t')\rho_1(\boldsymbol{k},t')\}\end{aligned}. \tag{8.61}$$

The two momenta variables in (8.61) are explicit functions of the retarded time t' on the right-hand side through the accelerative expressions

$$\begin{aligned}\boldsymbol{k}(t')&=\boldsymbol{k}(t)+\frac{e}{\hbar}\int_{t'}^{t}\boldsymbol{F}(t'')dt''\\ \boldsymbol{k}'(t')&=\boldsymbol{k}'(t)+\frac{e}{\hbar}\int_{t'}^{t}\boldsymbol{F}(t'')dt''\end{aligned}. \tag{8.62}$$

This is required due to the convolution integral and represents the acceleration of the particle *through* the collision process, thus accounting for the non-zero time duration of the collision itself [39]. The temporal scattering processes appear, for phonon scattering, as

$$S(\boldsymbol{k},\boldsymbol{k}';t,t')=Re\left\{\frac{2}{\hbar^2}\sum_q|M_q|^2 f(N)\delta_{\boldsymbol{k},\boldsymbol{k}'\pm q}e^{-\frac{i}{\hbar}\int_t^{t'}dt''\zeta(\boldsymbol{k},\boldsymbol{k}';t',t'')}\right\}, \tag{8.63}$$

where

$$F(N)=N_q+\frac{1}{2}\pm\frac{1}{2} \tag{8.64}$$

introduces the Bose–Einstein factors, with the upper sign for emission and the lower sign for absorption, and

$$\zeta(\boldsymbol{k}, \boldsymbol{k}';t', t'') = E(\boldsymbol{k}(t'')) - E(\boldsymbol{k}'(t'')) \pm \hbar\omega_{\boldsymbol{q}}. \tag{8.65}$$

In the semi-classical approach of the previous chapters, the last exponential in (8.63) represents the joint spectral densities of the two states. This would become the energy-conserving delta function if there was a restriction to remaining on the energy shells. The first term in the curly brackets of (8.60) refers to scattering *in* to the state with momentum k from all the possible states with momentum $\boldsymbol{k}'$. This process is weighted by the occupancy of the initial momentum through the density matrix at this momentum. On the other hand, the second term corresponds to scattering *out* of the state k to all the possible states with momentum $\boldsymbol{k}'$, and is weighted by the density matrix of this initial state. The temporal evolution of the scattering functions creates a barrier to generating an integral equation approach, but this can be overcome by the introduction of a secondary self-scattering process, just as in chapter 7 for EMC (obviously, the goal is to solve the path integral equation with EMC techniques). To do this, a new scattering function is introduced via

$$S^{*}(\boldsymbol{k}, \boldsymbol{k}';t, t') = S(\boldsymbol{k}, \boldsymbol{k}';t, t') + [\Gamma(t, t') - \Gamma_{\mathrm{out}}(t, t')\delta(\boldsymbol{k} \cdot \boldsymbol{k}')], \tag{8.66}$$

with

$$\begin{aligned} \Gamma_{\mathrm{out}}(t, t') &= \sum_{\boldsymbol{k}'} S(\boldsymbol{k}, \boldsymbol{k}';t, t') \\ \Gamma(t, t') &= \Gamma_0 \delta(t - t'). \end{aligned} \tag{8.67}$$

The terms in the square bracket of (8.66) make no contribution to the scattering integral, but the term in Γ_{out} plays a very important role. This term projects out the contribution due to uncompleted out-scattering processes, and the retardation in the out-scattering term is accounted for here rather than in the equivalent term in (8.61). With these specific terms, (8.61) can be written as

$$\left[\frac{\partial}{\partial t} + \frac{e}{\hbar}\boldsymbol{F} \cdot \frac{\partial}{\partial \boldsymbol{k}} + \Gamma_0\right]\rho_1(\boldsymbol{k}, t) = \int_0^t dt' \sum_{\boldsymbol{k}'} S^{*}(\boldsymbol{k}, \boldsymbol{k}';t, t')\rho_1(\boldsymbol{k}', t'). \tag{8.68}$$

Now, this can be approached precisely as the Boltzmann equation was in section 7.1 to yield a path integral. The result is

$$\begin{aligned} \rho_1(\boldsymbol{k}, t) = &\int_0^t dt' e^{-\Gamma_0(t-t')} \\ &\times \int_0^{t'} dt'' \sum_{\boldsymbol{k}'} S^{*}(\boldsymbol{k}, \boldsymbol{k}';t', t'')\rho_1(\boldsymbol{k}', t'') \end{aligned}, \tag{8.69}$$

where the momenta are functions of t'' through (8.62). In the limit of long times and instantaneous collisions and no quasi-particle effects from the field, this result reduces to that found for the Boltzmann transport equation and the EMC approach in chapters 6 and 7. The result (8.69) is known as the Barker–Ferry equation [40].

The evolutionary properties that are inherent in (8.62), and thus in (8.69), lead to what is known as an intra-collisional field effect [41, 42]. This effect is very important when the field is large or when the collision duration is a significant fraction of the mean time between collisions, and has previously been analyzed for steady-state transport in uniform, time independent fields [31, 43]. In the steady-state case, the intra-collisional field effect induces a broadening and skewing of the usual energy-conserving delta functions [44]. This leads, for example to a lowering of the threshold for phonon emission due to acceleration during the collision, a lowering of the scattering strength at high fields, and an early inter-valley transfer and impact ionization [45] at high energies.

The path integral (8.69) is the basis for simulations of the transport with EMC techniques, as described in chapter 7. Here, however, the intra-collisional field effect and off-shell corrections are easily added to the EMC. This approach first appeared for studying high field transport in SiO_2, and in studies of the mechanisms by which dielectric breakdown would occur in this material [46, 47]. These two approaches both used the polar optical phonons of SiO_2 as the main scattering process, but also included the acoustic phonons of low energy. However, it was found that the polar optical phonons could not stabilize the distribution function above fields of about 2–3 MV cm^{-1} [48]. If no other process occurred, this would lead to a much lower breakdown field for the oxide than is observed experimentally, due to runaway due to the inability of the polar modes to fully relax the carriers at high energy. To solve this problem, an additional process was introduced. Fischetti [46] introduced a high energy acoustic process which involved Umklapp scattering across the zone boundary. On the other hand, Porod and Ferry [47] introduced non-equivalent inter-valley scattering with the second set of valleys lying some 2–3 eV above the principal minimum of the conduction band. Both of these processes are successful in controlling the distribution function and pushing the breakdown field somewhat above 7–10 MV cm^{-1}. The experiments and knowledge of the properties of the oxide at high energies are so limited that it is not possible to select which of the two processes actually is the most important [49], as both lead to basically the same velocity-field curve and the same threshold for impact ionization. This is because both are high energy scattering processes that limit the velocity and prevent runaway, and they appear to do so with equal effectiveness. In figure 8.8, the simulated velocity-field curve is shown for an oxide conduction effective mass of $0.5m_0$, using the presence of satellite valleys to limit the runaway [47]. Experimental data from Hughes is also shown for comparison [50, 51].

8.4 Nonequilibrium Green's functions

In chapter 6, the Boltzmann equation was discussed. There, one had a nearly equilibrium situation in which the distribution function was a Maxwellian or a Fermi–Dirac at a possible temperature above that of the lattice. Then, there was the true hot carrier system in which the distribution was unknown and had to solved via one of several possible methods. With Green's functions, the system is quite the same. In equilibrium, one has the normal Green's functions which are assumed to be

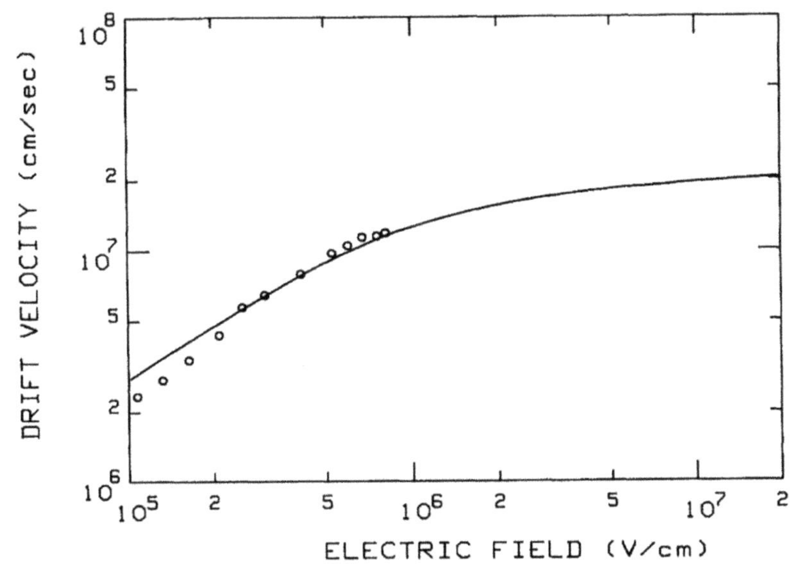

Figure 8.8. The results of an EMC simulation of transport in SiO_2 (solid line), with an effective mass of $0.5m_0$ and an acoustic deformation potential of 25 eV. Satellite valleys lying ~3 eV above the conduction band minima are assumed. The circles are data from Hughes [50, 51]. Reprinted with permission from Porod and Ferry [47], copyright (1985) by the American Physical Society.

at low temperature so that the Fermi energy is a sharp cutoff and one thinks of electrons being excitations above the Fermi energy and holes as being the empty state that arises below the Fermi energy. Of course this can be carried over to semiconductors without much difficulty. Then, there are the Matsubara, or temperature, Green's functions which are a totally different beast [2, 16], with temperature entering through the substitution leading to (8.41) where imaginary time transforms to a temperature. Finally, there is the situation in which the system is well out of equilibrium, and new equations are required to determine just what the hot carrier distribution really is in this state. The same is true for Green's functions, and the new functions that must be found to describe this distribution are the nonequilibrium Green's functions (NEGF) [52, 53], which are based upon original work by Schwinger [54, 55]. In this section, the equilibrium Green's functions and NEGF will be described along with some of their applications. But, some problems that limit the applicability of NEGF will also be discussed. In general, the Green's functions are far more difficult to use, especially when anisotropic scattering centers, such as the Coulombic impurity and polar optical phonon scattering, are present.

8.4.1 Equilibrium Green's functions

In the density matrix, only a single time variable appears in the problem, as it was assumed that the two wave functions were to be evaluated at the same time. In essence, these two approaches build in correlations in space, but do not consider that there may be correlations in time. On the other hand, the kernel (8.39) possesses both times, and this will lead to the concept of the Green's functions [16]. In general, we separate the kernel into a term which is forward in time ($t > t_0$) and a term which is backward in time ($t < t_0$), by changing the order of the two wave functions. Then, we combine the two, which produces a commutator relationship, and define the *retarded* Green's function as

$$G_r(r, r'; t, t_0) = -i\Theta(t - t_0)\langle\{\psi(r, t), \psi^\dagger(r', t_0)\}\rangle. \tag{8.70}$$

In this equation, several new notations have been included. First, the $\Theta(x)$ is the Heavyside step function which is 1 for $x > 0$ and zero otherwise. Then, the angle brackets have been added to indicate an ensemble average, which corresponds to a sum over the appropriate basis states, while the curly brackets denote an *anti*-commutator relation where the normal negative sign is replaced with a positive sign, which is the normal case for fermions, and may be written as

$$\{c, c^\dagger\} = cc^\dagger + c^\dagger c = 1. \tag{8.71}$$

Of course, for the wave functions, the value of unity is actually $\delta(r - r')\delta(t - t_0)$. By the same token, there exists an advanced Green's function that is the adjoint of the retarded Green's function and is written as

$$G_a(r, r'; t, t_0) = i\Theta(t_0 - t)\langle\{\psi^\dagger(r', t_0), \psi(r, t)\}\rangle. \tag{8.72}$$

In addition, the wave functions are now *field operators*, which means that they operate on an empty world like creation and annihilation operators; e.g., they incorporate fermion creation and annihilation operators within their definitions. That is, if the wave functions are described by (8.38), then the coefficients c are no longer simple constants, but are these creation and annihilation operators [2, 16], and the resulting field creation operator creates and excitation *with that wave function*. For example, the term $\psi(r, t)\psi^\dagger(r', t_0)$ may be considered to create an excitation above the Fermi level at r' and t_0, which then propagates to r, t where it is annihilated. So, each of the two Green's functions contains a pair of these processes, in which one is the reverse of the other.

The general non-interacting ground state Green's function must satisfy two forms of the Schrödinger equation, in terms of the different sets of variables. The equations of motion are given as

$$\begin{aligned} \left[i\hbar\frac{\partial}{\partial t} - H_0(r) - V(r)\right]G^0(r, r'; t, t_0) &= \hbar\delta(r - r')\delta(t - t_0) \\ \left[-i\hbar\frac{\partial}{\partial t_0} - H_0(r') - V(r')\right]G^0(r, r'; t, t_0) &= \hbar\delta(r - r')\delta(t - t_0) \end{aligned}. \tag{8.73}$$

For the moment, it will be assumed that the Green's function here is a function only of the differences in position and time, and that there is no potential in the system. Then, the two equations above can be Fourier transformed in these difference variables as

$$G^0(k, \omega) = \frac{\hbar}{\hbar\omega - E(k)}. \tag{8.74}$$

The bare Hamiltonian in (8.73) produces the energy as a function of momentum from the Fourier transform of the *difference* in positions, while the frequency arises from the Fourier transform of the *difference* in times. (In equilibrium, it is assumed that the system is homogeneous and globally time independent.) Neglecting the potential means that a free propagating Green's function for a homogeneous system with no boundaries is being considered in (8.74). Now, if the inverse of this Fourier transform is desired, the retarded and advanced Green's functions can be produced by introducing a convergence factor described by η. Hence, the retarded and advanced bare (or ground state) Green's functions arise from (8.74) as

$$G^0_{r,a}(k, \omega) = \frac{\hbar}{\hbar\omega - E(k) \pm i\eta}. \tag{8.75}$$

The inverse transform in time can be written as

$$G^0_{r,a}(k, t) = \int \frac{d\omega}{2\pi} G_{r,a}^{\ 0}(k, \omega) e^{-i\omega t}. \tag{8.76}$$

The integration proceeds along the real frequency axis. But, as for any integration involving a complex variable, it has to be closed to obtain a full contour integration. for $t > 0$, the contour has to close in the lower half plane, in order to contain the proper pole. On the other hand, for $t < 0$, the contour has to close in the upper half plane. The introduction of the convergence factors in (8.75) are addressed by this closure, as they move the pole off the real axis. Hence, closing in the lower half plane requires the upper sign, and

$$G^0_r(k, \omega) = \frac{\hbar}{\hbar\omega - E(k) + i\eta}, \quad G^0_a(k, \omega) = \frac{\hbar}{\hbar\omega - E(k) - i\eta}. \tag{8.77}$$

Naturally, these two functions are conjugates of one another.

The integral (8.76) leads to a number of interesting and important properties that we associate with the Green's functions. First, there is an important property of complex functions such as (8.77). When η is vanishingly small, as it is for these two functions where it serves only as a convergence factor, we can write the functions as

$$\frac{1}{x \pm i\eta} = P\frac{1}{x} \mp i\pi\delta(x). \tag{8.78}$$

Here, P denotes the principal part of the function (and the resulting integral). The delta function yields the on-shell relationship between the energy $\hbar\omega$ and the momentum-dependent kinetic energy $E(\mathbf{k})$. Hence, for the non-interacting Green's functions, the delta function arises from

$$2\pi\delta(\hbar\omega - E(k)) = -\mathrm{Im}\{G^0_r(k, \omega) - G^0_a(k, \omega)\}. \tag{8.79}$$

When interacting Green's functions are used, such as when scattering is added, the convergence factors will become the self-energies, and (8.80) will be replaced by the spectral density function, introduced in (8.5), as

$$A(k, \omega) = -\mathrm{Im}\{G_r^0(k, \omega) - G_a^0(k, \omega)\} = i[G_r^0(k, \omega) - G_a^0(k, \omega)]. \tag{8.80}$$

8.4.2 The interaction representation and impurity scattering

To proceed beyond the ground state of the non-interacting Green's functions (8.77), one needs to develop an approximation technique for the interacting Green's functions. The standard approach is the interaction representation, which provides a method for developing the perturbation series. Unfortunately, this often leads to multiple, messy integrals. Fortunately, the Feynman diagrams give us an elegant and pictorial method of looking at the various terms in the perturbation series [56, 57]. As a result, they also give us a much more intuitive method of examining the various terms. It is impossible to include all possible perturbative terms. The number of these terms is just too large. In addition, one must be sure that the perturbation series actually converges, although it is not always possible to do so. This presents a problem in some cases, where it is necessary to push ahead while not being confident that the approach actually converges to the correct, or even to any, answer. This has, on occasion, led to unphysical results. In practice, only certain sets of terms, are retained, such as those based leading to the Hartree energy. The method of generating the perturbation series has come to be known as the interaction representation.

At the start, the wave function is given as a condensed form of (8.38), where the wave function is written in terms of an initial value as

$$\psi(r, t) = e^{-iH(t-t_0)/\hbar}\psi(r, t_0). \tag{8.81}$$

Here, the operators appearing in the Hamiltonian are not explicit functions of time (although this will change when one deals with the electron–phonon interaction). Normally, the Heisenberg representation occurs when the basis functions are independent of time, and time variation is generated for the operators $O(t)$ via the form

$$\frac{\partial O(t)}{\partial t} = i/\hbar[O(t), H], \; O(t) = e^{iHt/\hbar}O(0)e^{-iHt/\hbar}. \tag{8.82}$$

In the interaction representation, it is assumed that there is a base Hamiltonian that contains the ground state Hamiltonian H_0 and a perturbing potential $V(t)$, and the time variation arises from (8.82). Now, (8.81) may be rewritten, for an arbitrary initial time, as

$$\psi(r, t) = e^{iH_0(t-t_0)/\hbar}e^{-iH(t-t_0)/\hbar}\psi(r, t_0). \tag{8.83}$$

It may be recognized that this defines a unitary operator, which after some algebra on the expansion, can be used to define the perturbation series from [16]

$$U(t, t_0) = T\left[\exp\left(-\frac{i}{\hbar}\int_{t_0}^{t} V(t')dt'\right)\right]. \tag{8.84}$$

An important point here, is that the operator U is only unitary if the potential is real. If it contains imaginary components, then problems arise. Earlier, in this chapter, the scattering matrix was discussed. It will appear here as the S-matrix, which is defined from (8.84) as

$$S(t, t') = U(t, t_0)U^{\dagger}(t', t_0) = T\left[\exp\left(-\frac{i}{\hbar}\int_{t'}^{t} V(t'')dt''\right)\right]. \tag{8.85}$$

The operator T that appears in these last two equations is the time-ordering operator. In the series expansion of the exponential, there will be products of the potential V at different times. The time-ordering operator arranges so that the earliest time potential is on the right-hand side of the product and the latest time potential is on the left, and with all times ordered according to this requirement. In addition, the terms must arrange according to Wick's theorem in which each term has a creation and an annihilation operator, with the latter to the right, the so-called operator-ordering [58].

It is very difficult to start the Green's function at $t = 0$ and then activate the perturbation at that time. This is because the state of the system at $t = 0$ is not known. It could be in some sort of quasi-equilibrium which may not be the normal ground state. For example, a p-n junction diode just sitting on the bench is in a steady-state when no bias is applied. But neither of the two regions, n and p, are in equilibrium. This is because of the built-in potential and charge dipole that would not exist, if the two materials were in equilibrium. So, while it may be thought as being in some type of thermal quasi-equilibrium, it is not thermodynamic equilibrium. At $t = 0$, it is not clear just what the state of the system really is, so it will denote it by an undefined state |0>; that is, the empty bra indicates that it is not really known what the state is. However, with the scattering operator, it is possible to say some things about this state. For example, when the perturbation is turned on in the infinite past,

$$\begin{aligned} \psi(0) &= S(0, -\infty) \mid 0\rangle \\ \psi(t) &= S(t, -\infty) \mid 0\rangle \\ \psi(\infty) &= S(\infty, -\infty) \mid 0\rangle \end{aligned} \; . \tag{8.86}$$

In these expressions, the state on the far right of each expression is the true ground state of the system that occurs in the infinite past. Using these expressions, and some simple algebra, the general Green's function can be written in a properly normalized way as

$$G(r, r'; t, t') = \frac{\langle 0 \mid T[\psi(r, t)\psi^{\dagger}(r', t')S(\infty, -\infty)] \mid 0\rangle}{\langle 0 \mid S(\infty, -\infty) \mid 0\rangle}. \tag{8.87}$$

Consider the simple case termed the 'empty band.' For this, the annihilation operator acting upon the ground state produces zero, or

$$\psi(r, t) \mid 0\rangle = 0. \tag{8.88}$$

This implies the Green's function must consist only of the retarded time ordering, where the creation operator is to the right, and occurs earlier, and the retarded Green's function must be given by (8.70). The angular brackets in the function now refer to the ensemble averaging denoted in (8.87).

The power of these Green's functions lies in the knowledge that they remain near equilibrium and can be referred to a true ground state in the infinite past and in the infinite future. When the system is actually in an unknown state with the perturbation turned on at an arbitrary time, this assurance is lost. This will be one of the main problems when attention is turned to the NEGF.

Ionized impurities create a set of charge centers that interact with the carriers via the Coulomb potential. This interaction is a form of scattering which contributes to the resistivity (or conductivity, as the case may be). The impurity has an associated Coulomb potential that is long-range in nature. This long-range interaction is often cut off either by assuming that it is screened in one way or another. These impurities were discussed in chapter 6 without going into any great detail. In the present discussion, the perturbative treatment of impurities will be the focus to demonstrate the approach. The contribution of the impurity potential to the energy is illustrated in (8.73), in which

$$V(r) = \int d^3r' \sum_j V\left(r - r_j - r'\right) n(r'), \tag{8.89}$$

where the r_j are the sites at which the impurities are situated. This is the total system energy, and the effect on transport will arise when the derivative of the total energy is taken with respect to the density n.

The expansion of the S-matrix in the scattering operator for the impurities leads to an infinite series of terms, which change the equilibrium state. The higher-order terms are usually broken up by the use of Wick's theorem (the possible products of the operators will appear as all possible sums of pairs of creation and annihilation operators, which satisfy the time-ordering operator) [16, 58], and this leads to a diagram expansion. The formation of the Green's functions in the S-matrix expansion is accompanied by an averaging process over the equilibrium state (this is actually coupled to an average over the impurity configuration as well, which is discussed below). The equilibrium state must be renormalized in this process, and this causes a cancellation of all disconnected diagrams that are canceled by the equivalent terms in the denominator of (8.87). The result is an expansion in only the connected diagrams. Typical diagrams for the impurity scattering are shown in figure 8.9(a). Here the non-interacting Green's functions ($G^0_{r,\,a}$) are shown in the light-green lines, and the impurity Coulomb interaction is denoted by the dashed red lines. The thicker green line on the left-hand side is now the interacting Green's function, in which the self-energy will replace the convergence factor. The argument is usually that one must carry out an averaging process over the position of the impurities, since the end result (at least in macroscopic samples) should not depend upon the unique distribution of these impurities. Yet, *it is well known today that, for example, a modern MOSFET may well be affected by the precise position of the*

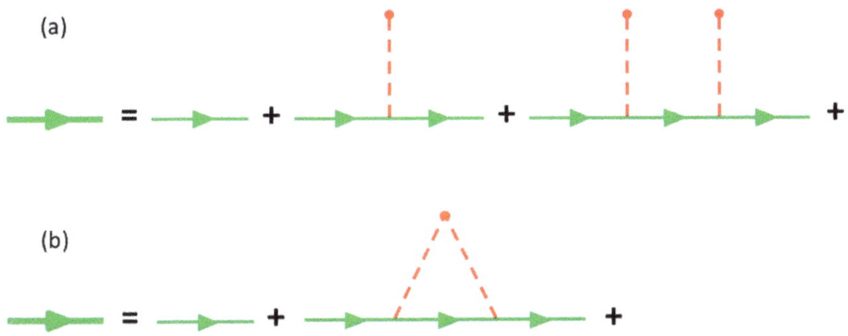

Figure 8.9. (a) The perturbation series for the Green's function in the presence of impurity scattering. The interactions with individual impurities are indicated by the red lines. (b) Reduced interaction after impurity averaging. The double red line represents the interaction as it appears in the Fermi golden rule, with the two lines corresponding to the square magnitude of the matrix element.

impurity atoms [59]. Moreover, the universal conductance fluctuations observed at low temperature in mesoscopic systems are very sensitive to the exact configuration of the impurities [14, 60]. For these reasons, in the semi-classical world, the impurities have come to be treated in real space with molecular dynamics as discussed in section 7.4.

However, the perturbative approach requires the magnitude squared matrix elements for the Fermi golden rule. And, this now requires impurity averaging. The impurity averaging may be understood by noting the summation over the impurity positions that appear in (8.89). What will be required is that the two matrix elements must have momenta that add to zero (which explains the adjoint matrix element in the calculation), so that the impurity has, on average, zero total momentum effect on itself. This condenses the diagrams to the simpler one shown in figure 8.9(b). Additional terms will have two, three, etc, versions of the pair of red lines with the included Green's function. A little algebra will show that it can be resummed to write the Green's function as

$$G_{r,a}(k,\,\omega) = G^0_{r,\,a}(k,\,\omega) + G^0_{r,\,a}(k,\,\omega)\Sigma_{r,a}(k,\,\omega)G_{r,a}(k,\,\omega), \tag{8.90}$$

Which can be rewritten as Dyson's equation

$$G_{r,a}(k,\,\omega) = \frac{G^0_{r,\,a}(k,\,\omega)}{1 - G^0_{r,\,a}(k,\,\omega)\Sigma_{r,a}(k,\,\omega)} = \frac{1}{G^0_{r,\,a}(k,\,\omega)^{-1} - \Sigma_{r,a}(k,\,\omega)}. \tag{8.91}$$

The self-energy that has been introduced into these last two equations is the central part of figure 8.9(b), and is shown in figure 8.10(a). The internal Green's function in (8.90) is the full Green's function (8.91). In many cases, where the number of impurities is small, the scattering is weak, so that the self-energy can be approximated by using only bare Green's function. A slightly better approach is to keep only the lowest-order diagram, but to use the full Green's function, solving the overall problem by iteration. This approach is termed the *self-consistent Born approximation.* The real part of the self-energy gives a correction to the

Figure 8.10. (a) The self-energy diagram for determining the effect of impurity scattering. (b) The lowest-order bubble for computing the conductivity.

single-particle energy $E(\boldsymbol{k})$. This *dresses* the energy due to the interaction with the impurities, as discussed in section 8.1. Such dressing gives a momentum-dependent shift in the energy, which also leads to a change in the effective mass of the particle. The imaginary part of the self-energy represents the dissipative interaction of the scattering processes. While this was originally included in an ad hoc manner with the parameter η, it can now be fully determined. The sign of the imaginary part of the self-energy is important to determine the imaginary part represents the retarded or the advanced Green's function.

To evaluate the self-energy the full Green's function is retained in the latter term, so that the process must be iterated. However, only the lowest-order term in the self-energy will be retained and the internal Green's function is replaced with the bare equilibrium Green's function. It is done as the impurity scattering is weak, so that the form becomes

$$\Sigma_{r,a}(k,\,\omega) = \frac{N_i}{\hbar^2\Omega}\sum_q |\,V_i(q)\,|^2\,G^0_{r,\,a}(k-q,\,\omega). \tag{8.92}$$

The factor $1/\hbar^2$ comes from the pre-factor of the S-matrix terms and assures us that (8.92) retains the proper dimensionality. The factor N_i is the total number of impurities in the semiconductor, but together with the volume Ω gives the average density of impurities n_i. Only a single interaction appears in (8.92), rather than the scattering by the various assortment of impurities, a result known as *impurity averaging*. This single interaction is multiplied by the number of impurities to arrive at the total scattering strength. This impurity averaging pairs up the scattering diagrams, as shown in figure 8.10(a), but it assumes that the position of the impurity is irrelevant, which may be somewhat dubious! It is pertinent also to note that the impurity potential is usually screened by the free carriers, known as a vertex correction it changes the strength of the red dot in figure 8.10(a). It is easiest to assume a δ-function potential (in real space), which means that the momentum change in scattering is small compared to the screening momentum, or $V(\boldsymbol{q}) = V_0$, and $V_0 \sim e^2/\varepsilon_s q_D^2$. However, the form shown in (8.92) will be retained as it is desired to produce a higher-order correction that modifies this term.

As the imaginary part of the self-energy is of primary interest, this can be obtained by recognizing that η is small, so that we retain only the imaginary part of the bare Green's function. The real parts of the self-energy are no more than a shift of the energy scale, and this should be quite small in the weak scattering limit. Then, the limit of small η leads to (8.78), and

$$G^0_{r,a}(k - q, \omega) \sim i\pi\delta(\hbar\omega - E). \tag{8.93}$$

The summation over q may be converted to an integral in spherical coordinates, which leads to

$$\Sigma_{r,a}(k, \omega) = \frac{i\pi n_i}{\hbar}\int_0^{2\pi}\int_0^{\pi}\frac{\sin\vartheta d\vartheta d\phi}{4\pi}\,|\,V_i(q)\,|^2 \rightarrow i\frac{\hbar}{2\tau(\omega)}. \tag{8.94}$$

Within the approximations used, the self-energy is now a function only of energy (through ω). There remains another important point here, and that is the fact that the computation leading to (8.94) ignores the factor of $(1 - \cos\vartheta)$ that arises even in the Boltzmann approach, as found in (6.12), and appears due to the fact that the simple *impurity scattering does not relax the momentum completely*. In essence, this factor arises because the Coulomb scattering is not isotropic, so similar problems will occur also with polar optical phonon scattering.

To understand the problem with impurity scattering, it is necessary to understand that the Coulomb force only deflects the carrier, without altering its energy or magnitude of the momentum. Hence, the carrier retains nearly all of its coherence throughout the scattering. One method of understanding this is to observe the interference effects retained by the carrier during scattering by the impurity. This coherence retention is well-known in the mesoscopic world by the observation of processes such as the Aharonov–Bohm effect, weak localization, and universal conductance oscillations [14]. This can be illustrated by a process much like Young's optical two-slit experiment [61]. This experiment can be replicated with electrons. A charged wire in a TEM creates what is known as a bi-prism, and electrons split as they pass this wire and form an interference pattern just as that of Young's photons [62]. Importantly, a single electron in the TEM, much like a single photon [63], will still lead to the interference pattern. That is, the single electron (or photon) must have both wave and particle properties, as recognized by Einstein [64]. For the purpose here, these observations mean that a single charged impurity will behave much like the bi-prism in the TEM, and a single electron passing by this impurity can create an interference pattern representing its wave-like self-interactions in diffracting around the impurity.

An analytical study of the coherence during an impurity scattering with a single impurity has been done by Barker, using the NEGF [65] (discussed below). In this work, it was observed that the diffraction of the single carrier produced as many has five peaks in the resulting pattern. It was clear that the interference arises from the matrix element itself for the scattering interaction. A more instructive simulation involved two electrons and two impurities, which provide a complicated wave interaction process that illustrates the same interference principle. In this latter study, a simple nanowire structure with two repulsive impurities embedded in it, as shown in figure 8.11 [66], was used. Here, the nanowires are 40 nm wide, and the

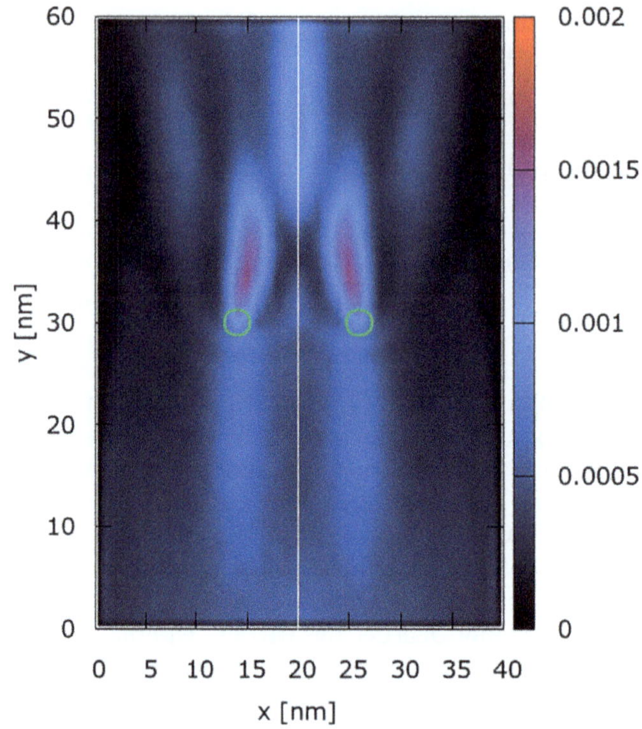

Figure 8.11. Quantum electron density after 200 fs of the initial minimum uncertainty wave packet is injected at $x = 20$, $y = 0$. The green circles are equal energy lines at 0.175 eV of the repulsive impurity potential. Reprinted from Weinbub *et al* [66], under the creative commons open access provisions.

area of interest is some 60 nm long. The impurities are located at the two green circles, which are constant energy lines around the impurities. The electrons are continually injected at $y = 0$, $x = 20$ nm, and absorbed completely at $y = 60$ nm. These electrons are simulated by a Wigner function wave packet (Wigner functions are covered more fully below). The continual injection allows the study of the steady-state inference of the electrons and impurities. The electrons are injected at an energy of 0.14 meV, which is above the second transverse sub-band in the nanowire. This, plus the non-local interaction with the impurities leads to the injected carriers forming the two beams apparent in the figure in the region before the impurities (the carriers travel from bottom to top in the figure). The carrier 'density' from $| \psi(x, y) |^2$ is plotted in the figure. It is clear that the carriers are diffracted by the impurities, and then interfere with themselves (or each other) downstream from them. This interference would be present even if only a single impurity were present, as was shown by Barker, mentioned above. The interference peak evolves all the way to the exit of the nanowire. These interference peaks are certainly reminiscent of the two-slit experiment. The five beams are precisely the same as those using the NEGF [65].

The importance of these various effects—the interference, the Aharonov–Bohm effect, weak localization, and universal conductance oscillations, as well as the diffraction discussed just above—lies in the fact that it is just quite likely that perturbation theory will not work well with impurities. It is known that it doesn't work well for a simple system like a harmonic oscillator in a constant electric field [67]. The message is therefore, that when the number of impurities is not very large, their effect should not be treated perturbatively, but treated realistically in real space. Otherwise, significant parts of the interaction with the carriers will be missed, especially with quantum transport, and these interactions can directly alter the nature of the system/device being considered.

8.4.3 Conductivity and the Bethe–Salpeter equation

As discussed in chapter 6, the semi-classical theory of conductivity is based upon the Boltzmann transport equation that describes a one-particle distribution function. This is the normal approach to study transport in semiconductors, particularly because the response of semiconductor systems for the full quantum interactions is not necessary. The essential point is the Markovian behavior of the scattering processes and the Gaussian nature of the distribution function. A Gaussian has no higher-order cumulants, so that one need consider only single-particle Green's functions. Morover, coherence of the wave function is fully destroyed in each collision. As we have seen in the preceding section, this is not the case with impurity, or polar optical phonon, scattering. A more general approach for quantum transport is the Kubo formula. In this method, the conductance is computed from the current–current correlation function [68]. This approach describes transport with only an assumption that the current is linear in the applied electric field. Then, the current (or an ensemble averaged current) is given by a correlation function as

$$\begin{aligned}\langle \mathbf{J}(r, t)\rangle &= \langle \mathbf{j}(r, t)\rangle \\ &\quad - i\Theta(t - t')\int dt' \int d^3r' \langle [\mathbf{j}(r, t), \mathbf{j}(r', t')]\rangle \cdot \mathbf{A}(r', t')'\end{aligned} \tag{8.95}$$

where $\mathbf{A}$ is the vector potential. In the scattering limit, the ballistic response (which leads to plasma oscillation effects in the dielectric function) can be ignored, and the conductance is given by the current–current correlation function as shown. The current operator is given by standard quantum mechanics as

$$\mathbf{j}(r, t) = \frac{ie\hbar}{2m^*}\{\psi^\dagger(r, t)\nabla\psi(r, t) - [\nabla\psi^\dagger(r, t)]\psi(r, t)\}. \tag{8.96}$$

In the Fourier transform version, this becomes

$$\mathbf{j}(k, t) = \frac{e\hbar \mathbf{k}}{m^{*2}}\psi^\dagger(k, t)\psi(k, t). \tag{8.97}$$

The Fourier transformed conductivity then becomes

$$\sigma_{ij} = \left(\frac{e\hbar}{m^*}\right)^2 \sum_{k,k'} k_i k'_j G_2(k, t; k't')\delta(E_k - E_F). \tag{8.98}$$

The subscripts define the coordinate axes. Here, the new function G_2 is *the two-particle Green's function.* The last delta function assures that the conductivity is measured at the Fermi surface. A two-particle Green's function actually contains four field operators as

$$G_2(k, t; k't') = -\langle T[\psi^\dagger(k', t')\psi^\dagger(k, t)\psi(k, t)\psi(k', t)]\rangle. \tag{8.99}$$

The above treatment leading to the two-particle Green's function (homogeneous conductance and the *zero-frequency* static conductivity) may be rewritten in Fourier transform form as the product of two functions, giving a convolution integral in Fourier space. Since we are interested in the static result, only the frequency integral survives. The integration over momentum is already contained in (8.98). This result is the lowest-order approximation to the two-particle Green's function and appears as

$$G_2(k, k') \cong \int \frac{d\omega}{2\pi} G_r(k, \omega) G_a(k', \omega). \tag{8.100}$$

The two Green's functions form a bubble diagram, as shown in figure 8.10(b). Here, the Green's functions are the vectors $\boldsymbol{k}$ and $\boldsymbol{k}'$ and the integration is over the primed coordinate, or alternatively over the exchanged momentum $\boldsymbol{q}$. Moreover, this bubble is going to be horizontal as the times at the beginning and end are different. It is important to recognize that the result (8.98) is a d.c. result; the factor ω in the integral is not the applied frequency, but the energy $E/\hbar$. The two-function loop diagram of figure 8.10(b) is a particle–hole propagator (the two lines point in different time directions). The electron is excited above the Fermi energy and is called a quasi-particle; its characteristic energy is measured from the Fermi energy itself. This is represented by the retarded Green's function. The hole is a quasi-particle that exists below the Fermi energy; its energy is measured downward from the Fermi energy, and is described by the advanced Green's function, flowing backward in time (which is measured as motion to the left).

The conductivity may now be determined, although only the diagonal conductivity will be retained under the assumption that the current is parallel to the electric field. Hence, $i = j$, and the two Green's functions have to be at the same momentum as each involves only a single momentum and they come together at both ends of the bubble, as shown in figure 8.10(b). Then,

$$\sigma = \left(\frac{e\hbar^2}{m^*}\right)^2 \sum_k k_i^2 \int \frac{d\omega}{2\pi} \frac{1}{\hbar\omega - E_k - i\hbar/2\tau} \frac{\delta(E_k - E_F)}{\hbar\omega - E_k + i\hbar/2\tau}. \tag{8.101}$$

The frequency integral given here can be evaluated quickly by the method of residues. Then, using $k_i^2 = k^2/3$, the conductivity is finally found to be

$$\sigma = \frac{1}{3}\left(\frac{e\hbar^2}{m^*}\right)^2 \sum_k \frac{k^2\tau}{\hbar^2} \delta(E_k - E_F) = \frac{ne^2\tau}{m^*}, \tag{8.102}$$

with

$$n = \frac{1}{\pi^2}\left(\frac{2m^* E_F}{\hbar^2}\right)^{3/2}. \tag{8.103}$$

This result is at low temperatures for a strongly degenerate electron gas. This is the usual low-frequency result described by the Drude formula. It is not the usual conductivity we find in Boltzmann transport theory, for example in chapter 6. The discussion in chapter 6 for the impurity mobility includes a factor $(1 - \cos \vartheta)$, which is missing in (8.102). In this latter result, τ is a scattering time and not a relaxation time. It also must be noted that it is a result in which the two Green's functions are evaluated in isolation from one another; there is no interaction between the two. If the angular variation that appears semi-classically is to be recovered, higher-order terms in which the two Green's functions interact with one another must be considered.

The most important higher-order corrections are those in which the impurities connect the two Green's functions that appear in (8.100). Typical new diagrams that appear are shown in figure 8.12. The diagrams all have four interaction lines from either a single or a pair of impurities. In this complex situation, the two-particle Green's function cannot be treated simply by figure 8.10(b) or (8.100). In general, the two-particle Green's function is frequency dependent, although the static conductivity does not have this frequency dependence. The first diagram (upper left of figure 8.12) leads to a succession of impurity interactions that cross between the two sets of retarded and advanced Green's function. These are often referred to as 'ladder' diagrams. The second diagram contains two impurity interactions in which the lines cross one another. These are difficult to handle, but can be un-nested by reversing the lower set of retarded functions in time. This latter result is the set of diagrams used for weak localization [18]. The bottom set of diagrams in figure 8.12 are multiple scattering processes from a single impurity. This also is difficult to include in normal calculations, but have been found to occur in semiconductor transport when the impurities are included in real space by molecular dynamics [69], as such it is hard to argue that they can be ignored.

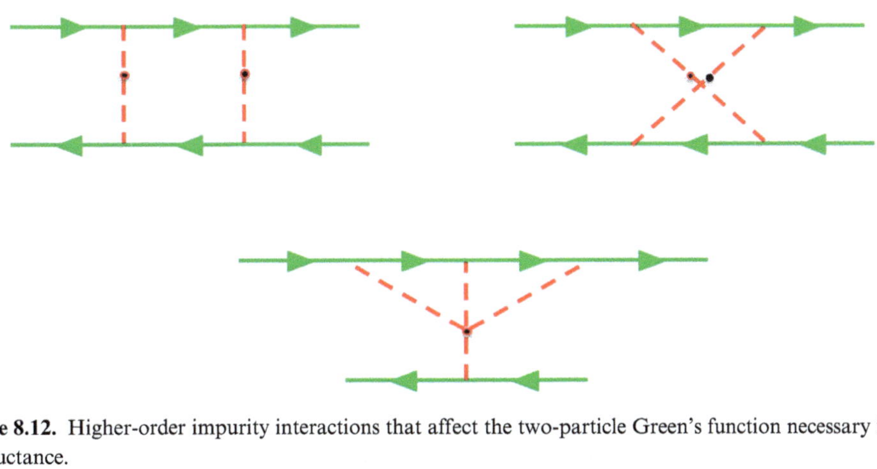

Figure 8.12. Higher-order impurity interactions that affect the two-particle Green's function necessary for the conductance.

Nevertheless, it has become usual to treat only the ladder diagrams, for which the the static conductivity, in the isotropic limit, is [70, 71]

$$\sigma = \frac{1}{3}\left(\frac{e\hbar^2}{m^*}\right)^2 \int \frac{d\omega}{2\pi} F(\omega), \tag{8.104}$$

where

$$\begin{aligned} F(\omega) &= \sum_{\boldsymbol{k},\boldsymbol{k}'} \boldsymbol{k} \cdot \boldsymbol{k}' G_2(\boldsymbol{k}, \boldsymbol{k}', \omega) \\ &= \sum_{\boldsymbol{k}} G_r(\boldsymbol{k}, \omega) G_a(\boldsymbol{k}, \omega) \left\{ k^2 \right. \\ &\quad \left. + \sum_{\boldsymbol{k}'} \boldsymbol{k} \cdot \boldsymbol{k}' \Lambda(\boldsymbol{k}, \boldsymbol{k}', \omega) G_r(\boldsymbol{k}', \omega) G_a(\boldsymbol{k}', \omega) \right\} \end{aligned} \quad . \tag{8.105}$$

In the scalar form here, the two wave vectors satisfy a dot product. The first term in the curly brackets leads to the results obtained above in which the higher-order terms were ignored. The new term is the second one in the curly brackets. The two Green's functions are those on the far right of the second diagram in figure 8.13. Internal Green's functions are part of the kernel Λ. The quantity Λ is referred to as the irreducible scattering vertex. This kernel can be evaluated by the various terms of figure 8.10(b) and various approximations result. The ladder terms shown as the upper left diagram in figure 8.12 can be summed nicely. It is apparent that the dot product in the second term is going to give rise to the missing angular term. The particular form, illustrated in figure 8.13, gives the leading terms. With the full two-particle Green's function, the result (8.104) is known as the Bethe–Salpeter equation [70, 72].

In the above approximation, the key contributions to the scattering vertex Λ arise from the ladder diagrams. These impurity interactions do not transport energy (the scattering from impurities is elastic), so the upper and lower Green's functions are at the same energy. In the static case, the two frequencies also are the same. In the following, only the lowest-order correction is kept and this is the sum of these ladder diagrams of the upper left diagram in figure 8.12. The kernel for the single scattering pair that spans the upper and lower Green's function is given by

$$\Gamma_0 = \frac{n_i}{\hbar^2} \mid V(q) \mid^2 . \tag{8.106}$$

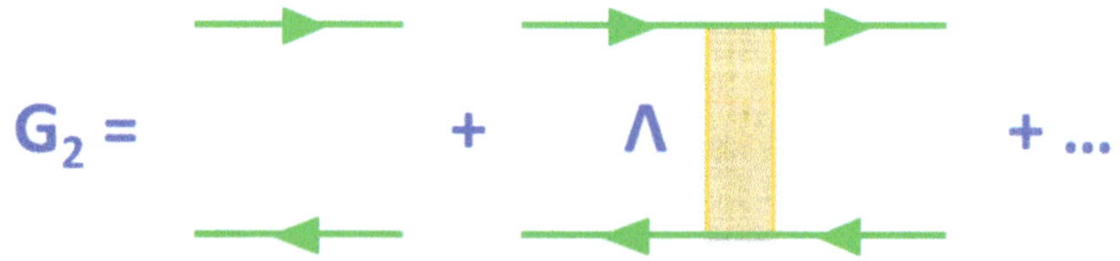

Figure 8.13. The leading correction term to the two-particle Green's function as represented in (8.104). The scattering vertex Λ is approximated by the ladder diagrams—the upper left set in figure 8.12.

The scattering kernel Λ can be treated separately from the output pair of Green's functions. The series of ladder diagrams can be written as

$$\Lambda = \Gamma_0 + \Gamma_0\Pi\Gamma_0 + \Gamma_0\Pi\Gamma_0\Pi\Gamma_0 + \cdots = \frac{\Gamma_0}{1 - \Pi\Gamma_0}. \tag{8.107}$$

The Green's function polarization term is defined as

$$\Pi = \int \rho(E)dE \frac{\hbar}{\hbar\omega - E_k - i\hbar/2\tau} \frac{\hbar}{\hbar\omega - E_k + i\hbar/2\tau} = -2\pi\hbar\tau\rho(\hbar\omega). \tag{8.108}$$

The last pair of Green's functions, on the right end of the second term in figure 8.13, adds another factor of Π to the summation in (8.107). Using the properties of the various terms, (8.105) can now be rewritten as

$$F(\omega) = \sum_k k^2 G_r(\boldsymbol{k}, \omega) G_a(\boldsymbol{k}, \omega) \left\{ 1 + \cos\vartheta \frac{\Gamma_0\Pi}{1 - \Pi\Gamma_0} \right\}. \tag{8.109}$$

Generally, an assumption is made that the product $\Pi\Lambda_0$, is large compared to unity, so that one arrives at the term involving just the cosine of the scattering angle with its sign changed. Carrying out the frequency integration in (8.104) leads to the Boltzmann result

$$\sigma = \frac{1}{3}\left(\frac{e\hbar}{m^*}\right)^2 \sum_k k^2 \tau (1 - \cos\vartheta)\delta(E_k - E_F) = \frac{ne^2\tau_m}{m^*}. \tag{8.110}$$

This result, with the approximations used, provides a direct connection with the semi-classical result for impurity scattering obtained from the Boltzmann equation in chapter 6. The complications certainly are greater here, but this suggests an approach that can be used to treat other diagrammatic terms for new effects that are not contained in the Boltzmann equation. Boltzmann's 'molecular chaos' assumption avoids all processes that would lead to two-particle Green's functions. In the above, the angular dependence of the fraction that appears as the second term in the curly brackets of (8.109) provides the needed conversion from a simple scattering time to a momentum relaxation time. On the other hand, if the scattering matrix element does not depend upon $\boldsymbol{q}$, which is the case in many phonon scattering processes, then the ladder correction terms do not contribute as the angular variation integrates to zero. This simple fact is often overlooked in semi-classical treatments, as this latter angular variation often is put in by hand, and its origin is not fully appreciated.

8.4.4 NEGF

The NEGF has been widely applied to simulations of semiconductor devices, despite concerns over the applicability of many of the approaches, while one of the most extensive introductions has been given by Danielewicz [73] and its use has been reviewed [74–76]. While an exhaustive coverage is outside the scope of this work, an introduction and review of some typical examples will be presented for clarity.

Nevertheless, a number of problems, some of which have been mentioned above, exist which suggest that many studies using NEGF have overstepped the mark. For example, there are many studies claiming to use NEGF to study devices in the ballistic limit, when it is widely known that in this limit the NEGF reduces to the equilibrium Green's functions examined above [77], and the Landauer formula. Hence, continued use of NEGF in ballistic transport cases is often done more for the hype of the name rather than for the necessity of the approach.

To begin, the limits on (8.84) need to be examined. In general, these limits are written with the understanding (made clear in the scattering matrix itself and the normalization of the Green's function) that

$$\begin{aligned} t &\rightarrow \infty, \\ t' &\rightarrow -\infty. \end{aligned} \tag{8.111}$$

The first of these two limits is unacceptable. It assumes that one knows what will occur in the infinite future and thus violates causality. In particular, the distribution in the future is not the equilibrium form, and that also suggests the first limit is invalid. The second of the two limits is problematical as well. To invoke such a limit implies that the system evolution smoothly evolves from the initial, equilibrium state of the far past. It is not clear that this procedure is valid either, as engineers like to turn the device on at (preferably) $t = 0$. Nevertheless, this is an approach useful for equilibrium conditions; the failure of this process will be discussed later. How to handle the upper limit of the integral in the unitary operator will be treated first.

To avoid the need for $t \rightarrow \infty$ in the perturbation series, a new time path for the real-time functions was suggested by Blandin *et al* [78] (see figure 8.14). (There may have been others who have suggested this approach, but this work seems to be blamed most often.) The evolution begins with a thermal Green's function at $t = t_0 - i\hbar\beta$, where β is the inverse of the thermal energy as previously introduced. The path then flows into the nonequilibrium, non-interacting Green's function at t_0. It then flows forward in time to the maximum of (t, t'), at which point it returns *backward* in time to t_0. The limit $t_0 \rightarrow -\infty$ is often invoked when one is not interested

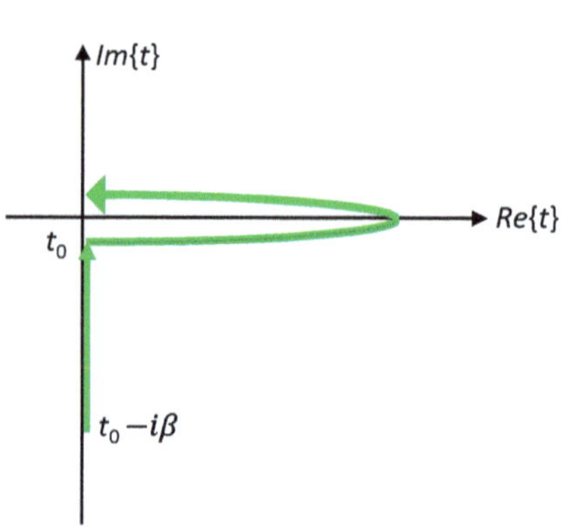

Figure 8.14. Contour along which the ordered time evolution evolves in evaluation of Green's function products in NEGF.

in the initial condition (as in most far-from-equilibrium systems). The Green's functions that arise when both time arguments are on either the upper branch (backward in time) or the lower branch (forward in time) are straightforward using (8.90). When the two time arguments are on different branches of the time contour, two new correlation functions must be defined [53, 79].

In chapter 6, it was necessary to solve the Boltzmann in order to find the nonequilibrium distribution function. Here, the nonequilibrium distribution function is similarly unknown and must be found. The two new correlation functions allow this to be achieved. The first of these two is the 'less than' function

$$G^{<}(r, r'; t, t') = i\langle \psi^{\dagger}(r', t')\psi(r, t)\rangle. \tag{8.112}$$

The second new function is the 'greater than' function

$$G^{>}(r, r'; t, t') = -i\langle \psi(r, t)\psi^{\dagger}(r', t')\rangle. \tag{8.113}$$

These are correlation functions so that the operator products produce c-numbers (non-operators).

The original two equilibrium Green's functions in (8.90) and the two new correlation functions, now give four Green's functions. These are all that one needs, as the latter two will relate to the far-from-equilibrium distribution function. (While this is true to lowest order, higher-order multi-particle Green's functions may be needed as well, just as was described in section 8.4.3.) In practice, two further Green's functions are found to be useful. These can be obtained from the previous four, so are merely combinations of the previous ones. These are the time-ordered Green's function

$$\begin{aligned} G_t(r, r'; t, t') &= i\langle T[\psi^{\dagger}(r', t')\psi(r, t)]\rangle \\ &= \Theta(t - t')G^{>}(r, r'; t, t') + \Theta(t' - t)G^{<}(r, r'; t, t') \end{aligned} \tag{8.114}$$

and the anti-time-ordered function

$$G_{\bar{t}}(r, r'; t, t') = \Theta(t' - t)G^{>}(r, r'; t, t') + \Theta(t - t')G^{<}(r, r'; t, t'). \tag{8.115}$$

In addition, all of the Green's functions can be connected via the equations

$$\begin{aligned} G_r &= G_t - G^{<} = G^{>} - G_{\bar{t}} = \Theta(t - t')(G^{>} - G^{<}) \\ G_a &= G_t - G^{>} = G^{<} - G_{\bar{t}} = -\Theta(t' - t)(G^{>} - G^{<}) \end{aligned}. \tag{8.116}$$

There is a major difference between these nonequilibrium Green's functions and the equilibrium ones. The ensemble average, indicated by the angular brackets in (8.87) involve the equilibrium distribution function. In these new nonequilibrium functions, this angular average no longer signifies the averaging used previously. Instead, the brackets indicate an average needs to be performed over the available states of the far-from-equilibrium system, with these states weighted by the far-from-equilibrium distribution function. This new distribution function will be described later, as this far-from-equilibrium distribution is usually the most difficult part of the nonequilibrium problem that must be solved. In quantum transport, this problem is far more difficult, due to the off-shell effects, and has been solved in only a few cases. Usually this is achieved under stringent approximations. Some approaches have used atomistic approaches to provide the use of realistic band structures [80–82], phonon spectra [83], strain effects [84], interface roughness [85]. These atomistic approaches put the individual atoms on their lattice and compute the band structure for the material and then use this in transport simulations.

As there are two times in the correlation functions (8.112) and (8.113), a pair of equations of motion for these functions is needed, one for variation on each time scale. The added complications of having four functions means these equations will be more complicated than either the simple Boltzmann equation or the simple equilibrium forms used previously. Since only four of the various functions introduced above are actually independent, Keldysh [86] introduced a general method with a single matrix Green's function. Therefore, our equations reduce to a single matrix equation to be solved. The overall effort required is not reduced, but the equations in which these functions are described are simplified, at least mathematically. Each Green's function is defined in terms of two field operators. Each of these two field operators can be on the upper line or the lower line of the overall. The Keldysh matrix is a 2×2 matrix. If one refers to the equations (8.90) and (8.112)–(8.116), the rows of the matrix are defined by the operator $\psi(r,t)$. That is, row one of the matrix corresponds to when t is on the lower path, while row two is defined when t is on the upper path of the trajectory in figure 8.14. Similarly, the operator $\psi^{\dagger}(r', t')$ defines the columns of the matrix. Column one is defined when t' is on the lower portion of the path, while column two is defined when t' is on the upper portion of the path. Thus, for example, we can write the 11 element by assuming *contour ordering* (time advances on the contour, which means it increases on the lower branch and decreases on the upper branch as described above) is just the time-ordered function (8.114), and so on. The 12 element gives us the 'less than' function. The contour-ordered matrix may now be written as

$$\mathbf{G}_C = \begin{bmatrix} G_t & G^{<} \\ G^{>} & G_{\bar{t}} \end{bmatrix}. \tag{8.117}$$

In an attempt to simplify this further, one *does a rotation in Keldysh space* [86]. The first step is subtract G_t from each term in the matrix, and then multiply through with a minus sign, so that the new Keldysh matrix is

$$\mathbf{G}'_C = \begin{bmatrix} 0 & G_a \\ G_r & G_K \end{bmatrix}, \tag{8.118}$$

where

$$G_K = G^{<} + G^{>}. \tag{8.119}$$

From (8.118), a rotation removes some of the degeneracy of the various elements, separating the retarded and advanced functions as well as simplifying the matrix itself. The non-unitary matrices used in this rotation are

$$\mathbf{L} = \frac{1}{\sqrt{2}}\begin{bmatrix} 1 & -1 \\ 1 & 1 \end{bmatrix}, \mathbf{L}^{\dagger} = \frac{1}{\sqrt{2}}\begin{bmatrix} 1 & 1 \\ -1 & 1 \end{bmatrix}, \boldsymbol{\sigma}_z = \begin{bmatrix} 1 & 0 \\ 0 & -1 \end{bmatrix}. \tag{8.120}$$

With these matrices, the Keldysh matrix becomes

$$\mathbf{G}_K = \mathbf{L}\boldsymbol{\sigma}_z\mathbf{G}'_C\mathbf{L}^{\dagger} = \begin{bmatrix} G_r & G_K \\ 0 & G_a \end{bmatrix}. \tag{8.121}$$

Admittedly, this is not much of a change, but now the Keldysh function is the only non-diagonal term, so that equations of motion should be easier. (It should be noted that G_K is the matrix element (8.119) and $\boldsymbol{G}_K$, which is in bold, is the actual matrix (8.121).)

But, there are further problems here. Most modern approaches to the use of NEGF base their work on that of Keldysh [86]. As will be seen later, the scattering self-energies depend on the correlation functions $G^{<}$ and $G^{>}$ and on the greater and lesser Green's functions of the phonons, which account for the occupancy of the phonon states and depend on the phonon energies. While this approach has been widely used, Keldysh himself uses the S-matrix from equilibrium theory that *works only upon the assumption that the system is close to equilibrium*, so that the normal techniques can be used. Keldysh assumes that the interaction representation, in which the scattering-derived interaction representation is assumed to be a unitary translation operator. But, this assumption fails when the energies contain self-energies so that the Hamiltonian is no longer Hermitian. More importantly, Danielewicz has pointed out that correlated initial conditions also make the S-matrix questionable at best [73]. The equivalent equations of motion obtained by Kadanoff and Bahm have been solved for correlated initial conditions [87, 88].

Again, most approaches for devices use only the Dyson equation to provide a re-summation of interaction terms, as in (8.107), into a simple self-energy. With anisotropic scattering, such as that by impurities or polar phonons, one needs to utilize the Bethe–Salpeter equation to completely evaluate the scattering. The onset of nonequilibrium phonon distributions, proper treatment of the scattering in many cases, and full use of both the Bethe–Salpeter equation with vertex corrections, increase tremendously the method's computational effort. It is then easy to see that particle-based approaches, such as the Wigner functions, discussed below allow a quicker route to the solution, as well as being more applicable than NEGF in many cases, particularly when inhomogeneous and transient transport is desired.

Subject to the above caveats, the non-interacting forms of these Green's functions will satisfy the equations

$$\begin{gathered} \left[i\hbar\frac{\partial}{\partial t} - H_0(r) - V(r)\right]\mathbf{G}_{K0} = \hbar\boldsymbol{I} \\ \left[-i\hbar\frac{\partial}{\partial t'} - H_0(r') - V(r')\right]\mathbf{G}_{K0} = \hbar\boldsymbol{I} \end{gathered}. \tag{8.122}$$

Here, $\boldsymbol{I}$ is the identity matrix (unity on the diagonal and zero off the diagonal). The potential has been assumed to be local, hence only a function of a single position.

The zero subscript on the Green's function indicates the non-interacting form. Scattering generally uses the S-matrix as discussed and this leads to the self-energies that enter these equations in the interacting situation. Equation (8.122) is modified to

$$\begin{aligned}\left[i\hbar\frac{\partial}{\partial t} - H_0(r) - V(r)\right]\mathbf{G}_K = \hbar\boldsymbol{I} - \boldsymbol{\Sigma}\mathbf{G}_K \\ \left[-i\hbar\frac{\partial}{\partial t'} - H_0(r') - V(r')\right]\mathbf{G}_K = \hbar\boldsymbol{I} - \mathbf{G}_K\boldsymbol{\Sigma}\end{aligned}, \tag{8.123}$$

where the self-energy matrix is given by

$$\boldsymbol{\Sigma} = \begin{bmatrix}\Sigma_r & \Sigma_K \\ 0 & \Sigma_a\end{bmatrix}. \tag{8.124}$$

In general, it is a stretch to expect the self-energy to preserve the Keldysh form shown here. If the non-unitary transformations defined in and below (8.120) were applied to the convolution functions represented by the last terms in (8.123) the results would differ greatly. Moreover, care must be taken in correctly formulating the self-energy expressions in the contour-ordered Green's functions. The product terms (of the self-energy and the Keldysh matrix) involve at least four wave functions, and necessarily a two-particle Greens' function (that has been ignored) must be considered. Therefore, many effects are not included here.

The discussion of the last paragraph suggests that the self-energy and the Green's function in Keldysh form in the last terms on the right of (8.123) are not at all obvious. Generally, the unit matrix on the right side of the equation is not appropriate, as the element in the 22 position should be $-\delta(t - t')\delta(r - r')$ [89]. The transform matrix $\boldsymbol{\sigma}_z$ converts this to a proper identity matrix which the remaining two transformation matrices leave unchanged. The various transformations seem to be chosen for easy application rather than mathematical integrity. For the product of the self-energy and the Green's function, however, it has already been pointed out that the application of the non-unitary transformations in Keldysh space fails. The problem lies in defining just what the terms in the two matrices $\boldsymbol{\Sigma G}$ and $\boldsymbol{G\Sigma}$ should be. The problem arises from the convolution integral over the internal time variable in these terms. The problems lie in the time integration, and not in the spatial integration. Consider the product

$$C(t, t') = \int dt'' A(t, t'')B(t'', t'). \tag{8.125}$$

The evaluation commonly used is due to Langreth [90]. Consider the case for $C^<$. Here, t is on the bottom leg of the contour, while t' is on the return path of the contour. There are two possibilities. Langreth first deformed the contour by taking the contour at time t and pulling it back to t_0, and then returning it out to *time* t'. Now there are two time loop 'spikes,' and the result depends on which loop the time t'' resides. Langreth separated the integral into two parts, one for t'' on each of the two new paths. He refers to these as the sub-contours C_1 and C_2, respectively. Then, (8.125) is written as

$$C^<(t, t') = \int_t^{-\infty} dt'' A(t, t'')B(t'', t') + \int_{-\infty}^{t} dt'' A(t, t'')B(t'', t'). \tag{8.126}$$

In the first integral over the C_1 contour, B is the less than function, as its arguments always fold around the C_2 contour. The function A is the anti-time-ordered function, but since $t > t''$, the theta function inherent in this last function gives us just the less than function. In the second integral over the C_2 contour, A is the time-ordered function, and yields just the greater than function. Thus, these two integrals become

$$C^<(t, t') = \int_{-\infty}^{t} dt''\Theta(t - t'')(A^> - A^<)B^< = \int_{-\infty}^{t} A_r B^< dt''. \tag{8.127}$$

It is also possible to deform the contours not from t, but from t', which makes the A term always give the less than function, and then the two paths contribute to yield the advanced function for B. These two possibilities must be combined to give the general result. Integrals such as this allow us to write the product formulas as (with the integration assumed, but not shown) [90]

$$\begin{aligned} C^< &= A_r B^< + A^< B_a,\ C_r = A_r B_r, \\ C^> &= A_r B^> + A^> B_a,\ C_a = A_a B_a. \end{aligned} \tag{8.128}$$

This is known as the Langreth theorem, and similar results can be achieved for triple products.

As mentioned, the Langreth theorem is consistently used when NEGF approaches are adapted. But, there are some problems with it. First, only simple one-particle Green's functions are included here, as if Wick's theorem carried over to the nonequilibrium case. But, as mentioned, even Keldysh worried about this when he resorted to the S-matrix which is only for near-equilibrium. The role of the two-particle Green's functions has not been explored in this context. Secondly, all of the trajectories discussed assume that the perturbation approach derives from the equilibrium distribution function and that the S-matrix will remain unitary for this purpose. But, this doesn't even work for the semi-classical case with the Boltzmann equation. Once one obtains the path integral in chapter 7, there is no mention of the initial condition—it is forgotten. The nonequilibrium distribution arises from a balance between the driving forces and the relaxation forces, if this balance can be achieved. There is no reason to believe that the situation will be different for NEGF. In fact, Prigogine has emphasized that the transition from equilibrium to the far-from-equilibrium system involves a phase transition that at least breaks time-reversal symmetry [91]. Again, the system that is being studied, within its environment, seeks a steady-state, far-from-equilibrium stable state that balances the driving forces and the dissipative forces. During the transient response toward this stable state, if it exists, the system may evolve though a number of intermediate phases; e.g., homogeneous, inhomogeneous, linear, nonlinear, etc. When the forces are removed, the system response does not reverse its course through these different phases, but seeks a relaxation toward the equilibrium steady-state from which it initially deviated. It is well-known in nonlinear systems that these paths are quite different, and the same is true in semiconductor devices. The excitation process generates entropy [92]. The relaxation does not remove this entropy from the system but generates even more as dissipation still occurs. The failure of Wick's theorem has already been experienced with the need for the Bethe–Salpeter equation above. Thus, there is no rational reason to believe that these trajectories, such as that of figure 8.14, have any more than faith to make them germane to the task. Basically, these approaches, such as the Langreth theorem, are mathematical approaches that are mathematically correct, but they have never been properly connected to the physics that is being studied. In spite of the caveats and problems cited in the last paragraph, the fact is that this is the most well-developed approach for NEGF and has been widely adopted.

As previously, it is now necessary to proceed to the conductivity and current in the case of NEGF. For this, we still use the results of the Kubo formula, as given in (8.95) and (8.96). Using some different choices of variables for the multiple internal integrations, the current is written in terms of the real-time Green's functions is

$$\begin{aligned}\langle J(r,\,t)\rangle = -\frac{ie^2\hbar}{4m^{*2}}\lim_{r',t'\to r,t}\left(\frac{\partial}{\partial r}-\frac{\partial}{\partial r'}\right)\int dt_s\int d^3s\lim_{s',t'_s\to s,t_s}\left(\frac{\partial}{\partial s}-\frac{\partial}{\partial s'}\right)\\ \times[G_r(r,\,s';\,t,\,t'_s)G^<(s,\,r';\,t_s,\,t')+G^<(r,\,s';\,t,\,t'_s)G_a(s,\,r';\,t_s,\,t')]\mathbf{A}(s,\,t_s)\end{aligned} \quad (8.129)$$

As discussed above, each Green's function product should in reality be a two-particle Green's function. However, this has been expanded into the pair lowest-order Green's functions, as suggested by the Keldysh approach. This is an approximation, and higher-order Green's function products certainly will be needed to treat certain physical processes, just as was done for the conductivity previously. Nevertheless, the result (8.129) can be Fourier transformed for a homogeneous system, at least on the average scale of the response functions, giving (after a summation over spins)

$$\begin{aligned}\sigma(k,\,\omega) = -\frac{e^2\hbar}{m^{*2}\omega}\int\frac{d^3k'}{(2\pi)^3}\int\frac{d\omega'}{2\pi}k'\cdot k\\ \times[G_r(k',\,\omega')G^<(k'-k,\,\omega'-\omega)\\ +G^<(k',\,\omega')G_a(k'-k,\,\omega'-\omega)]\end{aligned}. \quad (8.130)$$

This form differs somewhat from that found in (8.98), beyond just the real-time functions that appear here. First, the higher-order two-particle Green's functions have not been included, so this formulation is that of the lowest-order Green's functions. Even the simple form of (8.130) has not been utilized very much in transport calculations, when NEGF is used. Nevertheless, the conductance here is a double integral over the current–current correlation function, just as was used in the near-equilibrium situation.

As discussed above, one eventually needs the distribution functions themselves. For this, the ansatz [93] introduces the nonequilibrium distribution function $f(\omega)$ from the retarded and advanced Green's functions to be

$$G^{<}(k, \omega) = if(\omega)[G_r(k, \omega) - G_a(k, \omega)]. \tag{8.131}$$

This ansatz is known to be correct in the zero temperature formulation. However, it is not clear that it still holds in the nonequilibrium case [17], and has been the subject of much discussion, with some modifications. With this ansatz, (8.131) can be rewritten as

$$\begin{aligned}\sigma(k, \omega) = -\frac{e^2\hbar}{m^{*2}\omega}\int\frac{d^3k'}{(2\pi)^3}\int\frac{d\omega'}{2\pi}k'\cdot k[&G_a(k'-k, \omega'-\omega)f(\omega') \\ &- G_r(k', \omega')G_r(k'-k, \omega'-\omega)f(\omega'-\omega) \\ &+ G_r(k', \omega')G_a(k'-k, \omega'-\omega)\{f(\omega')-f(\omega'-\omega)\}]\end{aligned} . \tag{8.132}$$

The first two products in the square brackets if one changes the frequency variables as $\omega'' = \omega' - \omega$, and uses the fact that $\sigma^*(k, \omega) = \sigma(k, -\omega)$ and $G_r = G_a^*$. Thus, we are left with only the last term in the square brackets. For low frequencies (and the long-time limit of $\omega = 0$ will be taken immediately), the distribution function can be expanded about ω', so that

$$\sigma(k, \omega) = -\frac{e^2\hbar}{m^{*2}\omega}\int\frac{d^3k'}{(2\pi)^3}\int\frac{d\omega'}{2\pi}k'\cdot kG_r(k', \omega')G_a(k'-k, \omega'-\omega)\frac{\partial f(\omega')}{\partial\omega'}. \tag{8.133}$$

Finally, at low frequencies (and for homogeneous material), this is reduced to

$$\sigma(k, 0) = -\frac{e^2\hbar}{m^{*2}}\int\frac{d^3k'}{(2\pi)^3}\int\frac{d\omega'}{2\pi}k'\cdot k\,|\,G_r(k', \omega')\,|^2\,\frac{\partial f(\omega')}{\partial\omega'}. \tag{8.134}$$

In the case of low temperatures, the form used in previous chapters is recovered. Here, the derivative of the distribution function is replaced with the negative of a delta function at the Fermi energy. The summation over the momentum states counts the number of states that contribute to the conductivity. This gives the carrier density at the Fermi energy for low temperature. The quantity $|\,G_r(k', \omega')\,|^2$ represents the transmission through the system, whether in general, or for a set of modes, such as for the Landauer formula. Even if there is no transverse variation, integration over the longitudinal component of the wave vector will produce the difference in the Fermi energies at the two ends of the samples (as in, for example, the quantum Hall effect). The point is that the reduction of the less than function has allowed a separation of the density–density correlation function from the current–current correlation function.

In general, the expression (8.134) remains an approximation, in that $|\,G_r(k', \omega')\,|^2$ is really a lowest-order representation of the actual polarization that appears in the conductivity bubble. This was addressed earlier in connection with figure 8.13, and the extended version that arises with consideration of the Bethe–Salpeter equation is to interchange the two integrals and replace the integration over momentum with

$F(\omega')$ with the corresponding value in (8.105) including the factor of 3 arising from the dot product of the two momenta. Thus, the extended conductivity can be expressed as

$$\sigma(k, \omega) = -\frac{e^2\hbar}{m^{*2}} \int \frac{d^3k'}{(2\pi)^3} \int \frac{d\omega'}{2\pi} \Pi(k', \omega') \frac{\partial f(\omega')}{\partial \omega'}, \tag{8.135}$$

where Π is the polarization that is represented by the integrand of (8.105)

$$\begin{aligned} \Pi(k, \omega) = G_r(k, \omega) G_a(k, \omega) \Big\{ & k' \cdot k \delta(k' - k) \\ & + \int \frac{d^3k'}{(2\pi)^3} \frac{k' \cdot k}{k^2} \Lambda(k' - k) \Pi(k', \omega') \Big\}. \end{aligned} \tag{8.136}$$

The actual form (8.105) replaces the polarization in the last term with the approximation of the first term, but the full Bethe–Salpeter equation has this integral equation that must solved.

It is important to point out that the form of the distribution function that appears in (8.135) still must be determined by a balance between the driving forces and the dissipative forces if such a balance exists. The less than Green's function must be determined in the nonequilibrium system, and the distribution function extracted from it, just as the distribution function must be determined for the Boltzmann approach introduced in chapter 6. Moreover, it is also important to note once again that the entire development of this section has assumed a homogeneous system and only the difference in the two times has been retained. Generally, this implies that center-of-mass quantities are used:

$$\begin{aligned} R &= \frac{1}{2}(r + r'), \qquad r_d = r - r', \\ T &= \frac{1}{2}(t + t'), \qquad t_d = t - t'. \end{aligned} \tag{8.137}$$

Then, the momentum vector arises from the Fourier transform of r_d and the frequency arises from the Fourier transform of t_d. The average time and position have fallen out of the problem but they have not been totally neglected. Inhomogeneous systems have been treated more carefully [94, 95], and particularly for devices as discussed in the next section. One final warning caveat lies in the observation that correlation effects in NEGF can lead to unphysical effects that tend to drive the system toward homogeneity [96].

8.4.5 NEGF in hot carrier devices

There have been several attempts to acquire a quantum equivalent to the Boltzmann transport equation. One of the best of these was to obtain the equivalent of (6.10) by using a gradient expansion of the Green's functions around their average position. That is, the center-of-mass type transformation (8.137) is used, and the system expanded about R and T [53]. With this, the less than Green's function can be found to be a product of the spectral density and an approximation to the distribution function similar to (6.10), except that the scattering time τ is replaced by a polarization function that is found from an integral equation similar to that of (8.136), but where the G_rG_a term is replaced by the spectral density [89]. Jauho reviewed the use of the NEGF formalism for hot carriers, and discussed how to move to the Wigner function (discussed below) and to treat inhomogeneous systems [97]. Bertoncini *et al* observed that, in a high electric field, the energy band appeared as in a triangular potential well, and introduced Airy transforms to uncouple the product terms for the scattering functions [98, 99]. This latter work showed that the retarded function had the same general form as in equilibrium, a result found earlier [89], and the less than function could be written as

$$G^{<}(k, s, s', \omega) = G_r(k, s, \omega)G_a(k, s', \omega)\Sigma^{<}(k, s, s', \omega), \tag{8.138}$$

where s and s' are the variables for the Airy transform of z and z', respectively, *in the direction along the field.* The momentum k is a two-dimensional variable in the plane perpendicular to the field. With the assumption that $s = s'$, this could be unfolded into the form of the above ansatz (8.131) as [25]

$$G^{<}(k, s, \omega) = A(k, s, \omega)f(k, s, \omega), \tag{8.139}$$

with

$$f(k, s, \omega) = \frac{\Sigma^{<}(k, s, \omega)}{2\,\mathrm{Im}\{\Sigma_r(k, s, \omega)\}}. \tag{8.140}$$

The important point of this transform is that the spectral function $A(k, s, \omega)$ clearly showed a shift and a broadening of the single peak with the electric field, a clear exhibition of the intra-collisional field effect discussed in connection with (8.69) above. Since the distribution function also appears within the self-energies, just as in the semi-classical form discussed in chapter 6, (8.140) is an integral equation that has to be solved for the distribution function [25, 99].

NEGF was used to treat optical excitation of semiconductors by Kurokawa and Takemori [100]. Here, ultra-short laser pulses (a double-pulsed experiment) were studied with NEGF, taking care to treat the Coulomb interactions up to high order, since this is the dominant scattering at low temperatures. Because the sample is excited inhomogeneously (the thickness is not small compared to the absorption length), the random phase approximation has to be performed in real time, taking care to include the two-particle Green's functions. Bryskin and Kleinert studied high electric field transport in the direction normal to the interfaces of a superlattice [101]. Here, they found important effects of the intra-collisional field effect as well as shifts in the bound state energies due to the field.

Mori and Hamaguchi examined the role of phonon scattering in current fluctuations in GaAs quantum wires [102]. Beginning with NEGF, they eventually transform to a distribution given as the Wigner function. Surprisingly, they find that the average energy of the carriers is not effectively lowered by the optical phonon scattering.

The use of ARPES for studying hot carriers was discussed in chapter 4, where it was especially useful for observations of intervalley transfer in semiconductors. Recently, a full theory of the process has been developed using NEGF [103]. In this work, the authors developed the formalism for the spectral function and used it to study the time resolved ARPES process. This technique is used for a study of the thermalization of hot carriers in graphene. The results of the theory and those from the experiment are shown in figure 8.15. The electronic band structure of graphene, in the region of the K point, is shown in panel (a) of the figure, while the calculated ARPES signals are shown in panels (b)–(e). Differential signals calculated from the excited and equilibrium signals are shown in (f) and (g) and integrated signals are compared with the experiments in the remaining panels. The experiments use a pump and probe technique, as discussed in chapter 4, where the excitation is at 8 J m^{-2} and 3.46 J m^{-2}. The time delay in the figure is measured from the switch off of the pump, although the signals are taken to be 30 fs Gaussian pulses. With the sample at 300 K, the pump raises the carrier temperature to more than 2550 K and 1930 K, respectively, already at $t = 0$. Altogether, the theory allows for study of the time resolved ARPES in systems driven out of equilibrium by a strong laser pulse.

Semiconductor devices are continuing to downscale, and have progressed through FinFETs to nanowires or nanosheets by 2021 [104]. Since the first FinFETs [105], the transport channel has become of interest for quantum transport, as the sizes are

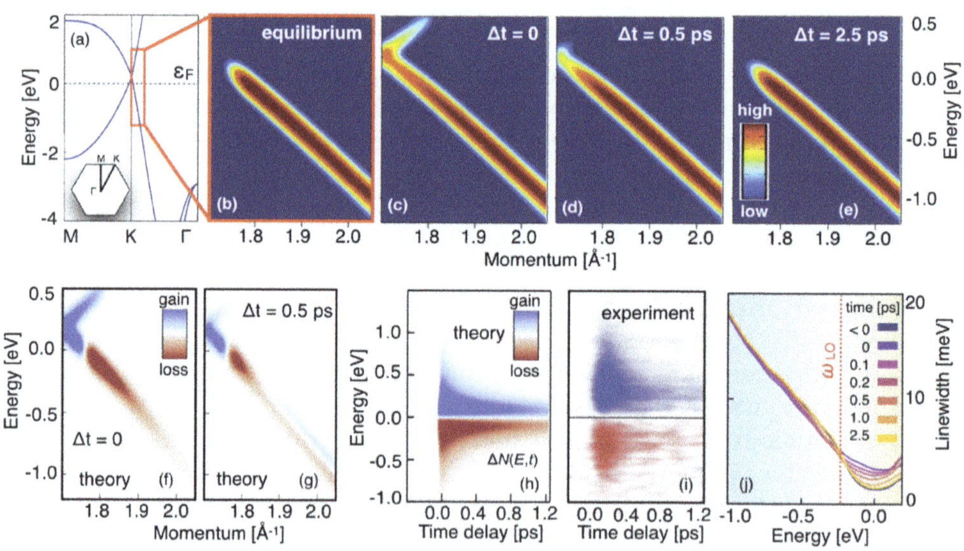

Figure 8.15. (a) Band structure of graphene near the K point. The Fermi energy is 0.2 eV below the Dirac point. (b–e) ARPES signals calculated along the K–Γ line (red box in (a)) at equilibrium and for different delays after the pump signal. Calculated differential signals for two delays (f,g), in which the equilibrium signal is subtracted. Differential pump–probe signals (integrated over momentum) for theory (h) and experiment (i). Electron line widths are shown in (j). Reprinted with permission from Caruso *et al* [103], copyright (2020) by the American Physical Society.

of the order of the de Broglie wavelengths. The study of quantum transport in semiconductor devices is, of course, older, although it has become a significant part of device simulation in the last few decades. The earliest developments were to use classical approaches, such as the EMC methods, and to insert quantum corrections, such as the quantum potential [106]. The idea of the quantum potential derives from the work of Madelung [107] and Kennard [108], the latter of which suggested that quantum wave packets representing particles would travel under normal classical forces plus any quantum potential in the system. The form of the quantum potential was rediscovered later by Bohm [109]. If the Schrödinger wave function is written as

$$\psi(r, t) = A(r, t)e^{iS(r, t)/\hbar}, \tag{8.141}$$

where both the amplitude A and the action S are real functions, then the quantum potential (energy) can be written as [106–108]

$$V_Q(r, t) = -\frac{\hbar^2}{2m^*A}\frac{\partial^2 A}{\partial r^2} = -\frac{\hbar^2}{2m^*\sqrt{n}}\frac{\partial^2 \sqrt{n}}{\partial r^2}. \tag{8.142}$$

The last form follows from the general assumption that the local density is given by the magnitude squared of the wave function, hence $n = A^2$. One of the earliest applications of the quantum potential in a device was to study resonant tunneling diodes [110]. It was then used in field effect transistors, as noted [105]. But, the resonant tunneling diode became the fruit fly for studies of quantum transport, because of the availability of considerable experimental data [111–113].

The second quantum effect to be added to semi-classical device simulation was to account for the non-zero size of the wave packet that represented an individual carrier. This led to the introduction of an effective potential [114], which was acquired by smoothing the classical potential found from Poisson's equation by the carrier's Gaussian wave packet. Such an effective potential had been discussed quite early by Wigner [115], and by Feynman and Kleinert [116].

A form of the effective potential for the interface of a MOSFET is illustrated in figure 8.16. The effective potential achieves two important quantum effects in the device: (1) it moves the maximum carrier density away from the interface as indicated by the set-back d, and (2) it raises the lowest energy of the carriers away from the classical minimum, thus providing the quantization energy in the well ΔE_c. Finally, the third quantum effect that has been included in semi-classical simulations is the collision broadening, which accounts for the off-shell discussion at the start of this chapter. In this approach, a distribution of such broadening is used to break the energy-conserving delta function in each collision [117, 118].

All three of these effects will be automatically included within the NEGF treatment of a device. An important non-quantum effect that produces quantum interference is the discreteness of impurity atoms in the device. It has already been pointed out that the exact position of an impurity, and its potential, dramatically affects device performance, even in the classical regime. The importance of this was pointed out by Keyes some time ago [119]. Discrete impurities were introduced into semi-classical simulations by Wong and Taur [120], and the fact that a single

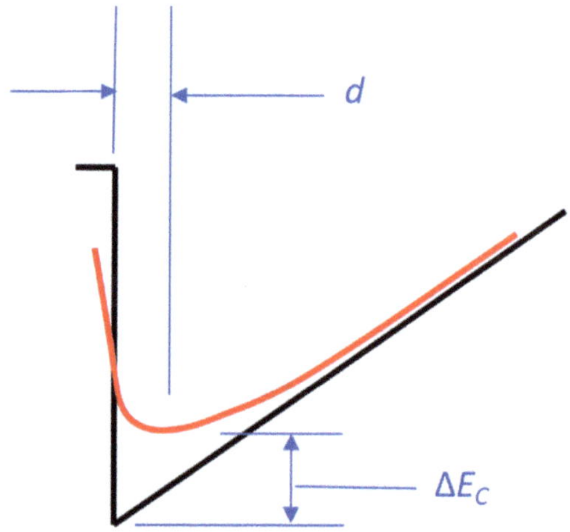

Figure 8.16. The effective potential (red) for the confinement of carriers in the normal triangular quantum well (black) at the oxide interface in a Si MOSFET.

impurity can affect the device performance was shown by various simulations [59]. That this impurity also can produce quantum effects has already been demonstrated with figure 8.11. Thus, the inclusion of atomistic doping, where the exact positions of the impurities are used without resorting to perturbation theory, has become extremely important in modern device simulation.

Now consider the problem of simulation of very small semiconductor devices (or other structures). In the case that the transverse dimensions are small, it is likely that the band structure determined in large volumes is no longer appropriate. Here, the analytic band, effective mass approach used previously (such as with the scattering matrix and Lippmann–Schwinger equation approach to obtain the transmission discussed above) often is no longer applicable. It is not the calculation of the transport that is no longer applicable, but the energy structure of the system under consideration that is likely to be modified by the small transverse dimensions. Thus, modern approaches even use atomistic methods to calculate the proper band structure, either by empirical pseudo-potential approaches or by more compact tight-binding methods. One of the first modern simulation packages using tight-binding was for the study of transport in molecules [121]. A somewhat more famous energy structure program is SIESTA, which also uses localized atomic orbitals (but numerically simulated ones) [122, 123]. This program has been extended to computation of the conductance via NEGF in TranSIESTA [124, 125]. A similar approach is used in the FIREBALL code [126, 127]. While these approaches use real space atomic orbitals, these are density-functional theory approaches. A different approach uses semi-empirical methods with atomic orbitals in which orbital overlap functions are empirically adjusted to fit to experimental band structures, and transport is accomplished with NEGF. One such semi-empirical approach is NEMO, which began with studies on the resonant tunneling diode [128]. MOSFETs were studied later using an advanced version of NEMO [129], as well as other equivalent formulations [130]. Other NEGF approaches appeared about the same time, but without the atomistic treatment of either impurities or the band structure [131, 132].

In spite of the presence of these packages, many groups have developed their own NEGF codes for simulation of semiconductor devices. For example, studies of the efficacy of single molecule transistors have used NEGF [121, 133]. The quantum effects in short-channel MOSFETs have also been examined [130, 134]. The performance of MOSFETs as analog devices has also been investigated [135], and the effect of remote impurities (in the oxides) on Si nanowire FETs also has been examined [136]. More recently, the performance of nanowires in several materials has been discussed [137]. In figure 8.17, the nanowire structure for this latter work is shown in panel (a), in this case a wire of 2.2 × 2.2 nm^2. It is a gate all-around structure in which the device is coated on all sides with an oxide. The gate is applied to the central 6 nm section. The authors investigate the role of the self-energy, which is shown in panel (b). As may be seen, the real part gives little effect except in the sub-threshold regime. This panel is for a general nanowire that is not in the device configuration. In figure 8.18, the current spectra is shown for a FET with a 24 nm source region and a gate voltage of 0.4 V. The position (in energy) of the lowest sub-band due to the transverse quantization is shown by the white dashed line. The quasi-Fermi level is shown as the lower magenta line indicating the amount of bias applied to the drain. The upper red line is the 'current energy,' which is essentially the integral of the product of current and energy over the cross-section of the device. It may be seen there is a lot of energy at the peak of the sub-band (near the peak of the electric field in the device, which is given as the negative derivative of the Fermi energy).

It was pointed out relatively early that structures such as FinFETs would be effective in moving the carriers away from the surfaces into the bulk [138]. This will be even more true in the case of nanowires. This is shown in figure 8.19 for a Si

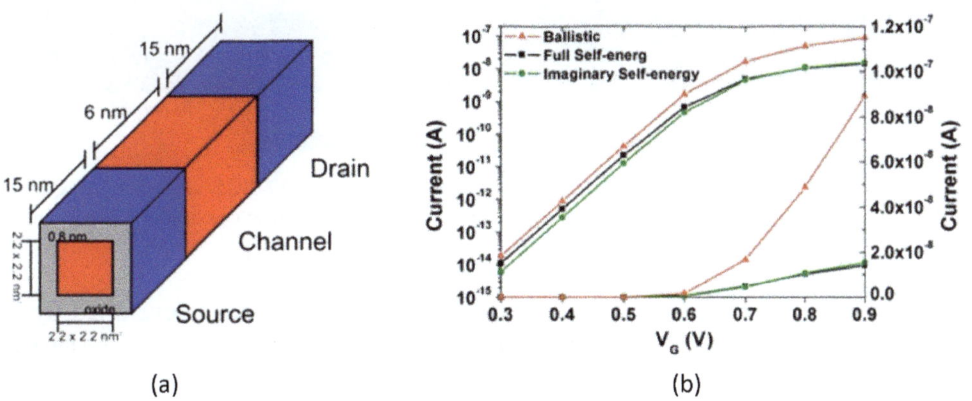

Figure 8.17. (a) General shape of the nanowire considered, although the source and drain lengths were varied in the study. The gate is applied in the central (red) region. (b) Currents through this nanowire for various values of the self-energy. The imaginary curve (green) neglects the real part of the self-energy. The ballistic properties diverge greatly once the transistor is turned on. Reprinted from Martinez *et al* [137], under the creative commons CCY open access license.

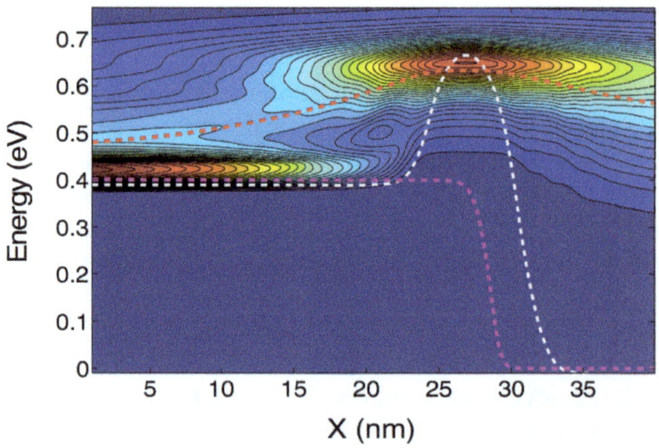

Figure 8.18. The current spectra for the device of figure 8.17(a). The lowest sub-band is shown with the white dashed line, the quasi-Fermi level with the (lower) magneta line, and the current energy with the (upper) red line. Reprinted from Martinez *et al* [137], under the creative commons CCY open access license.

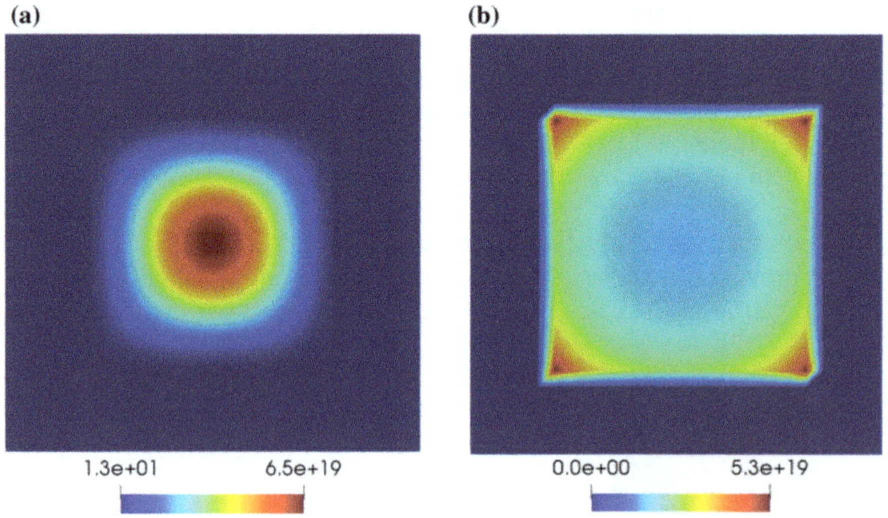

Figure 8.19. Carrier density across the transverse section of the nanowire. (a) NEGF simulation. (b) Classical drift-diffusion simulation. Reprinted from Berrada *et al* [139], under the creative commons CCY open access license.

nanowire of 3×3 nm^2 using a newer NEGF simulation package developed by the Glasgow group [139]. The panel on the right (b) illustrates the location of the carriers in the classical case, where they are drawn to the higher surface fields at the corners of the gate all-around structure. In this case, the nanowire channel (gate length) is 20 nm with 0.4 V on the gate and 0.6 V on the drain. In panel (a), the equivalent charge distribution, for the same biases, is shown for the NEGF simulation. It is clear that the carriers are moved away from the surface channels and into the bulk of the nanowire.

As the device world moves from nanowires to nanosheets, which are nanowires whose width is larger than their height, inevitably this leads to the ultimate nanosheet represented by monolayer compounds. Graphene was at the heart of such a study quite early but suffers from the lack of a band gap. However, using bilayer graphene can lead to the opening of a gap that would be suitable for FETs [140]. Recently, interest has shifted to the transition-metal di-chalcogenides (TMDCs) which have a relatively wide band gap and can lead to reasonable device characteristics, and are natural for simulations with NEGF [141, 142].

NEGF has been applied to a number of optical devices as well. These include quantum cascade lasers [143] and nitride LEDs [144]. In addition, it has been applied to study the concept of hot carrier quantum well solar cells [145] and nanowire versions [146]. These later works have become extremely important in their respective communities.

8.5 Wigner functions

So far only quantum functions, such as the density matrix (8.42), that are a function only of times and the position variables have been studied. In classical physics, one has the Boltzmann equation which is a function of position, momentum, and time, so a phase space formulation of the transport problem follows naturally. It is often very convenient to describe the system with a phase space formulation even in quantum mechanics, although introductory texts don't often mention this possibility. Such a formulation exists and was proposed by Wigner [115]. To approach this, the center-of-mass and difference coordinates are defined through (8.137), although r_d will be replaced by s, as is usual in the Wigner function. Like the density matrix, the Wigner function has only a single time, so that t, $t' \equiv t$. Then, the density matrix may be transformed on the difference coordinate as

$$f_W(R, t) = \frac{1}{h^3}\int ds \rho\left(R + \frac{s}{2}, R - \frac{s}{2}, t\right) e^{ip\cdot s/\hbar}. \tag{8.143}$$

It may be noted that this is related to the less than Green's function of (8.113) in the case in which the two times in the latter are equal. Using the transformation (8.137) and the Liouville equation for the density matrix, the equation of motion for the Wigner function becomes

$$\frac{\partial f_W}{\partial T} + \frac{p}{m^*}\cdot\frac{\partial f_W}{\partial r} + \frac{1}{i\hbar}\left[V\left(R + \frac{s}{2}\right) - V\left(R - \frac{s}{2}\right)\right] f_W = 0 \tag{8.144}$$

in the case for which there is no scattering. Using the displacement operators in position, this can be rewritten as

$$\frac{\partial f_W}{\partial T} + \frac{p}{m^*}\cdot\frac{\partial f_W}{\partial r} - \frac{1}{i\hbar}\left[V\left(R + \frac{i\hbar}{2}\frac{\partial}{\partial p}\right) - V\left(R - \frac{i\hbar}{2}\frac{\partial}{\partial p}\right)\right] f_W = 0, \tag{8.145}$$

or, by separating the two potential terms into separate integrals and reversing s in the second, it may be written in the more usual form

$$\frac{\partial f_W}{\partial T} + \frac{p}{m^*}\cdot\frac{\partial f_W}{\partial r} - \frac{1}{h^3}\int dP W(R, P) f_W(R, p+P) = 0$$
$$W(R, P) = \int ds \sin\left(\frac{s\cdot P}{\hbar}\right)\left[V\left(R+\frac{s}{2}\right) - V\left(R-\frac{s}{2}\right)\right] \tag{8.146}$$

The form (8.145) is useful for using a Taylor series to expand all functions around R as was shown by Wigner [115]. On the other hand, the last form (8.146) is useful when one adds a scattering function and wants to develop a path integral for numerical trajectory approaches with particles [147, 148].

In general, the Wigner function is not a positive definite function. This is a consequence of the uncertainty relationship. On the other hand, if the Wigner function is integrated over either position or momentum, a positive definite function gives the probability functions (square magnitude of the wave functions) in position or momentum. In fact the negative excursions of the function exist over phase space regions whose volume is of order $\hbar^3$, so that smoothing of the function over a volume corresponding to the uncertainty principle will produce a positive definite function. The Wigner function is always positive definite in the ground state appropriate to equilibrium. However, the onset of negative excursions of the Wigner function are always viewed as the entrance of correlation/entanglement into the problem [149]. Such correlation or entanglement is connected to off-diagonal elements of the density matrix, and these lead to oscillatory behavior in the Wigner function with negative excursions. In this regard, modern approaches to the Wigner function provide a distinct formulation that is recognized as equivalent to normal operator-based quantum mechanics, but provides a different perspective [150]. Because of the phase-space nature of the Wigner distribution, it is possible to identify where quantum corrections enter a problem by comparing with the classical version. Wigner himself derived an effective quantum potential that can be used as a correction term for the potentials in thermodynamics [115].

Consider a simple Gaussian wave packet, representing a single carrier, which may be written as

$$\psi(R) = \frac{1}{(2\pi\sigma^2)^{1/4}} e^{-R^2/4\sigma^2 - ik_0R}. \tag{8.147}$$

The last term in the exponential gives motion to the wave packet so that it moves at a velocity $v_g = \hbar k_0/m^*$. This wave function can be used in the Wigner formulation (8.143) to give the result, in one dimension,

$$f_W = \frac{1}{h} e^{-2\sigma^2(k-k_0)^2 - R^2/4\sigma^2}. \tag{8.148}$$

This is a positive definite function, which is appropriate for what is basically a ground state function. Unlike the Green's functions, the Wigner function can be used directly from the equation of motion (8.146), either with or without scattering. The scattering integral is where the problem lies, as it is quite difficult to handle [149, 151]. Consequently, many approximations are used which basically boil down to modifying the normal semi-classical scattering functions to account for off-shell effects and collisional broadening.

8.5.1 Simulations with the Wigner function

The resonant tunneling diode (RTD) was conceived by Esaki and his colleagues [111], but became much more interesting after it was demonstrated that oscillations in the THz regime could be expected [112]. In this device, two tunneling barriers (typically GaAlAs) separated by a quantum well (typically GaAs) are used, and the bound state in the well provides the resonant level for tunneling. For a given potential structure, such as found in the RTD, the tunneling coefficient can be calculated as a function of the incident energy using an exact formulation or the WKB approximation. This can then be used for a simple analytical model, but this is not a good approach to use, as the tunneling probability will change with bias applied across the structure. Better simulations have been developed which solve Poisson's equation for a self-consistent potential, and coupling this with a transport equation using the Wigner function [13, 152, 153].

In one approach, the initial condition for the Wigner function at the source and drain ends of an RTD are generated by computing the ensemble set of wave functions, and the density matrix, from proper scattering states [12]. Plane waves are used to build up the density matrix. These plane waves are chosen with the Fermi–Dirac probability function; that is the number of plane waves at any incident energy is proportional to the value of this function at that energy. For states which are incident from the left of the barrier (these are waves traveling in the positive x-direction, with $k > 0$ in one dimension), the scattering states become [11]

$$\psi_k(x) = \begin{cases} \dfrac{1}{\sqrt{2\pi}}[e^{ikx} + R(k)e^{-ikx}], & x < x_{\min}, \\ \dfrac{1}{\sqrt{2\pi}}T(k)e^{ikx}, & x > x_{\max}. \end{cases} \tag{8.149}$$

The quantities $x_{\min}$ and $x_{\max}$ are the left and right sides of the barrier–well region. $R(k)$ and $T(k)$ are the reflection and transmission coefficients for the barrier–well structure, respectively. Using these scattering states, the density matrix is a summation over bound states and an integration over propagating states, and then the Wigner function is generated numerically from this density matrix [12]. A similar set is generated for waves coming from the right.

Solving the Wigner equation of motion then requires discretization of the function on a grid in both real space and momentum space. This discretization puts some limits on the range of the variables, that arise because the grid is a lattice and leads to periodic effects just as in band theory. When the Wigner function is discretized with a spatial increment Δx, the momentum space becomes periodic with period $\pi\hbar/\Delta x$. In order to consider both positive and negative momentum, the range of the grid is taken to be

$$-\frac{\pi}{2\Delta x} < k \leqslant \frac{\pi}{2\Delta x}. \tag{8.150}$$

The next major problem is that the Wigner equation of motion includes the first spatial derivative of the Wigner function. It is convenient split the overall grid into two regions, one for positive (right-going) momentum and one for negative (left-going) momentum. Then, a first-order upwind differencing scheme [154] can be used to propagate the Wigner distribution to the outgoing boundary along the characteristic direction as

$$\frac{\partial f_W}{\partial x} = \begin{cases} \frac{1}{\Delta x}[f(x_i) - f(x_{i-1})], & k > 0, \\ \frac{1}{\Delta x}[f(x_{i+1}) - f(x_i)], & k < 0. \end{cases} \tag{8.151}$$

The time step is required to satisfy the Courant–Friedrichs–Lewy condition $\Delta x/\Delta t < v_g$. With a device like the resonant tunneling diode (RTD), the self-consistent potentials in the system must also be updated every time step. This is handled by solving Poisson's equation at each time step. The self-consistent Wigner distribution is first calculated for the equilibrium state with no bias applied but with the presence of the ionized donors. Knowledge of the Fermi energy from the temperature and doping level, as well as the equilibrium temperature itself, is sufficient to determine this equilibrium state. Then, the bias can be ramped up, and down to recover the equilibrium state.

A typical structure is a GaAs device with embedded GaAlAs barriers, which provide 0.3 eV barrier height [13]. Each of the barriers and the quantum well are 5 nm in extent, and the regions outside the barriers are doped to 10^{18} cm^{-3}. The I–V curve shows the typical peak and valley with the peak current at ~0.28 V applied to the drain end of the device. The Wigner function at this bias is shown in figure 8.20. In the simulations, $\Delta x = 0.25$ nm and $\Delta t \sim$ a few femtoseconds. Scattering was instituted via the relaxation time approximation with a scattering time appropriate to a mobility of 3000 cm^2 Vs^{-1}. Steady-state was achieved within 0.7 ps in all bias cases. Once steady-state is reached at each bias level, the current through the device

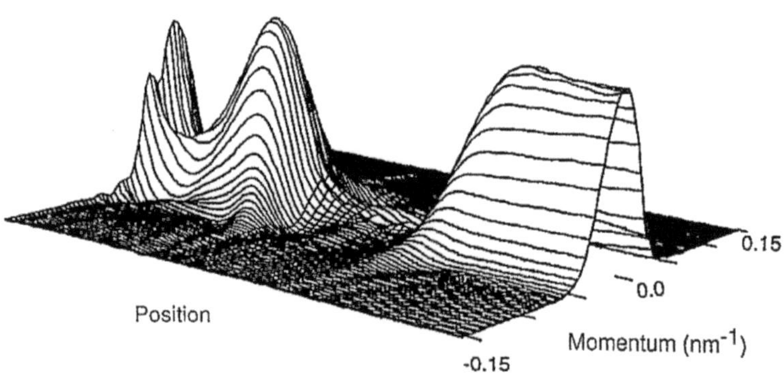

Figure 8.20. The steady-state Wigner function at the peak of the I–V curve (~0.28 V). The small values near the source (left) end arise from the formation of a 'triangular' potential well between the source and the first tunneling barrier. This leads to depletion of the carriers in this region. Reprinted from Kluksdahl *et al* [13], copyright (1989) by the American Physical Society.

can be determined. This produces a peak current density ~0.18 mA cm^{-2} and a peak-to-valley ratio of 2:1, all at a temperature of 300 K.

More recently, the Wigner function was used to study a resonant-tunneling nanowire field effect transistor [155], with interesting results. Here, the active material was GaAs, with AlGaAs barriers, and a wrap-around gate was used on the cylindrical nanowires. The gates covered the RTD segment which was the active channel. The two barriers were 2.8 nm and 2.26 nm in thickness, while the well width was varied from 2.5 to 7.2 nm. The bodies of the nanowires varied from 2.0 to nm in radius. The authors found that the on/off ratio was more controlled by the gate than the drain.

8.5.2 Wigner Monte Carlo

It is possible to take the kinetic equation for the density matrix, and convert this kinetic equation into a path variable equation, just as was done for the Boltzmann equation in chapter 7. Then, the various orders of scattering, which contribute to the overall integral equation, may be sampled by a Monte Carlo process. This same approach has been developed for the Wigner equation of motion [156–158]. That is, the Wigner equation of motion is converted into an integral equation. Then the Neumann series representing the integrand that is developed over many scattering processes (again for a backward propagation form of solution) is solved by a Monte Carlo sampling process, in the same manner as discussed in chapter 7. The only changes are methods to incorporate off-shell effects, such as caused by the collision duration, and a method to handle the Wigner potential W in (8.146).

A number of early steps toward simulation of the Wigner equation have appeared. In one approach, the simulation uses particles to handle the transport, but the quantum effects of tunneling are introduced by solving for the energy dependent tunneling probabilities and then using these to decide whether or not a particle passes the barrier [159]. That is, the tunneling is treated as a scattering process, as has been suggested by Barker [160]. Another approach used the particles to simulate the Wigner equation of motion and study a process by which particles could be created (multi-particle production) [161]. In this approach, they weighted the individual particles, by a process that was introduced earlier in classical Monte Carlo to study rare events [162]. While these approaches bring particles into the simulation, they don't fully solve the quantum coherence problem in quantum transport. On the other hand, it is known that the Wigner equation of motion will bring coherence and interference into the solutions, as was demonstrated in figure 8.11.

The idea of using particle trajectories dates at least to the work of Kennard [108], who asserted that, with the Schrödinger equation, the carriers would follow their classical potentials plus the effects produced by any quantum potential present. Thus, an EMC approach represents the Wigner function by a collection of particles that follow the same paths as for the Boltzmann function of chapter 7. The new difficulties are related to the Wigner potential as mentioned. The concept of a Wigner path is thus useful as it provides a representation of the quantum evolution, and is quite useful for numerical simulation of the Wigner equation of motion. Such a path is the path followed by a small sample of the Wigner function as it evolves through phase space (these are the particles used). Moreover, this path evolves quite like a classical particle, except for scattering and for rapidly varying potentials.

A version termed the weighted Monte Carlo approach to simulations of the Wigner function transport was mentioned above [162]. However, there are some differences here, especially as paths with negative weights have to appear to represent the negative parts of the Wigner function [163]. Since some of the Monte Carlo particles carry these negative weights, a problem arises as to how to keep the density at its nominal value, as there may well be more simulation particles than real particles. This is achieved through the concept of a *maximal envelope*. Simply speaking, the maximal envelope at a point in space is the momentum distribution of the magnitude of the Wigner function at that point in space. For example, when simulating a Gaussian Wigner wave packet impinging upon a barrier, it is necessary to have enough particles present to account for the large negative distributions that can occur at the barrier. Thus, a significant fraction of the associated particles carry negative weight as a means of giving these negative values of the Wigner distribution. These negative weights are a different approach to representing the phase interference that leads to these negative excursions of the Wigner function. Now, the Wigner function at a point in phase space can be described by a sum over the particles as

$$f_W(r, k) = \sum_{i=1}^{N} A_i \delta(r - r_i)\delta(k - k_i). \tag{8.152}$$

N is the number of simulation particles and A_i is the weight, or *affinity*, attached to each of these particles. Absorbing boundary conditions on the exit boundaries from the device may also be instituted in a very easy manner. An important aspect of the affinity is its sign and magnitude change during the simulation. The non-local Wigner potential determines the change in the affinity that each particle possesses. That is, the wave nature of the particles is maintained through the variation of each particle's affinity. In figure 8.21, the Wigner function is shown for a Gaussian wave packet moving through a single tunneling barrier [164]. In this structure, based upon the normal GaAs RTD, only a single potential barrier is present with width of 3 nm. At the time of the image, the packet has traveled about halfway through the entire tunneling process, and one can observe the transmitted and reflected Gaussian packets. In addition, there is a rapidly oscillating (in space) contribution that has positive and negative portions. This portion of the overall Wigner function represents the coherence (and memory) between the transmitted and reflected packets. These latter two packets are entangled and this entanglement will remain present until it is broken up by some decoherence process that occurs, such as scattering. This approach also has been used for RTDs [165].

The weight and the affinity are really artificial numerical quantities whose purpose is to simulate the quantum phase interference that occurs in the Wigner function during real propagation within a semiconductor device. A melding of these

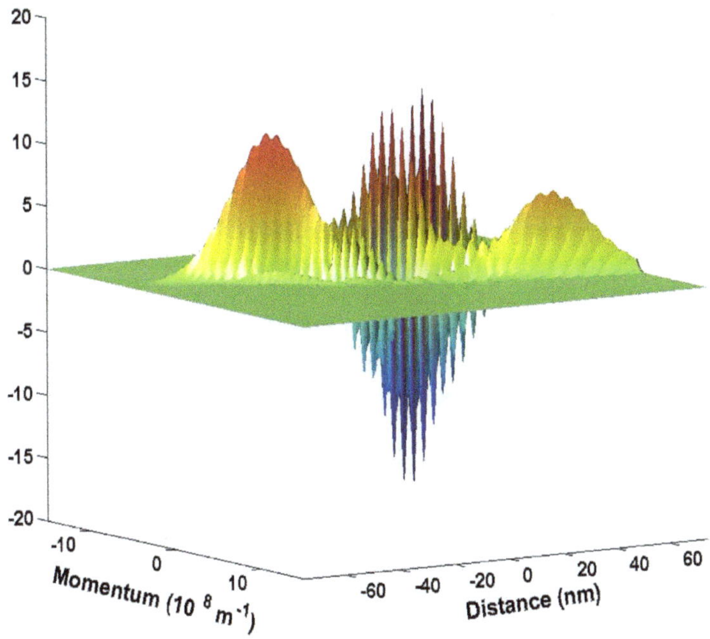

Figure 8.21. Gaussian wave packet after it has interacted with a single potential barrier of 3 nm width and 0.3 eV height in a GaAs tunneling structure. The transmitted and reflected packets have been fully formed as can be seen. The correlation/entanglement between these two packets is the rapidly oscillating part. Reprinted from Shifren and Ferry [164], copyright (2002) with permission from Springer.

two concepts was pursued in another approach [166]. In this approach, the weights are accompanied with signs associated with them. This led to high variance in the results, so that a weight decomposition approach, which limits the value of a weight by storing part of it on a phase space grid, was necessary. This method significantly improved the variance during simulations with this approach. The variance problem finally led to the adoption of a pure sign convention [167, 168]. This creates a model in which the Wigner function is considered to include a Boltzmann-like scattering term, but which now includes a generation term. The quantum information then is carried by the sign of the quasi-particles. This approach is most useful when treating the interaction with the non-local potential as a scattering event. When a scattering event from the potential occurs, two new particles are created, one with the momentum increased by q and one where it is reduced by q, with q determined randomly from the probability distribution of the potential's spatial Fourier transform. The sign on one of these new particles is taken to be the same as the incident particle, while the sign on the other is the opposite. These signs are taken into account in each averaging process that is used to find average values. Equivalent particles with opposite signs annihilate one another when they meet in phase space. This approach was used for figure 8.11.

The general affinity method has been adopted by others to simulate a variety of MOSFETs. These include double-gate MOSFETs of 6 nm gate length [169, 170]. It has also been used to study planar MOSFETs with a quantum corrected EMC [171].

8.5.3 Scattering in Wigner functions

As remarked above, many approximations are used to include collision effects and scattering within the Wigner approach. These usually boil down to modifying the normal semi-classical scattering functions to account for off-shell effects and collisional broadening. Levinson has developed a formal approach to scattering [151], and his results tend to reinforce this idea, as the main differences from normal semi-classical scattering are just these major effects. In quantum transport, scattering from a phonon, for example, does not have to be local, the event may spread over a small spatial distance. This arises as the collision is not instantaneous, but takes a few femtoseconds to be completed [39]. Hence, due to the intra-collisional field effect, in which the particle continues to be accelerated by the field during this short time, the particle also moves a bit due to the particle velocity. In addition to the non-local scattering, the loss of a one-to-one correspondence between energy and momentum, there is also a broadening of the energy conservation that arises from replacement of the delta function by the spectral density function.

Here, an approach based upon Wigner paths, uses weighted particles evolving forward in time. Each path will have several collisions occurring during the simulation time. There is an alternative approach based upon the backward Monte Carlo method, where particles evolving backward in time are associated with the Levinson and the Barker–Ferry (see section 8.3.3) models of electron–phonon interaction. The latter goes one step further than the two-time Levinson model by taking into account the finite lifetime of the particle states due to the scattering process [40]. The homogeneous Levinson and Barker–Ferry equations have been generalized to account for the spatial evolution in a quantum wire. Effects related to the initial build-up of the energy-conserving delta function and the evolution of the Wigner state during the process of collision—the intra-collisional field effect—are observed. According to a classical model, the need for energy and momentum conservation causes the fastest components of the state to travel ballistically. In the quantum models, uncertainty arising from the collision durations leads to broadening in a manner in which there are electrons which gain energy and can move faster than in the ballistic case.

As a result of the fact that this is the evolution of a quantum system, the scattering integrands assume complex values as a result of the quantum interference and the fact that the system is subjected to the uncertainty principle. The negative parts of the Wigner function are direct indications of the regions where uncertainty is important. An extended version of the weighted Monte Carlo [162], or the equivalent affinity approach [163], is now used to generate a set of Wigner paths. The sum over these paths is a numerical estimate of the Wigner function itself. In order to generate a Wigner path, there is an entire sequence of random variable selections that must be made [172]. The most important aspect of this is that the collision is a two-step process. That is, the Fermi golden rule involves the magnitude squared matrix element. Each matrix element is one step in the process. Thus, in the EMC for the Wigner function, one-half the phonon momentum is passed to the path at each of the two time steps. For a virtual process, in which the scattering is not completed, this one-half of the momentum is then withdrawn at the second vertex, so no net momentum is passed to the path. If the scattering process is completed, however, the net momentum change is $\hbar q$, just as in the classical case. The passing of half the momentum at each vertex can be seen in the Levinson formulation [151]. So, this is a correct interpretation of the manner in which the scattering events are inserted into the paths. Then, a phase factor is added to the particle, which is proportional to

$$\cos\left[q(x_1 - x_2) + \omega_q(t_1 - t_2)\right]. \tag{8.153}$$

Despite this, there is still the problem of identifying the two positions and the two times in (8.153); e.g., identifying the position change and time change during the collision. That is, the scattering is non-local and involves a time duration. While it is not something that is often discussed with the Wigner function, the momentum in the Wigner formulation arises from the difference in the two positions of the wave functions. The momentum is connected to the direction normal to the main diagonal in the density matrix. Hence, the time duration is connected to that required to transit between the two positions at the corresponding momentum. These different variables are thus tied together in a type of self-consistent manner. Many methods have been suggested to include the collisional displacement as a modification to the use of semi-classical scattering within the Monte Carlo method [173–175]. In the case of [173], it was suggested to replace the energy-conserving delta function with the joint spectral density, represented by a Lorentzian line centered around the classical energy conservation point. This was not very successful due to the very long tails of this function, which led to arbitrarily high energies. To properly include the role of the finite collision duration, we need to account for two effects: (1) the collision duration itself and the resulting energy shift, and (2) the position change during the collision.

The collision duration itself is usually only a few femtoseconds [39]. In this time, the change in position that occurs as a result of the collision duration may be written approximately as [176]

$$\delta x = \left[v(t) + \frac{e\boldsymbol{F}\cdot\boldsymbol{v}(t)}{m^* v(t)}\frac{\tau_c}{2}\right]\tau_c. \tag{8.154}$$

In this expression, $\boldsymbol{F}$ is the electric field and τ_c is the collision duration. The first term is just simple displacement due to the particle's motion. The second term represents the intra-collisonal field effect. During the interaction, there is an overlap between the two wave packet at the two positions separated by δx. The two wave functions are Wigner packets which may be taken to be Gaussian in shape, which leads to an overlap that varies as

$$I(\delta x) = \frac{1}{\sqrt{4\pi\sigma^2}}e^{-(\delta x)^2/4\sigma^2}, \tag{8.155}$$

where the broadening σ is taken to be the thermal de Broglie wavelength

$$\sigma \sim \sqrt{\frac{\hbar^2}{12 m^* k_B T}} \sim 2\ \text{nm}. \tag{8.156}$$

Then the probability for a given shift is

$$P(\delta x) = | I(\delta x) |^2 = \frac{1}{4\pi\sigma^2} e^{-(\delta x)^2/2\sigma^2}. \tag{8.157}$$

The direction of the shift is given by the relation between the velocity and the electric field. This provides a method of incorporating the displacement in time and space when using the semi-classical (instantaneous) approximation for the scattering process. When the collision duration is known, one can estimate the position shift from (8.154) and then evaluate the probability for such a shift from (8.157). As the approach is for the Monte Carlo, a decision on whether or not to accept the shift can be made naturally, just as for a self-scattering event. In a sense, this physical shift, and the evolution of the Wigner function during the collision, introduces a non-Markovian behavior in the equation of motion [177].

With the above approach, both collisional broadening and nonlocality of the scattering are both incorporated, even though the motion is largely classical during the collision [176]. Generally, the collision duration may be considered to be exponentially distributed, and then may be estimated by

$$\tau'_c = -\tau_c \ln(r), \tag{8.158}$$

where τ_c is the collision duration determined from direct calculations [39] and r is a random number between 0 and 1. This duration determined from (8.158) is then used in (8.154) to determine the change in positon during the scattering. This duration also gives the change in energy from the field during the collision which provides the broadening. In figure 8.22, the average value of the displacement during the collision is plotted for an RTD, with the tunneling structure centered around

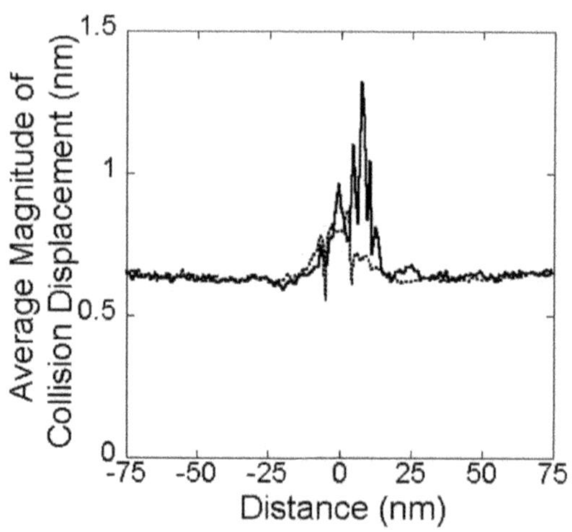

Figure 8.22. Root-mean-square displacement during a phonon scattering event as a function of distance through an RTD. The tunneling structure is centered around $x = 0$. The solid curve is for a bias of 0.5 V, while the dotted curve is the absence of any bias. Reprinted from Shifren and Ferry [176], copyright (2003) with permission from Elsevier.

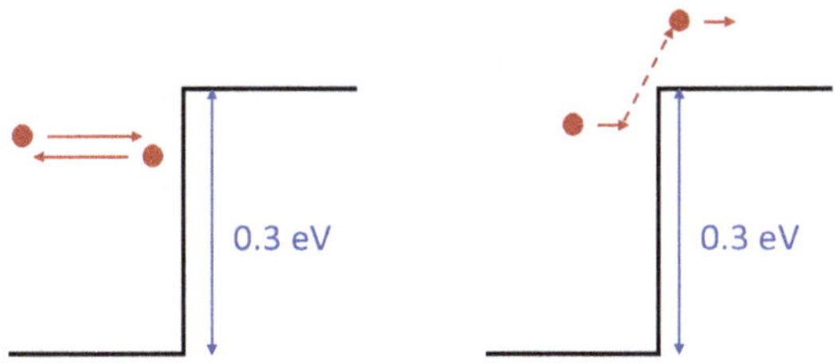

Figure 8.23. In the left panel, a normal interaction occurs in which the particle approaches the barrier and may be either reflected (as shown) or penetrates the barrier (tunnels). But, in the right panel, it absorbs a phonon, which gives it sufficient energy to surpass the barrier during the collision, yet must lose the barrier energy and winds up with less kinetic energy.

$x = 0$. The RTD structure contains 3 nm barriers of 0.3 eV height and a 5 nm well. It may be seen that, away from the tunneling region, the r.m.s. displacement is about 0.6 nm, although the average displacement is zero, as positive and negative displacement (with and opposite to the field) occur. But, in the tunneling region, the r.m.s. displacement can be much larger. This is because the field is considerably larger around the tunneling structure, where most of the potential drop occurs. And, in this region, the displacement is peaked in the direction of the electric field.

The increase shown in figure 8.22 near the tunneling region can lead to some interesting and unexpected results. Consider, for example, the structure in figure 8.23, which represents a region near the tunneling barrier. In panel (a), a particle is approaching a barrier of modest height. If there is no collision, the particle will likely be reflected backwards, but there is a small probability that it will penetrate the barrier. This particle could also absorb or emit a phonon, without the presence of quantum effects during the collision, it still will likely be reflected. In panel (b), the effect of a particle undergoing the intra-collisional field effect near the barrier is shown. If the field is sufficiently high, the particle may gain enough energy to surpass the barrier and wind up in a state which goes directly over the barrier. This may be accompanied by a positional displacement, which is indicated in the figure. During this collision, the particle may emit or absorb a phonon. If it emits a phonon, then it will wind up with a lower kinetic energy in its final state. The carrier may also absorb a phonon and still wind up with a lower kinetic energy in its final state. The difference in these two processes is a difference in the energy that must be absorbed from the electric field during the collision. Thus, the energy conservation requirement becomes

$$E_{k\pm q} - E_k \pm \hbar\omega_q = V(x) - V(x + \delta x). \tag{8.159}$$

That is, if the positional displacement takes it into the barrier region, yet the absorbed energy exceeds the barrier height, then the particle has to lose the potential energy represented by the barrier.

8.5.4 Wigner function in devices

In recent device studies, the signed particle Monte Carlo approach has been used to study a number of small semiconductor devices, including the role of interface roughness [178]. They also used a potential decomposition to improve the method [179]. In this approach, the non-local potential is decomposed into a classical part and a quantum mechanical non-local part. The classical part gives simply the classical electric field that enters the Boltzmann equation. The quantum part is then the difference between the actual non-local potential and the classical potential and is treated by the scattering method discussed above for signed particles. Scattering by the phonons and the quantum potential both involve the generation of particles and the use of the particle sign to impart quantum interference.

Device studies started with a Si nanowire transistor using the full Wigner function [180]. In this work, the transport along the channel is solved with a one-dimensional Wigner function approach, while the transverse quantization is handled by a two-dimensional Schrödinger solver. A relaxation time approximation form for the scattering within the channel was assumed. Now, one has to be careful with quantum simulations in short devices. The existence of longitudinal resonances has already been discussed in connection with figure 8.5. These longitudinal resonances have also appeared in treatments with the Wigner function and are genuine characteristics [181]. In the Wigner function case, as with the earlier case, this is a sign of coherence between the source and drain Wigner functions. Further simulations of quantum wires, complete with dissipation have appeared from Barraud [182, 183]. In this latter work, the author used a gate-all-around format on the nanowire and showed that dissipation works well to break up correlation between the input and output.

Some of the earliest work with Wigner functions in devices was for multi-quantum well lasers [184]. Here, the authors studied the role of the nonequilibrium carriers on the gain in these quantum well lasers. In later work, a quantum Boltzmann equation was developed from the Wigner equation of motion and used to study a single quantum well laser [185, 186]. Finally, dissipative quantum transport in the superlattices used for quantum cascade lasers has been studied with the Wigner function [187].

Wigner functions were used to study the behavior of cylindrical MOSFETs with the presence of acoustic phonons and impurities [188]. The study of the Wigner function showed that, in spite of the scattering, current flowed through the device in a narrow stream. This may be seen in figure 8.24. Panel (a) shows the situation without scattering, while panel (b) illustrates the scattering case. Variations tend to be at low momentum near the center of the channel. Similar results were later obtained for a planar MOSFET [189] and for a double-gate MOSFET [190]. In these latter cases, surface roughness scattering was also included. Transport in a double-gate MOSFET was studied in the quasi-ballistic case in order to compare solutions from NEGF and Wigner function approaches [191], with little difference being observed. Finally, the electron dynamics in a nanoscale transistor were compared for classical Boltzmann dynamics and Wigner function dynamics [192]. As mentioned above, the general affinity method has been adopted by others to simulate a variety of MOSFETs. These include double-gate MOSFETs of 6 nm gate length [169, 170]. It has also been used to study planar MOSFETs with a quantum corrected EMC [171].

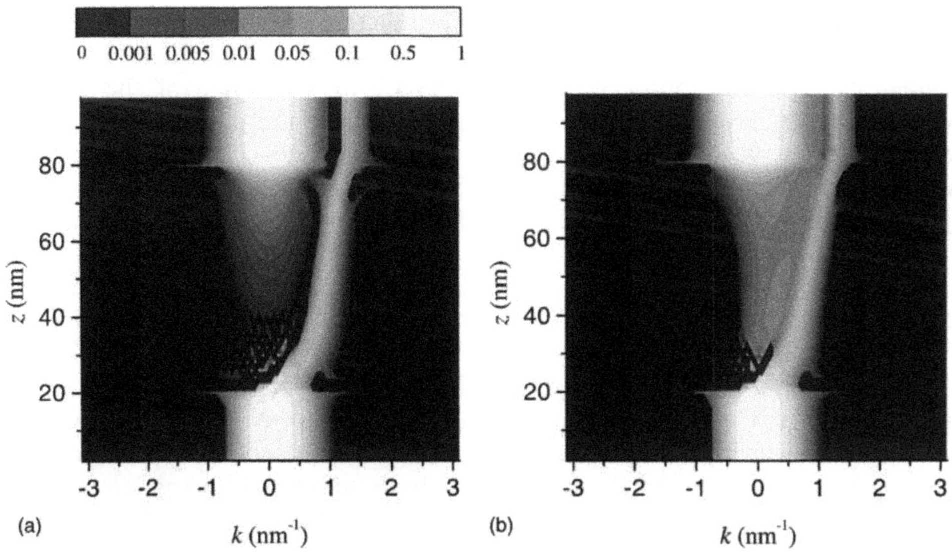

Figure 8.24. Contour plots of the Wigner function for the lowest sub-band in the quantum wire FET with a gate voltage of 1 V and a drain voltage of 0.3 V. The channel length is 60 nm. (a) No scattering is present. (b) Scattering by acoustic phonons and impurities is present. Reprinted from Balaban *et al* [188], copyright (2002) with permission from Elsevier.

8.6 Some final comments

Transport in a semiconductor can be thought of as being a simple concept. However, in reality, the semiconductor is a very complex system, and the simulation of the transport also may be a very complex endeavor involving full simulation of the processes by which it is created as well as simulation of its operation and resulting characteristics [193]. This complexity arises not merely from the need for a deeper understanding of the transport phyiscs of hot carriers, but also because these devices interact with their environment. From a physical point of view, a *device* comprises an active semiconductor region, which is open to the environment in which it is embedded. This connection to the environment may be through a set of contacts, or through interactions with the phonon structure of the lattice in which the device is embedded, or through other types of interactions, as mentioned below. The central feature is that the device dynamics cannot be treated in isolation and must be considered in conjunction with this environment [194], including the array of other devices and interconnects nearby [195]. This is especially true in the quantum regime [196].

Quantum transport in these complex systems requires a complex quantum description, which takes into account coherent phenomena, as well as dissipative processes of scattering, with both modification by the interaction with the environment as well as by the onset of new quantum effects, such as the intra-collisional field effect. There are many approaches to study the quantum effects, as has been detailed in this chapter. The author's preference is for the Wigner function, as perhaps was evident in the preceding section. The use of the Wigner function is particularly useful in studies of complex systems, as the Wigner function may explicitly illustrate the important quantum effects and entanglement that is a signature of quantum interactions. Particle approaches and Monte Carlo simulation of the quantum Wigner function provide an efficient approach to the study of such methods. It becomes clear how to study each process with this approach and to establish the importance of the specific process in the behavior of the overall system. The Wigner function also allows one to clearly identify the quantum effects, particularly the entanglement that arises between different parts of the quantum system.

However, the world mostly has gone a different direction, with NEGF becoming increasingly applied in practice. But, there are problems with NEGF, both conceptually and with implementation in real devices. These have been discussed in section 8.4, and should not be ignored, if for no other reason than that they call into question the accuracy of present approaches to NEGF. Moreover, the assumption of merely using quantum approaches for the channel region, and coupling to the source and drain 'contacts' by a simple self-energy term loses a great deal of important effects. To pursue this approach neglects the fact that these 'contacts' are not simple regions in their own right. The effects that can occur in these regions affect the performance and observable behavior of the entire system.

References

[1] Kohanoff J 2006 *Electronic Structure Calculations for Solids and Molecules* (Cambridge: Cambridge University Press)

[2] Ferry D K 2017 *Quantum Transport in Semiconductors* (Singapore: Pan Stanford)

[3] Landauer R 1957 *IBM J. Res. Develop.* **1** 223

[4] Landauer R 1970 *Phil. Mag.* **21** 863

[5] Duke C B 1969 *Tunneling in Solids* (New York: Academic)

[6] Kroemer H 1968 *IEEE Trans. Electron Dev.* **15** 819

[7] Ando T 1991 *Phys. Rev.* B **44** 8017

[8] Usuki U, Saito M and Takatsu M *et al* 1995 *Phys. Rev.* B **52** 771

[9] Gilbert M J and Ferry D K 2004 *J. Appl. Phys.* **95** 7954

[10] Gilbert M J, Akis R and Ferry D K 2005 *J. Appl. Phys.* **98** 094303

[11] Merzbacher E 1970 *Quantum Mechanics* 2nd edn (New York: Wiley)

[12] Kriman A M, Kluksdahl N C and Ferry D K 1987 *Phys. Rev.* B **36** 5953

[13] Kluksdahl N C, Kriman A M, Ferry D K and Ringhofer C 1989 *Phys. Rev.* B **39** 7720

[14] Ferry D K 2020 *Transport in Semiconductor Mesoscopic Devices* 2nd edn (Bristol: IOP Publishing)

[15] Akis R, Ferry D K and Bird J P 1996 *Phys. Rev.* B **54** 17705

[16] Fetter A L and Walecka J D 1971 *Quantum Theory of Many-Particle Systems* (New York: McGraw-Hill)

[17] Kadanoff L P and Baym G 1962 *Quantum Statistical Mechanics* (Reading, MA: Benjamin)

[18] Ferry D K, Goodnick S M and Bird J P 2009 *Transport in Nanostructures* 2nd edn (Cambridge: Cambridge University Press)

[19] Kotylar R, Obradovic B and Matagne P *et al* 2004 *Appl. Phys. Lett.* **84** 5270

[20] Gilbert M J and Ferry D K 2005 *IEEE Trans. Nanotechnol.* **4** 355

[21] Akis R, Gilbert M and Ferry D K 2006 *J. Phys. Conf. Series* **38** 022

[22] Kohn W and Luttinger J M 1957 *Phys. Rev.* **108** 590
[23] Siegel J L and Argyres P N 1969 *Phys. Rev.* **178** 1016
[24] Argyres P N and Siegel J L 1973 *Phys. Rev. Lett.* **31** 1397
[25] Ferry D K 1991 *Semiconductors* (New York: Macmillan)
[26] Nakajima S 1958 *Prog. Theor. Phys.* **20** 948
[27] Zwanzig R 1960 *J. Chem. Phys.* **33** 1338
[28] Mori H 1965 *Prog. Theor. Phys.* **33** 423
[29] Pottier N 1983 *Physica* **A117** 243
[30] Barker J R 1980 *Physics of Nonlinear Transport in Semiconductors* ed D K Ferry, J R Barker and C Jacoboni (New York: Plenum) pp 127–51
[31] Barker J R 1978 *Sol.-State Electron.* **21** 197
[32] Balescu R 1975 *Equilibrium and Nonequilibrium Statistical Mechanics* (New York: Wiley)
[33] Knezevic I and Ferry D K 2003 *Phys. Rev.* E **67** 066122
[34] Knezevic I and Ferry D K 2005 *J. Comp. Electron.* **3** 359
[35] Knezevic I and Ferry D K 2004 *Semicond. Sci. Technol.* **19** S220
[36] Ferry D K, Barker J R and Grubin H L 1981 *IEEE Trans. Electron Dev.* **28** 905
[37] Kurosawa T 1966 *J. Phys. Soc. Jpn.* Suppl. **21** 424
[38] Rees H D 1972 *J. Phys. C* **5** 641
[39] Bordone P, Vasileska D and Ferry D K 1996 *Phys. Rev.* B **53** 3846
[40] Barker J R and Ferry D K 1979 *Phys. Rev. Lett.* **42** 1779
[41] Thornber K K 1971 *Phys. Rev.* B **3** 1929
[42] Barker J R 1973 *J. Phys. C* **6** 2663
[43] Ferry D K 1980 *Physics of Nonlinear Transport in Semiconductors* ed D K Ferry, J R Barker and C Jacoboni (New York: Plenum) pp 577–88
[44] Ferry D K, Kriman A M, Hida H and Yamaguchi S 1991 *Phys. Rev. Lett.* **67** 633
[45] Bude J, Hess K and Iafrate G J 1992 *Phys. Rev.* B **45** 10958
[46] Fischetti M 1984 *Phys. Rev. Lett.* **53** 1755
[47] Porod W and Ferry D K 1985 *Phys. Rev. Lett.* **54** 1189
[48] Fitting H-J and Friemann J-U 1982 *Phys. Stat. Solidi (a)* **63** 349
[49] Ferry D K 1988 *The Physics and Technology of Amorphous* SiO_2 ed R A B Devine (New York: Plenum) pp 365–73
[50] Hughes R C 1973 *Phys. Rev. Lett.* **30** 1333
[51] Hughes R C 1978 *Sol.-State Electron.* **21** 251
[52] Baym G and Kadanoff L P 1961 *Phys. Rev.* **124** 287
[53] Kadanoff L P and Baym G 1962 *Quantum Statistical Mechanics* (New York: W. A. Benjamin)
[54] Schwinger J 1951 *Proc. Nat. Acad. Sci.* **37** 452
[55] Schwinger J 1951 *Proc. Nat. Acad. Sci.* **37** 455
[56] Feynman R P 1949 *Phys. Rev.* **76** 749
[57] Feynman R P 1949 *Phys. Rev.* **76** 769
[58] Wick G C 1950 *Phys. Rev.* **80** 268
[59] Khan H R, Vasileska D and Ahmed S S *et al* 2004 *J. Comp. Electron.* **3** 337
[60] Liu B, Akis R and Ferry D K *et al* 2016 *J. Phys. Cond. Matter* **28** 135302
[61] Young T 1807 *Course of Lectures on Natural Philosopy and Mechanical Arts* (London: J. Johnson) L. XXXIX
[62] Akashi T, Harada K and Kasai H *et al* 2002 *Appl. Phys. Lett.* **81** 1922

[63] Taylor G I 1909 *Proc. Cambridge Phil. Soc.* **14** 417
[64] Einstein A 1909 *Phys. Z.* **10** 817
[65] Barker J R 2003 *J. Comp. Electron.* **2** 153
[66] Weinbub J, Ballicchia M and Nedjalkov M 2018 *Phys. Stat. Sol. RRL* **12** 1800111
[67] Ferry D K 2021 *Quantum Mechanics* 3rd edn (Boca Raton, FL: CRC Press) sec. 7.2.2
[68] Kubo R 1957 *J. Phys. Soc. Jpn.* **12** 570
[69] Joshi R P and Ferry D K 1991 *Phys. Rev.* B **43** 9734
[70] Enz C P 1992 *A Course on Many-body Theory Applied to Solid-State Physics* (Singapore: World Scientific Press)
[71] Abrikosov A A, Gor'kov L P and Dzyaloshinskii I Y 1965 *Quantum Field Theoretical Methods in Statistical Physics* (New York: Pergamon)
[72] Bethe H and Salpeter E 1951 *Phys. Rev.* **84** 1232
[73] Danielewicz P 1984 *Ann. Phys.* **152** 239
[74] Spicka V, Velický B and Kalvová A 2005 *Phys. Rev.* E **29** 154
[75] Spicka V, Velický B and Kalvová A 2005 *Phys. Rev.* E **29** 175
[76] Spicka V, Velický B and Kalvová A 2005 *Phys. Rev.* E **29** 196
[77] Meir Y and Wingreen N S 1992 *Phys. Rev. Lett.* **68** 2512
[78] Blandin A, Nourtier A and Hone D W 1976 *J. Physique* **37** 369
[79] Rammer J and Smith H 1986 *Rev. Mod. Phys.* **58** 323
[80] Lousier M, Schenk A, Fichtner W and Klimeck G 2006 *Phys. Rev.* B **74** 205323
[81] Maassen J, Harb M and Michaud-Rioux V *et al* 2013 *Proc. IEEE* **101** 518
[82] Martinez A, Kalna K and Sushko P V *et al* 2009 *IEEE Trans. Nanotechnol.* **8** 159
[83] Zhou W X and Chen K Q 2014 *Sci. Repts.* **4** 7150
[84] Vasileska D, Prasad C, Wieder H H and Ferry D K 2003 *J. Vac. Sci. Technol.* B **21** 1903
[85] Vasileska D, Eldridge T, Bordone P and Ferry D K 1998 *VLSI Des.* **6** 21
[86] Keldysh L V 1965 *Sov. Phys. JETP* **20** 1018
[87] Semkat D, Kremp D and Bonitz M 1999 *Phys. Rev.* E **59** 1557
[88] Garny M and Müller M M 2009 *Phys. Rev.* D **80** 085011
[89] Mahan G D 1992 *Quantum Transport in Semiconductors* ed D K Ferry and C Jacoboni (New York: Plenum Press) pp 101–40
[90] Langreth D C 1976 *Linear and Nonlinear Electron Transport in Solids* ed J T Devreese and E van Doren (New York: Plenum Press)
[91] Prigogine I 1981 *From Being to Becoming* (San Francisco, CA: W. H. Freeman)
[92] Zubarev D 1974 *Nonequilibrium Statistical Thermodynamics* (Berlin: Springer)
[93] Lipavský P, Ŝpiĉka V and Velický B 1986 *Phys. Rev.* B **34** 3020
[94] Dahlen N E and van Leeuwen R 2007 *Phys. Rev. Lett.* **98** 153004
[95] Dahlen N E and van Leeuwen R 2009 *J. Chem. Phys.* **130** 224101
[96] Stan A 2016 *Phys. Rev.* B **93** 041103
[97] Jauho A-P 1989 *Sol.-State Electron.* **32** 1265
[98] Bertoncini R, Kriman A M and Ferry D K 1989 *Phys. Rev.* B **40** 3371
[99] Bertoncini R, Kriman A M and Ferry D K 1989 *Phys. Rev.* B **41** 1390
[100] Kurokawa Y and Takemori T 1999 *Physica B* **272** 306
[101] Bryskin V V and Kleinert P 1999 *J. Phys. A* **32** 2731
[102] Mori N and Hamaguchi C 1994 *Semicond. Sci. Technol.* **9** 941
[103] Caruso F, Novko D and Draxl C 2020 *Phys. Rev.* B **101** 035128

[104] https://newsroom.ibm.com/2021-05-06-IBM-Unveils-Worlds-First-2-Nanometer-Chip-Technology,-Opening-a-New-Frontier-for-Semiconductors
[105] Hisamoto D, Hu C and Bokor J *et al* 2000 *IEEE Trans. Electron Dev.* **47** 2320
[106] Zhou J-R and Ferry D K 1992 *IEEE Trans. Electron Dev.* **39** 473
[107] Madelung E 1926 *Z. Phys.* **40** 322
[108] Kennard E H 1928 *Phys. Rev.* **31** 876
[109] Bohm D 1952 *Phys. Rev.* **85** 166
[110] Grubin H L and Kreskovsky J P 1989 *Sol.-State Electron.* **32** 1071
[111] Chang L L, Esaki L and Tsu R 1974 *Appl. Phys. Lett.* **24** 593
[112] Sollner T C L G, Goodhue W D and Tannenwald P E *et al* 1983 *Appl. Phys. Lett.* **43** 588
[113] Goldman V J, Tsui D C and Cunningham J E 1987 *Phys. Rev. Lett.* **58** 1256
[114] Ferry D K, Akis R and Vasileska D 2000 *Proc. Int. Electron Dev. Mtg* (New York: IEEE Press) p 287
[115] Wigner E 1932 *Phys. Rev.* **40** 749
[116] Feynman R P and Kleinert H 1986 *Phys. Rev.* **A34** 5080
[117] Kim K, Mason B and Hess K 1987 *Phys. Rev.* B **36** 6547
[118] Register L F and Hess K 2000 *J. Appl. Phys.* **87** 303
[119] Keyes R W 1977 *Science* **195** 1230
[120] Wong H-S and Taur Y 1993 *IEEE Electron. Dev. Mtg.* (New York: IEEE Press) p 705
[121] Di Carlo A and Lugli P 1995 *Semicond. Sci. Technol.* **10** 1673
[122] Ordejón P, Artacho E and Soler J M 1996 *Phys. Rev.* B **53** 10441
[123] Soler J M, Artacho E and Gale J D *et al* 2002 *J. Phys. Cond. Matt.* **14** 2745
[124] Brandbyge M, Mozos J-L and Ordejón P *et al* 2002 *Phys. Rev.* B **65** 165401
[125] Stokbro K, Taylor J, Brandbyge M and Ordejón P 2003 *Ann. N. Y. Acad. Sci.* **1006** 212
[126] Sankey O F and Niklewski D J 1989 *Phys. Rev.* B **40** 3979
[127] Demkov A A, Ortega J, Sankey O F and Grumbach M P 1995 *Phys. Rev.* B **52** 1618
[128] Bowen R C, Klimeck G and Lake R K *et al* 1997 *J. Appl. Phys.* **81** 3207
[129] Jovanovic D 2002 *Int. Conf. Modeling Simul. Microstruc.* (Computational Publishers) p 478
[130] Xia T, Register L F and Bannerjee S K 2003 *IEEE Trans. Electron Dev.* **50** 1511
[131] Yu Z, Yergeau D W and Dutton R W *et al* 2001 *Proc. SPIE* **4600** 117
[132] Mori N and Hamaguchi C 2003 *J. Comp. Electron.* **2** 119
[133] Damle P, Rakshit T, Paulsson M and Datta S 2002 *IEEE Trans. Nanotechnol.* **1** 145
[134] Takeda H, Mori N and Hamaguchi C 2003 *J. Comp. Electron.* **2** 119
[135] Jiménez D, Iñíguez B, Suñé J and Sáenz J J 2004 *J. Appl. Phys.* **96** 5271
[136] Poli S, Pala M G and Poiroux T 2009 *IEEE Trans. Electron. Dev.* **56** 1191
[137] Martinez A, Price A and Valin R *et al* 2016 *J. Comp. Electron.* **15** 1130
[138] Shishir R and Ferry D K 2008 *J. Comp. Electron.* **7** 305
[139] Berrada S, Carrillo-Nunez H and Lee J *et al* 2020 *J. Comp. Electron.* **19** 1031
[140] Khaliji K, Noei M and Tabatabaei S M *et al* 2013 *IEEE Trans. Electron Dev.* **60** 2464
[141] Song X-J, Xu L-C and Bai H-F *et al* 2017 *J. Appl. Phys.* **121** 144505
[142] Shadman A, Rahmen E and Khosru Q D M 2017 *Superlatt. Microstruc.* **111** 414
[143] Haldas G, Kolek A and Pierscinska *et al* 2017 *Optics Quantum Electron.* **49** 22
[144] Shedbalkar A and Witzigmann B 2018 *Optics Quantum Electron.* **50** 67
[145] Aeberhard U 2019 *Semicond. Sci. Technol.* **34** 094002
[146] Fast J, Aeberhard U, Bremner S P and Linke H 2021 *Appl. Phys. Rev.* **8** 021309
[147] Samson J H 2003 *J. Phys. A* **36** 10673

[148] Sels D, Brosens F and Magnus W 2013 *Physica A* **392** 326
[149] Ferry D K and Nedjalkov M 2018 *The Wigner Function in Science and Technology* (Bristol: IOP Publishing)
[150] Dias N C and Prata J N 2004 *Ann. Phys.* **313** 110
[151] Levinson I B 1970 *Sov. Phys. JETP* **30** 362
[152] Ravaioli U, Osman M A and Pötz W *et al* 1985 *Physica B + C* **134B** 36
[153] Frensley W R 1987 *Phys. Rev.* B **36** 1570
[154] Shokin Y I 1983 *The Method of Differential Approximation* (Berlin: Springer)
[155] Lee J-H, Shin M and Seo J H 2019 *J. Appl. Phys.* **125** 174502
[156] Rossi F, Jacoboni C and Nedjalkov M 1994 *Semicond. Sci. Technol.* **9** 934
[157] Nedjalkov M, Dimov I, Rossi F and Jacoboni C 1996 *Math. Comp. Modeling* **23** 159
[158] Bordone P, Pascoli M and Brunetti R *et al* 1999 *Phys. Rev.* B **59** 3060
[159] Jensen K L and Buot F A 1991 *IEEE Trans. Electron Dev.* **38** 2337
[160] Barker J R 1985 *Physica* **134B** 22
[161] Bialas A and Krzywicki A 1995 *Phys. Lett.* B **354** 134
[162] Rossi F, Poli P and Jacoboni C 1992 *Semicond. Sci. Technol.* **7** 1017
[163] Shifren L and Ferry D K 2001 *Phys. Lett.* A **285** 217
[164] Shifren L and Ferry D K 2002 *J. Comp. Electron.* **1** 55
[165] Shifren L, Ringhofer C and Ferry D K 2003 *IEEE Trans. Electron Dev.* **50** 769
[166] Nedjalkov M, Kosina H and Selberherr S 2004 *Large Scale Scientific Computing* ed I Lirkov *et al* vol 2907 (Heidelberg: Springer) p 178
[167] Nedjalkov M, Kosina H, Ungersboeck E and Selberherr S 2004 *Semicond. Sci. Technol.* **19** S226
[168] Nedjalkov M, Kosina H and Selberherr S *et al* 2004 *Phys. Rev.* B **70** 115319
[169] Querlioz D, Saint-Martin J and Do V-N *et al* 2006 *IEEE Trans. Electron Dev.* **5** 737
[170] Querlioz D, Do V-N and Saint-Martin J *et al* 2008 *Phys. Stat. Solidi (c)* **5** 150
[171] Sampredo C, Gámiz F, Godoy A and Jiménez-Molinos F 2007 *J. Comp. Electron.* **6** 41
[172] Jacoboni C, Bertoni A, Bordone P and Brunetti R 2001 *Math. Comp. Simul.* **55** 67
[173] Reggiani L, Lugli P and Jauho A P 1987 *Phys. Rev.* B **36** 6602
[174] Pascoli M, Bordone P, Brunetti R and Jacoboni C 1998 *Phys. Rev.* B **58** 3503
[175] Bertoni A, Bordone P, Brunetti R and Jacoboni C 1999 *J. Phys. Cond. Matt.* **11** 5999
[176] Shifren L and Ferry D K 2003 *Phys. Lett.* A **306** 332
[177] Nettel S and Beck H 2014 *Physica B* **436** 91
[178] Sellier J M, Nedjalkov M, Dimov I and Selberherr S 2013 *J. Appl. Phys.* **114** 174902
[179] Sellier J M, Nedjalkov M, Dimov I and Selberherr S 2015 *Math. Comp. Simul.* **107** 108
[180] Yamada Y, Tsuchiya H and Ogawa M 2009 *IEEE Trans. Electron Dev.* **56** 1396
[181] Ferrari G, Bordone P and Jacoboni C 2006 *Phys. Lett.* A **356** 371
[182] Barraud S 2009 *J. Appl. Phys.* **106** 063714
[183] Barraud S 2011 *J. Appl. Phys.* **110** 093710
[184] Tsuchiya H, Hiyashi Y and Miyoshi T 1996 *Physica B* **227** 411
[185] Weetman P and Weetman P 2003 *Microwave Opt. Technol. Lett.* **38** 369
[186] Weetman P and Wartak M S 2003 *J. Appl. Phys.* **93** 9562
[187] Jonasson O and Knezevic I 2015 *J. Comp. Electron.* **14** 879
[188] Balaban S N, Pokatilov E P and Fomin V N *et al* 2002 *Sol.-State Electron.* **46** 435
[189] Croitoru M D, Gladilin V N and Fomin V N *et al* 2003 *J. Appl. Phys.* **93** 1230
[190] Croitoru M D, Gladilin V N and Fomin V N *et al* 2004 *J. Appl. Phys.* **96** 2305

[191] Jiang H and Cai W 2010 *J. Comp. Phys.* **229** 4461
[192] Sellier J M, Amoroso S M and Nedjalkov M *et al* 2014 *Physica A* **398** 194
[193] Lorenz J K, Bär E and Clees T *et al* 2011 *IEEE Trans. Electron Dev.* **58** 2227
[194] Ferry D K, Nedjalkov M and Weinbub J *et al* 2020 *Entropy* **22** 1103
[195] Ferry D K, Grondin R O and Reich R K *et al* 1984 *The Physics of Submicron Devices* ed H L Grubin *et al* (New York: Plenum Press) p 267
[196] Zurek W H 2003 *Rev. Mod. Phys.* **75** 715